Sulphur in Plants

Sulphur in Plants

Edited by

Yash P. Abrol

Department of Environmental Botany,
Faculty of Science,
Hamdard University,
New Delhi, India

and

Altaf Ahmad

Department of Environmental Botany,
Faculty of Science,
Hamdard University,
New Delhi, India

SPRINGER-SCIENCE+BUSINESS MEDIA, B.V.

A C.I.P. Catalogue record for this book is available from the Library of Congress.

ISBN 978-90-481-6276-5 ISBN 978-94-017-0289-8 (eBook)
DOI 10.1007/978-94-017-0289-8

Printed on acid-free paper

CONTENTS

PREFACE

Sulphur (S) plays a pivotal role in various plant growth and development processes being a constituent of sulphur-containing amino acids, cysteine and methionine, and other metabolites viz., glutathione and phytochelatins, co-factor of enzymes which contribute to stress repair and amelioration of heavy metal toxicity. Besides, a number of S-containing components are biologically active and, thus, a source for use as medicinal value.

The basic global issue before the agricultural scientist and world community is to evolve cultivars and develop methodologies for efficient use of inputs to enhance agricultural productivity. This is particularly true of the developing countries which are going to see maximum rise in population with changing food demands and declining availability of land. Amongst the inputs, nutrients play a crucial role. The major requirement is for N, P and K followed by several micro-nutrients. In this context reports of world-wide S deficiency in the agricultural systems are relevant. The reasons are many. Broadly speaking reduction in S emission, use of S-free N, P and K fertilizers and higher biomass production contributed the maximum. Despite the need for sulphur as an essential plant nutrient and the substantial returns expected from its use, very little attention has been given to fill the gap between supply and demand of S.

During the recent past, there has been spurt in research activity in various aspects of S utilization by crops. These range from an understanding of the global S cycle to the biotechnological approaches to improve the utilization efficiency of not only S but also its role in balanced use of fertilizers. This is to contribute to efficient use of nutrients and, thus, development of sustainable agricultural practices.

In this book, the editors have attempted to put together the research contributions by experts in their respective fields to elaborate the various aspects of sulphur to be considered from the agricultural point of view.

In chapter 1, a general introduction of the S, its global inventory and the expected changes concerning its turnovers on a global scale within the next 15 years are discussed comprehensively; the S cycle of upland, wetland and limnic ecosystems are examined, and S balances at different scales from landscape over sub-field down to plant level are presented.

Chapter 2 deals with the food requirement scenario keeping in view the increase in population, income level, urbanization and, thus, the changing food habits, and need to remedy the under-nourishment and malnutrition prevailing amongst the teeming millions in some regions of the world. This is followed by discussion on prevalence of S deficiency, its causes and the present and future scenario of fertilizer demand and supply.

Chapter 3 deals with managing sulphur (S) in agroecosystems. In the first part of this chapter global biogeochemical S cycle, S sources and transformation processes in the environment are dealt. The second part deals specifically with the management of S in agroecosystems.

Chapter 4 summarizes what is known about the uptake and transport of sulphate in plants. The physiological and biochemical background on sulphate transport characteristics are summarized together with a critical appraisal of current molecular approaches in this research area. A model of a 'highly regulated circuit' controlling sulphate uptake and assimilation, mediated by feedback loops involving key metabolites of cysteine biosynthesis has been proposed.

In chapter 5, the newly obtained information for the current view on the pathway of assimilatory sulphate reduction are presented. A brief historical account of investigations of sulphate assimilation is followed by biochemical and molecular characterisation of the enzymes and their regulation. Finally, a summary of the remaining open questions and directions for future research is given.

Chapter 6 deals with the sulphur distribution and redistribution in vegetative and generative plants. In the first part, evidence for the reallocation of S, factors that influence reallocation changes in mobility and sinks with respect to plant development, and mechanisms for loading sulphate and organic-S into the phloem are discussed. In the second part, pattern of distribution and redistribution of sulphur in vegetative plants in monocots and dicots, pathways for the importation of S into developing grains in monocots and dicots and forms of S imported by developing cereal grains and incorporated into grain proteins are given.

Chapter 7 deals with the biosynthesis of cysteine and methionine.

Chapter 8 focuses on the recent advances in the elucidation of glucosinolate biosynthesis. Finally, future research need on identification of regulators of glucosinolate biosynthesis to make it possible to metabolically engineer glucosinolate profiles in a tissue specific manner to improve disease resistance and nutritional value is given.

Chapter 9 and 10 deal with ecophysiological and molecular aspects of metallothionine (MTs) and phytochelatins (PCs). In chapter 9, emphasis is placed on their role in metal detoxification, metal specific genetically defined tolerance (hypertolerance) and the possible use of PCs as biomarkers of heavy meal toxicity, and chapter 10 reviews our current understanding of the biosynthesis, expression, regulation and functions of MTs and PCs, drawn from a range of physiological, biochemical, genetic and molecular biological approaches to their study.

In chapter 11 occurrence, localization, coding gene, metabolism and function of a higher plant sulpholipid, sulphoquinovosyl diacylglycerol (SQDG), are given in 1-5 sections. In 6th sections, a number of different stress responses were considered, and tried to evaluate whether SQDG can play a role in adaptation. The future looks very interesting for research on this unique membrane lipid!

Chapter 12 discusses the role of thiol compounds (with special emphasis on GSH redox system) in the defense systems and their involvement in plant responses to environmental stress.

In chapter 13, an extensive review of genetic engineering strategies used to modify stress tolerance is presented. Main focus is given on the changes in the content and redox state of glutathione, a major low-molecular-weight thiol-containing compound and a scavenger of reactive oxygen species.

In chapter 14 detail mechanisms of H_2S emission and utilization by plants that will help to develop strategies for usefulness of the channelling of extra sulphur into specific sulphur pool in order to improve crop quality by using molecular biology tools are described.

Chapter 15 deals with the consequences of SO_2 influx in plants.

Chapter 16 describes the very important role plants play in providing the protein required in animal diets. Mechanisms by which the plant regulates the relative abundance of these two protein classes are proposed. Also, efforts to enhance the relative abundance of the sulphur amino acids in seed storage proteins are described.

Chapter 17 is concerned specifically with impact of S on processing quality of wheat. An overview of the distribution of S deficiency, its amelioration, effect of other plant nutrients on S uptake is also given.

Chapter 18 examines the effect of sulphur on yield and quality of oilseed rape and also describes the relationship between sulphur nutrient status within the plant and disease development.

Chapter 19 synthesizes the available information on the responses of field crops, both in terms of yield and quality to S fertilization in monoculture and cropping systems that are prevalent in different states of India, and the role of S in optimizing crop production and reducing environmental risks.

Chapter 20 deals with the interaction of S with other nutrients which is relevant to considering/developing strategies for balanced use of fertlizers.

Chapter 21 discusses the biological profile of some of the sulphur compounds of plant origin, and rationale for their possible use in the form of medicine.

Thus, this book has presented an authorative review of present status of knowledge of sulphur and its availability for crop production and quality, and identified further research areas to be explored. We believe that this book, written by experts in different areas of research, will be useful for under-graduates, graduates, professors, scientists in biology and agronomy.

We are thankful to the authors for their time and efforts. We express our sincere gratitude to Mr. Siraj Hussain (IAS, Vice-Chancellor) and Professor Muhammad Iqbal (Head, Department of Environmental Botany) of Hamdard University for their constant encouragement and for providing us necessary facilities to effectively carry out this work. Financial supports from Indian National Science Academy (under Honorary Scientist Scheme) and Department of Science and Technology (under SERC Fast Track Young Scientist Scheme), Govt. of India are gratefully acknowledged.

Yash P Abrol and Altaf Ahmad
The Editors

CONTRIBUTORS

Malik Z ABDIN
Centre for Biotechnology, Faculty of Science, Hamdard University, New Delhi, 110062, India

Yash P ABROL
Department of Environmental Botany, Faculty of Science, Hamdard University, New Delhi, India

Madhoo AGRAWAL
Department of Botany, Banaras Hindu University, Varanasi 221 005, India

Altaf AHMAD
Department of Environmental Botany, faculty of Science, Hamdard University, New Delhi, 110 062, India

Mohammad ALI
Department of Pharmacognosy & Phytochemistry, Faculty of Pharmacy, Hamdard University, New Delhi, 110 062, India

John W ANDERSON
Botany Department, La Trobe University, Bundoora, Victoria 3083, Australia

Milkha S AULAKH
Department of Soils, Punjab Agricultural University, Ludhiana, 141004, India

Anna BŁASZCZYK
Institute of Biochemistry and Biophysics, Polish Academy of Sciences, ul. Pawinskiego 5A, 02-106 Warsaw, Poland

Elke BLOEM
Institute of Plant Nutrition and Soil Science, Federal Agricultural Research Centre (FAL), Bundesallee 50, D-38116 Braunschweig, Germany

Elaine J BOOTH
Scotaland Agriculture College (SAC), Craibstone Estate, Bucksburn, Aberdeen, Scotland, AB21 9YA, UK

Peter BUCHNER
Agronomy and Environment Division, IACR-Rothamsted, Harpenden, Hertfordshire, AL5 2JQ, UK

Christopher S COBBETT
Department of Genetics, The University of Melbourne, Parkville, Victoria, 3052, Australia

Melissa A FITZGERALD
Yanco Agricultural Institute, NSW Agriculture, Private Mail Bag, Yanco, NSW, 2703, Australia

Dali GAGANIDZE
Institute of Biochemistry and Biophysics, Polish Academy of Sciences, ul. Pawinskiego 5A, 02-106 Warsaw, Poland

Erich GLAWISCHNIG
Plant Biochemistry Laboratory, Department of Plant Biology, The Royal Veterinary and Agricultural University, 40, Thorvaldsensvej, DK-1871 Frederiksberg C, Denmark

Dieter GRILL
Institute of Plant Physiology, University of Graz, Schubertstraße 51, 8010 Graz, Austria

Gábor GULLNER
Plant Protection Institute, Hungarian Academy of Sciences, 1525 Budapest, POB 102, Hungary

Barbara A HALKIER
Plant Biochemistry Laboratory, Department of Plant Biology, The Royal Veterinary and Agricultural University, 40, Thorvaldsensvej, DK-1871 Frederiksberg C, Denmark

Silvia HANEKLAUS
Institute of Plant Nutrition and Soil Science, Federal Agricultural Research Centre (FAL), Bundesallee 50, D-38116 Braunschweig, Germany

Kashif-ul-HAQ
Department of Pharmacognosy & Phytochemistry, Faculty of Pharmacy, Hamdard University, New Delhi, 110 062, India

John L HARWOOD
School of Biosciences, Cardiff University, PO box 911, Cardiff Cf10 3us, U.K.

Malcolm J HAWKESFORD
Agronomy and Environment Division, IACR-Rothamsted, Harpenden, Hertfordshire, AL5 2JQ, UK

Laura HOPKINS
Agronomy and Environment Division, IACR-Rothamsted, Harpenden, Hertfordshire, AL5 2JQ, UK

Jonathan R HOWARTH
Agronomy and Environment Division, IACR-Rothamsted, Harpenden, Hertfordshire, AL5 2JQ, UK

John IMSANDE
Plant Genetics Group., Department of Agronomy, Iowa State University, Ames, Iowa 50011, USA

Muhammad IQBAL
Department of Environmental Botany, Faculty of Science, Hamdard University, New Delhi, 110062, India

Arshad JAMAL
Centre for Biotechnology, Faculty of Science, Hamdard University, New Delhi, 110062, India

Kacper KAZIMIERCZUK
Institute of Biochemistry and Biophysics, Polish Academy of Sciences, ul. Pawinskiego 5A, 02-106 Warsaw, Poland

Ishrat KHAN
Centre for Biotechnology, Faculty of Science, Hamdard University, New Delhi, 110062, India

Nuzhat KHAN
Centre for Biotechnology, Faculty of Science, Hamdard University, New Delhi, 110062, India

Tamás KÖMIVES
Plant Protection Institute, Hungarian Academy of Sciences, 1525 Budapest, POB 102, Hungary

Stanislav KOPRIVA
Institute of Forest Botany and Tree Physiology, Georges-Köhler-Allee 053, D-79085 Freiburg, Germany

Anna KOPRIVOVA
Department of Plant Biotechnology, Faculty of Biology, Schänzlestr. 1, D-79104 Freiburg, Germany

Karunachand C LAKKINENI
201-2300 Grenet, St-Laurent, Montreal, QC, Canada H4L 4Y9

Frantz LISZEWSKA
Institute of Biochemistry and Biophysics, Polish Academy of Sciences, ul. Pawinskiego 5A, 02-106 Warsaw, Poland

Finlay MACRITCHIE
Department of Grain Science and Industry, Kansas State University, Manhattan, KS 66506, USA

Michael D MIKKELSEN
Plant Biochemistry Laboratory, Department of Plant Biology, The Royal Veterinary and Agricultural University, 40, Thorvaldsensvej, DK-1871 Frederiksberg C, Denmark

Hamid A NAEEM
Department of Grain Science and Industry, Kansas State University, Manhattan, KS 66506, USA

Masaaki NOJI
Department of Molecular Biology and Biotechnology, Graduate School of Pharmaceutical Sciences, Chiba University, Yayoi-cho 1-33, Inage-ku, Chiba 263-8522, Japan

Oene OENEMA
Wageningen University and Research Center, Alterra, P.O.Box 47, NL-6700 AA Wageningen The Netherlands.

Alexander A OKANENKO
National Taras Shevchenko University, Plant Physiology & Ecology Department, Kyiv, 01033, Ukraine

NS PASRICHA
Postah Research Institute, Gurgaon, Haryana, India

Romke POSTMA
Nutrient Management Institute NMI, Haagsteeg 2b, NL-6708 PM Wageningen (RP), The Netherlands.

Kazuki SAITO
Department of Molecular Biology and Biotechnology, Graduate School of Pharmaceutical Sciences, Chiba University, Yayoi-cho 1-33, Inage-ku, Chiba 263-8522, Japan

Henk SCHAT
Department of Ecology and Physiology of Plants, Vrije Universiteit, Faculty of Earth and Life Sciences, De Boelelaan 1085, 1081 HV Amsterdam, The Netherlands

Ewald SCHNUG
Institute of Plant Nutrition and Soil Science, Federal Agricultural Research Centre (FAL), Bundesallee 50, D-38116 Braunschweig, Germany

Agnieszka SIRKO
Institute of Biochemistry and Biophysics, Polish Academy of Sciences, ul. Pawinskiego 5A, 02-106 Warsaw, Poland

F Else C SNELLER
Department of Chemistry and Ecotoxicology, National Institute of Inland Water Management and, Wastewater Treatment (RIZA), P.O. Box 17, 8200 AA Lelystad, The Netherlands

Michael TAUZ
Institute of Plant Physiology, University of Graz, Schubertstraße 51, 8010 Graz, Austria

Joc AC Verkleij
Department of Ecology and Physiology of Plants, Vrije Universiteit, Faculty of Earth and Life Sciences, De Boelelaan 1085, 1081 HV Amsterdam, The Netherlands

Kerr C WALKER
Scotaland Agriculture College (SAC), Craibstone Estate, Bucksburn, Aberdeen, Scotland, AB21 9YA, UK

ABBREVIATIONS

γ ECS	γ-glutamyl-cystcinc synthasc
γGluCys	γ-glutamylcysteine
ADP	Adenosine diphosphate
AOS	active oxygen species
APK	APS kinase
APR	APS reductase
APS	adenosine 5'-phosphosulphate
APSST	APS sulphotransferase
APX	ascorbate peroxidase
APXs	Ascorbate peroxidases
ATPS	ATP sulphurylase
BSO	buthionine sulphoximine
CAT	catalase
CDPK	calcium dependent protein kinase
COS	carbonylsulphide
CP	chlorophyll-protein
CS_2	carbon disulphide
CHS	chalcone synthase
CYP	cytochrome P450
Cys	cysteine
DAG	diacylglycerol
DAP	diammonium phosphate
DGDG	digalactosyldiacylglycerol
DIIA	dchydroascorbatc
DMBA	9,10-Dimethyl 1,2-Benzanthracene
DMDS	dimethyl disulphide
DMS	dimethyl sulphide
dpa	days post anthesis
ESP	epithiospecifier protein
GC	gas chromatography
GCS	γ-glutamylcysteine synthetase
GPX	glutathione peroxidase
GR	glutathione reductase
GS	glutathione synthase
GSH	glutathione
GSH-S	glutathione synthetase
GSL	glucosinolate
GSNO	nitroso-glutathione
GST	glutathione-S-transferases

GUS	β-glucuronidase
H_2O_2	hydrogen peroxide
H_2S	hydrogen sulphide
hGSH	homoglutathione
HMW	high molecular weight
HMW-GS	high molecular weight glutenin subunits
hmPC	hydroxylmethyl PC
hPCs	homo-PCs
HSO_3^-	bisulphite
IAA	indole-3-acetic acid
IAAld	indole-3-acetaldehyde
IAN	indole-3-acetonitrile
IAOx	indole-3-acetaldoxime
ICP-OES	inductively coupled plasma-optical emission spectroscopy
kDa	kilo Daltons
LMW	low molecular weight
LMW-GS	low molecular weight glutenin subunits
MAPK	mitogen activated protein kinase
MDA	malondialdehyde
MDHA	monodehydroascorbate
Me	methyl
Met	methionine
MGDG	monogalactosyldiacylglycerol
MIC	minimum inhibitory concentration
MMT	*S*-adenosylmethionine methyltransferase
MOP	muriate of potash
MS	Mass Spectrometry
MSDs	membrane spanning domains
MTs	metallothioneins
MV	methyl viologen
O_2	superoxide radical ion
OAS	*O*-acetyl-L-serine
OAS-TL	O-acetylserine(thiol) lyase
OTC	L-2-oxothiazolidine-4-carboxylic acid
PAL	phenylalanine ammonia lyase
PAPS	adenosine 3'-phosphate 5'-phosphosulphate
PCs	phytochelatins
PhG	phenylglyoxal
PHGPX	phospholipid hydroperoxide glutathione peroxidase
PHLOH	phospholipid alcohol
PHLOOH	phospholipid peroxide
PLA_2	lipid-activated phospholipase A_2
PxV	potato potevirus
ROS	reactive oxygen species

QTL	quantitative trait locus
S-GT	UDPG:thiohydroximic acid glucosyltransferase
ST	sulphate transporter
TPX	thioredoxin peroxidase
UPP	unextractable polymeric protein

S. HANEKLAUS, E. BLOEM AND E. SCHNUG

THE GLOBAL SULPHUR CYCLE AND ITS LINKS TO PLANT ENVIRONMENT

Abstract. The fundamental importance of sulphur as one of the three philosophic essentials goes back to alchemy. The production of sulphuric acid was a gauge for the industrial state of a nation at the start of the 20th century. Less than 80 years later sulphur was stigmatised as the yellow poison after being identified as one of the major reasons for the acid rain problem. Later, after clean air acts came into force, severe sulphur deficiency became a major nutrient disorder in agricultural production. An insufficient sulphur supply reduces crop productivity, diminishes crop quality, impairs the nitrogen utilisation efficiency and thus increased the undesired nitrogen losses to the environment. Understanding the mechanisms and processes involved in the sulphur cycling at different scales and its links to the plant environment is an important contribution to sustainable agricultural production and environmental protection. Significant advances in plant nutrition and soil science can be expected by establishing site-specific sulphur balances for improving fertilizer strategies, by identifying regulatory mechanisms in the sulphur metabolism of plants for enhancing the natural resistance against pathogens, and by determining the sulphur cycling in the edaphon for promoting soil life.

1. INTRODUCTION

Sulphur (S) is ubiquitous abundant on earth and statistically S is available in adequate amounts to satisfy globally plant growth. Nevertheless S deficiency is an important nutrient disorder in agricultural production on all continents. Besides a decrease in crop productivity (Schnug and Haneklaus, 1998) and negative impacts on crop quality (Schnug, 1993; Schnug et al., 1993a; Schnug, 1997), a higher susceptibility of crops to certain diseases was observed (Schnug et al., 1995; Bourbos et al., 2000). Additionally, the nitrogen (N) utilisation efficiency is closely linked to the S supply; on average every kg of S missing to satisfy the plant S demand results in a potential N loss of 15 kg N to the environment (Schnug et al., 1993b).

Plants have been classified biogeochemically (Duvigneaud and Denayer-De Smet, 1968), physiologically (Kinzel, 1972) and ecologically (Grime, 1979) on the basis of their demand and accumulation of S. Based on Grime's classification system, Ernst (1997) characterized the S metabolism of different plant species and mechanisms of adaptation to S rich environments. Drastic reductions in the S supply, for instance after clean air acts coming into force presumably not only have a significant effect on agricultural, but also other ecosystems. For the diagnosis of the plant nutritional S status, soil and plant analysis, and site-specific S balances are usually applied. A molecular approach has been proposed in order to identify plant genes whose expression is regulated by the S status, to couple the promoters of these genes with fluorescent reporter genes in transgenic plants for registering online the S status in the field (Hawkesford, 2002). The major problem of this procedure is that the variability of the S supply is inversely related to the scale of observation, so that a transfer of

Yash P. Abrol and Altaf Ahmad (eds.), Sulphur in Plants, 1—28.

measurements from molecular to field level is fiction.

Balancing S inputs and outputs in agricultural ecosystems proved to be a difficult tool to predict the S requirement of crops, because some sources such as the direct assimilation of sulphur dioxide (SO_2) and the mineralization from organic matter have been misinterpreted or strongly overestimated, while others such as the soil water regime, have been drastically underestimated. Soil hydrology has been identified as one of the key parameters governing S dynamics in the soil S so that this feature is crucial for a realistic modelling of the S supply of crops in agricultural ecosystems (Bloem, 1998; Haneklaus et al., 2002).

A direct uptake of S gases such as hydrogen sulphide (H_2S) and SO_2 from the atmosphere takes place; concentrations that are higher than 100 μg m^{-3} for SO_2 (Saalbach, 1984) and 46 μg m^{-3} for H_2S (De Kok, 1990) are, however, phytotoxic. Mean SO_2 and H_2S concentrations are 10 μg m^{-3} and 0.07 μg m^{-3}, respectively in rural areas and 100 μg m^{-3} and 0.35 μg m^{-3}, respectively in industrial zones (De Kok, 1990). Results from fumigation experiments delivered a contribution of SO_2 to the plant S supply of 25 to 80% (Lauenroth et al., 1983; Fiedler and Thakur, 1984), which is not representative because of specific experimental conditions such as extraordinary high SO_2 concentrations combined with a low S supply in the culture substrate. But in areas with high S depositions, as for instance those at close distance (3 - 16 km) to coal fired plants took up 10% to 44% of S from the atmosphere (Noggle, 1979). In general, however, the uptake of gaseous S compounds contributes no more than 1 kg S ha^{-1} yr^{-1} to the total S balance, assuming an assimilation rate equally high as that of CO_2 (Schnug, 1988). The S cycle can be studied at different scales with the largest being the globe and the smallest referring to individual plants in their habitat. In this chapter a general, qualitative and quantitative inventory of S in the environment is given in section 2.1 and 2.2. Global S pools and fluxes are presented in section 2.2.1 and 2.2.2 and expected changes of the S turnover within the next 15 years are discussed comprehensively. In section 2.2.3 the S cycle of upland, wetland and limnic ecosystems is examined. In section 3, S balances at different scales from landscape over sub-field down to plant level are presented. Based on today's knowledge future research needs are pointed out in section 4.

2. GLOBAL ASPECTS OF THE SULPHUR CYCLE

2.1 Qualitative Aspects

2.1.1 Inorganic and organic sulphur species

Sulphur is bound in valence states ranging from +6 in sulphate to -2 in sulphides. There are numerous inorganic and organic S compounds (section 2.1; Table 1) abundant in different spheres on earth.

Important inorganic and organic S compounds of anthropogenic, geogenic and biogenic origin, which are found in atmosphere, hydrosphere, lithosphere and biosphere, are listed in table 1. The key objectives of this chapter focus on quantitative aspects of the S cycle. Detailed information on individual S species is given for instance by Weast et al. (1964), Senning (1982) and Oae and Okuyama (1992).

Table 1. Relevant inorganic and organic S compounds in atmosphere, hydropshere, lithosphere and biosphere

Inorganic compounds		**Organic compounds**	
Name	**Formula**	**Name**	**Functional group**
Hydrogen sulphide	H_2S	*Reduced*	
sulphate	SO_4^{2-} plant: lipid sulphates minerals:$CaSO_4*2H_2O$, $Na_2SO_4*10\ H_2O$, $MgSO4*H_2O$, $BaSO_4$, $SrSO_4$	carbonyl sulphide carbon disulphide thiol/mercaptanes sulphide (dimethyl sulphide, DMS)	COS CS_2 R-SH R-S-R
thiosulphate	$S_2O_3^{2-}$	disulphide (dimethyldisulphide, DMDS)	R-S-S-R
sulphite	SO_3^{2-}	Polysulphide	$R-S_x^-$
polythionate	$S_xO_6^{2-}$	*Oxidised*	
polysulphide	S_x^{2-}	sulphoxide sulphone	R_2SO R_2SO_2
sulphur dioxide	SO_2	Sulphonic acid	$R-SO_3H$
sulphide	minerals: PbS, ZnS, FeS, FeS_2, Fe_3S_4, HgS, $CuFeS_2$, FeAsS, Cu_2S, MoS_2, Ag_2S, Sb_2S_3, As_4S_4	Thiosulphate sulphonium ion sulphate ester	$R-S_2O_3^-$ R_3S^+ $R-O-SO_3^-$
elemental S	S_8		

2.2 Quantitative Aspects

The global S pools comprise of atmospheric, hydrogenic, biogenic and lithosperic S reservoirs (section 2.2.1). Temporal changes of S in different pools are for instance caused by volcanic activities, underground coal seam fires, burning of biomass and industrial activities, and will particularly affect regional S budgets and fluxes (section 2.2.2).

2.2.1 Global sulphur pools

Global S pools and annual S fluxes are shown in figure 1. Major terrestrial S pools are volcanoes and rocks with 50-106 and 2.4-1010 Tg S and significant marine reservoirs are oceanic sediments and sea water with 3-108 and 1.3-109 Tg S, respectively. Substantially, the S reservoir of soil and land biota (3-105 Tg S), lakes (300 Tg S), soils (120-103 Tg S) and atmosphere (5 Tg S) is lower. Values for S bound in plant biomass vary between 0.6-103 Tg S (Trudinger, 1979) and 8.5-103 Tg S (Galloway, 1996). S inherent to the world population constitutes the smallest S reservoir with 0.84 Tg S. It can be expected that by 2025 with an estimated world population of 8 billion people about 1.1 Tg S will be bound in human bodies. Clark (2002) estimates an intake of 310 g S yr^{-1} to be sufficient to satisfy the human S demand. Sulphur represents about 0.25% (140 g S) of the total body weight (Clark, 2002). This equals an annual release of 0.008 Tg S at actual mortality rate of 54-106 people (Anon, 2002a).

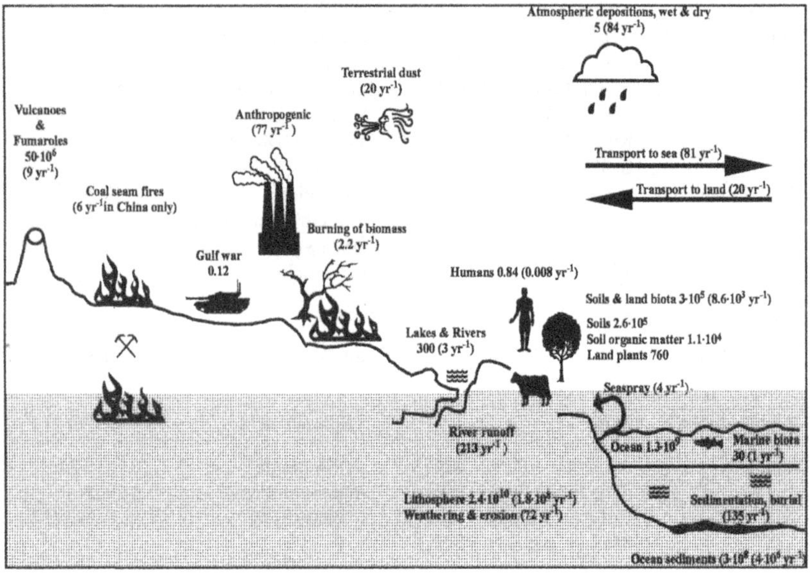

Figure 1. Global S pools (Tg S) and fluxes (Tg S yr^{-1}) (Sources: Eriksson, 1960; Krauskopf, 1967; Trudinger, 1979; Freney and Williams, 1983; Freney et al., 1983; Brimblecombe et al., 1989; Bates et al., 1992; Galloway, 1996; Reeburgh, 1997; Omar et al., 1998; Anon, 2002b; Clark, 2002; Greenpeace, 2002)

Volatile sulphur compounds in the atmosphere. Industrialisation has been accompanied by increased levels of sulphite and sulphur dioxide which contributed to the phenomenon of acid rain. 1.5 µg SO_2 g^{-1} lowers the pH of rain water from 5.7 to 4.0 (Agripedia, 2002). The sulphate content in rain water varied between 3 and 12 mg SO_4 L^{-1} (Van Loon, 1973; Taylor et al., 1983) before clean air acts came into force and were reduced by these measures to 1-3.8 mg SO_4 L^{-1} a decade later (Klemm and Gray, 1982).

Aerosols with a diameter of < 2.5 µm contain 0.3-23 µg SO_4-S m^{-3} and aerosols with a diameter >2.5 µm showed a mean S content of 0.2 - 0.5 µg S m^{-3} (Taylor et al., 1983). In 1993, mean annual SO_2 concentrations were as low as 0.5 µg S m^{-3} in Ireland and as high as 5 µg S m^{-3} in eastern countries of Europe including East Germany, Southern Poland, Czech Republic, Slovakia and North Eastern part of Austria (Larssen and Hagen, 1996). Peaks of single measurements showed values of > 500 µg S m^{-3} (Larssen and Hagen, 1996).

Sulphur compounds in the hydrosphere. Volatile S gases are produced biologically (CH_3SCH_3, CS_2, CH_3SH, CH_3SSCH_3) and/or phytochemically (H_2S, COS) in the surface water of the ocean (Andreae, 1986). Dimethyl sulphide (DMS) concentrations vary between 0.5 to 5 nmol L^{-1} and are being strongly dependent on region and season (Bates et al., 1987). In comparison the DMS concentration in freshwater is only about 0.1-1.0 nmol L^{-1} (Nriagu and Holdway, 1989). Carbon disulphide (CS_2) concentrations are about 16 pmol L^{-1} and carbonyl sulphide (COS) concentrations range from 10 to 100 pmol L^{-1} (Bates et al., 1992).

As sulphate is the most stable S compound under oxidising conditions, it is the predominant S form in soils and aerated water. Cloud water concentrations given in literature vary between 77 µg SO_4^{2-}-S L^{-1} (Scott and Laulainen, 1979) and 250 µg SO_4^{2-}-S L^{-1} (Petrenchuk and Drozdova, 1966). The geogenic background value of sulphate in surface water is ≤ 8.3 mg SO_4^{2-}-S L^{-1} (UBA, 2001). The S concentration of freshwater and sea water ranges between 3 up to 30 mg S L^{-1} in lakes (Health Canada, 1994) and 900 mg S L^{-1} in ocean (Boudreau and Westrich, 1984; Capone and Kiene, 1988; Andreae and Jaeschke, 1992; Giblin and Wieder, 1992). Long-term studies (1988-2000) of annual sulphate fluctuations in two Irish rivers revealed a range of 3-12 and 2-120 mg SO_4^{2-}-S L^{-1}, respectively (NIVA, 2001). Lake water sulphate content significantly decreased from 1989 to 1991 on 67 out of 98 test sites in Northern America and Europe (NIVA, 2001). Typical sulphate concentrations in natural groundwater vary between 5-100 mg SO_4^{2-}-S L^{-1} (Strebel and Boettcher, 1985; Bloem, 1998). The S concentration in aquifers is closely related to geological formations (LaMoreaux and Tanner, 2001) and the variation of sulphate in mineral water being correspondingly wide with concentrations of 0.03 mg S L^{-1} to 18.33 g S L^{-1} (Geiser, 2002). Fontes et al. (1989) found increasing sulphate concentrations of groundwater with depth in granitic batholith at Stripa, Sweden. Worthington and Ford (1995) stated that high sulphate contents of spring water was related to base level and thermal, while limestone springs with low sulphate content were bound to higher elevations above base level and were not thermal. They also found a direct relationship between sulphate concentration and spring temperature, and an inverse relationship with discharge. The S source is presumably S minerals which were dissolved by circulating meteoric water in deep zones by which tectonic fractures of initially 0.1 mm aperture were widened to at least 0.2 mm (Worthington and Ford, 1995). The critical threshold for S as sulphate in drinking water is for example 83 mg SO_4^{2-}-S L^{-1} in the EU and 167 mg SO_4-S L^{-1} in Canada (Health Canada, 1994). Higher intake of sulphate rich mineral water may cause adverse effects. Recommended thresholds for sulphate in drinking water for cattle range from 83 mg SO_4^{2-}-S L^{-1} in The Netherlands to 667 mg SO_4^{2-}-S L^{-1} in USA (Oude Elferink and Meijer, 2002).

Sulphur in the pedosphere. The concentration of S in parent materials ranges from 0.026% to 1% S with igneous rocks ≤ metamorphic rocks ≤ magmatic rocks of upper continental crust << limestones < sedimantary rocks (sulphides) < shales < sedimentary rocks (sulphates) << coal (Bowen, 1966; Wedepohl, 1968; Friend, 1973). The typical range of S in agricultural soils of humid and semi-humid regions is 100 to 500 µg g^{-1}, or 0.01 to 0.05% S. This equals 224 to 1,120 kg S ha^{-1} in the Ah-horizon (Stevenson, 1986). The total S content of soils may be as low as 20 µg g^{-1} (0.002%) in highly leached and weathered soils of humid regions or as high as 35,000 μg g^{-1} (3.5%) in marine marsh soils and up to 50,000 μg g^{-1} (5%) in calcareous and saline soils of arid and semi-arid regions (Stevenson, 1986). Tropical soils generally contain low amounts of S due to their low organic matter content. Under humid conditions Haehndel and Isermann (1993) determined sulphate concentrations of 592 to 7,006 kg SO_4^{2-}-S ha^{-1} in sub-soils and values of 67 to 233 mg SO_4^{2-}-S L^{-1} in the soil solution at the seam of the mean groundwater table.

The C:N:S ratio varies depending on parent material, leaching and fertilisation, and is on average 130:10:1.3. It may be as narrow as 90:8:1 on mollisols or as wide as 200:12:1 on more leached spodosols (Agripedia, 2002). The C:S ratio in soils from different regions in the world ranged from 85 on cultivated land up to 250 on grassland (Stevenson and Cole, 1999). The N:S ratio on these sites varied between 6.7 and 11.1 (Stevenson and Cole, 1999). The critical C:S ratios for mineralisation and immobilization are 200:1 and 400:1, respectively (Agripedia, 2002).

The edaphon constitutes about 5% of the total organic matter in soils (Topp, 1981). The Ah horizon of soils typically comprises soil biota in the following amounts (dry matter m^{-2} in a no-till farm soil): bacteria (50 g), fungi (100 g), amoeba (5 g), nematodes (0.2 g), arthropods (0.5 g) and worms (1-20 g) (Seafriends, 2002). Farm stock amounts to 50 g m^{-2} and is, thus, comparable to bacteria and earth worms while fungi outweigh this number considerably (Seafriends, 2002).

Sulphur bacteria have S globules in their cells and are capable of oxidising/reducing sulphide/sulphate. These comprise the following groups: thiosulphate bacteria, denitrifying thiosulphate bacteria, sulphate reducing bacteria and saprophytic bacteria (Ellis, 1932). It was Winogradsky (1887) who found out that Beggiatoa used H_2S as an energy source and developed the concept of autotrophy, and its relationship to natural cycles.

Microbial biomass is in the range of 146 - 968 μg g^{-1} soil (Roembke et al., 2002) with S content of 928-1355 μg S g^{-1} (Saggar et al., 1981). Sulphur bound in microbial biomass amounts to about 1-3% of the total organic S in agricultural soils (Saggar et al., 1981; Chapman, 1987; Wu et al., 1994). Hu et al. (2002) found higher values with 4.4 to 4.9% in non-vegetated soils and 5.2 and 8.8% in vegetated soils. The turnover of the soil microbial biomass is fundamental for the incorporation of SO_4^{2-}-S into soil organic matter (O'Donnell et al., 1994).

Anon (2002c) determined an active fungal biomass of 17.9 μg g^{-1} and a total fungal biomass of 63 μg g^{-1} in an agricultural soil, but also higher values of 1750 μg g^{-1} were found (Gisi, 1990). The S content of fungi species varied between 852 μg g^{-1} and 2772 μg g^{-1} (Saggar et al., 1981). On agricultural soils between 3 and 11.4 g (dry matter) and on grassland up to 46.3 g m^{-2} (dry matter) of biomass in the form of lumbricidae was found (Roembke et al., 2002). These findings indicate that lumbricidae supposedly bind a distinct amount of S.

2.2.2 Global sulphur fluxes

The pedosphere is a border phenomenon of biosphere, hydrosphere and lithosphere on the terrestrial surface. In figure 1 values for total annual S fluxes are presented. Estimates on global S fluxes and S species may differ widely because of transferring regional measurements to a global scale, analytical methods employed and temporal and spatial variations caused by climatic and microbiological conditions (Andreae and Jaeschke, 1992).

Atmospheric sulphur fluxes. Together with the industrialisation in the 18th century the technical meaning of sulphur and particularly sulphuric acid increased continuously (Rutherford, 2002). The production of sulphuric acid was a gauge for the industrial state of a country at the start of the 20th century. In the pre-industrial era anthropogenic S emissions were negligible and S depositions were only 43 Tg S yr^{-1} (Andreae, 1986). Now-a-days, the combustion of fossil fuels accounts for 80 to 85% of the total anthropogenic S emissions (Whelpdale, 1992). Bates et al. (1992) point out the necessity to evaluate S cycling on a regional scale, because most S species have an atmospheric lifetime in the order of days (Summers and Fricke, 1989; Wojcik and Chang, 1997) and larger scales make it difficult to assess the climatic and environmental significance. The same authors found that the northern hemisphere is strongly influenced by anthropogenic S emissions which provide a global input of about 80 Tg S; in this region only 8% of the total S emissions come from natural sources (Dignon and Hameed, 1989; Bates et al., 1992). In the southern hemisphere, volcanic and oceanic emissions make up 58% of the total S emissions (Bates et al., 1992; Guttikunda et al., 2001). Ichikawa et al. (2001) expect an increase of SO_2 emissions in East Asia by 34% until 2030 under current legislation options; in some eastern parts of China the same authors expect a doubling of S depositions. In 1995, China emitted 23.7-109 t of SO_2 and about one third of the country is already affected by acid rain (Hao et al., 2001) with tremendous economic effects (Feng et al., 2002). Bouwman and van Vuuren (1999) assume a global increase of 3% in anthropogenic SO_2 depositions from 1990 to 2015 while they expect global S depositions on terrestrial surface to remain unchanged in this period.

Biomass burning is another major contributor to SO_2 emissions of which 90% is caused by humans (Hileman, 1990) and about 80% of it occurs in the tropics during the dry season (Dignon and Penner, 1991). About 8700 Tg biomass (DM) is burnt annually which contributes 1.44 to 2.94 Tg S to the global emissions (Andreae, 1991; Bates et al., 1992).

Surface and underground coal seam fires are severe problems in North and South America, Australia, Africa and Asia, and only in China about 6 Tg S yr^{-1} is emitted by this source (Greenpeace, 2002). Perspectives for 2050 with a view to the development of SO_2 emissions compared to those of 2000 are as follows: +23% for Europe, +108% for Asia, -14% for USA, +78% for China and +243% for India (Matsuoka, 1992). S fluxes may change rapidly. An impressive example is the abrupt reduction of anthropogenic S emissions after breakdown of the former East Germany at the beginning of the 1990's. Earlier, the German Democratic Republic alone delivered nearly the same amount of SO_2 as all the neighbouring countries put together to West-Germany (Schnug, 1998). In Germany SO_2 emissions declined after re-unification by 84% to 831·103 t SO_2 yr^{-1} and must be further reduced to 520·103 t SO_2 yr^{-1} in 2010 according to EU guidelines (UBA, 2002).

Another example is the gulf war. During the Iraqi invasion and occupation of Kuwait, one of the worst ever man-made environmental disasters took place (Al-Ajmi,

1991; USGO, 1992; Al-Awadhi et al., 1992). The blazing and gushing oil wells (more than 700) spewed vast amounts of oil to the land, estimated at over 60 million barrels (Al-Attar, 1994), forming about 300 oil lakes that covered over 49 km^2 of land surface. The amount of oil lost due to fires and oil flows has been calculated to about 1.0-1.5 billion barrels (Petroleum Economist, 1992), ending up in an S emission of about 0.12 Tg S during just one year (Fig. 1).

Hydrospheric sulphur fluxes. The global S flux was estimated to be in the range of 131 Tg S yr^{-1}, of which 46 - 85 Tg S yr^{-1} is of anthropogenic origin. Constant S fluxes are, for instance, caused by the discharge from springs. For example, the Rhume spring in the German Harz mountain area in Northern-Germany (51.592°N; 10.304°E) delivers each year 7092 t of S bound to gypsum (Herrmann, 1969) which is twice the amount of S sold in the entire country as potassium-sulphate fertilisers and theoretically enough to fertilise 10% of all oilseed rape cropped in Germany with S.

Husar and Husar (1985) determined the water and S run-off for 54 large river basins in the world. The mean S concentration in all rivers was 3.1 mg S L^{-1}. Mean values of 0.5 mg S L^{-1}, 8 mg S L^{-1} and 100 mg S L^{-1} were found in African, European and North American rivers, respectively. The S run-off increased with industrialisation to maximum values of > 2 g S m^{-2} yr^{-1} in the studies areas; the natural background value is 1-3 mg S m^{-2} yr^{-1} (Husar and Husar, 1985).

Fluxes of organic and inorganic sulphur compounds between atmosphere, biosphere, pedosphere and hydrosphere. Semi-constant S fluxes are those by volcano emissions, exchange between hydrosphere and atmosphere and by industry (Fig. 2). Reduced S gases emitted to the atmosphere are rapidly oxidised before they are deposited so that global fluxes consider sulphate and sulphur dioxide (Andreae and Jaeschke, 1992).

In general, the flux of S gases from soils increases exponentially with surface temperature and moisture content (Staubes et al., 1986; Goldan et al., 1987; Lamb et al., 1987; Melillo and Steudler, 1989). Marsh and tideland S emissions are the major contributors to biogenic DMS and H_2S emissions (Bates et al., 1992). Up to 51 Tg DMS and 15 Tg S H_2S are released from the hydrosphere in total (Fig. 2). The sea to air flux is proportional to the air-sea concentration difference (Andreae and Jaeschke, 1992). The possible role of organic sulphide gases in the biosphere for S based cooling, and its possible ameliorating effect on climate is discussed by Shaw (1983).

Under humid conditions, the soil water regime is characterised by S losses through leaching over winter, but this S flux has so far not been quantified on a regional scale. In section 3 special attention is paid to the significance of soil hydrology to the S balance for agricultural sites.

2.2.3 Sulphur transformation processes in ecosystems

S cycle implies several gaseous species, ions, solid phases, poorly soluble minerals such as pyrite and organic compounds (section 2.2.1 and 2.2.2). Abiotic redox reactions lead to the conversion of SO_2 to sulphuric acid in the atmosphere, the precipitation of FeS_2 under anoxic conditions and its solubilisation in the presence of oxygen. In comparison, the dissimilatory sulphate reduction to H_2S under anoxic conditions is a microbially

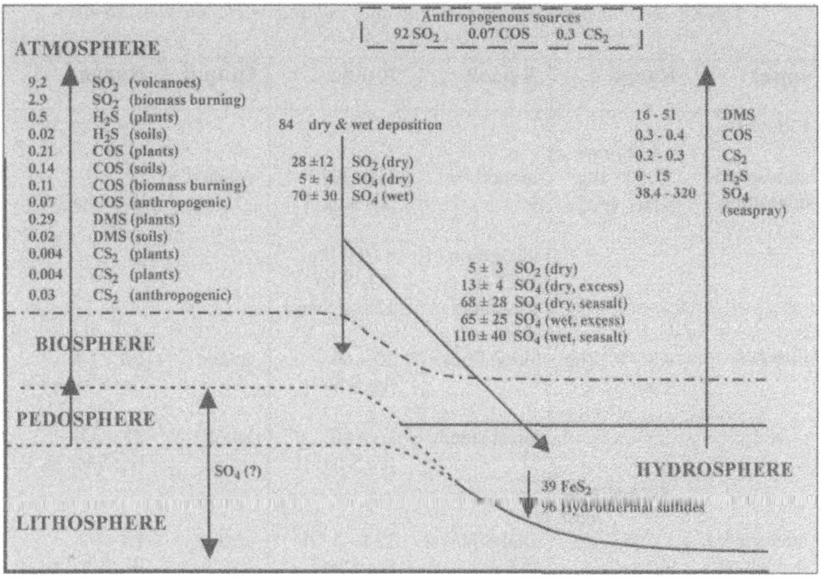

Figure 2. Interactions and fluxes (Tg S yr^{-1}) of S compounds in the global biogeochemical S cycle. (Sources: Brimblecombe et al., 1989; Kelly and Smith, 1990; Andreae and Jaeschke, 1992; Bates et al., 1992; Pham et al., 1995)

mediated redox process. Additionally, mass transfer processes such as diffusion and volatilisation affect S cycling rates. In the following sections the main transformation processes and differences of the S cycle in upland, wetland and limnic ecosystems are highlighted, pool sizes outlined and S fluxes between different ecosystems presented. In table 2 characteristic figures are given for S pools and fluxes in upland, wetland and limnic ecosystems.

Upland ecosystems. In typical forest and agricultural ecosystems oxic conditions predominate throughout the year. Atmospheric S depositions are the major S input, while losses can be contributed to volatilisation, leaching and runoff (Table 2). Sulphur input through weathering usually contributes less than 5% to the total S input (Mitchell et al., 1986). In agricultural ecosystems only the bulk deposition is relevant for the S nutrition of crops (see section 1). In comparison, important S inputs in forests consist of wet and dry deposition, through-fall, stem-flow and interception (Evans et al., 1997). Mitchell et al. (1992) reported regional differences of wet S inputs in the range of two orders of magnitude (1.6 kg ha^{-1} yr^{-1} in Australia and >80 kg S ha^{-1} yr^{-1} in Poland). Likens et al. (1990) found out that monitoring through-fall and runoff S in forest ecosystems provides reliable information about the total S depositions. A beech forest, for instance, intercepts about 50 kg S ha^{-1} yr^{-1} whereas this value is as high as 85 kg S ha^{-1} yr^{-1} for a spruce forest (Ellenberg et al., 1986). The quantification of these parameters of the S balance is strongly influenced by the measurement techniques (Vet

Table 2. Sulphur pools and fluxes in upland, wetland and limnic ecosystems

Input	Range	S pools	Range	Output	Range
UPLAND					
	Forest				
atmospheric deposition	37-50 [kg S ha^{-1} yr^{-1}]	mineral soil S	310–3070 [kg S ha^{-1}]	seepage	37 - 41 [kg S $ha^{-1} yr^{-1}$]
		adsorption of SO_4^{2-}-S	8–700 [kg SO_4-S ha^{-1}]		
		microbial biomass	12 [kg S ha^{-}]		
litterfall	4.8 – 6.3 [kg S ha^{-1} yr^{-1}]	forest floor	20 – 60 [kg S ha^{-1}]	uptake	6.4 - 7.6 [kg S $ha^{-1} yr^{-1}$]
		forest stand	20 – 60 [kg S ha^{-1}]	runoff	21 – 34 [kg S $ha^{-1} yr^{-1}$]
	Agricultural land				
atmospheric deposition	12-21 [kg S ha^{-1} yr^{-1}]	total S in soil (A_h)	224 – 1120 [kg S ha^{-1} yr^{-1}]	uptake	13 – 42 [kg S $ha^{-1} yr^{-1}$]
ground water	0 – 295 [kg S ha^{-1} yr^{-1}]			leaching	30 – 80 [kg S $ha^{-1} yr^{-1}$]
minerali-zation	10 – 30 [kg S ha^{-1} yr^{-1}]			gaseous losses	0.2 – 3 [kg S $ha^{-1} yr^{-1}$]
WETLAND					
	Freshwater				
inflow	3.2 - 641[kg S ha^{-1} yr^{-1}]	SO_4^{2-}-S concentra-tion in water	30 – 500 [µM SO_4-S]	discharge	10 – 70% [% of input]
		burial in sediments	0.03–32 [kg S ha^{-1} yr^{-1}]	H_2S	0.01 – 26 [kg S ha^{-1} yr^{-1}]
				DMS	0.004 - 1.8 [kg S ha^{-1} yr^{-1}]
	Marine				
tidal flux	0.0031 – 1.6 [Tg S ha^{-1}]	SO_4^{2-}-S concentra-tion in water	1 – 40 [mM SO_4-S]	Max. export through creeks	192 [g S m^{-2} yr^{-1}]
		burial in sediments	32 – 320 [kg S ha^{-1} yr^{-1}	H_2S	0.009 – 6020 [kg S ha^{-1} yr^{-1}]
				DMS	0.00 – 8.8 [kg S ha^{-1} yr^{-1}]

LAKES atmosphere, streams, ground-water	6.7 – 13.5 [kg S ha^{-1} yr^{-1}]	lake water	1 – 1400 [µg SO_4-S g^{-1}]	gaseous losses	~0.05 [kg S ha^{-1} yr^{-1}]
		accumula-tion in sediments	1.6 – 25.6 [kg S ha^{-1} yr^{-1}	streams	Inflow conc. = outflow conc.

Sources: **uplands:** Stevenson (1986), Hultberg and Grennfelt (1992), Sah and Meiwes (1993), Feger (1995), Preuschoff (1995), Bloem (1998), Pritzel et al. (2001); **wetlands:** Howarth et al. (1983), Cook et al. (1990), Giblin et al. (1990), Giblin and Wieder (1992); **lakes:** Cook et al. (1990), Cook and Kelly (1992)

et al., 1988), the calculation method (Driscoll et al., 1998), the moisture content of the canopy and the ground level vegetation (Padro, 1994).

Acid sub-soils, e.g. under forest vegetation and peat soils show pH values < 5, which are far below the acidity levels of fertile agricultural soils. Large storage capacities for adsorbed sulphate exist on such sites because the pH value of the soil is lower than the zero point of charge (ZPC) resulting in the hydration of metal oxides and, thus, a positive surface (Curtin and Syers, 1990; Ajwa and Tabatabai, 1995). Although, adsorbed sulphate plays only a minor role in the direct S nutrition of agricultural crops it may contribute positively to the S balance of the whole surrounding landscape (see section 3.1). On a German forest site about 700 kg S ha^{-1} in the form of sulphate was adsorbed (Feger, 1995). Sah and Meiwes (1993) found that on an acidic beech soil about 20% (8 kg S ha^{-1} yr^{-1}) of the total S input is adsorbed, while no sulphate adsorption was determined on a calcareous soil.

Soils that have a low pH value and a high S adsorption contribute to the acidification of lakes. Yet, it is not clear if the acidification process is reversible and if it is, how long it will take (Christophersen et al., 1990; Skeffington and Brown, 1992; Bishop and Hultberg, 1995; Gbondo-Tugbawa and Driscoll, 2002; Moffat et al., 2002). On soils with a low S adsorption capacity a significant decrease of acidification is expected within 5-10 years after reduction of atmospheric S depositions (Wright et al., 1988; van Dijk et al., 1992; Skjelkvåle et al., 2001). In contrast, on soils with a high sulphate adsorption capacity this process presumably takes decades (Skeffington and Brown, 1992; Alewell, 2001) because of the slow release of adsorbed sulphate. Compared to forest soils, agricultural soils usually show pH values >5.5 so that sulphate adsorption is negligible and leaching of S during winter being the major process. Under humid conditions, S losses through leaching vary between 20 to 120 kg S ha^{-1} yr^{-1} with an average of 50 to 60 kg S ha^{-1} yr^{-1} (Table 2; Schachtschabel et al., 1998).

The total S content of a forest floor varies between 11 to 150 kg S ha^{-1}, while the S pool of the corresponding mineral soils may be as high as 310 to 3070 kg S ha^{-1} (Table 2; Mitchell et al., 1992). In hardwoods of the northern hemisphere up to 64 kg S ha^{-1} are bound in the biomass, while corresponding values for the southern hemisphere are as low as 0.3-2.6 kg S ha^{-1} (Turner and Lambert, 1979). Feger (1995) found that spruce forest stands accumulate 37-55 kg S ha^{-1} and that the annual S uptake is only 4 to 5 kg S ha^{-1} yr^{-1} (Table 2; Feger, 1995).

In agricultural systems S uptake and off-take are significantly higher than in forest ecosystems and the significance of their regional expansion for the S balance of landscapes is discussed in section 3.1. In table 3, S uptake of forests and agricultural crops are compared. The reason for the extremely low S content bound in *Pinus radiata* is the low biomass production of only 315 kg ha^{-1} together with extreme S deficiency (Table 3; Lambert and Turner, 1977).

Table 3. Sulphur bound in biomass of forests and S uptake of trees and agricultural crops

Upland ecosystem	**S deposition**	**S bound biomass (kg ha^{-1})**	**S uptake (kg ha^{-1} yr^{-1})**	**Litterfall (kg ha^{-1} yr^{-1})**
Forests				
Pinus radiata [a]	none	0.34	-----	----
Dediduous oak (*Quercus pyrenaica*) [b]	Low	19.6	3.7	
Norway Spruce (*Picea abies*)[c]	high	37 – 55	4 – 5	2.4 – 3.0
Beech forest (*Fagus sylvatica* L)[d+e]	high	68.1	6.4 – 7.6	4.8 – 6.3
Agricultural crops	**S supply**	**S off-take (kg S mg^{-1})**	**S uptake (kg ha^{-1} yr^{-1})**	**S in residues (straw) (kg ha^{-1})**
Winter wheat (*Triticum vulgare*)[f]	sufficient	3 – 5	30 – 43	7 – 26
Oilseed rape (*Brassica napus*)[g]	sufficient	15 – 20	45 – 100	15 – 51
Sugar beet (*Beta vulgaris*)[f]	sufficient	0.21– 0.44[1)]	13 – 18	8 – 15
Rice (*Oryza sativa* L.)[h]	sufficient	3.2	11 – 20	—

Sources: (a) Lambert and Turner (1977); (b) Quilchano et al. (2002); (c) Feger (1995); (d) Sah and Meiwes (1993); (e) Meiwes and Khanna (1981); (f) Bloem (1998); (g) Schnug (1988); (h) Katyal et al. (1997); [(1)]calculated per fresh weight of the beets

Filner et al. (1984) calculated a worldwide S emission from plants of 7.4 Tg S yr^{-1}, while Winner et al. (1981) came to a value of 54 Tg S yr^{-1}. These estimates seem to be too high when compared to the sum of S released by plants in the form of different S gases (Fig. 2).

Globally, Crutzen (1983) determined an annual S emission of < 4 Tg S yr^{-1} in the form of H_2S, $(CH_3)_2S$ and CH_3SH from agricultural fields, 24 Tg yr^{-1} from wetlands and estimated an emission rate of 16 Tg S yr^{-1} from tropical forests. Vegetation is the largest sink and source for/to atmospheric carbonyl sulphide (COS) and DMS, respectively (Fall et al., 1988; Bates et al., 1992). Gaseous S emissions from forests (vegetation and soil) in temperate regions are in the range of 0.16-0.53 kg S ha^{-1} yr^{-1} and, therefore, of minor significance for the S balance (Andreae and Jaeschke, 1992). Emissions from annual crops such as corn and soybean are mainly DMS, while deciduous trees emitted H_2S and DMS in similar amounts (Andreae and Jaeschke, 1992). Generally the emission of S gases increases with temperature and illumination (Fall et al., 1988). Root injury, high levels of atmospheric SO_2 and increased levels of soil water sulphate or bisulphite also promote the release of H_2S (Andreae and Jaeschke, 1992).

Total S emissions from agricultural soils show diurnal and seasonal variations (Kanda et al., 1995; Yang et al., 1998), increasing with the use of organic manure (Yang et al., 1998) and nitrogen fertilisers (Kanda et al., 1995), while decreasing with oxygen availability (Chapman et al., 1996). S emission rates of 40 - 95 g S ha^{-1} were reported for paddy rice fields in China (Yang et al., 1998), 56-170 g S ha^{-1} for fields where maize and wheat were grown (Kanda et al., 1995) and 7.5 g S ha^{-1} for peat bogs (Chapman et al., 1996). In comparison, Adams et al. (1981) calculated S emission rates of 2 kg S ha^{-1} yr^{-1} for coastal wetlands, 3 kg S ha^{-1} yr^{-1} for inland soils high in organic matter and 150 g S ha^{-1} yr^{-1} for characteristic inland soils.

Wetland ecosystems. The worldwide acreage of wetland ecosystems is about 5.8x108 ha (Giblin and Wieder, 1992). Saline and brackish wetlands comprise approximately 9% of that area, while freshwater wetlands cover the remaining area. Acid sulphate soils can be assigned to wetlands since they are strongly influenced by the soil water regime and are mentioned separately in this section because of their globally increasing relevance. Turner et al. (1990) have described wetlands as poorly drained areas transitional between terrestrial and aquatic systems, where the water table is periodically at or near the surface, or the land is covered by water. In wetland ecosystems sulphate reduction and the formation of metal sulphide are dominant processes. The soils are often high in organic matter (Cowardin et al., 1979). Under saturated conditions, wetland soils are oxic only at the surface (Howes et al., 1981). The transition from oxic to anoxic zones depends on soil drainage, water flow rates and plant species that deliver oxidants (Dacey and Howes, 1984; Bloem et al., 1995).

Sulphur in wetland soils is mainly abundant in the form of reduced inorganic S minerals (pyrite, iron monosulphides and elemental sulphur), organic S forms and sulphate. In salt marshes and brackish swamps reduced inorganic sulphur minerals are dominating; in freshwater wetlands mainly reduced organic S compounds are found (Casagrande et al., 1980). The most important processes in wetland ecosystems are the microbial dissimilatory sulphate reduction, the evolution of reduced sulphur compounds and the formation of reduced sulphides such as pyrite. Oxidation of S produces acidity and may affect the use of these drained soils for agriculture. In the course of the dissimilatory sulphate reduction fermentative micro-organisms break down organic matter into smaller organic molecules (e.g. lactate, acetate, alcohols) that are oxidised to CO_2 by sulphate-reducing bacteria using sulphate as terminal electron acceptor. Variation in sulphate reduction rates over time and between sites can be attributed to three major factors: temperature, substrate availability and partitioning between oxic and anoxic respiration (Giblin and Wieder, 1992). Sulphate reduction rates from salt marsh pans and creeks are generally distinctly lower than those from vegetated areas indicating the promoting influence of roots, rhizomes and possibly root exudates. Availability of organic carbon in marine sediments, and sulphate concentration in freshwater systems are the limiting factors of sulphate reduction.

Reported S reduction rates varied between 1 and 75 mol S m^{-2} yr^{-1} (Howarth and Teal, 1979; Oenema, 1988; Swieder, 1988) in salt marsh soils and 0.1 to 17.2 mol S m^{-2} yr^{-1} in freshwater wetlands (Brown and MacQueen, 1985; Spratt and Morgan, 1990; Wieder et al., 1990). Sulphide, which is produced through sulphate reduction, can be either deposited as metal sulphide (Berner, 1982) or it can be oxidised to elemental S or sulphate. Sulphide can also be methylated, both chemically and biologically to volatile organic compounds. A variety of volatile S compounds is produced in wetlands including hydrogen sulphide (H_2S), methylmercaptan (CH_4SH), dimethyl sulphide (DMS), dimethyl disulphide (DMDS), methanethiol (MT), carbonylsulphide (COS) and carbon disulphide (CS_2) (Lomans et al., 2002). A part of these biogenic gases is emitted to the atmosphere where they are oxidised to sulphur dioxide and sulphate. These S containing gases contribute to the acidification of rainfall (Andrea and Jaeschke, 1992) and some of them are relevant for global warming (Lomans et al., 2002; see section 2.2.2).

H_2S fluxes range from 0.013 to 68,721 µg S m^{-2} h^{-1}, DMS fluxes from 0.3 to 328 µg S m^{-2} h^{-1} and other S gases from <0.1 to 100 µg S m^{-2} h^{-1} (Giblin and Wieder, 1992). The contribution of these gases to the annual S fluxes is, however, only small (Streudler and Peterson, 1985; Giblin and Wieder, 1992).

Annual gaseous S emissions from saline wetlands are about 1000 times greater than from freshwater wetlands (Adams et al., 1981). Ivanov (1977) estimated a total rate of H_2S production in marine sediments of 900 Tg S yr^{-1}. In comparison, the global H_2S emission of soils is estimated to be in the range of 0.02 Tg S (Fig. 2). The annual amount of sulphide buried as pyrite in sediments is only 39 Tg S yr^{-1} (Berner, 1982). These findings imply that the major part of sulphide produced in marine sediments is lost by gaseous emissions, where it is re-oxidised to sulphate by dissolved oxygen (Jørgensen, 1978; Berner, 1982).

Acid sulphate soils. Acid sulphate soils describe natural sediments and soils that contain iron sulphides (pyrite, iron mono-sulphide or their precursors). The exposure of sulphide in these soils to oxygen, for example by drainage or excavation causes the production of sulphuric acid. So-called actual acid sulphate soils contain highly acidic soil layers (pH = 4) due to the aeration of the soil. Wang and Luo (2002) measured pH values between 2.8 and 3.5 in acid sulphate soils and calculated a lime requirement of up to 2687.5 t ha^{-1} for adjusting the soil pH to an optimum level. In comparison, potential acid sulphate soils have not been exposed to air yet and are, therefore, not acidic (Anon, 2002d), a fact of prime relevance with a view to changes in land use. Acid sulphate soils may have adverse side-effects on environment, economy and human health (Anon, 2002d).

Limnic ecosystems. Lakes account for only 0.01% of the water on the earth surface, but they are among other important sources and sinks for water (Anon, 2002e). Streams, precipitation and groundwater discharge are the main factors of water inflow, while evaporation, groundwater recharge and discharge to streams drain the lake. The chemistry of freshwater lakes is closely linked to the water temperature (Anon, 2002e). Relevant parameters of S inputs are wet and dry atmospheric S depositions, inflow of sulphate and organic S via groundwater and rivers and litter-fall from surrounding forests (David and Mitchell, 1985; Mitchell et al., 1992; Urban et al., 2001). In northern temperate regions, the export of S from terrestrial ecosystems is in balance with atmospheric inputs and the latter directly related to the sulphate concentration of lake water (Sullivan et al., 1988). Usually the sulphate discharge is slightly smaller than the inflow. Seepage lakes, however, retain 40 to > 80% of the added sulphate (Lin et al., 1987; Marnette et al., 1993), supposedly due to the extended contact time of water and sediments (Kelly et al., 1987; Evans et al., 1997).

Giblin et al. (1990) have reported S accumulation rates of 160 to 2,560 mg S m^{-2} d^{-1} in different lake sediments. The burial and release of S from sediments is of ecological interest, because fixation of S in lake sediments increases alkalinity and release of S from lake sediments increases acidity (Urban and Monte, 2001). This means that in regions with high atmospheric S depositions the lake alkalinity production is substantially enhanced through increased bacterial sulphate reduction (Cook et al., 1986; Evans et al., 1997).

Sulphate concentrations in lakes vary between 0.7 and 640 mg SO_4 L^{-1} (Table 2; Cook and Kelly, 1992). The highest values were found in pit lakes from mine drainage. In marine systems the sulphate reduction is not limited by sulphate because of the high concentration of 900 mg S L^{-1} in seawater (see section 2.2.1). In contrast, in lakes a sulphate concentration of < 96-192 mg SO_4 L^{-1} limits the reduction (Ramm and Bella, 1974; Lovley and Klug, 1986; Kelly and Rudd, 1989).

In lakes with extremely low sulphate concentrations such as Lake Victoria in Africa (< 160 µg SO_4-S L^{-1}) turnover times of sulphate were as short as 10 - 14 days (Lehman and Branstrator, 1994). In eutrophic, hard water lakes with high sulphate concentrations turnover times are in the order of years (King and Klug, 1982).

Gaseous losses of 0.05 to 0.5 kg S ha^{-1} yr^{-1} were measured at the surface of deeper and shallow unstratified lakes, respectively (Table 2; Cook and Kelly, 1992). Sulphate in lakes is conveyed by diffusion and bio-turbation to anoxic zones; here, bacterial sulphate reduction releases hydrogen sulphide. Hydrogen sulphide may either react to form metal sulphides, or diffuse to oxic zones where it is oxidized chemically or by chemoautotrophic or photoautotrophic S-oxidising bacteria to S0, Sx2- (polysulphides) and sulphate (Luther and Church, 1992). Between 2.3 and 23.4 kg SO_4-S ha^{-1} yr^{-1} is reduced by bacteria (Cook and Kelly, 1992). In acid pit lakes values of 144 SO_4-S kg ha^{-1} day^{-1} were measured (Herlihy et al., 1987).

Total S budgets are difficult to set up for lakes as it is complicated to quantify sedimentation of organic S, retention of sulphides, mineralisation rates and oxidation of reduced S species (Rudd et al., 1986). Cook and Kelly (1992) calculated a natural S input of 6.7 to 135 kg S ha^{-1} yr^{-1} and an outflow of 3.5 to 129 kg S ha^{-1} yr^{-1} for 7 lakes.

Pit lakes. As soon as open-pit mining ends, inflow of groundwater and runoff from adjacent land leads to the development of pit lakes. Pit lakes store significant amounts of water and may have serious effects on the regional groundwater quality. In eastern Germany, 118 of the largest pit lakes have a volume of 6.5 km^3 and are, thus, higher than all larger natural lakes in the northern part of the country (Geller, 2000). Pit lakes are deeper than natural lakes and, therefore, differ from them in stratification, chemical properties and the activity of sulphate reducing bacteria. The water of pit lakes at metal-sulphide and coal mines is strongly acidic (pH 2.5 - 3.5; Geller, 2000), high in sulphate and eventually rich in dissolved heavy metals and metalloids (Obermann et al., 2000). Sulphate concentrations of 1,400 to 6,000 mg SO_4 L^{-1} were determined (Obermann et al., 2000; Strauch et al., 2000; Tempel et al., 2000; USGS, 2002). The following heavy metal concentrations were measured: 0.29-0.59 mg As L^{-1} to 96 mg As L^{-1} in the epilimnion (Tempel et al., 2000), 25 mg Fe L^{-1} (USGS, 2002), 0.46 mg Se L^{-1} and 8.1 mg U L^{-1} (Paulson, 2001). Sulphate reducing bacteria are important for neutralising acidic lake water (see above) and precipitating heavy metals. Bioremediation, therefore, focuses on optimum growth conditions for sulphate reducing bacteria which imply adjustment of the pH and the supply of additional carbon sources (USGS, 2002).

3. SULPHUR CYCLING AT DIFFERENT SCALES

The primary interest in nutrient balances on any spatial scale is the determination of nutrient surpluses and deficits. S balances on a national scale for instance reveal that China and India are the countries with the strongest and increasing S deficit in agriculture against the background of an over-proportionally increasing demand for food production (Messick, 1997). Messick (1997) estimates that in 2005 the S deficit will be 5.5 Tg S in Asia, 2.2 Tg S in China, 1.6 Tg S in India, 1.3 Tg S in Africa, 1.2 Tg S in North America, 0.9 Tg S in Latin America and 0.5 Tg S in Western Europe.

The S supply of agricultural crops is closely related to landscape characteristics and principle interactions are outlined in section 3.1. S fertilisation to agricultural crops based on S balances needs to consider all relevant S input and output. This implies that traditional S balances are extended by soil hydrological S fluxes (section 3.2).

3.1 Sulphur Cycling in Landscapes

S input and off-take in agro-ecosystems may have a strong influence on neighbouring ecosystems. S is often in excess in natural systems, even if the atmospheric S input is low because the turnover of S is much lower in these than in agricultural systems. Natural vegetation, land without plant production and forests show a positive S balance even when atmospheric inputs are low (section 2.2.3, Table 2; Eriksen et al., 1998). The reasons for differences in the S balance between agricultural and forest sites were outlined in section 2.2.3. Under humid conditions, forests can be described as S accumulating areas, while agricultural farmland is S consuming. Agricultural crops may benefit from this S pool, if groundwater reservoirs of both ecosystems are connected. Consequently, in forest rich landscapes the risk of S deficiency in agricultural crops is reduced. The consequences of deforestation for the S cycling are summarised by Vitousek (1983) as follows: sulphate leaching is the most likely dominant form of S loss; harvest and biomass burning remove S rapidly and erosion of organic S-containing compounds may be significant; a rapid decline in soil and stream water sulphate concentrations is possible.

3.2 Site-Specific Sulphur Cycling and Sulphur Balances for Agriculture

The S status of an agricultural site is difficult to assess, because soil analysis is not reliable due to the high spatial and temporal variability of sulphate in soils (Schnug and

Haneklaus, 1998; Bloem et al., 2001; Haneklaus et al., 2002). Sources and sinks commonly included in the S balance are inputs by depositions from atmosphere, fertilisers, plant residues and mineralisation, and outputs by S off-take and S losses due to leaching. A frequent problem when establishing such simple S balance is that the budget does not correspond with the actual S supply. The reason is that under temperate conditions it is rather the spatio-temporal variation of hydrological soil properties which controls the plant available sulphate-S content. Thus, a site-specific S budget which includes such information is a more promising way for the evaluation of the S supply. The major soil properties and external S sources affecting the amount of plant available sulphate in the soil are shown in figure 3.

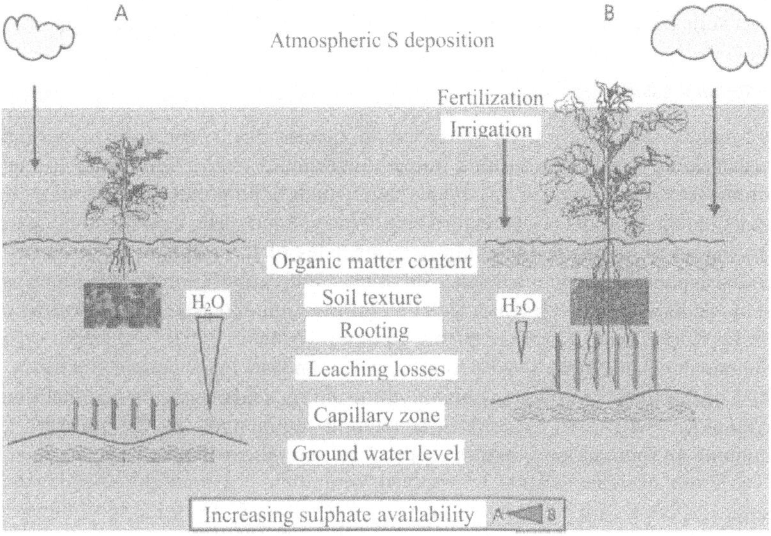

Figure 3. Overview and efficacy of essential impact factors modifying plant available sulphate in soils. (Source: Haneklaus et al., 2002)

The significance of plant available soil water as source and storage for S has been disregarded or underestimated so far. But, as a matter of fact, especially under humid growing conditions, plant available soil water is the largest contributor to the S balance (Bloem, 1998). Leaching and import from sub-soil or shallow groundwater sources (Bloem, 1998; Eriksen et al., 1998) can change the amounts of sulphate available for plants within a very short time. Groundwater is a large pool for S, because S concentrations of 5-100 mg S L^{-1} (see section 2.2.1) are common in surface near groundwater (Isermann, 1993; Bloem, 1998). There are three ways how groundwater contributes to the S nutrition of plants (see also Fig. 3):

- There is a direct S input if the groundwater level is only 1-2 m below surface, which is sufficient to cover the S requirement of most crops, as plants can utilise the sulphate in the groundwater directly by their root system.

- Groundwater which is used for irrigation can supply up to 100 kg S ha^{-1} to the crop (Preuschoff, 1995; Schlichting, 1996; Bloem, 1998; Pedersen et al., 1998), but irrigation water will only contribute significantly to the S supply if applied at the start of the main growth period of the crop.
- The capillary rise of groundwater under conditions of a water saturation deficit in the upper soil layers leads to an S input. This process is highly dependent on climatic conditions.

The S supply of a crop increases with the amount of plant available water or shallow groundwater. The higher the water storage capacity of a soil, the less likely are losses of water and sulphate-S by leaching, and the greater is the pool of porous water and also the more likely is an enrichment of sulphate-S just by subsequent evaporation. Thus, heavier soils have a higher charging capacity for sulphate-S than lighter ones.

3.3 Sulphur Cycling in Plants

The smallest scale of S cycling is that in plants. Plants are able to assimilate atmospheric SO_2 and H_2S, but this fraction contributes only a minor way to their S nutrition (see section 1 and 2.2.3). Plants satisfy their S requirement predominately by sulphate uptake by roots (Hawkesford et al., this volume). The next metabolic step is the reduction of sulphate. Metabolic intermediates of the sulphate assimilation are activated adenosine phosphosulphate, sulphite, sulphide and S containing amino acids such as methionine and cysteine. Detailed information about metabolic pathways is given by Kopriva and Koprivova (this volume). Cysteine is the first stable organic compound in which S, taken up by roots or leaves, appears in the plant metabolism. Up to 90% of the total S in plants is bound in the amino acids, cysteine and methionine (Giovanelli, 1990). All reduced S containing compounds in the primary (e.g. methionine and glutathione) and the secondary (e.g. glucosinolates) metabolism are derived from cysteine. Plants co-evolved with their environment and developed strategies to cope with excessive S (Ernst, 1993). During the reduction of sulphate, taken up by plant roots, significant amounts of S are released to the atmosphere, of which H_2S is the predominant form (Schroeder, 1993). It has been suggested that the release of H_2S regulates, homeostatically the size of the cysteine pool and, thus, maintains it at low level because of its cytotoxicity (Rennenberg, 1984).

Brassica plants also developed mechanisms to cope with an insufficient S supply. The bio-recyclisation of glucosinolates is a fundamental part of the ecological survival strategy of *Brassica* species, because the dissemination of these plants relies on a large number of small seeds with high concentrations of S containing amino acids. A proposed model for the regulation of the glucosinolate and myrosinase system's dependence on the S supply is given by Schnug and Haneklaus (1994).

4. FUTURE RESEARCH NEEDS

There are still various open questions in relation to the S cycling and its links to plant environment. From an agricultural point of view the following research topics are of prime interest:

Quantification of S fluxes in soil micro-pores. In section 3.2 the significance of the water storage capacity of soils for the S supply of crops was mentioned. This raises new questions concerning the physical S cycling on small scales, in particular the influence of sulphate retention in micro-pores on soil water dynamics. The retention of S in surface soils is strongly dependent on pH and mineralogy (see section 3.2). Rapid leaching may, however, also occur on tropical soils by preferential flow under non-equilibrium conditions that usually occur during and shortly after rainfall, and under humid conditions it is possible that considerable amounts of sulphate are stored in soil water fractions which are less available to plants. Combining the convection-dispersion equation for the evaluation of solute flow with the mobile-immobile model (Van Genuchten and Wierenga, 1976; Vogeler et al., 1998) is supposed to significantly improve the prognosis of S fluxes and, thus, to modelling the soil S balance for different soils and climatic conditions.

S cycling in the edaphon. The edaphon is an active part of mobilisation and immobilisation processes in soils (section 2.2.1). However, quantitative relationships between the microbial immobilisation of inorganic S, the turnover of soil microbial biomass S and subsequent formation of organic S as well as the extent of availability of these S fractions for plants have not been determined experimentally so far. There is also a lack of knowledge about the S metabolism of nematodes, arthropods and worms and their interactions with the S cycle of plants. Particularly, the significance of *Lumbricidae* on the S transformation processes in soils is of interest, because they bind significant amounts of S (section 2.2.1), their activity enhances the turnover of soil organic matter, and promotes soil life and soil structure (Topp, 1981).

Volatile S emissions of plants. The mechanisms for sulphur induced resistance (SIR) of plants against pests and diseases include an increased synthesis of natural components, the degradation of glycosides, the synthesis of new components and hypersensitive response (Haneklaus et al., 2001). So far quantitative data about volatile S emissions of plants are scarce and there is only circumstantial evidence of a link to the S nutritional status (section 3.3). H_2S may be released prior or after cysteine formation (Giovanelli, 1990) but the question is still open as to which enzymes catalyse the release of H_2S. The studies of Burandt et al. (2001) revealed that genotypical differences existed for L-cysteine desulfhydrase and O-acetylserine(thiol)lyase activity as well as total S status among 10 different lines of oilseed rape. An increasing total S content of the plants was associated with decreasing L-cysteine desulfhydrase and increasing O-acetyl serine(thiol)lyase activity. The verification of causal relationships between S supply, release of H_2S and resistance against pathogens could be a breakthrough for applying SIR as a means of biological plant protection.

5. REFERENCES

Adams DF, Farwell SO, Pack MR and Robinson E (1981) Biogenic sulfur gas emissions from soils in Eastern and Southeastern United States. *J Air Poll Control Assoc* **31:** 1983—2009

Agripedia (2002). Introductory Soil Microbiology. http://www.ca.uky.edu/Agripedia/classes/pls566/class22.htm

Ajwa HA and Tabatabai MA (1995) Metal-induced sulfate adsorption by soils: I. Effect of pH and ionic strength. *Soil Sci* **159:** 32—42

Al-Ajmi D (1991) Effects of mixing height and wind speed on the dispersion of air pollution in Kuwait. Harvard Conference on Kuwait Oil-Fire Impact, Boston, Massachusetts

Al-Attar MH (1994) An integrated approach for over-coming the adverse environmental impacts inflicted upon Kuwait during the Iraqi occupation. Proceedings of International Symposium on Environment, Budapest

Al-Awadhi N, El-Nawawy A and Al-Daher R (1992) In-situ and on-site bioremediation of oil-contaminated soil in Kuwait. Proceedings of the Third World Academy of Sciences (TWAS), Kuwait

Alewell C (2001) Predicting reversibility of acidification: The European sulfur story. *Water Air Soil Pollut* **130:** 1271—1276

Andreae MO (1986) The ocean as a source of atmospheric sulfur compounds. In: Buat-Menard P (ed) The Role of Air-sea Exchange in Geochemical Cycling, pp 331—362. Reidel, Dordrecht

Andreae M (1991) Biomass burning: its history, use, and distribution and its impact on environmental quality and global climate. In: Levine JS (ed) Global Biomass Burning: Atmospheric, Climatic and Biosphere Implications, pp 3—21, MIT Press, Cambridge

Andreae MO and Jaeschke WA (1992) Exchange of sulphur between biosphere and atmosphere over temperate and tropical regions. In: Howarth RW, Stewart JWB and Ivanov MV (eds) SCOPE 48 Sulphur Cycling on the Continents: Wetlands, Terrestrial Ecosystems and Associated Water Bodies, pp 27—66. John Wiley & Sons, Chichester

Anon (2002a) http://www.demographie.de/info/epub/pdfdateien/SWITodesursachen

Anon (2002b) http://www.asl-associates.com/content3.doc

Anon (2002c) http://www.co.marion.or.us/mpubwork/parks/bonesteele/soilfoodweb.htm

Anon (2002d) Planning and managing development involving acid sulfate soils. DRAFT State planning policy including guidelines, Queensland Government, http://www.ipa.qld.gov.au/docs/sppacidsulfsoil.pdf

Anon (2002e) Lecture No 20 Pit lakes. http://www.cs.umt.edu/GEOLOGY/classes/Geol431/lectur20htm

Bates TS, Cline JD, Gammon RH and Kelly-Hansen SR (1987) Regional and seasonal variations in the flux of oceanic dimethylsulfide to the atmosphere. *J Geophys Res* **92:** 2930—2938

Bates TS, Lamb BK, Guenther A, Dignon J and Stoiber RE (1992) Sulfur emissions to the atmosphere from natural sources. *J Atmos Chem* **14:** 315—337

Berner RA (1982) Burial of organic carbon and pyrite sulfur in the modern ocean: its geochemical and environmental significance. *Am J Sci* **282:** 451—473

Bishop KH and Hultberg H (1995) Reversing acidification in a forest ecosystem: The Gårdsjoen Covered Catchment. *Ambio* **24:** 85—91

Bloem E (1998) Schwefel-Bilanz von Agraroekosystemen unter besonderer Beruecksichtigung hydrologischer und bodenphysikalischer Standorteigenschaften. Dissertation TU Braunschweig. Landbauforschung Voelkenrode Sonderheft 192

Bloem E, Luettmann B and Giani L (1995) Jahreszeitliche Schwefeldynamik semisubhydrischer Salzmarschen. *Z Pflanzenernaehr Bodenk* **158:** 251—256

Bloem E, Haneklaus S, Sparovek G and Schnug E (2001) Spatial and temporal variability of sulphate concentration in soils. *Commun Soil Sci Plant Anal* **32:** 1391—1403

Boudreau BP and Westrich JT (1984) The dependence of bacterial sulfate reduction on sulfate concentration in marine sediments. *Geochim Cosmochim Acta* **48:** 2503—2516

Bourbos VA, Skoudridakis MT, Barbopoulou E and Venetis K (2000) Ecological control of grape powdery mildew (*Uncinula necator*). http://www.landwirtschaft-mlr.badenwuerttemberg.de/la/lvwo/kongress/SULFUR.html

Bouwman L and van Vuuren D (1999) Global assessment of acidification and eutrophication of natural ecosystems. *UNEP/DEIA&EW/TR* 99—106

Bowen HJM (ed) (1966) Trace Elements in Biochemistry. Academic Press, London

Brimblecombe P, Hammer C, Rodhe H, Ryaboshapko A and Bourton CF (1989) Human influence on the sulphur cycle. In: Brimblecombe P and Lein AY (eds) Evolution of the Global Biogeochemical Sulphur Cycle, pp 77—121. John Wiley & Sons, New York

Brown KA and MacQueen JF (1985) Sulphate uptake from surface water by peat. *Soil Biol Biochem* **17:** 411—420

Burandt P, Papenbrock J, Schmidt A, Bloem E, Haneklaus S and Schnug E (2001) Genotypical differences in total sulfur contents and cysteine desulfhydrase activities in *Brassica napus* L. *Phyton* **41:** 75—86

Capone DG and Kiene RP (1988) Comparison of microbial dynamics in marine and freshwater sediments: Contrasts in anaerobic carbon catabolism. *Limnol Oceanogr* **4:** 725—749

Casagrande DJ, Gronli K and Sutton N (1980) The distribution of sulfur and organic matter in various fractions of peat: Origins of sulfur in coal. *Geochim Cosmochim Acta* **44:** 25—32

Chapman SJ (1987) Microbial sulphur in some Scottish soils. *Soil Biol. Biochem.* **19:** 301—305

Chapman SJ, Kanda KI, Tsuruta H and Minami K (1996) Influence of temperature and oxygen availability on the flux of volatile sulphur compounds from wetlands: a comparison of peat and paddy soils. *Soil Sci Plant Nut* **42** 279—288

Christophersen N, Neal C, Seip HM and Stone A (1990) Hydrochemical models for simulation of present and future short term changes in stream chemistry: development and status. In: Mason BJ (ed) The Surface Water Acidification Programme, pp 97—106. Cambridge University Press, Cambridge

Clark TJ (2002) Sulfur. http://www.vitamins-minerals.tv/minerals/sulfur.htm

Cook RB, Kelley CA, Schindler DW and Turner MA (1986) Mechanisms of hydrogen ion neutralization in an experimentally acidified lake. *Limnol Oceanogr* **31:** 134—148

Cook RB, Kreis RG, Kingston JC, Camburn KE, Norton SA, Mitchell MJ, Fry B and Shane LCK (1990) Paleolimnology of McNearney Lake: an acidic lake in northern Mitchigan. *Paleolimnology* **3:** 13—34

Cook RB and Kelly CA (1992) Sulphur cycling and fluxes in temperate dimictic lakes. In: Howarth RW, Stewart JWB and Ivanov MV (eds) SCOPE 48 Sulphur Cycling on the Continents: Wetlands, Terrestrial Ecosystems and Associated Water Bodies, pp 145—188. John Wiley & Sons, Chichester

Cowardin LM, Cater V, Gollet FC and Laroe ET (1979) Classification of wetlands and deepwater habitats of the United States. FWS/OBS-79/31, US Fish and Wildlife Service

Crutzen PJ (1983) Atmospheric interactions - homogenous gas reactions of C, N, and S containing compounds. In: Bolin B and Cook RB (eds) The Major Biogeochemical Cycles and Their Interactions, SCOPE 21, pp 67—112. John Wiley & Sons, Chichester

Curtin D and Syers JK (1990) Extractability and adsorption of sulfate in soils. *J Soil Sci* **41:** 305—312

Dacey JW and Howes BL (1984) Water uptake by roots controls water table movement and sediment oxidation in short *Spartina* marsh. *Science* **224:** 487—489

David MB and Mitchell MJ (1985) Sulfur constituents and cycling in water, seston, and sediments of an oligotrophic lake. *Limnol Oceanogr* **30:** 1196—1207

De Kok LJ (1990) Sulfur metabolism in plants exposed to atmospheric sulfur. In: Rennenberg H, Brunold C, De Kok LJ, Stulen I (eds) Sulphur Nutrition and Sulphur Assimilation in Higher Plants; Fundamental, Environmental and Agricultural Aspects, pp 111—130. SPB Academic Publishers, The Hague

Dignon J and Hameed S (1989) Historic emissions of sulfur and nitrogen oxides from 1860 to 1980. *J Air Poll Con Assoc* **39:** 180—186

Dignon J and Penner JE (1991) Biomass burning: a source of nitrogen oxides in the atmosphere. In: Levine JS (ed) Global biomass burning, pp 370—375. MIT Press, Cambridge

Driscoll CT, Likens GE and Church MR (1998) Recovery of surface waters in the northeastern U.S. from decreases in atmospheric depositions of sulfur. *Water Air Soil Pollut* **105:** 319—329

Duvigneaud P and Denayer-De Smet S (1968) Essai de classification chimique (éléments minéraux) des plantes gypsicoles du Bassin de l'Ebre. *Bull Soc r Bot Belg* **101:** 279—291

Ellenberg H, Mayer R and Schauermann J (eds) (1986) Oekosystemforschung. Ergebnisse des Sollingprojektes 1966-1986. Ulmer Verlag, Stuttgart

Ellis D (ed) (1932) Sulphur bacteria - a monograph. Longmans, Green and Co., London

Eriksen J, Murphy MD and Schnug E (1998) The soil sulphur cycle. In: Schnug E (ed) Sulphur in Agroecosystems, pp 39—74. Kluwer Academic Publishers, Dordrecht

Eriksson E (1960) The yearly circulation of chloride and sulfur in nature; meteorological, geochemical and pedological implications. Part II. *Tellus* **12:** 63—109

Ernst WHO (1993) Ecological aspects of sulfur in higher plants: the impact of SO_2 and the evolution of the biosynthesis of organic sulfur compounds on populations and ecosystems. In: De Kok LJ, Stulen I, Rennenberg H, Brunold C and Rauser WE (eds) Sulfur Nutrition and Assimilation in Higher Plants, pp 295—313. SPB Academic Publishers, The Hague

Ernst WHO (1997) Life-history syndromes and the ecology of plants from high sulphur-habitats. In: Cram WJ, De Kok LJ, Stulen I, Brunold C and Rennenberg H (eds) Sulphur Metabolism in Higher Plants, pp 131—146. Backhuys Publishers, Leiden

Evans HE, Dillon PJ and Molot LA (1997) The use of mass balance investigations in the study of the biogeochemical cycle of sulfur. *Hydrological Processes* **11:** 765—782

Fall R, Albritton DL, Fehsenfeld FC, Kuster WC and Goldan PD (1988) Laboratory studies of some environmental variables controlling sulfur emissions from plants. *J Atmos Chem* **6:** 341—362

Feger KH (1995) Solute fluxes and sulfur cycling in forested catchments in SW Germany as influenced by experimental (NH4)2SO4 treatments. *Water Air Soil Pollut* **79:** 109—130

Feng Z, Miao H, Zhang F and Huang Y (2002) Effects of acid deposition on terrestrial ecosystems and their rehabilitation strategies in China. *J Environ Sci* **14:** 227—233

Fiedler, HJ and Takur S (1984) Zur Schwefelernaehrung der Waldbaeume und blattanalytischen Bewertung ihres Ernaehrungszustandes. *Beitr f d Forstwirtschaft* **18:** 81—86

Filner P, Rennenberg H, Sekiya J, Bressan RA, Wilson LG, Le Cureux L and Shimel T (1984) Biosynthesis and emission of hydrogen sulfide by higher plants. In: Koziol MJ and Whatley FR (eds) Gaseous Air Pollutants and Plant Metabolism, pp 291—312. Butterworth, Stoneham

Fontes J-C, Fritz P, Louvat D and Michelot J-L (1989) Aqueous sulphates from the Stripa groundwater system. *Geochim Cosmochim Acta* **53:** 1783—1789

Freney JR and Williams CH (1983) The sulphur cycle in soil. In: Ivanov MV and Freney JR (eds) The Global Biochemical Sulphur Cycle. pp 129—201. John Wiley & Sons, Chichester

Freney JR, Ivanov MV and Rodhe H (1983) The sulphur cycle. Section 2.5. In: Bolin B and Cook RB (eds) The Major Biogeochemical Cycles and Their Interactions, 56—61. John Wiley & Sons, New York

Friend JP (1973) The global sulfur cycle. In: Rasool SI (ed) Chemistry of the Lower Atmosphere, 177—201. Plenum Press, New York

Galloway JN (1996) Anthropogenic mobilization of sulfur and nitrogen: Immediate and delayed consequences. *Annu Rev Energy Environ* **21:** 261—292

Gbondo-Tugbawa SS and Driscoll CT (2002) Evaluation of the effect of future controls on sulfur dioxide and nitrogen oxide emissions on the acid-base status of a northern forest ecosystem. *Atmos Environ* **36:** 1631—1643

Geiser PM (2002) http://www.mineralwaters.org

Geller W (2000) Schwefelsaure Bergbaurestseen. DFG-Programmkonzeption Geotechnologien, Leitprojekt VIII "Stoffkreislaeufe", 30.08.2000, Hannover

Giblin AE, Likens GE, White D and Howarth RW (1990) Sulfur storage and alkalinity generation in New England lake sediments. *Limnol Oceanogr* **35:** 852—869

Giblin AE and Wieder RK (1992) Sulphur cycling in marine and freshwater wetlands. In: Howarth RW, Stewart JWB and Ivanov MV (eds) SCOPE 48 Sulphur Cycling on the Continents: Wetlands, Terrestrial Ecosystems and Associated Water Bodies, pp 85—124. John Wiley & Sons, Chichester

Giovanelli J (1990) Regulatory aspects of cysteine and methionine biosynthesis. In: Rennenberg H, Brunold C, De Kok LJ and Stulen I (eds) Sulphur Nutrition and Sulphur Assimilation in Higher Plants; Fundamental, Environmental and Agricultural Aspects, 33—48. SPB Academic Publishers, The Hague

Gisi U (ed) (1990) Bodenoekologie. Thieme Verlag

Goldan PD, Kuster WC, Albritton DL and Fehsenfeld FC (1987) The measurement of biogenic sulfur emissions from soils and vegetation: three sites in the eastern United States revisited. *J Atmos Chem* **5:** 439—467

Greenpeace (2002) Minenfeuer - Wenn die Erde Feuer faengt. Greenpeace Magazin 2/00

Grime P (ed) (1979) Plant Strategies and Vegetation Processes. John Wiley & Sons, Chichester

Guttikunda SK, Thongboonchoo N, Arndt RL, Calori G, Carmichael GR and Streets DG (2001) Sulfur deposition in Asia: seasonal behavior and contributions from various energy sources. *Water Air Soil Pollut* **131:** 383—406

Haehndel R and Isermann K (1993) Soluble nitrogen and carbon in the underground of sports fields. Proceedings of 7th International Turfgrass Research Conference, Palm Beach, Florida, pp 1—6

Haneklaus S, Bloem E and Schnug E (2002) Sulfur In: Rattan Lal (ed) Encyclopedia of Soil Science, pp 1282—1288. Marcel Dekker Inc., New York

Haneklaus S, Bloem E and Schnug E (2002) The significance of sulphur induced resistance (SIR) for sustainable agricultural production. Proceedings of 13th International Reinhardsbrunn Symposium "Modern Fungicides and Antifungal Compounds III", Friedrichroda, pp 365-371

Hao J, Wang S, Liu B and He K (2001) Plotting of acid rain and sulfur dioxide pollution control zones and integrated control planning in China. *Water Air Soil Pollut* **130:** 259—264

Hawkesford MJ (2002) Smart plant technology for sensing crop nutritional status. http://www.iacr.bbsrc.ac.uk/aen/nutacquisit.htm

Health Canada (1994) Les sulfates. http://www.health-canada.ca/ehp/dhm/catalogue/dpc_pubs/rqepdoc_appui/sulfates.pdf

Herlihy AT, Mills AL, Hornberger GM and Bruckner AE (1987) The importance of sediment sulfate reduction on the sulfate budget of an impoundment receiving acid mine drainage. *Water Resources Res* **23:** 287—292

Herrmann A (1969) Die geologische und hydrologische Situation der Rhumequelle am Suedharz. Abh. Karst- und Hoehlenkunde, Reihe B, Heft 1, Muenchen. http://www.mediatec.de/speleo/german/de/spring/rhume.html

Hilemann B (1990) Biomass burning: Environment hurt more than thought. *Chem Engineering News* **26:** 4—5

Howarth RW and Teal JM (1979) Sulfate reduction in a New England salt marsh. *Limnol Oceanogr* **24:** 999—1013

Howarth RW, Giblin A, Gale J, Peterson BJ and Luther GW (1983) Reduced sulfur compounds in the pore waters of a New England salt marsh. *Ecol Bull* (Stockholm) **35:** 135—152

Howes BL, Howarth RW, Valiela I and Teal JM (1981) Oxidation - reduction potentials in salt marsh: Spatial patterns and interactions with primary production. *Limnol Oceanogr* **26:** 350—360

Hu Z, Haneklaus S, Wang S, Xu C, Cao Z and Schnug E (2003) Comparison of mineralization and distribution of soil sulfur fractions in the rhizosphere of oilseed rape and rice. *Commu Soil Sci Plant Anal* (in press)

Hultberg H and Grennfelt P (1992) Sulphur and seasalt deposition as reflected by throughfall and runoff chemistry in forested catchments. *Environ Pollut* **75:** 215—222

Husar RB and Husar JD (1985) Regional river sulfur runoff. *J Geophy Res* **90:** 1115—1125

Ichikawa Y, Hayami H, Sugiyama T, Amann M and Schoepp W (2001) Forecast of sulfur deposition in Japan for various energy supply and emission control scenarios. *Water Air Soil Pollut* **130:** 301—306

Isermann K (1993) Loeslicher N, Sulfat-S und (DO)C im (un-)gesaettigten Untergrund von Porengrundwasserleitern bei unterschiedlicher Landbewirtschaftung/Duengung. *Mitt Dt Bodenkd Ges* **71:** 141—144

Ivanov MV (1977) Influence of microorganisms and microenvironment on the global sulfur cycle. In: Krumbein WE (ed) Environmental Biogeochemistry and Geomicrobiology, The Aquatic Environment: pp, 47—61. Ann Arbor Science Publisher, Ann Arbor, MI

Jørgensen BB (1978) A comparison for methods for the quantification of bacterial sulfate reduction in coastal marine sediments. III. Estimation from chemical and bacteriological field data: *Geomicrobio J* **1:** 49—64

Kanda KI, Tsuruta H and Minami K (1995) Emissions of biogenic sulfur gases from maize and wheat fields. *Soil Sci Plant Nut* **41:** 1—8

Katyal JC, Sharma KL and Srinivas K (1997) Sulphur in Indian Agriculture, Proceedings of TSI/FAI/IFA Symposium on Sulphur in Balanced Fertilisation, KS-2/1-KS-2/12

Kelly CA and Rudd JWM (1989) Reply to comment by Urban and Baker. *Limnol Oceanogr* **34:** 1146—1148

Kelly DP and Smith NA (1990) Organic sulfur compounds in the environment. *Adv Microbial Ecol* **11:** 345—385

Kelly CA, Rudd JWM, Hesslein RH, Schindler DW, Dillon PJ, Driscoll CT, Gherini SA and Hecky RE (1987) Prediction of biological acid neutralization in acid-sensitive lakes. *Biogeochem* **3:** 129—140

King GM and Klug MJ (1982) Comparative aspects of sulfur mineralization in sediments of a eutrophic lake basin. *Appl Environ Microbiol* **43:** 1406—1412

Kinzel H (1972) Biochemische Pflanzenphysiologie. *Ver Verbr Naturwiss Kenntnisse Wien* **112:** 77—98

Klemm RF and Gray JML (1982) A study of the chemical composition of particulate matter and aerosols over Edmonton. Report No. RMD 82/9, prepared for the Research Management Division by the Alberta Research Council

Krauskopf KB (ed) (1967) Introduction to Geochemistry. McGraw Hill, New York

Lamb B, Westberg H, Allwine G, Bamesberger L and Guenther A (1987) Measurement of biogenic sulfur emissions from soils and vegetation: application of dynamic enclosure methods with Nautsch filter and GC/FPD analysis. *J Atmosph Chem* **5:** 469—491

Lambert MJ and Turner J (1977) Dieback in high site quality *Pinus radiata* stands- the role of sulphur and boron deficiency. *NZ J For Sci* **7:** 333—348

Larssen S and Hagen LO (1996) Air Quality in Europe, 1993- A pilot report. European Topic Centre on Air Quality, European Environment Agency, pp123

LaMoreaux PE and Tanner JT (eds) (2001) Springs and Bottled Waters of the World. Springer Verlag, Berlin

Lauenroth WK, Milchunas DG and Dodd JL (1983) Response of a grassland to sulfur and nitrogen additions under controlled SO_2 exposure. *Environ Exp Bot* **23:** 339—346
Lehman JT and Branstrator DK (1994) Nutrient dynamics and turnover rates of phosphate and sulfate in Lake Viktoria, East Africa. *Limnol Oceanogr* **39:** 227—233
Likens GE, Bormann FH, Hedin LO, Driscoll CT and Eaton JS (1990) Dry deposition of sulfur: a 23-year record for the Hubbard Brook Forest ecosystem. *Tellus* **42B:** 319—329
Lin JC, Schnoor JL and Glass GE (1987) Ion budgets in a seepage lake. In: Hites RA and Eisenreich SJ (eds) Sources and Fates of Aquatic Pollutants, pp 209—227. American Chemical Society, Washington, DC
Lomans BP, van der Drift C, Pol A and Op den Camp HJM (2002) Microbial cycling of volatile organic sulfur compounds. *Cell Mol Life Sci* **59:** 575—588
Lovley DR and Klug MJ (1986) Model for the distribution of sulfate reduction and methanogenesis in freshwater sediments. *Geochim Cosmochim Acta* **50:** 11—18
Luther III GW and Church TM (1992) An overview of the environmental chemistry of sulphur in wetland systems. In: Howarth RW, Stewart JWB and Ivanov MV (eds) SCOPE 48 Sulphur Cycling on the Continents: Wetlands, Terrestrial Ecosystems and Associated Water Bodies, pp 125—144. John Wiley & Sons, Chichester
Marnette ECL, Houweling H, van Dam H and Erisman JW (1993) Effects of decreased atmospheric deposition on the sulfur budgets of two Dutch moorland pools. *Biogeochem* **23:** 119—144
Matsuoka Y (1992) Future projections of global anthropogenic sulfur emissions and their environmental effects. *Environ Sys Res* **22:** 359—368
Meiwes KJ and Khanna PK (1981) Distribution and cycling of sulphur in the vegetation of two forest ecosystems in an acid rain environment. *Plant Soil* **60:** 369—375
Melillo JM and Steudler PA (1989) The effect of nitrogen fertilizer on the COS and CS_2 emissions from temperate forest soils. *J Atmos Chem* **9:** 411—417
Messick D (1997) Sulphur in world agriculture: the international and commercial situation. Proceedings of Symposium on Sulphur in Balanced Fertilisation, New Dehli, KS-1, pp 1—9
Mitchell MJ, David MB, Maynard DG and Telang SA (1986) Sulfur constituents in soils and streams of a watershed in the Rocky Mountains of Alberta. *Can J For Res* **16:** 315—320
Mitchell MJ, David MB and Harrison RB (1992) Sulphur dynamics of forest ecosystems. In: Howarth RW, Stewart JWB and Ivanov MV (eds) SCOPE 48 Sulphur Cycling on the Continents: Wetlands, Terrestrial Ecosystems and Associated Water Bodies, pp 215—254. John Wiley & Sons, Chichester
Moffat AJ, Kvaalen H, Solberg S and Clarke N (2002) Temporal trends in throughfall and soil water chemistry at three Norwegian forests, 1986-1997. *For Ecol Management* **168:** 15—28
NIVA (2001). National presentations from the 16th meeting of the ICP Waters programme task force in Riga, Latvia. ICP-Waters Report 63/2001
Noggle JC (1979) Sulfur accumulation by plants; the role of gaseous sulfur in crop nutrition. In: Shriner DS, Richmond CR and Lindberg SE (eds) Atmospheric Sulfur Deposition - Environmental Impact and Health Effects, pp 289—297. Ann Arbor Science Publisher, Ann Arbor, MI
Nriagu JO and Holdway DA (1989) Production and release of dimethyl sulfide from the great lakes. *Tellus* **41B:** 161—169
Oae S and Okuyama T (eds) (1992) Organic Sulfur Chemistry: Biochemical Aspects, CRC Press, Boca Rotan, Florida
Obermann P, Wisotzky F, Kuever J and Detmers J (2000) Quantifizierung und Modellierung des geogenen, biogenen und antropogenen Schwefelkreislaufs in Deutschland. DFG-Programmkonzeption Geotechnologien, Leitprojekt VIII "Stoffkreislaeufe", Hannover
O'Donnell AG, Wu J and Syers JK (1994) Sulphate-S amendments in soil and their effects on the transformation of soil sulphur. *Soil Biol Biochem* **26:** 507—514
Oenema O (1988) Early diagenesis in recent fine-grained sediments in the Eastern Scheldt. Ph.D. Dissertation, University of Utrecht, The Netherlands
Omar SAS, Briskey E, Misak R and Asem AASO (1998) The Gulf War impact on terrestrial environment of Kuwait: An overview. 1st Int. Conf. on Addressing Environmental Consequences of War - Legal, Economic and Scientific Perspectives, Washington, DC
Oude Elferink SJWH and Meijer GAL (2002) Sulfur-containing compounds in drinking water for cattle (Risico's van zwavelverbindingen in drinkwater voor runderen). ID TNO Diervoeding, Postbus 65, 8200 AB, Lelystad. http://www.precisievoeding.nl/documenten/ID_rapp2000-009.pdf
Padro J (1994) Observed characteristics of the dry deposition velocity of O3 and SO2 above a wet deciduous forest. *Sci Total Environ* **147:** 395—400

Paulson O (2001) Draft: Bioremediation of the sweetwater pit lake. http://www.greenworldscience.com/pitpapertext.htm

Pedersen CA, KnudsenL and Schnug E (1998) Sulphur fertilisation. In: Schnug E (ed) Sulphur in Agroecosystems, pp 115—134. Kluwer Academic Publishers, Dordrecht

Petroleum Economist (1992) The Oil Fires Story.

Petrenchuk O and Drozdova VM (1966) On the chemical composition of cloud water. *Tellus* **18:** 280—286

Pham M, Mueller J-F, Brasseur GP, Granier C and Megie G (1995) A three-dimensional study of the tropospheric sulfur. *J Geophysical Res* **100:** 26061—26092

Preuschoff M (1995) Untersuchungen zur Schwefelversorgung von Weißkohl an zwei Loeßstandorten. PhD, University Hanover, Verlag Ulrich E. Grauer, Stuttgart

Prietzel J, Weick C, Korintenberg J, Seybold G, Thumerer T and Treml B (2001) Effects of repeated (NH4)2SO4 application on sulfur pools in soil, soil microbial biomass, and ground vegetation of two watersheds in the, Black Forest/ Germany. *Plant Soil* **230:** 287—305

Quilchano C, Haneklaus S, Gallardo JF, Schnug E and Moreno G (2002) Sulphur balance in a broadleaf, non-polluted, forest ecosystem (central-western Spain). *For Ecol Management* **161:** 205—214

Ramm AE and Bella DA (1974) Sulfide production in anaerobic microcosms. *Limnol Oceanogr* **19:** 110—118

Reeburgh WS (1997) Figures summarizing the global cycles of biogeochemically important elements. *Bull Ecol Soc Am* **78:** 260—267

Rennenberg H (1984) The fate of excess sulphur in higher plants. *Annu Rev Plant Physiol* **35:** 121—153

Roembke J, Beck L, Foerster B, Fruend HC, Horak F, Ruf A, Rosciczweski C, Scheurig M and Woas S (2002) Boden als Lebensraum fuer Mikroorganismen. Eine Literaturstudie. http://www.uvm.baden-wuerttemberg.de/bofaweb/berichte/tbb04/tbb04.htm

Rudd JWM, Kelly CA and Furutani A (1986) The role of sulfate reduction in the long term accumulation of organic and inorganic sulfur in lake sediments. *Limnol Oceanogr* **31:** 1281—1291

Rutherford (2002) Schwefel. Chalogene VI. Hauptgruppe. http://www.uniterra.de/rutherford/ele016.htm

Saalbach E (1984) Die Bedeutung der in der Atmosphaere enthaltenen Schwefelverbindungen fuer die Schwefelversorgung landwirtschaftlicher Nutzpflanzen. *Angewandte Botanik* **58:** 147—156

Saggar S, Bettany JR and Stewart JWB (1981) Measurement of microbial sulphur in soil. *Soil Biol Biochem* **13:** 493—498

Sah SP and Meiwes KJ (1993) Sulfur inputs and outputs for two European beech forests growing on different soil substrates. *Can J For Res* **23:** 1626—1630

Schachtschabel P, Blume H-P, Bruemmer G, Hartge KH and Schwertmann U (eds) (1998). Lehrbuch der Bodenkunde. 14. Auflage, Ferdinand Enke Verlag Stuttgart

Schlichting M (1996) Der Sulfatgehalt des Grundwassers in Abhaengigkeit von bodennutzungsspezifischen Stoffeintraegen und dessen Bedeutung fuer die Nutzung des Grundwassers zur Feldberegnung - Beispiel Fuhrberger Feld. Master thesis, University of Hannover, Hannover

Schnug E (1988) Quantitative und qualitative Aspekte der Diagnose und Therapie der Schwefelversorgung von Raps (*Brassica napus* L.) unter besonderer Beruecksichtigung glucosinolatarmer Sorten. Habilitationsschrift (DSc thesis), Agrarwiss. Fakultaet der Christian-Albrechts-Universitaet zu Kiel

Schnug E (1993) Physiological functions and environmental relevance of sulfur-containing secondary metabolites. In: De Kok LJ, Stulen I, Rennenberg H, Brunold C and Rauser WE (eds) Sulfur Nutrition and Assimilation in Higher Plants, pp 179—190. SPB Academic Publishers, The Hague

Schnug E (1997) Significance of sulphur for the quality of domesticated plants. In: Cram WJ, De Kok LJ, Stulen I, Brunold C and Rennenberg H (eds) Sulphur Metabolism in Higher Plants, pp 109—130. Backhuys Publishers, Leiden

Schnug E (1998) Air pollution and global change - impact on agriculture. In: De Kok LJ and Stulen I (eds) Response of Plant Metabolism to Air Pollution, pp 15—22. Backhuys Publishers, Leiden

Schnug E and Haneklaus S (1994) Sulphur deficiency in *Brassica napus* - biochemistry, symptomatology, morphogenesis. Landbauforschung Voelkenrode, FAL Braunschweig, Sonderheft 144

Schnug E and Haneklaus S (1998) Diagnosis of sulphur nutrition. In: Schnug E (ed) Sulphur in Agroecosystems, pp 1—38. Kluwer Academic Publishers, Dordrecht

Schnug E, Haneklaus S and Murphy DPL (1993a) Impact of sulphur supply on the baking quality of wheat. *Aspects Appl Biol* **36:** 337—345

Schnug E, Haneklaus S and Murphy DPL (1993b) Impact of sulphur fertilization on fertilizer nitrogen efficiency. *Sulphur Agric* **17:** 8—12

Schnug E, Haneklaus S, Booth E and Walker KC (1995) Sulphur and stress resistance in oilseed rape. Proceedings of the 9th International Rapeseed Congress "Rapeseed Today and Tomorrow", Cambridge, Vol. 1 of 4, pp 229—231

Schroeder P (1993) Plants as sources of atmospheric sulfur. In: De Kok LJ, Stulen I, Rennenberg H, Brunold C and Rauser WE (eds) Sulfur Nutrition and Assimilation in Higher Plants, pp 253—270. SPB Academic Publishers, The Hague

Scott BC and Laulainen NS (1979) On the concentration of sulphate in precipitation. *J Appl Meteorology* **18:** 138—147

Seafriends (2002) Soil ecology http://seafriends.org.nz/enviro/soil/ecology.htm

Senning A (ed) (1982) Sulphur in organic and inorganic chemistry. Marcel Dekker Inc., New York

Shaw GE (1983) Bio-controlled thermostasis involving the sulfur cycle. *Climatic Change* **5:** 297—303

Skeffington RA and Brown DJA (1992) Timescales of recovery from acidification: Implications of current knowledge for aquatic organisms. *Environ Pollut* **77:** 227—234

Skjelkvåle BL, Tørseth K and Aas W (2001) Decrease in acid deposition - recovery in Norwegian waters. *Water Air Soil Pollut* **130**b 1433—1438

Spratt Jr HG and Morgan MD (1990) Sulfur cycling in a cedar dominated freshwater wetland. *Limnol Oceanogr* **35:** 1586—1593

Staubes R, Ockelmann G and Georgii HW (1986) Emissions of biogenic sulfur compounds from various soils. Proceedings of 2nd International Symposium on Biosphere-Atmosphere Exchange, Mainz, Federal Republic of Germany, pp 16—21

Stevenson FJ (ed) (1986) Cycle of Soil Carbon, Nitrogen, Phosphorus, Sulfur, Micronutrients. John Wiley & Sons, New York

Stevenson FJ and Cole MA (eds) (1999) Cycles of soil Carbon, Nitrogen, Phosphorus, Sulfur and Micronutrients. 2nd edi. John Wiley & Sons, New York

Strauch G, Glaeßer W, Knoeller K, Schlosser D and Trettin R (2000) Sulfatreduktion in bergbaulich gestoerten Sedimenten - hydrochemische Langzeitprozesse im Grundwasser. DFG-Programmkonzeption Geotechnologien, Leitprojekt VIII "Stoffkreislaeufe", Hannover

Strebel O and Boettcher J (1985) Einfluß von Bodennutzung und Bodennutzungsaenderungen auf die Stoffbilanz eines reduzierten Aquifers im Einzugsgebiet eines Foerderbrunnens. *Wasser Boden* 37

Streudler PA and Peterson BJ (1985) Annual cycle of gaseous sulfur emissions from a New England *Spartina alterniflora* marsh. *Atmos Environ* **19:** 1411—1416

Sullivan TJ, Eilers JM, Church MR, Blick DJ, Eshleman KN, Landers DH and DeHaan MS (1988) Atmospheric wet sulphate deposition and lakewater chemistry. *Nature* **331:** 607—609

Summers PW and Fricke W (1989) Atmospheric decay distances and times for sulphur and nitrogen oxides estimated from air and precipitation monitoring in eastern Canada. *Tellus* **41B:** 286—295

Swieder KT (1988) Transformations of sulfur compounds in salt marsh sediments. M.S. Thesis. SUNY at Stony Brook

Taylor GS, Baker MB and Charlson RJ (1983) Heterogenous interactions of the C, N, and S cycles in the atmosphere: the role of aerosols and clouds. In: Bolin B and Cook RB (eds) The Major Biogeochemical Cycles and their Interactions, SCOPE 21, pp 115—141. John Wiley & Sons, Chichester

Tempel RN, Shevenell LA, Lechler P and Price J (2000) Geochemical modeling approach to predicting arsenic concentrations in a mine pit lake. *Appl Geochem* **15:** 475—492

Topp W (1981) Biologie der Bodenorganismen. UTB, Heildelberg

Trudinger PA (1979) The biological sulfur cycle. In: Trudinger PA and Swaine DJ (eds) Biogeochemical Cycling of Mineral-Forming Elements, pp 293—313. Elsevier Science Publisher, Amsterdam

Turner J and Lambert MJ (1979) Sulfur nutrition of forests. In: Shriner et al. (eds) Atmospheric Sulfur Depositions - Environmental Impact and Health Effects, pp 321—333. Ann Arbor Science Publisher, Ann Arbor, MI

Turner RS, Cook RB, Van Miegrot H, Johnson DW, Elwood JW, Bricker OP, Lindberg SE and Hornberger GM (1990) Watershed and lake processes affecting surface water acid-base chemistry. NAPAP Report 10, in Acidic Deposition: State of Science and Technology. National Acid Precipitation Assessment Program, Washington, DC

UBA (2001) Wasserwirtschaft in Deutschland. Teil 2-Gewaesserguete oberirdischer Binnengewaesser. BMU

UBA (2002) Umweltdaten Deutschland 2002. UBA Berlin, pp 56

United States General Accounting Office (USGO) (1992) Efforts to address health effects of the Kuwait oil fires. GAO/HRD 92-50

Urban NR and Monte AE (2001) Sulfur burial in and loss from the sediment of Little Rock Lake, Wisconsin. *Can J Fish Aqua Sci* **58:** 1347—1355

Urban NR, Sampson CJ, Brezonik PL and Baker LA (2001) Sulfur cycling in the water column of Little Rock Lake, Wisconsin. *Biogeochem* **52:** 41—77

USGS (US Geological Survey) (2002) In situ remediation of acidic pit-lake water by stimulation of sulfate reducing bacteria. Water resources research grant proposal. http://water.usgs.gov/wrri/00grants/MTlake.html

van Dijk HFG, Boxmann AW and Roelofs JGM (1992) Effects of a decrease in atmospheric depositions of nitrogen and sulfur on the mineral balance and vitality of a *Scots pine* stand in The Netherlands. *For Ecol Management* **51:** 207—215

Van Genuchten MT and Wierenga PJ (1976) Mass transfer studies in sorbing porous media. I. Analytical solutions. *Soil Sci Am J* **40:** 473—480

Van Loon JC (1973) Toronto's precipitation analyzed for heavy metal content. *Water Pollut Control* **111:** 38

Vet RJ, Sirois A, Jeffries DS, Semkin RG, Foster NW, Hazlett P and Chan CH (1988) Comparison of bulk, wet-only and wet-plus-dry deposition measurements at the Turkey Lakes Watershed. *Can J Fish Aqua Sci* **45:** 26—37

Vitousek PM (1983) The effects of deforestation on air, soil, and water. In: Bolin B and Cook RB (eds) The Major Biogeochemical Cycles and their Interactions, SCOPE 21, pp 223—246. John Wiley & Sons, Chichester

Vogeler I, Scotter DR, Clothier BE and Tillman RW (1998) Anion transport through intact soil columns during intermittent unsaturated flow. *Soil Tillage Res* **45:** 147—160

Wang J and Luo S (2002) Sulfur and its acidity in acid sulfate soil of Taishan costal plain in southern China. *Commu Soil Sci Plant Anal* **33:** 579—593

Weast RC, Selby SM and Hodgman CD (1964) Handbook of Chemistry and Physics. A Ready-Reference Book of Chemical and Physical Data. 45th edi, The Chemical Rubbe Co., Cleveland Ohio

Wedepohl KH (1968) Chemical fractionation in the sedimentary environment. In: Ahrens LH (ed) Origin and Distribution of the Elements. Vol. 30 of Earth Sciences. Pergamon Press, London

Whelpdale DM (1992) An overview of the atmospheric sulphur cycle. In: Howarth RW, Stewart JWB and Ivanov MV (eds) Sulphur Cycling on the Continents: Wetlands, Terrestrial Ecosystems and Associated Water Bodies, SCOPE 48, pp 5—26. John Wiley & Sons, Chichester

Wieder RK, Yavitt JB and Lang GE (1990) Methane production and sulfate reduction in two Appalachian peatlands. *Biogeochemistry* **10:** 81—104

Winner WE, Smith CL, Koch GW, Mooney HA, Bewley JD and Krouse HR (1981) Rates of emission of H2S form plants and patterns of stable sulphur isotope fractionation. *Nature* **289:** 672—673

Winogradsky S (1887) Ueber Schwefelbakterien. *Botanische Zeitung* **XLV:** 489—507

Wojcik GS and Chang JS (1997) A re-evaluation of sulfur budgets, lifetimes, and scavenging ratios for eastern north America. *J Atmos Chem* **26:** 109—145

Worthington SRH and Ford DC (1995) High sulfate concentrations in limestone springs: an important factor in conduit initiation? *Environ Geology* **25:** 9—15

Wright RF, Lotse E and Semb A (1988) Reversibility of acidification shown by whole catchment experiments. *Nature* **334:** 670—675

Wu J, O'Donnell AG, He ZL and Syers JK (1994) Fumigation-extraction method for the measurement of soil microbial biomass-S. *Soil Biol Biochem* **26:** 117—125

Yang Z, Kong L, Zang J, Wang L and Xi S (1998) Emission of biogenic sulfur gases from Chinese rice paddy fields. *Sci Total Environ* **224:** 1—8

6. ACKNOWLEDGEMENT

The authors wish to express their sincere gratitude to Herbert Daybell (Agrimedia, Bottesford, UK) for the linguistic revision of this paper.

7. AFFILIATIONS

Institute of Plant Nutrition and Soil Science, Federal Agricultural Research Centre (FAL), Bundesallee 50, D-38116 Braunschweig, Germany
For correspondence - *Ewald Schnug*: pb@fal.de

N.S. PASRICHA AND Y.P. ABROL

FOOD PRODUCTION AND PLANT NUTRIENT SULPHUR

Abstract. There has been a tremendous increase in food grain production mainly through developments in crop science and production technology over the last four decades. However, there is increasing concern as to whether development in science and technology will be adequate to meet the increasing demands of the growing population. To produce and provide food needed for those additional 77 million people annually is possible but an uphill task. The lion's share (75 to 80%) of future food production growth will have to come from productivity increases. Further challenge would be not only to meet the food requirement from limited resources but it has to be produced in a changed demand structure in a most sustainable manner keeping in view the fragile environments especially in the resource poor countries of Asia and Africa.

Much of the increase in food grain production is attributable to use of high yielding cultivars and increased use of nitrogenous, phosphatic and potassic fertilizers. Switch over and rapid acceptance by the growers of high analysis fertilizers following World War II greatly increased the crop demands for some secondary and micronutrients. With in the last 4 decades, response to S application had been reported from most of the agricultural lands of the world. The extent and severity of S-deficiencies can be expected to increase in future because of expanding use of S-free fertilizers (high analysis N, P and K) and pesticides, increasing crop yields due to intensification that makes greater demands on soil nutrients, and implementation of air pollution control schemes. S-deficiency in crop plants work to the disadvantage of consumers by reducing quality and producers by limiting the yield, reduces use efficiency of other inputs and increases cost of production.

With increased crop production, S removal in crop harvests has increased nearly four fold from 4 mt in 1960 to current total crop removal the world over to about 15 mt which is expected to increase to over 20 mt when the world food production will be 40-50% higher by the year 2020. Rain, irrigation water, crop residues and manures alone will, thus, not be able to meet the tremendous S needs of the crops, which means this has to be supplemented with S-fertilizers. Total world production of all forms of S (elemental, pyrites and other forms) make about 60 mt, a major portion of which is used in fertilizer industry. Therefore, S for use as fertilizer will not be that easily available. This will lead to increase in the gap between S needs as estimated from crop harvest and the amount of S added as S-fertilizers. This is not a good trend and can become a cause of major worry when malnutrition due to S-deficiency becomes a problem in human and animal health. The only solution to such a situation is to use S-bearing N, P or K fertilizers wherever possible or alternately use other sources such as elemental S or gypsum.

1. INTRODUCTION

Food from agriculture supports around six billion people that inhabit this planet today. The population is expected to increase to about 7.2 and 8.3 billion, respectively, by 2015 and 2030. The major increase is expected to be in the developing countries. This will be despite the continuous decline in growth rates. There will be about 77 million added each year to the already burgeoning population (FAO, 2002). This scenario requires increased food production. This will certainly have to come through intensification of cultivation as there are limited land resources. This implies more pressure on soil and water resources, and is of particular concern because in many regions there are already "exhaustion symptoms" because of unsustained use of natural

Yash P. Abrol and Altaf Ahmad (eds.), Sulphur in Plants, 29—44.

resources (Abrol et al., 2002). For their sustained use, management of soil fertility with efficient utilization of available water resources will be a pre-requisite. Thus, additional production will be highly input sensitive, and mineral fertilizer will play a key role in realizing the potential of future crop cultivars evolved by wide breeding and transgenesis besides conventional breeding.

In a holistic scenario, mineral fertilizers accounted for around 30% of total increase in crop production in the last four decades during which world cereal production doubled (Roy, 2001). Now, there will be a change in the food requirement as it is expected that around 40 per cent of the world's population will be living in urban areas by the year 2020. For example, there will be more consumption of wheat than rice at the expense of coarse grain and the demand for vegetables and fruits will increase. Moreover, quality aspects have to be taken into consideration. The driving force for this increased quality food production is determined by improvement in nutritional status of millions who are under-nourished/malnourished on the one hand and by rising incomes of the people in the developing countries. Mineral fertilizers will play a vital role and their use has to be done in a more efficient manner. There is a possibility of some misuse of fertilizers, however, when trying to compensate poor growth caused by excessive losses on unsuitable soils or lack of certain nutrients (Luc Maene, 2002). Efficient plant nutrient use requires balanced fertilization, optimization of nutrient replenishment and minimum of nutrient losses to the environment.

Importance of S as a plant nutrient is becoming more imminent due to its effect on crop productivity and quality. In the past, inputs of S from atmospheric deposition and fertilizers were more than sufficient to meet the requirement in industrially developed countries of Western Europe and North America. In Asia and other developing world, the amount of S made available from mineralization of soil organic S was generally sufficient because of the low level of crop harvests. Even low level of SO_2 deposition from atmosphere because of industrial backwardness was no hamper for crop production. However, rapid population growth necessitated increased food production by adopting high yielding, fertilizer responsive, short duration crop cultivars grown intensively under irrigated conditions. Due to increased harvest levels, demand for all nutrients including S increased and the soil supply could no longer keep pace with the increased demand. This has resulted in wide spread S-deficiency and responses to its application. Meanwhile, traditional S containing fertilizers such as ammonium sulphate and single super phosphate are no longer widely used, although still more than 75% of total S for crop nutrition comes from these two sources. Their use is decreasing further with increased level of nutrients applied from S-free high analysis mineral fertilizer. Thus, the gap between total amount of S removed in crop harvests and that applied through fertilizers is increasing with time.

Major portion of 60 million tonnes of total S produced annually worldwide goes to the fertilizer industry for the manufacture of phosphate fertilizers. Although demand and supply scenario shows a surplus in the world market, this does not mean that adequate amount of S is used as a plant nutrient. In order to check stock piling of S produced, it is necessary to make its use in fertilizer formulations to meet the increasing needs of this element in crop production.

In this chapter, food requirement scenario keeping in view the increase in population, income level, urbanization and, thus, the changing food habits, and need to remedy the under-nourishment and malnutrition prevailing amongst the teeming millions in some regions of the world is discussed. This is followed by discussion on prevalence of S deficiency, its causes and the present and future scenario of fertilizer demand and supply.

2. FOOD REQUIREMENT AND POPULATION GROWTH

Several economic, social and political factors will continue influencing trends in food supply, demand, trade and prices. Factors with potentially significant effect on future development in world food situation besides population and income growth are urbanization, rate of increase in food production due to technological changes and productivity growth, prices of commodities and response of supply and demand to prices. Of all these factors, population and income growth are important in determining the food supply and demand balances in future.

The latest medium projection of the United Nation show that 6.1 billion people of 2000 will grow to 7.2 billion in 2015 and 8.3 billion in 2030 heading towards 9.3 billion 2050. In absolute number it amounts to current annual addition of 77 million i.e. a new Germany each year. Even by 2045-2050, there will be increments to the levels of 43 million each year. Because of large population base all these increases will be in the developing countries. The above scenario is despite that falling rate in population growth. The growth will be at an average rate of 1.1 per cent up to 2030 compared to 1.7 per cent a year over the last 30 years. Subsequent rate will be around 0.6 per cent. While in the developing world the rate will continue to be 0.7 per cent it will be 0.0 per cent for industrial countries and negative rate for transition countries (Table 1).

Table 1. Population projection

	Population (millions)				
	1979-81	**1997-99**	**2015**	**2030**	**2050**
World	4430	5900	7207	8270	9322
Developing countries	3259	4595	5858	6910	7987
Industrial countries	789	892	951	979	985
Transition countries	382	413	398	381	349
	Population growth (% per annum)				
	1979-99	**1989-99**	**1979-99 to 2015**	**2015-2030**	**2030-2050**
World	1.6	1.5	1.2	0.9	0.6
Developing countries	1.9	1.7	1.4	1.1	0.7
Industrial countries	0.7	0.7	0.4	0.2	0.01
Transition countries	0.5	0.1	-0.2	-0.3	-0.4

Source: FAO 2002: World Agriculture: towards 2015-2030

The declining growth rate will reduce the growth in demand of agricultural products. It will be still in the range of 1.7 to 2.2 per cent in developing and 0.4-0.7 in developed ones. This is to primarily meet the nutritional requirement of the millions who are undernourished. The scenario is not, however, uniform. In China, the average food consumption has reached an average of 3040 kcal (1997-99). In India with average figure being around 2500 kcal and population increase of 1 per cent a year over the next 20 years. Most of the undernourished are in the developing world with the highest population being in Asia and the sub-Saharan regions (Table 2)

Table 2. Calorie consumption and undernourished people

	CALORIE CONSUMPTION (kcal/calorie/day)				
	1961-63	**1979-81**	**1997-99**	**2015**	**2030**
World	2283	2552	2803	2940	3050
Developing countries	1960	3212	2681	2850	2980
Industrial countries	2891	3135	3380	3440	3500
Transition countries	3154	3389	2960	3060	3180

	UNDERNOURISHED million people				**(% of population)**			
	1990-92	**1997-99**	**2015**	**2030**	**1990-92**	**1997-99**	**2015**	**2030**
World		815				14		
Developing countries	816	777	610	443	20	17	11	6
Industrial countries		11				1		
Transition countries		27				6		

Source: FAO, 2002: World Agriculture: towards 2015/2030

The future economic growth will rise by 1.9% on per capita between 2000 and 2015, which will be higher than the 1.2% seen in 1990s. Closely related to population and income changes is the transformation of demographic patterns. The most vital demographic characteristic, particularly for projecting the kind and quantity of future food needs is the fast growing rate of urbanization besides fast growing economies in the developing countries. Approximately, 44% of the population of developing countries is residing in urban areas in 2000, up from 22% in 1960 and 30% in 1980. This has significant effect on food demand structure. Urbanization accelerates dietary transition from basic staples such as sorghum, millets, maize and root crops to fine cereals, fruits, vegetables, livestock products and processed foods. Cereal production is expected to grow by one billion tonnes by 2030. In developing countries dietary changes have already taken place during the period 1964-66 to 1997-99 and expected to alter still further by 2030 (Fig. 1). Most growth will come from yield increases and yield growth rates will be faster in developing countries especially for wheat and rice. Expansion of area under cultivation will not be significant source of future growth in cereal production as a whole. Food quality becomes more and more important to meet requirement of urban population. Plant nutrients applied as mineral fertilizers, particularly those which have a glaring effect on the food quality cannot be ignored under such situations. In this context, sulphur plays a significant role.

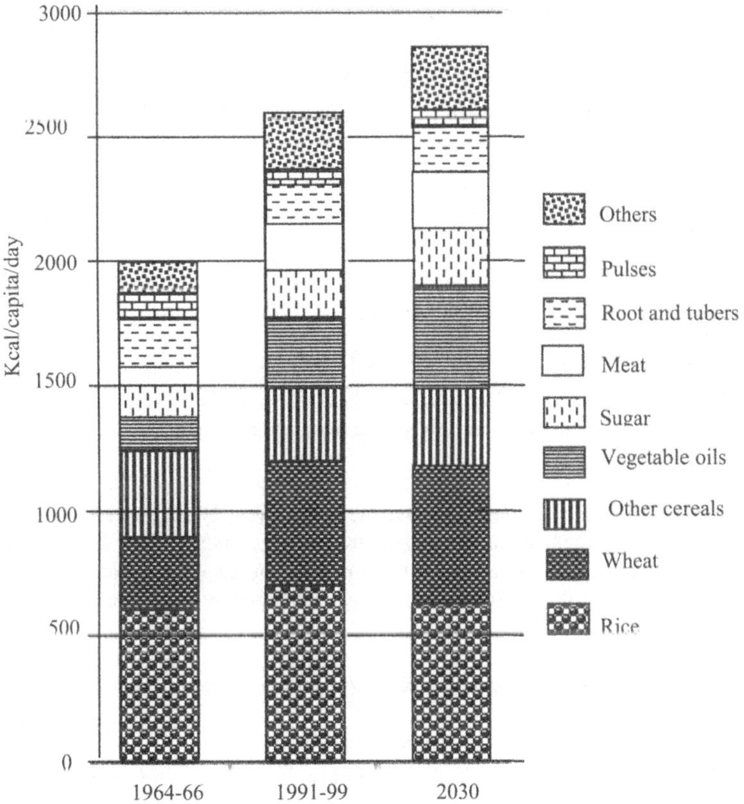

Figure 1. Dietary change in developing countries, 1964-66 to 2030

3. DEMAND FOR LIVESTOCK PRODUCTS AND GRASSLAND PRODUCTIVITY

Demand for livestock products will increase by almost 60% in developing countries from 16.5 kg yr^{-1} $capita^{-1}$ in 1990 to 26.4 kg $year^{-1}$ $capita^{-1}$ by the year 2020 while this increase will be only 5.5% for the developed countries where present average per capita meat consumption of 80 kg yr^{-1} is nearly four times higher than that in developing countries.

According to the recent FAO projection, meat consumption at global level will increase by about two fold. The consumption will be in the range of 55.0-66.0 in East Asia and Latin America/Caribbean. The present consumption in South Asia, particularly India is not high and is not expected to reach even the present levels of consumption in the other regions due to influence of dietary preference favouring less meat than in other societies (FAO, 2000) (Table 3).

Table 3. Change in the commodity consumption of food

	meat, carcass weight (kg/person/year)					
	1964/66	**1974/76**	**1984/86**	**1995/97**	**2015**	**2030**
World	24.1	27.4	30.7	34.7	40.2	43.8
South Asia*	3.9	3.9	4.3	5.5	8.2	11.8
East Asia	8.7	10.0	17.0	33.3	47.2	55.0
Sub-Sharan Africa	9.9	9.5	10.2	9.7	11.6	13.6
North east/ North Africa	11.9	13.7	20.5	20.0	26.6	32.0
Latin America/ Carribean	31.7	35.6	39.7	28.5	57.8	66.0

*Due to influence of dietary preference favouring less meat than in other societies
Source: FAO, 2000

Grasslands are major form of land uses the world over. These cover about 70% of agricultural land and 25% of the total land area (FAO, 1996) (Table 4). The role of grasslands has been to provide feed for the production of milk, meat and fibre by ruminant animals with at least 80% of the energy consumed by ruminants being derived from these sources.

Table 4. Permanent grasslands in different regions, 1994

Regions	**Permanent grasslands**	
	As % Agricultural Area	**As % total area**
Africa	84	30
North & Central America	57	17
South America	82	28
Oceania	89	51
Asia	67	34
Europe	33	16
Erstwhile Russian Federation	40	5
World Total	70	26

Source: FAO (1996) statistics

There is a lot of variation in grasslands as well as their management and productivity. The quantity of dry matter harvested by domesticated herbivores probably varies from 0.1 to around 20 t ha^{-1}. Grasslands have predominated in areas not suited for cropping and those where human population pressure is low (e.g. parts of South Africa) or level of affluence permit the devotion of land to animal population, whilst importing crop foods (e.g. parts of north-east Europe).

Jarvis and Aarts (2000) reviewed the efficiency of N transfer process involved in intensive dairy farming systems. Limits to what is considered to be technically achievable are indicated, but it is clear that performance in practice is highly variable and generally much lower than that which is technically possible. In order that the grasslands capture N with greater efficiency, it is necessary to have adequate S content in the soil. In the wake of increasing concern for clear environment especially in the developed world, S accretions from the atmosphere are decreasing resulting in poor efficiency of N in grasslands and range crops. Research has a major contribution to make by providing tools for more precise nutrient management. Models of S transformation processes through soil, plant, animal and atmosphere can be developed to indicate precise requirements for fertilizer S to sustain the desired rate of herbage growth with optimum N/S ratio for assimilation of N in plants and ruminants that feed on these herbage.

For example, botanical composition of grassland can be influenced by fertilizer practice. Knowledge of these competitive interactions can be incorporated into forage management. S deficiency may promote the disappearance of legumes while increases in the proportion of native clover species have been observed when S-containing fertilizers were applied in mixed communities (Jones, 1974). Lange (1998) suggested that S affects the leguminous plant species through its effect upon nitrogen fixation by *Rhizobium* micro-organisms. Even moderate S-deficiency diminished the nodule formation and nitrogenase activity.

In developing countries, with moderate population pressure, there are good opportunities to improve grassland productivity by introducing high yielding grass and legume species with improved nutrient management. Assimilation of N into animal proteins would be more efficient in the presence of adequate S in the plant tissue. It has been reported that natural grasslands commonly give beef live weight gains around 50 kg ha^{-1} annually. With improved management, that can be increased to 150 kg ha^{-1} but introduction of improved grasses and legumes with adequate nutritional management, it can increase output to over 1000 kg ha^{-1}. In South- and East Asia, there is substantial market development for milk and milk products (Wilkins, 2001). This is stimulating interest in introduction of high quality forages. In order to be competitive with cropping for direct consumption, sown grasses and legumes need to be adequately fertilized with S along with N, P, K and trace elements like Zn, Cu, B, Mo etc.

Dietary requirements of S in ruminants are reported to range from 0.1 to 0.21% of diet dry matter (NRC, 1985). Where rations are composed of maize grain and its silage, adequate S may not be provided to the animals. Animals feeding on grasses may show S-deficiency, as many of the grasses are low in S content in comparison to legumes and cruciferous plants. While in grasses the S content as percentage of dry matter is around 0.18, it range from 0.24-0.30 and 1.0-1.7 in legumes and crucifers, respectively (Kanwar and Mudahar, 1986). Since S nutrition in the animal system is related to N nutrition, therefore, it is more relevant to express S needs in terms of N:S ratio rather than S content alone. N:S ratio of 15:1 or narrow has been found to be adequate for S nutrition (Pasricha and Fox, 1993).

Sulphur containing amino acid, methionine, is essential for animal nutrition, as it can not be synthesized in the animal system. It acts as a precursor of other S-containing amino acid like cysteine and cystine, and has an important role in non-ruminant nutrition and productivity. S-amino acids requirements are especially high in growing pigs and poultry. Therefore, adequate S fertilization of forage crops and grasslands is important for increased livestock productivity.

4. WORLD FERTILIZER CONSUMPTION

The world fertilizer consumption increased by 7% from 130 mt in 1985 to 141 mt in 2000. Increase in fertilizer N alone accounted for 20% while consumption of both P and K decreased by 3 and 14%, respectively during the last 15 years. The growth rate in fertilizer use is higher in developing countries than the developed countries. For example, fertilizer consumption has a record level of 37.9 mt in China and 18.3 mt in India, the two population giants. However, rapid increase in fertilizer use especially in China @ 300 kg/ha ($N+P_2O_5+K_2O$) has attracted criticism. The use ratio of mineral fertilizers has been highly imbalanced in favour of N. This results in low use efficiency. Balanced fertilization is both soil as well as crop specific. The needed nutrients have to be applied in recommended level depending upon soil test value and specific crop requirements. While consumption of N increased by more than 100%, the relative use of S as fertilizer at around 10 mt per annum remained more or less static during the last 30 years (Fig. 2).

The areas of intensification are experiencing the deficiency of sulphur more frequently (Jansson, 1994). Therefore, balanced fertilization in the intensification areas is not only to be limited to N, P and K but also include secondary nutrients like S and micronutrients like Zn. In the past, S deficiencies were masked to some extent by S

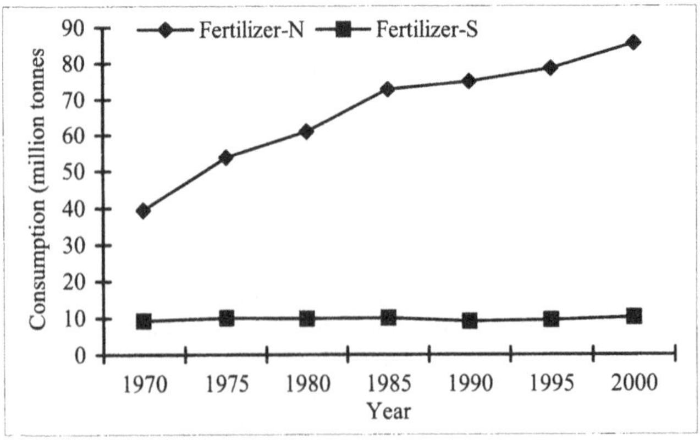

Figure 2. Trends in world consumption of fertilizer N vis-à-vis fertilizer S

inputs through precipitation, irrigation water, manures, crop residues, and S-containing nitrogenous and phosphatic fertilizers. Significant amount of S was also received as atmospheric S accrued with rainfall and direct absorption of SO_2. However, with more and more stringent environmental laws in many countries, S emissions have decreased drastically and this mission will continue to be implemented more vigorously in future in many countries. Thus, S contributions from rainfall are decreasing every day. Also, the use of S-bearing nitrogenous and phosphatic fertilizers has decreased drastically and their place has been taken by high analysis S-free fertilizers like urea, diammonium phosphate (DAP) etc. Sulphur deficiencies have, thus, become wide spread and will grow in future as more S is removed from the soil as a result of an increase in agricultural production by increased use of fertilizers, intensifying cropping systems with high yielding cultivars (Ceccotti and Messick, 1994).

5. DECLINE IN PRODUCTION AND CONSUMPTION OF S-CONTAINING MAJOR FERTILIZERS

A number of fertilizer materials, which were used for N, P or K content, also contained S. A fact often overlooked is that materials such as sulphate of ammonia, single super phosphate and potassium sulphate are priced largely on the basis of their N, P or K content, while no credit is given to their relatively high S content. The importance of these materials has diminished in the world-trade, as the economics of fertilizer distribution and use generally dictated the manufacture of materials with a high plant nutrient (N or P or K) content. This has often meant the exclusion of S from the product. Thus, in the world-trade, urea (46% N), DAP (48% P_2O_5) and muriate of potash (60% K_2O) have replaced sulphate of ammonia (SA; 20.5% N), single super phosphate (SSP; 16% P_2O_5) and Sulphate of potassium (SOP; 40% K_2O), respectively. Some single super phosphate plants are still operational in some parts of India and Brazil because of low investments cost, simple technology, availability and use of local skills, short gestation periods and high rates of capacity utilization: In 1950-51, 100% of P used in Indian agriculture was as single super phosphate. However, with time, total P consumption increased several fold from 8800 tonnes P_2O_5 in 1950-51 to 4.2 mt in 2000-01, the proportion of SSP dropped to near 10.4%. This is expected to drop further with government policy not to extend any subsidy on SSP in future. Similarly, use of sulphate of ammonia as a source of N was 100% in 1950-51 which dropped to negligible amount of 1.1% in 2000-01 (Table 5). However, in some countries like Brazil, sulphate of ammonia is still a major source of N. Sulphates of ammonia plants have some advantages and for that reason, synthetic sulphate of ammonia will continue to be produced on a local scale. About 55% are available as co-product in the manufacture of synthetic fibre intermediates such as caprolactam, acrylonitrite and methyl/methacylate. Since the sixties, the co-product sulphate of ammonia, in particular has been sold on the world N market in substantial quantities, although, its share is expectantly decreasing with increasing total N consumption (Table 6).

The most important source of co-product ammonium sulphate is caprolactam, which is used almost exclusively in the manufacture of nylon-6. It is widely used for making clothing, carpets, tires and as engineering plastics. The market price of sulphate of ammonia and single super phosphate over the years has been related more to the world market of urea and DAP prices than the market S prices. So far developments in the price of S had no influence on developments in the price of sulphate of ammonia and single super phosphate. Continuation of this situation will depend largely on the growth in consumption of sulphate of ammonia and SSP for use as an S and N; and S and P fertilizers rather than as N or P fertilizers alone.

Table 5. Trends in production and consumption of ammonium sulphates (AS) and single super phosphate in India

Year	AS Production (.000 t N)	Total N Consumption (.000 t N)	% of N use as AS	SSP Production (.000 t P_2O_5)	Total P_2O_5 Consumption (.000 t)	% of total P_2O_5 consumed as SSP
1950-51	28.8	29.0	100	9.8	8.8	100
1955-56	76.5	107.5	71.2	12.4	13.0	95.2
1960-61	81.8	211.7	38.6	52.4	53.1	98.6
1965-66	92.1	574.8	16.0	106.3	132.5	80.2
1970-71	125.5	1479.3	8.5	98.1	541.0	18.1
1975-76	125.3	2148.6	5.8	73.8	466.8	15.8
1980-81	89.4	3678.1	2.4	173.6	1213.6	14.3
1985-86	106.4	5660.8	1.9	341.8	2005.2	17.0
1990-91	124.3	7997.2	1.6	584.0	3221.0	18.1
1995-96	130.1	9822.8	1.3	512.0	2897.5	17.7
2000-2001	121.7	10,920.2	1.1	438.8	4214.6	10.4

Source: Fertilizer statistics 2000-2001, FAI

Worldwide consumption of S as a plant nutrient is estimated at 10 million tonnes per annum. Although their use as a source of major plant nutrients (N and P) has decreased drastically, but SSP and SA still continue to be the major source through which more than 8 mt of S is applied in crop production (Table 6). The use of SSP and SA will continue declining in future as well.

Table 6. World consumption of the three most important S containing fertilizers (2000)

Fertilizer	Nutrient consumption (million tonnes)	Total Nutrient consumption from all sources	% of total used as AS, SSP or SOP	S-consumption (million tonnes)
Sulphate of Ammonia	2.63 (N)	85.02(N)	3.09	3.05
Single Super Phosphate	6.04 (P_2O_5)	33.39 (P_2O_5)	18.09	5.32
Sulphate of Potassium	0.68 (K_2O)	22.21 (K_2O)	3.06	0.33

Source: Fertilizer statistics 2000-2001, FAI

6. ATMOSPHERIC SULPHUR ACCRETIONS AND PREVALENCE OF S DEFICIENCY

Global atmospheric S emissions are predicted to rise by 30 per cent beyond the year 2000 upon present day rates due to increased coal consumption (Wellburn, 1994). In a study, reported by Alcamo et al. (1995), emissions of SO_2 will range from 86 to 106.5 Tg a^{-1} in 2010 and from 139 to 279.5 Tg a^{-1} the year 2100 from the protocol and non-protocol scenario, respectively. In West Europe and North America, SO_2 emission increased from early part of the century to a peak in 1970s and since then has been decreasing due to adoption of pollution control measures (Whelpdale, 1992). In Europe, the amount of S acquired from the atmosphere has fallen from about 50 kg S ha^{-1} in 1990 to less than 10 kg S ha^{-1} (Mackenzie, 1995). In UK, SO_2 emission decreased from approximately 6.8 mt 1965 to under 4 mt in 1990. SO_2 emission in developing world is still increasing. In China, due to very high industrial development, S emission increased more than 25 fold during 1950-1990 with average annual increase of the 14%, whereas in India, the annual increase is still relatively low at 5% per annum (Table 7).

Table 7. Trends of SO_2 emissions in UK, China and India

Year	Sulphur Dioxide Emissions (kt/a)		
	UK	China	India
1950	6684	1048	600
1955	7724	2498	792
1960	6894	10,196	1088
1965	6806	6070	1398
1970	5748	9350	1606
1975	5216	12,952	2056
1980	2802	16,658	2404
1985	3488	23,026	3226
1990	3526	28,428	4386

Source: http:.//www.uea.ac.uk/su/~e490/su/INDIA.html

All India emissions are estimated at 3.54 Tg in 1990 and 4.64 Tg in 1995. This is attributed to increase in consumption of coal, high speed diesel and oil indicating expansion of transport and industry sector (Garg et al., 2001). Based on the RAINS-ASIA model SO_2 emission will show an increase of 321 per cent by 2020 (Panwar, 1998). Significance of atmospheric SO_2 as a source of S nutrition for the crops become obvious with the implementation of strict environmental regulations in West Europe. Appearance of S deficiency symptoms was reported for a number of crops. For more detail see Haneklaus et al. (this book).

S deficiency is becoming widespread throughout the world. While in the Western

Europe strict implementation of the environmental regulations to control SO_2 emission is primarily contributing to appearance of S deficiency (see earlier section), various factors for the developing countries are: increased use of high analysis, low S fertilizers with a view to meet the requirement of increasing population declining reserves of the soil, decreased use of S in pesticides and use of crop residues for food and fuel. An insufficient S supply affects yield and quality of the crop. Decrease in productivity by about 5% per year in Europe since 1990 has been reported (Mackenzie, 1995). Schnug (1998) reported from the findings on the impact of clean air on S deficiency that due to high mobility of SO_4^{2-} ions in soils, areas receiving sufficient amount of SO_2 deposits can become S-deficient after emissions are reduced. Relatively high NO_3^- content was reported in some crops. Moreover S deficiency in soil may increase the amount of unutilised N in soil which may be lost through volatilisation and leaching. A significant decrease in S content of wheat grown in England attributed to declining S accretions, affect nutritional as well as baking quality (McGrath and Zhao, 1995). Innumerable quality effects of S on forage, range crops and grasslands include Vitamin A, chlorophyll, protein, amino acids and cyanide content. S content affects natural resistance of plants against disease and insects which are important for animals.

7. PLANT NUTRIENT SULPHUR DEFICIT AND SUPPLY

The amount of total crop removal of S the world over is about 18 mt which is expected to increase to over 24 mt. The worldwide current deficit of 6 mt of S will grow to 11 mt by year 2010 and more than 13 million tonnes by the 2020 (Fig. 3). Major deficit is projected in Asia particularly the two big nations viz., China and India which account for 74% of Asia and 45% of the world (Fig. 4). This is for two reasons. First, intensification for increased crop production in future, especially China and India, where more than 30% of the cultivated lands representing major agricultural regions are already S-deficient according to experimental studies carried out by TSI and other agencies. Secondly, intensification of crop production has always been assigned with disproportion in favour of fertilizer N. There is, however, less environmental pressure regarding SO_2 emission. The situation is of concern as there are no specific programmes of S additions in the balance fertilization of the crops. In China, a part of imported KCl

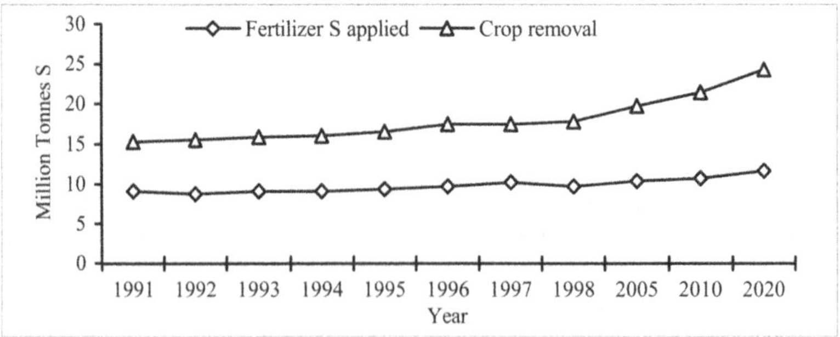

Figure 3. Projected world plant nutrients balance (Source: Messick and de Brey, 2001)

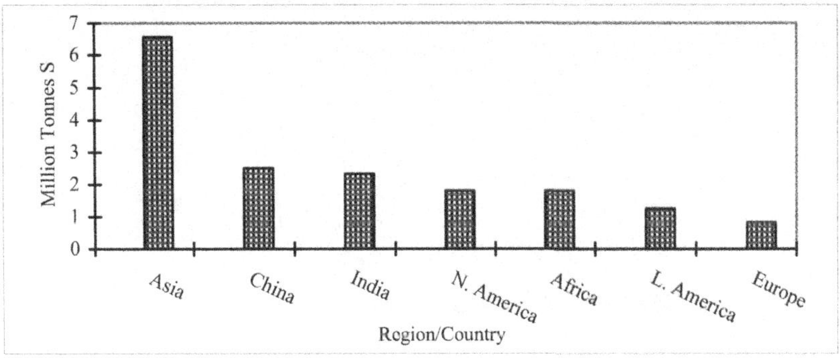

Figure 4. Region wise plant nutrient S deficit in 2010 (Source: Messick and de Bery, 2001)

is converted to K_2SO_4 by reacting with indigenously available Na_2SO_4. SOP, thus, obtained is used where crops demand both K and S. Drastic measures need to be taken at the government and public level to increased use of S in their fertilizer programme (Messick and de Brey, 2001).

8. WORLD DEMAND AND SUPPLY OF SULPHUR

World-wide huge reserves of S are available in the form of Frasch deposits, pyrite and sour natural gas (H_2S). Total of all forms of S (elemental S, pyrite and other forms) produced in 2000 was 60.1 mt. Elemental S, pyrite and other forms amount to 40.0, 5.44 and 14.7 mts, respectively (Fig. 5). Eighty one, sixty six and almost one hundred per cent of elemental S, total pyrite and other forms, respectively are used for making H_2SO_4 which is utilized as a raw material in the fertilizer industry. The proportion of recovered S is increasing while that of Frasch is on decline, and this scenario is to

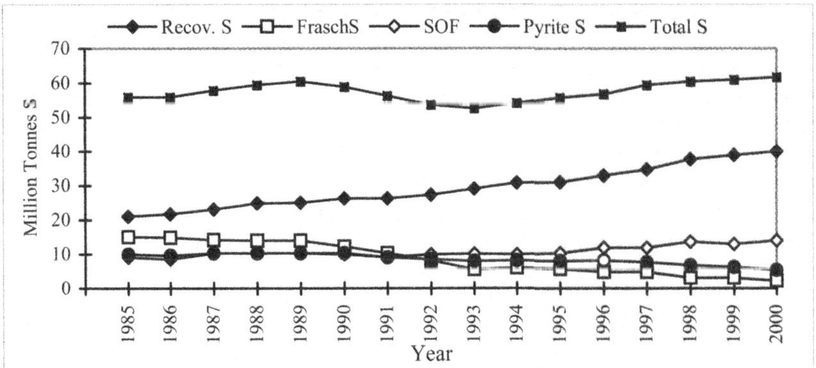

Figure 5. Production trends of different types of sulphur (Source: Luc Maene, personal communication)

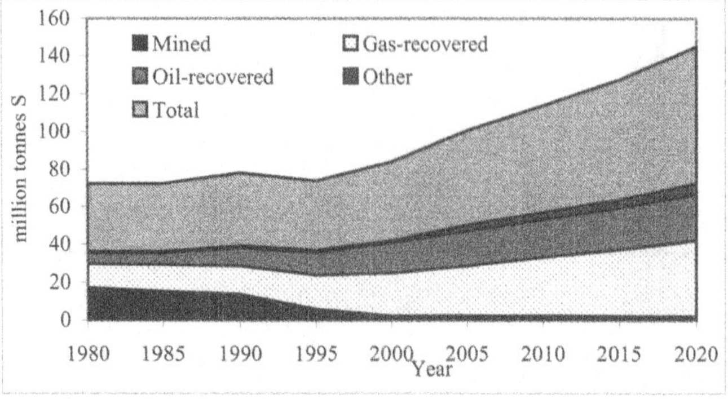

Figure 6. World elemental sulphur production and forecast (Source: Ian Service, Fertecon Research Centre, UK

continue in the future. For example, production and recovery of S from petroleum due to strict enforcement of antipollution regulations worldwide will increase. It is likely that there will be an addition of 15 mt per annum of recovered S production with possibility of further strong growth beyond 2010 (Fig. 6)

World trade in elemental S in 2000 was around 5 mt. Middle east, Western Canada, Former Soviet Union, Germany, France, U.S. and Mexico have substantial amounts of natural gas deposits. World Frasch S production peaked during the first half of the 1970's at around 17 million tonnes yr^{-1}, when recovered S from gas and oil overtook its production. Currently only a few Frasch mines remain operational. Recovery of S from natural gas processing gained momentum from 1950 and by mid 1975, it replaced the Frasch S and pyrites as world's major source of S.

Major exporters are Canada, Russia, S. Arabia, Japan and U.A.E. while the major importers are China, Morocco, USA, India, Tunisia and Brasil (Table 8). China has recently begun large-scale importation of elemental S to substitute domestic pyrite. Use of elemental S reduces the burden of transporting low-grade pyrite and has less adverse impact on the environment. It is assumed that 2 mt per annum sulphur-equivalent of pyrite consumption will remain in China in 2010 requiring an other 1.5 million tonnes per annum growth in elemental S consumption for pyrite substitution. Pyrite use is unlikely to be significant in 2020.

Projected worldwide balance of S in all forms is shown in Table 9. There is surplus supply of S. However, this does not mean that S supply is at ease but because of our ignorance of its use in balanced fertilization for optimum returns from this and other inputs in agriculture. Although, more S will be consumed in developing world as compared to developed world, still the trends for future demands for S in developing countries, where it is most needed, is projected to be low (Fig. 7). While the need for sulphur as a fertilizer is becoming widely accepted, the alternatives, such as adding gypsum or byproduct SA to a fertilizer formulation, are limiting the market for elemental S. Currently about 1 million tonnes of elemental sulphur are consumed to provide plant nutrient sulphur. The market is growing strongly, but it would take an 8%

growth to add another 1 million tonnes per annum by 2010. This is seen as maximum. The forecast for 2020 is put in at 4 million tonnes per annum.

Table 8. Ten major exporters and importers of elemental – sulphur in 2000

,000 tonnes in Elemental S-equivalent			
Countries	**S-Exports**	**Countries**	**S-Imports**
Canada	7,146	China	2,868
Russia	3,007	Morocco	2,455
S.Arabia	2,220	USA	2,450
Japan	1,063	India	1,807
UAE	1,056	Tunisia	1,599
Germany	991	Brazil	1,413
USA	749	Jordan	732
Poland	707	S.Africa	617
Iran	651	Israel	561
Kuwait	539	Mexico	543

Source: Luc Maene (personal communication)

Table 9. Forecasts of S-in-all-forms, demand and supply (mt)

Year	Demand		World Supply	Balance
	Developed countries	Developing countries		
2001	27.0	33.0	63.2	3.2
2002	28.8	33.7	64.8	2.3
2003	29.9	35.3	66.7	1.6
2004	30.2	37.6	69.3	2.4
2005	30.8	38.7	70.6	1.3

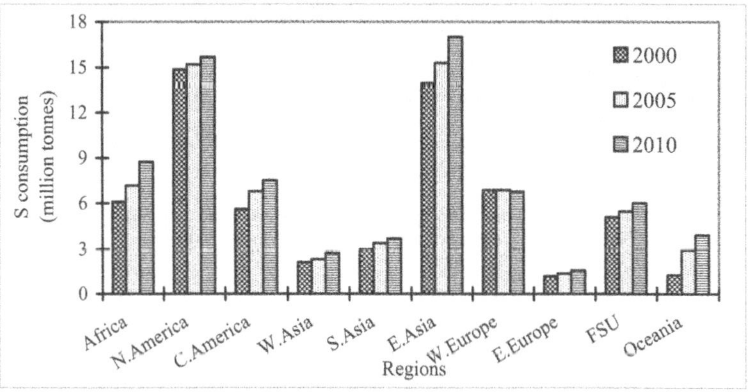

Figure 7. Region-wise S consumption and forecast (Source: TSI, Washington, DC)

9. REFERENCES

Abrol YP, Sangwan S and Tiwari MK (eds) (2002) Land Use-Historical Perspective – Focus on Indo-Gangetic Plains. Allied Publishers, New Delhi

Alcamo J, Krol M and Posch M (1995) An integrated analysis of sulphur emission acid deposition and climate change. *Water Air Soil Pollu* **85:** 1539—1550

Ceccotti S and Messick D (1994) Plant nutrient sulphur: A global review of crop requirements, supply, and environmental impact on nutrient balance. *Norwegian J Agric Sci* **15:** 7—25

FAO (1996) World Food Summit. Technical Background Documents. **2 & 3**: 6—11, 12—15. FAO, Rome

FAO (2000) Agriculture: Towards 2015/2030. Technical Interim Report, FAO Rome

FAO (2002) World Agriculture: Towards 2015/2030. Long Term Perspectives – The Outlook of Agriculture, FAO, Rome

Fertilizer Statistics (2001) Fertilizer Association of India, New Delhi

Garg A, Shukla PR, Bhattacharya S and Dadhwal VK (2001) Sub-region (district) and sector level SO_2 and NO_x emissions for India: assessment of inventories and mitigation flexibility. *Atmos Environ* **35:** 703-713

Ian Service (2002) Prospects for sulphur supply. Demand and Trade Wordwide. Fertecon Research Centre

Jansson H (1994) Sulphur status of soils – a global study. *Norwegian J Agric Sci* **15:** 27—20

Jarvis SC and Aarts HFM (2000) Nutrient management from a farming systems perspective. *Grassland Sci Euro* **5:** 363—373

Jones (1974) Fertilization of annual grasslands of California and Oregon. In: May DA (ed), Forage Fertilizer, pp 43—408. American Society of Agronomy, Madison, Wisconsin

Kanwar JS and Mudahar MS (1986) Fertilizer Sulfur and Food Production. Matinus Nizhoff/Dr W. Junk Publisher, Dordrecht

Lange A (1998) Einfluss du Schwefel-Versoegung auf die biologiche Stickshoff-Firuerung von Leguminosen. Ph.D. Thesis, University of Bonn

Luc Maene M (2002) Global potassium fertilizer situation: Current use and perspectives. In: Pasricha NS and Bansal SK (eds), Proceedings of International Symposium in Nutrient Management for Sustainable Crop Production in India, Potash Research Institute of India, Gurgaon, pp17—30

Mackenzie D (1995) Killing crops with cleanliness. *New Scientist,* 23 September, p 4

McGrath SP and Zhao FJ (1995) A risk assessment of sulphur deficiency in cereals using soil and atmospheric deposition data. *Soil Use Management* **11:** 110—114

Messick DL and de Brey C (2001) The global sulphur situation and outlook. In: Proceeding of the 51st Fertilizer Industry Round Table, St. Peter Beach, Florida

National Research Council (1985). Nutrient requirements of domestic animals: nutrient requirements of Sheep. National Academy of Sciences, National Academy Press, Washington, DC

Pasricha NS and Fox RL (1993) Plant nutrient sulphur in the tropics and sub-tropics. *Adv Agron* **50:** 209-269

Panwar TS (1998) Acid rains - case study for India (including RAINS-Asia model). Project Report: Acid rains and Emission Reduction in Asia Phase II, Asian Development Bank

Roy RH (2001) Fertilizer feeds the world. Proceeding of the Fertilizer Industry Federation of Australia Biennial Conference, Gold Coast, Australia

Schnug E (ed) (1998) Sulphur in Agro-Ecosystems. Kluwer Academic Publishers, Dordrecht

Wellburn A (1994) Air Pollution and Climate Change: The Biological Impact. 2nd edi. Longman Scientific and Technical, Essan

Whelpdale DM (1992) An overview of atmospheric sulphur cycle. In: Howarth RW, Stewart JWB and Ivanov MV (eds), Sulphur Cycling on the Continents, SCOPE 48, pp 5—26. John Wiley & Sons, Chichester

Wilkins RJ (2001) Facing the growing needs of mankind – Grasslands and rangelands. In: Nősberger J, Geiger HH and Struik PC (eds), Crop Science: Progress and Prospects, pp 63—80. CABI Publishing, Wallingford

10. AFFILIATIONS

Potash Research Institute of India, Gurgaon, Haryana (NSP); Department of Environmental Botany, Hamdard University, New Delhi (YPA), India

For correspondence – ypabrol@vsnl.com (YPA); priigurgaon@sify.com (NSP)

O. OENEMA AND R. POSTMA

MANAGING SULPHUR IN AGROECOSYSTEMS

Abstract. This chapter discusses the global biogeochemical sulphur (S) cycle and S pool, the role of nutrient management, in general, and of S management, in particular, in achieving agronomic and environmental objectives. Understanding the biogeochemical cycling of S is of crucial importance for S management in agroecosystems. Sulphur is involved in reduction-oxidation reactions and thereby transformed in different species, which have different reactivity and mobility in the environment. Unlike most other nutrients, plant available S is transported both through the atmosphere and with the hydrological cycle.

Atmospheric S deposition is the single largest external source of plant available S. Crop offtake and S leaching are the largest outputs of S in agroecosystems. The S supply from internal sources, i.e. net mineralisation of soil organic matter, crop residues and animal manure, is important for the match of S supply to S demand, but difficult to predict. Because of the huge variation in both S supply and S demand, managing S in agroecosystems has to be farm, site, soil, crop, and climate specific.

Management is the single most important production factor of agroecosystems, determining the economic and environmental performances of these systems. The focus of S management in agroecosystems is on matching the S supply to the S demand, i.e. the right amount of S in the right form at the right place at the right time. These management activities have to be considered in a whole farm perspective, to be able to achieve consistency in the general farm strategy and also to achieve a balanced supply for all essential nutrients.

Simple S budget models greatly facilitate the decision support process in S management. These models calculate all S inputs and all S outputs on a daily or weekly basis and thereby are capable of predicting whether or not fertiliser S is required. So far, S fertilisers still constitute a minor source of S in agroecosystems but forecasts indicate that the use of S fertiliser will double in the next two decades.

1. INTRODUCTION

The concern of nutrient management is the application of expertise and information about nutrients in guiding decisions and actions in agroecosystem. Basically, nutrient management is a set of specialised activities aimed at manipulating the nutrient sources and transformations of the agroecosystem, so as to achieve both economic and environmental targets (Beegle et al., 2000; Oenema and Pietrzak, 2002). The activities focus on the synchronisation and synlocalisation of plant-available nutrient sources to the demand of nutrients by crops and animals, i.e. the right nutrients, in right amounts, at the right time and at the right place. Thus, the basic question in nutrient management is "how to get the right nutrients in the right amounts at the right time and at the right place?"

The term 'nutrient management' has become common from the early 1980s onwards, when environmental effects of the intensification of agricultural production through increased applications of fertiliser and animal manure became evident. Nutrient management emphasises that both agronomic (crop yield, quality and soil fertility) and environmental objectives with respect to resource use and nutrient losses to the wider environment need to be considered. All nutrients and their sources need to be taken into account, i.e. nutrients in soil reserves, crop residues, animal manure, wastes,

Yash P. Abrol and Altaf Ahmad (eds.), Sulphur in Plants, 45—70.

atmospheric deposition, biological nitrogen fixation, and fertilisers. Though, there are 14 nutrient elements known to be essential for plants, and 18 for animals (Follett and Wilkenson, 1995; Marschner, 1995), the focus in nutrient management often is on nitrogen (N) and phosphorus (P). These nutrients are the main yield-limiting nutrients, and the losses of these nutrients from agroecosystems have also the largest impacts on the environment (Tilman et al., 2001). Efficiency of utilisation of these nutrients in crop and animal production also depends on the availability of other nutrients (e.g. Marschner, 1995; Whitehead, 2000), and neglecting other nutrients in nutrient management can be highly counter-productive.

This chapter deals with managing sulphur (S) in agroecosystems. Sulphur is an essential nutrients for both plants and animals. Shortages and excess of S limit crop and animal productivity. Thus, sulphur needs to be managed because its supply from 'natural' sources does not always match the S demand by crops and animals. In case of short supply, S has to be added via S containing fertilisers and/or organic soil amendments. When the supply is too large, measures have to be taken to decrease the supply. Usually, farmers can easily increase the S supply by applying S fertilisers, but they have limited possibilities to decrease the S supply from 'natural' sources. Agreements between (supra)national governments and industry can decrease the S emissions into the atmosphere and thereby the atmospheric deposition of S to agro-ecosystems, and thereby the supply to crops and animals. Evidently, management of S requires actions from both governments and farmers (Fig 1). Actions from governments are needed especially when the atmospheric S deposition, originating from industrial sources, is too large. Actions from farmers are needed when the S supply is too small to match S demand by crops and animals, but also when the S supply is too large.

The anthropogenic influence on the global S cycle has been enormous during the last two centuries, especially in the industrialised world (Brimblecombe et al., 1989).

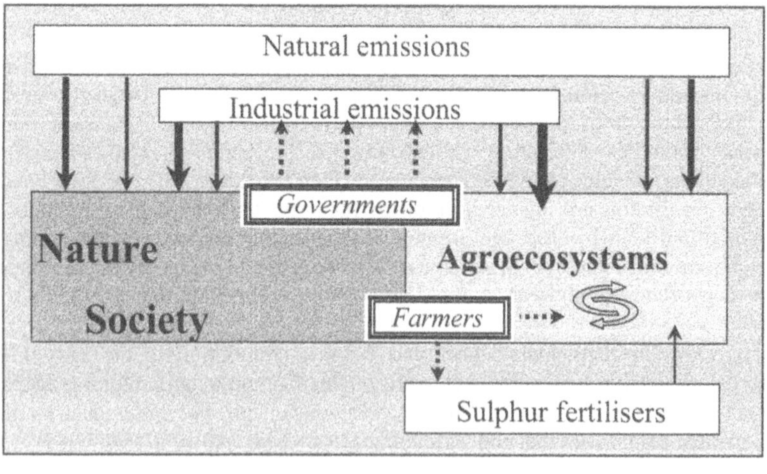

Figure 1. Analytical framework for the management of S in agroecosystems, in dependence of the management of industrial S emissions by governments and industry

Large amounts of sulphur dioxide (SO_2) have been and are emitted into the atmosphere through fossil-fuel combustion, mainly through the burning of S-containing coal and oil. Anthropogenic emissions of S into the atmosphere are now about twice as large as the global production of gaseous S compounds by bacteria (Smil, 1995). Because of human health effects and environmental effects associated with SO_2 in the atmosphere, governmental policies have been introduced from the 1970's onwards to decrease the emissions of SO_2 in the atmosphere. These policies have been highly successful in many industrialised countries, but the increasing quest for cheap energy from coal, notably by now rapidly industrialising countries in Asia, sets constraints to emission reduction targets. Evidently, the strategy for managing S in agroecosystems depends also on the (success of the) management of industrial S emissions by government and industry. The first part of this chapter deals with the global biogeochemical S cycle, S sources and transformation processes in the environment. The second part deals specifically with the management of S in agroecosystems.

2. SULPHUR IN THE ENVIRONMENT

2.1 Global Sulphur Reservoirs

The major global reservoirs of S near the earth surface are found in sedimentary rock, the ocean and evaporites derived from ocean water. The biosphere contains relatively little S. Most of the biospheric S is in dead organic matter buried in the soil (Table 1). Most of the S in rock is locked up in pyrite (FeS_2) with a mean residence time over geological era (Ga, 10^9 years). The S organic matter and atmosphere is actively involved in biogeochemical cycling processes and, therefore, has a much smaller mean residence time (days to years). The estimated amount of S in living plants is 600 Tg and in soil is 5000 Tg. For consumption, the world consumption of elemental S for industrial purposes was about 40 Tg per year in the last decade, i.e. only a fraction of S in soils and plants and also a fraction of the massive reserves of S associated with hydrocarbon deposits worldwide.

Table 1. Global reservoirs of S near the surface of the earth

Reservoir	S stock, 10^{15} g S
Ocean rocks	2400×10^3
Sedimentary rock	
Evaporites	5100×10^3
Other sedimentary rock	2700×10^3
Ocean/Sea	1280×10^3
Atmosphere	4×10^3
Fresh water and ice	9
Dead organic matter in soil	5
Live plants	0.6

Source: Schlesinger (1997)

2.2 Biogeochemical Sulphur Cycling

The cycling of S in the biosphere is strongly linked with the cycling of oxygen (O), carbon (C) and nitrogen (N). The fate of these elements depends on its cycle from inorganic compounds in water, soil, and atmosphere to living matter and back again. In this intricate interplay of living plants, animals and microorganisms with the environment, the elements are physically transferred from one compartment to another. At the same time, the elements are involved in complex sequences of oxidation and reduction reactions with transfers of energy, electrons and protons. S, N and C can appear as different species, depending on the oxidation-reduction status of the environment, and the reactivity and mobility of these different species can be very different. The double mobility of O, C, N and S species in water and air makes the cycling of O, C, N and S much faster than the cycling of mineral elements (Smil, 1995).

Sulphur is distributed over the globe mainly via atmosphere and surface waters. Humans have greatly increased the input of S into the atmosphere, and thereby, the transport of S over the globe during the last few centuries (Fig. 2).

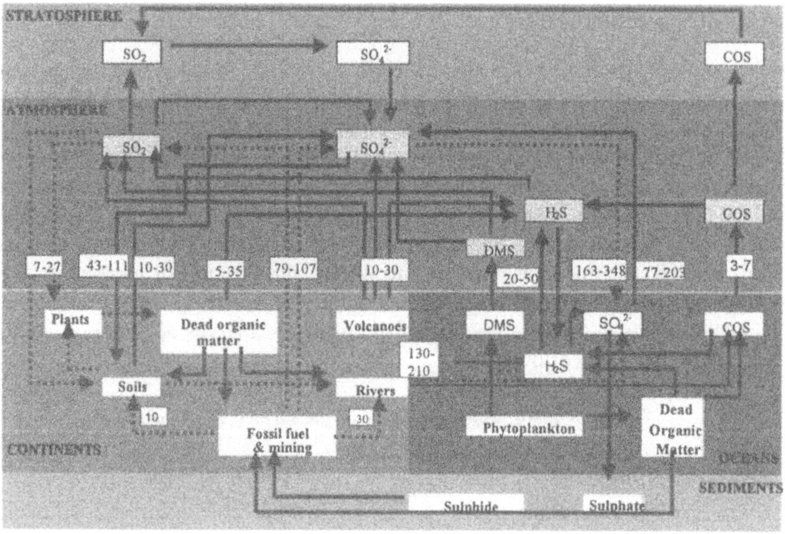

Figure 2. The global biogeochemical sulphur cycle. Arrows indicate the flows of S in Tg S per year. Solid arrows indicate natural S flows, dotted arrows indicate direct and indirect anthropogenic interference on the S flows. Diagram after Smil (1995), estimated fluxes (in boxes) after Brimblecombe et al. (1989). The estimated flow from 'fossil fuel & mining' to 'soils' after http://www.ifa.org.

Dominant sources of S in the atmosphere are seas (sea salt sprays and biogenic gases DMS, H_2S COS and CS_2), burning of biomass and fossil fuel (SO_2, SO_4^{2-}-S and H_2S), plants and decaying organic matter in swamps, marshes and excrements (SO_2, SO_4^{2-}-S and H_2S), and volcanoes (SO_2, SO_4^{2-}-S and H_2S). In the atmosphere, reduced S species are rapidly oxidised to SO_4^{2-}-S and washed out from the atmosphere to ocean and continents. Estimated total fluxes of S species into and out of the atmosphere are relatively uncertain because of the spatial and temporal variability. The mean flux from volcanoes into the atmosphere has been estimated at 10-30 Tg per year (Fig. 2), but single eruptions of volcanoes may eject as much as 10-50 Tg into the atmosphere in a relatively short period (Smil, 1995). Sea sprays are the largest source of S in the atmosphere, but most of the S is rapidly deposited again into the ocean; only about 10% of the S released in sea sprays reaches the continent. Burning of fossil fuels and mining activities release 79-107 Tg S into the atmosphere as SO_2 and SO_4^{2-}-S, of which most is deposited on and scavenged by the biosphere. Sulphur containing fertilisers add about 10 Tg S to soils, mainly via phosphate fertilisers. Rivers transport 130 to 200 Tg S per year from the continent to sea (Fig. 2). This flux is thought to have increased also due to human inference with the S cycle.

Evidently, the cycling of S in the biosphere is complex, involving many different S sources, transformation, transfer processes. Understanding this complexity is essential for managing S flows in agroecosystems. Proper management requires notion of the S sources, oxidation-reduction reactions, the linkages between O, C, N and S cycling, and of the different reactivity and mobility of S species. Understanding the global biogeochemical cycle of S has also enormous economic significance (Schlesinger, 1997). However, full treatise of the biogeochemical cycles and reactions of S is beyond the scope of this chapter; the reader is referred to Kellog et al. (1972), Ivanov and Freney (1983), Widdel (1988), Smil (1995), Brimblecome et al. (1989), Howarth et al. (1992); Maynard (1998), and references therein.

2.3 Sulphur in the Atmosphere

Sulphur concentrations in unpolluted atmosphere are low, ranging from 0.2 ppbv for SO_2 and H_2S (1 ppbv = 1 μl per m^3 ; 1 ppbv SO_2 ~ 2.9 μg SO_2 per m^3; 1 ppbv H_2S ~ 1.5 μg H_2S per m^3) to 0.5 ppbv for carbonyl sulphide (COS) (Kellog et al., 1972; Walker, 1977). The main reason for the low concentrations is the high reactivity of S-containing gases. All S gases are involved in complex chemical reactions, and rapidly transformed into predominantly sulphates, which are rapidly washed out the atmosphere to the earth surface. As a result, the mean residence time of S compounds in the atmosphere is small, ranging from <1 day for DMS, 1-2 days for H_2S, 1-10 days for SO_2, 4-15 days for SO_4^{2-}-S, 10-20 days for CS_2, and 2-5 year for COS. Mean transport distances are related to the mean residence time, and range from less than 200 km for DMS to 2000 km for SO_4^{2-}-S. Transport distances for COS reach global dimension.

Sulphur in the atmosphere originates from oceans and marshes, mainly in the form of DMS $(CH_3)_2S)$, COS, SO_4^{2-}-S containing aerosols and from industry and volcanoes mainly in the form of SO_2 (Fig. 2). In addition, there are small contributions from decaying organic wastes in the form of hydrogen sulphide (H_2S) and from plants emitting S compounds. Concentrations of S species in the atmosphere depend on distance of S emitting industrial sources, volcanoes and oceans. Concentration of SO_2 range from less than 1 ppbv (in areas remote from industry and traffic) to more than 70 ppbv (in the plumes of large SO_2 emitting sources, notably coal powered electricity generating plants) (Smil, 1995). Because of the damaging effects of high SO_2

concentrations on plants and human health, various international protocols and agreements have been made to decrease the emissions of SO_2 in the atmosphere drastically. An initial European Union limit was 120 μg SO_2 per m^3 (~34 ppbv SO_2) for annual median daily concentrations and the European Union guide value for mean annual SO_2 concentrations 40-60 μg SO_2 per m^3 (Saunders, 1985). Various subsequent international agreements have come up with more strict targets for the deposition of acidifying S compounds, and have strongly contributed to decrease in the emissions of SO_2 into the atmosphere, and subsequent S deposition (see also http://www.unece.org/env/lrtap). Figure 3 shows how total S emissions into the atmosphere have decreased in The Netherlands during the last two decades, following various policies and measures to decrease S emissions from power plants and industry. Evidently, this policy has been successful. The atmospheric deposition of S (both wet and dry) to soil ranges from about 1 kg per ha per year in remote areas to more than 100 kg per ha per year in heavily polluted urban and industrial areas. Loads as high as 100 kg S per ha per year likely have damaging effects on crop and animal productivity, and have to be decreased significantly.

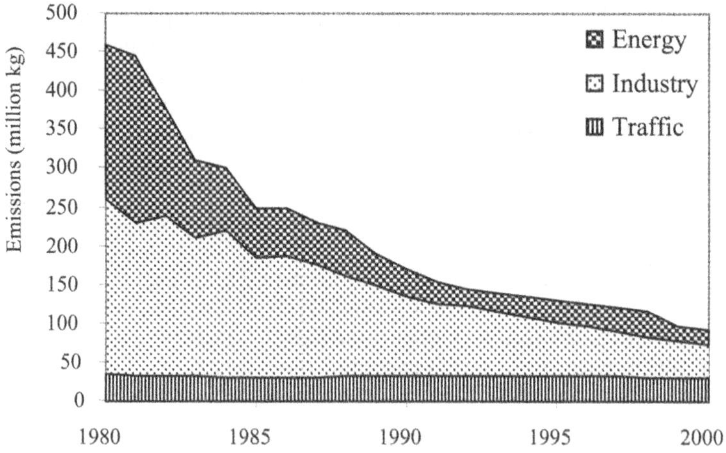

Figure 3. Changes in the emissions of SO_2 into the atmosphere in The Netherlands in the period 1980–2000, following policies and measures to decrease the emissions from industry and power plants. For comparison, the total S input via S fertilisers into agriculture in The Netherlands ranged between 10 and 20 million kg (10-20 Gg) in this period. Source: RIVM (2001)

2.4 Sulphur in Soils

Soils contain variable amounts of S depending on parent material, weathering stage and climate (leaching rate). Weathered soils derived from granite and sandstone usually contain little S (about 0.02 g per kg), while soils derived from sediments (shale, limestone) contain appreciable amounts of S (about 0.2 g per kg). Peat soils, acid sulphate soils and recently reclaimed soils from marine sediments may contain up to 10 g per kg or more. Evidently, S contents of surface soils may differ a thousand-fold.

Organic S is the dominant form of S in soils. Its content is related to the organic matter content (Eriksen et al., 1998). The organic S compounds in soils are derived from plant and animal residues. Amino acids and esters account for most of the S in organic matter. Net mineralisation, i.e. the release of inorganic S (primarily SO_4^{2-}-S) from the decomposition of S containing organic matter, occurs when the C:S ratio is less than about 200. Net immobilisation of S, i.e. the assimilation of SO_4^{2-}-S into microbial biomass, occurs when the C:S ratio is about 400 or more. At C:S ratios in between 200 and 400, there is essentially no net mineralisation or immobilisation of S in soils (Anderson, 1978; Tabatabai, 1986; Stevenson and Cole, 1999).

Dominant forms of inorganic S are sulphates in well-drained soils and sulphides in poorly drained anoxic soils. In arid regions, sulphates may occur as gypsum ($CaSO_4.2H_2O$) co-precipitated with calcium carbonate ($CaCO_3$), but in most well drained soils sulphates are only present as dissolved and adsorbed SO_4^{2-}-S. Soils containing aluminium and iron hydroxides are capable of adsorbing significant amounts of SO_4^{2-}-S. The adsorption increases as pH decreases due to the increasing number of positively charged sites with decreasing pH. Usually soil pH decreases with depth, and as a consequence the amount of adsorbed S is usually larger in the subsoil than in the topsoil, unlike the content of organic S. In anoxic soils, S occurs predominantly as pyrite (FeS_2), which is highly stable under reducing conditions. When exposed to oxic conditions, FeS_2 is rapidly oxidised to SO_4^{2-}-S by microorganisms. Elemental S (S^o) may be formed as temporary intermediate during the oxidation of FeS_2, but S^o is stable neither in reducing nor in oxidising environments. Soils gain S from external sources via atmospheric deposition and possibly via addition of fertilisers, animal manure, and various wastes, including sewage sludge. Soils are depleted via harvesting crops, leaching and erosion. The balance of S inputs and outputs, and the net mineralisation are important indicators for S management in agroecosystems, as further discussed in section 3 of this chapter.

2.5 Sulphur in Plants

It is not the purpose of this section to summarise the whole book, but rather to point at the small ranges in S contents of plant species wherein S is deficient, optimal and excessive. Plants acquire the required S mainly through the uptake of sulphate-S (SO_4^{2-}-S) by root from soil. Some amounts of S from the sulphur dioxide (SO_2) in the atmosphere may be obtained via the stomata of the plant leaves. Plants are exposed to a wide range in SO_2 concentration. Spells of ambient SO_2 above 70 ppbv or chronic concentrations above 20 ppbv are known to damage plants, causing tissue death and induced deficiency of cations (e.g. Kropff, 1989). Mosses, lichens and conifers are the most sensitive. Sulphur is an essential constituent of amino acids (i.e. methionine, cystine and cysteine) in proteins, and it occurs in plant constituents involved in enzymes, hormones and redox reactions. When S is deficient, the content of S containing amino acids in plant tissue are decreased, which in turn negatively affect yield and quality of the crop. Deficiency of S in legumes restricts biological N_2 fixation, and as a consequence, S-deficient legumes are also low in total nitrogen (N). Because of the co-variation of S content, protein content and yield, there is usually only a small variation in S content and N:S ratio in plants. In plant proteins, the ratio of N:S in terms of atoms is 36:1, which is equivalent to about 15:1 on a weight basis. This is close to the ratio of total N to total S that occurs in plants with adequate amounts of both elements. With advancing maturity and in senescing plants, S contents slowly decrease. When S supply is abundant, plants tend to accumulate inorganic S in the tissue in the form of SO_4^{2-}-S. Accumulation of SO_4^{2-}-S also occurs in plants in which protein synthesis is restricted by N deficiency. Some plant species contain relatively large amounts of S-containing secondary metabolites, such as glucosinolates.

The S content of leafy biomass of most agricultural crops typically varies in the narrow range of 1 to 3 g kg per kg dry matter. Among the families of crop plants, the S requirement increases in the order Gramineae < Leguminosae < Brassicaceae. This is also reflected in corresponding differences in the S content of their seeds: 1.5-1.9; 2.5-3.0 and 12-17 g per kg, respectively (Marschner, 1995). S contents of leafy biomass may be as low as 0.4 in S-deficient areas, and as high as 8 g per kg in S-abundant areas, depending also on plant species and plant variety. Woody biomass contains little protein and as a consequence, the S content of trees is about one order of magnitude less than that of leafy biomass of agricultural crops. Both contents of total S and SO_4^{2-}-S and N:S ratios have been suggested as indicators for the evaluation of the S nutrition status of plants and crops. Some studies were successful in deriving critical levels below or above which no response to S fertiliser application occurs (Syers et al., 1987). Other studies showed however, that the suggested critical levels were not uniformly applicable, but depending on species, variety, plant part, stage of maturity, nutrition status of other nutrients, and weather conditions, notably temperature and sunshine. Table 2 lists some reported values for critical contents of total S and SO_4^{2-}-S and of critical N:S ratios. Sulphur deficiency is usually manifested through visual symptoms of light green to yellow leaves that appear first along the veins of young leaves. S-deficiency is sometimes diagnosed as N-deficiency because of the similarity in symptoms. The number of reports showing S-deficiency in crop growth and increased crop yield due to S fertilisation have steadily increased over the last decade. This is related to the switch over to the usage of N and P fertilisers and pesticides low in S, and to decrease in S emissions into the atmosphere and atmospheric S deposition (Fig. 3). These decreases in S supply occurred concomitant with steady increases in crop S requirement because crop yields have steadily increased over the last decades. Generally, problems of S-deficiency in crops are more likely in the Southern Hemisphere than the Northern Hemisphere because of lower atmospheric S deposition.

2.6 Sulphur in Animals and Animal Manure

Animals acquire S via intake of S-containing feed (fodder) and drinking water. Sulphur is an essential constituent of body tissue, milk, egg, hair and wool. Body tissue contains on average 1.0-1.5 g S per kg body tissue (Anderson, 1978; Bowen, 1979; Whitehead, 2000), which is equivalent to 3-5 g S kg^{-1} dry weight. Milk contains about 0.3 g per kg, and wool and hair 25 and 38 g per kg, respectively. About 90% of the S in wool occurs in the form of cysteine/cystine with an N:S ratio of 5:1 (Underwood and Suttle, 1999).

Sulphur has several important roles in the metabolism of ruminant animals. It is essential for the synthesis of microbial biomass. Microorganisms in the rumen are able

Table 2. Reported ranges of values of critical S, SO_4^{2-}-S contents and N:S ratios in some plants

Plant species	Growth stage	Portion of plant	S content (g/kg)	SO_4^{2-}-S content (mg/kg)	N:S ratio
Barley	mid-late tillering	YEB	1.5-4.0		14
Canola /rape	Prior to flowering	YMB	3.5-4.7		19-30
Corn	Tasseling	Ear leaf	1.6-2.4	<400	17
Cotton	Early flowering	YMB	2.0-2.5		13
Oats	mid-late tilering	YEB	1.5-4.0		14-20
Ryegrass	Active growth	Young herbage	1.0-2.5	160-500	15-17
Peanut	Pre-flowering or flowering	YML	2.0-3.5		13-15
Sugar cane	Active growth	Top visible dewlap	1.2-1.3		17
Soybeans	Early flowering	YMB	2.0-4.0		16.5
White clover	Active growth	Young herbage	1.8-3.0	230	17-22
Lucerne	Early flowering	Whole shoot	1.0-2.2		11-17
Wheat	mid-late tillering	YEB/YMB	1.5-4.0	200-460	17

YEB = youngest emerged leaf blade; YMB = youngest mature leaf blade;
YML = youngest mature leaf
Sources: Reuter and Robinson (1997), Whitehead (2000)

to convert both organic and inorganic forms of S of feed to sulphides (S^{2-}, HS^-, H_2S), and to assimilate the sulphides into microbial protein. When S is deficient, the synthesis of microbial biomass is slowed down. This may limit the degradation of fibre in the rumen, decrease feed intake and impair animal productivity. Another metabolic function of S is to form sulphate conjugates with compounds such as phenols and indole. These compounds are derived from bacterial action in the digestive system and would otherwise be toxic to the animal. After conjugation with sulphate, they are less toxic, and excreted in the urine (Whitehead, 2000). Sulphides, formed in excess of the requirements for microbial biomass synthesis in the rumen, are absorbed into the blood and converted to SO_4^{2-}-S in the liver. Some of the SO_4^{2-}-S is then returned to the rumen and some is excreted in the urine. The recycling of S in the rumen enables S to be conserved, an important factor when the intake of S by the animal is low (Whitehead, 2000).

The supply of S for the synthesis of microbial protein is normally adequate when the N:S ratio of the diet is about 14:1, and the S content of the diet is 1.6 to 2 g per kg dry weight, depending on the protein content of the diet. Sheep have a slightly higher S requirement because of the high S content of wool. For optimum wool growth, the recommended S content in the diet is 2.5 g per kg (Underwood and Suttle, 1999).

Symptoms of S-deficiency in animals are not specific. Deficiency of S decreases the rate of feed digestion, utilisation of N and feed intake, which in turn lead to decrease in live-weight gain, and milk and wool production. Symptoms of excess S in the diet have been observed with S contents as low as 3.0 to 4.0 g per kg dry weight. Excess S in the diet leads to decrease in appetite and animal productivity, possibly due to sulphide toxicity. High sulphide contents in the digestive track decrease the motility of the rumen, impair respiration, and decrease the absorption of metal elements like copper and selenium. The latter may contribute to secondary effects like copper and selenium deficiency in the animal.

Animals utilise only a fraction of the plant nutrients contained in the animal feed. For S, only 5 to 30% of the amount of S in feed is transferred into animal products, i.e. meat, wool, milk and eggs. The remainder is excreted via faeces and urine. The proportion of the excreted S in urine depends on the S content of the feed. Usually, more S is excreted in the urine than in the faeces. However, it has been shown that the proportion of S in urine in sheep may range from 6% with herbage containing less than 1 g S per kg to 90% with herbage containing 5 g S per kg dry weight (Barrow, 1987). Hence, excess S in the diet is excreted in the urine. The content of S in the faeces is rather constant, about 3 g per kg dry weight, and only slightly increases with increase in S content and digestibility of the diet.

Because of the variable partitioning of S in the diet over animal products, urine and faeces, neither urine nor faeces has the same N:S ratio as the feed. The N:S ratio of faeces varies within a much smaller range than the N:S ratio of urine, because N and S contents of the urine are both highly variable, depending on both N and S contents of the diet. Most of the S in faeces is organically bound S, but some inorganic SO_4^{2-}-S is also present. The organic S consists mainly of undigested protein, microbial protein and ester sulphate. Urine contains both inorganic and organic S. The proportion of inorganic S increases with an increase in S content of the diet, but usually varies between 50-80%. During storage, some of the organically bound S in urine and faeces are transformed into inorganic S by bacterial breakdown, and as a result the inorganic S fraction increases with storage time. Small amounts of S may volatilise from animal excrements, mainly as hydrogen sulphide (H_2S) and methyl sulphide ($(CH_3)_2S$), but this loss is usually only a small fraction of the total S in stored animal manure. The composition of animal manure sampled in stables depends on animal type, but varies widely due to differences in animal ration, housing system, manure storage system and manure storage time. Data presented in Table 3 indicate that poultry slurry has a slightly wider C:S, N:S and P:S ratios than cattle and pig slurry.

2.7 Sulphur in Fertilisers

Approximately 90% of the total world market for elemental sulphur (40-42 Tg in 2000) is for the production of sulphuric acid. Most (95%) of this elemental S is 'recovered S', i.e. originates as a by-product from the processing of sour natural gas and sour crude oils. The other 5% come from the mining of naturally occurring S deposits and are known as 'mined S'. Forecasts indicate that the world elemental S consumption will increase to about 72 Tg S by the year 2020 (http://www.IFA.org). The expected increase is for the greater part related to the expected increase in the demands for phosphate and sulphur fertilisers.

The greater part of the sulphuric acid produced is used for the manufacture of phosphate fertilisers. In 2000, about 25 Tg of elemental S was consumed for the manufacture of phosphate fertilisers, and this amount will increase to about 42 Tg in the

Table 3. Composition (means and standard deviations) of cattle slurry, pig slurry and poultry slurry, sampled at a large number of farms in 1985. Number of samples ranged from 379 for poultry slurry to 436 for cattle slurry

Animal slurry	Dry matter (g/kg)	Organic C (g/kg)	N g/kg	P g/kg	S g/kg	C:S	N:S	P:S
Cattle slurry	96±21	39±9	4.9±0.8	0.9±0.3	0.7±0.2	58	7.2	1.3
Pig slurry	74±25	28±11	6.5±1.4	1.7±0.7	0.6±0.2	44	10.2	2.7
Poultry slurry	144±33	53±10	10.6±1.5	3.5±1.1	0.9±0.2	61	12.0	4.0

Source: Hoeksma (1987)

year 2020 (http//:www.IFA.org). Manufacture of S fertilisers (e.g. elemental S, ammonium sulphate, ammonium thiosulphate, etc.) consumed about 1 Tg of S in 2000, and this amount is expected to increase to about 4 Tg in the year 2020. The world total amount of S applied to agricultural soils via fertiliser has been estimated at about 10 Tg in 2000. The S is applied via ordinary superphosphate which contains 7 to 10% of P and 8 to 10 % of S as gypsum ($CaSO_4.2H_2O$), and via various other S containing fertilisers, many of which also contain other essential nutrients (elemental S, K_2SO_4, $(NH_4)_2SO_4$, $(NH_4)_2S_2O_3$, $CaSO_4.2H_2O$, $MgSO_4.7H_2O$). For many years, ordinary superphosphate has been a major S source, but during the last two decades ordinary superphasphate has been replaced by triple superphosphate with 19-23 % of P and essentially no S, and by ammonium phosphate fertilisers with also no or very little S. As a result, S addition to soils via fertilisers has decreased during the last two decades, notably in the eighties. At the same time, atmospheric deposition of S also decreased, especially in rich countries that successfully decreased S emissions from industry and power plants (Fig. 3). These two effects combined with the quest for more food by the growing population will increase the demand for S fertilisers in crop production in next decades.

3. MANAGING SULPHUR IN AGROECOSYSTEMS

3.1 General Framework of Nutrient Management

Basically, nutrient management is a series of six activities (Beegle et al., 2000; Oenema and Pietrzak, 2002):

0. Analysis, i.e. the analysis of requirements, pools and sources of nutrients, and the analysis of constraints set by labour, capital, policies and measures.
1. Decision making process, i.e. (i) development of options on the basis of the previous analyses, (ii) assessment of the consequences of the various options, and (iii) selecting the best option for achieving both agronomic and environmental targets.
2. Planning, i.e. working out in broad outline the things that need to be done and measured; when and where and how and with how much. This step results in the actual nutrient management plan.

3. Executions, i.e. implementation of the nutrient management plan in practice, taking into account actual environmental conditions, and best management guidelines and recommendations.
4. Monitoring, i.e. collecting data on yield and quality and nutrient losses.
5. Evaluation, i.e. verification and control of achievements relative to objectives.

This cycle of nutrient management activities has to be considered in a whole farm perspective (Fig. 4). These have a large impact on crop and animal productivity, and on the environmental performance of the agro-ecosystems.

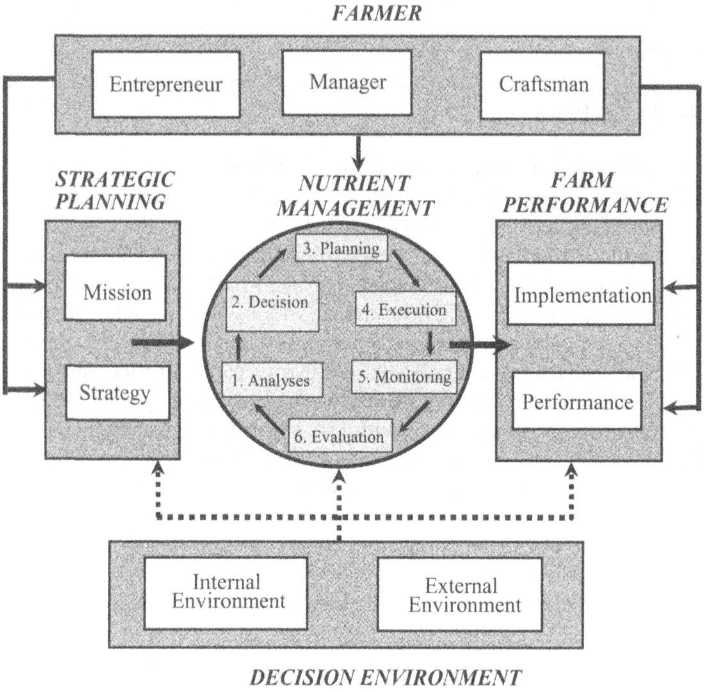

Figure 4. Analytical framework of nutrient management in a whole-farm perspective. The farmer is at the same time entrepreneur, manager and craftsman, and as a result is involved in strategic planning, management and implementation. The decision environment directly and indirectly influences the decisions and subsequent implementation steps. The internal environment deals with the agroecosystems itself, the external environment deals with market and technology developments, governmental policies and measures, suppliers and advisers, and the informal network. Sources: Oenema and Pietrzak (2002)

Therefore, the activities have to be consistent with the farm strategy and the long-term goals of the farm. A farm usually is a dynamic enterprise, which changes in response to the farmers' decisions upon developments, inside and outside the farm. At farm scale, management decisions take place first at strategic, than tactical and finally at operational levels (Beegle et al., 2000; Oenema and Pietrzak, 2002). *Strategic decisions* regard long-term agronomic and economic goals of the farm, i.e. decision dealing with investments and structural changes in the type, size and intensity of the farm. Strategic management is the 'art of formulating, implementing and evaluating decisions that enable the farm to achieve its objectives (David, 2001). Environmental targets have to be considered fully at strategic level, to be able to make nutrient management effective and efficient at tactical and operational levels. *Tactical management* deals with the implementation of strategic objectives; the annual nutrient management plan is established at this level. *Operational management* deals with the implementation of the nutrient management plan in practice on a day-by-day basis; when to do what and where in the field.

Nutrient management becomes effective and efficient when decisions at strategic management level guide decisions at tactical and operational management levels. Planning at tactical level is an essential activity of nutrient management, and has a major impact on the whole farm performance. Preparing the nutrient management plan, i.e. *analyses - decision making - planning* (Fig. 4), involves the following *detailed activities*:

1. Field and soil maps that indicate the location, size, morphology, soil types, streams and wells of the fields and its surroundings. Delineation of sensitive areas, i.e. areas and sites vulnerable for nutrient losses and sensitive sites that require specific protection, (sloping areas, sandy soils, nutrient-rich soils, sites near hedges, streams);
2. Soil tests that indicating the fertility status of the soil;
3. Crop rotation plan
4. Estimated crop yields and nutrient uptake by the crops, using historic records to indicate trends whenever possible. Calculation of nutrient input-output balances;
5. Identification of available nutrient sources at the farm, i.e. nutrients from soil, atmosphere, irrigation, groundwater seepage, crop residues, wastes, animal manure and fertilisers;
6. Soil, crop and weed management, i.e. what is the insect, weed and disease pressure, what happens with the stubble, roots and crop residues; how does it affect nutrient demand and nutrient supply;
7. Consideration of environmental restrictions due to legislation, and of requirements of processing industry and retailers with regards to production process and product quality;
8. Assessment of the best application methods of animal manure and fertilisers, e.g., surface spreading or soil injection, band application or broadcast application, single spreading or combined with pest management, all at once or split;
9. Assessment of the best time of manure and fertiliser applications, taking into account planting date and crop demand, climatic conditions, storage facilities and availability of contractor;
10. Assessment of recommended nutrient application rates; calculating the optimum amounts of nutrients to be applied to the various crops as function of site and crop, considering all possible constraints and nutrient sources at the farm; and
11. Planning, i.e. systematically working out the things that need to be done and measured; when and where and how and with how much. This step results in the

actual plan. It allocates the available nutrients in a way that maximizes the economic benefit, while minimizing the environmental impact and satisfying environmental limits.

Evidently, the detailed activities 1 to 7 are "*analyses*", and that in 8 to 10 are part of the *decision making process* (Fig. 4). Decision support systems or simple simulation models can greatly facilitate the decision making process by integrating the information from the various analyses and by providing information about results of alternative options. This holds especially for key activity 10; assessment of the amount of nutrients to be applied.

Crop and animal productivity, soil tests, and nutrient balances are in general the most important indicators for evaluation of the success of the nutrient management plan. Crop and animal productivity can be translated directly in economic terms while results of soil tests and nutrient balances provide information about nutrient use efficiency and the potential for nutrient losses. A deficit on the balance indicates that the nutrient stock of the soil is depleted while a surplus indicates enrichment, but also a potential for nutrient loss to the environment, depending on nutrient species. This desired relationships between soil test and nutrient balance would vary with nutrient species, environmental conditions and with the nutrient management strategy (soil depletion versus soil enrichment). At present, the evaluation of nutrient management occurs most often on single indicators (one by one), as quantitative relationships between soil test and nutrient balances, that would allow a more integrated evaluation of the nutrient management, are still lacking.

3.2 Framework of Sulphur Management

In large areas of the world, crop and animal production suffer from a low supply of S, notably in Australia, New Zealand and Canada, but increasingly also in countries in Africa, Asia, Europe and America. Sulphur has been mentioned the 'neglected element' (Eriksen et al., 1998), simply because for long the supply from natural sources (atmosphere, soil) and NPK fertilisers and pesticides (with S containing 'impurities') was sufficient to meet the crop requirement. But with the development of high quality fertilisers and pesticides without S impurities, the decline in S deposition (Fig. 3), and with the rapid increase in crop yields, the incidences of S deficiency in crops have become more frequent. As a result, there is an increasing need for S management.

The general framework of nutrient management discussed above in section 3.1 holds for all nutrients, including S. The objective of S management in agroecosystems is to get the proper amounts of S, in the proper form, at the proper place and time, so as to sustain crop and animal productivity. Indeed, current focus of S management is on supplying proper amounts of S for crop and animal production, and not on minimising S losses. The role of S dissipated from agroecosystems in the pollution of air, groundwater and surface waters is small. Leaching of excessive amounts of SO_4^{2-}-S may deteriorate the quality of drinking water extracted from groundwater and surface waters. Leaching of SO_4^{2-}-S may indirectly also contribute to eutrophication of surface waters, when the SO_4^{2-}-S is reduced to sulphide, and the sulphides liberate phosphate bound to iron-oxy-hydroxides in sediments. Though these possible effects do indicate that leaching losses must be minimised, there are as yet no strict environmental targets for concentrations of SO_4^{2-}-S in groundwater and surface waters, that would restrict S management for high crop and animal production.

Assessment of the amount of S to be applied is the core of the nutrient management related specifically to S. This assessment is most easily done by using a modelling approach (decision support system), given the complexity of the S cycle (section 2), and the specific effects of farm strategy (Fig. 4), crop, site, soil and climate in balancing S demand to S supply. Various simulation models have been developed over the last decades. Some of these models have a strong focus on the net mineralisation of organically bound S in soils, and on the linkages between C, N, P and S (Parton et al., 1988; McCaskill and Blair, 1990; Castellano and Dick, 1991; Delgado and Amacher, 1997). Other models focus on the prediction of fertiliser S requirement and on the best timing of the application (Sinclair, 1986, Postma, et al. 1999). Basically, these latter models calculate the changes with time in the concentration of SO_4^{2-}-S in the soil solution of the topsoil, according to simple input-output budget calculations. Inputs of S via atmospheric deposition, animal manure, fertilisers, compost, irrigation and net mineralisation, and outputs via uptake by the crop, erosion and net leaching are calculated on a daily or weekly basis (Fig. 5). The initial amount of SO_4^{2-}-S in the soil solution and the S mineralisation potential should be derived from soil tests, or from calculations using detailed models. The models need site-specific climate data (rainfall,

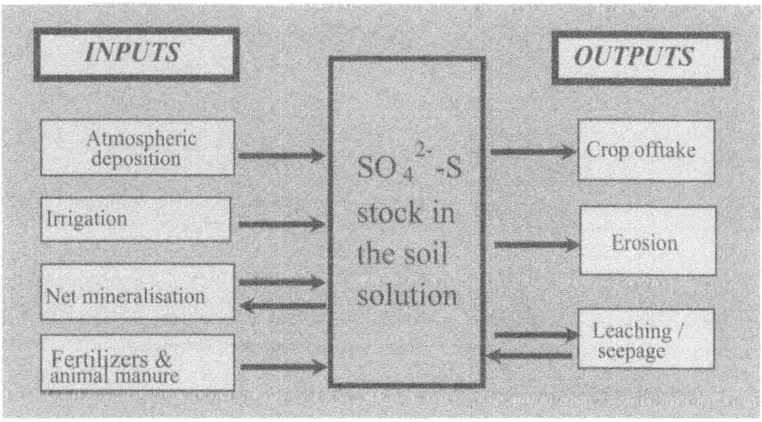

Figure 5. Schematic outline of the input-output budget model. At each time step considered, the concentration of SO_4^{2-}- S in the soil solution is calculated, and evaluated whether this amount can satisfy the requirement of the crop

evapotranspiration, temperature), soil-specific data (leaching), and crop-specific data (potential yield and S uptake). The models calculate whether or not additional fertiliser S is required, and how much. Evidently, such models provide a good basis for S fertiliser recommendations (amount, form, timing, and frequency), on which the farmer can make decisions.

The accuracy of the model predictions depends on the notion of the agroecosystem and the S cycle to be ascribed, the assumptions in the model and the reliability of the input data. Simple models usually require little input data but tend to be less site, soil, and climate specific than detailed models. The basic information required follows directly from the input-output budget model shown in figure 5. Estimates of atmospheric S inputs are usually available from environmental statistics, for different parts of countries. Inputs via irrigation can be calculated readily from the composition of the irrigation water. Estimates of net mineralisation are difficult to predict accurately, as these are highly soil specific (see section 3.3 below). Data about S uptake and crop offtake are available, but should be made farm specific. The minimum S requirement varies greatly between crops, both in total amount and the period of the growing season (Fig. 6). Estimates of S leaching can be obtained from data on rainfall (surplus), soil characteristics and sulphate retention in the soil; estimates of seepage also follow from data on groundwater fluctuations and groundwater composition. Losses via erosion can be estimated from rainfall (surplus), soil characteristics and morphological characteristics.

Evidently, evaluation of model prediction is essential, using the indicators suggested before, i.e. crop and animal productivity, soil tests, and nutrient balances. More specific indicators include plant S contents, N:S ratios and SO_4^{2-}-S contents (Table 2).

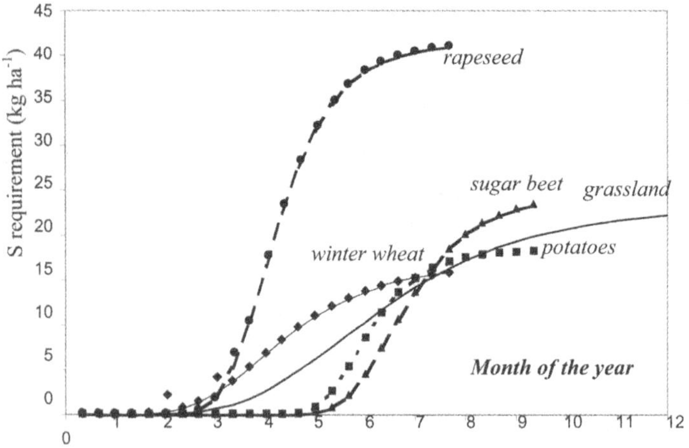

Figure 6. Course of the minimum S requirement of five dominant crops during the growing season in The Netherlands. Crops differ in total S uptake and in the period of maximum uptake. Source: Postma et al. (1999)

3.3 Assessing the S Supplying Capacity of the Soil

The soil is a large reservoir of S (see section 2.4), but only a fraction is directly available to plants. Various soil tests have been developed to estimate the plant available amounts of S in soil. These tests have been developed by individual research workers, and reflect, in part, the level of knowledge at the time of development of methods (Houba et al., 2000). The basics of soil testing are that nutrients are extracted from soil samples, and the amounts extracted are related to crop performance. Relating soil tests to the response of the crop mostly proceeds empirically via field experimentation. As the results of the field experiments depend on field management, soil type, crop, and climate, relationships between soil tests and crop response often show a wide scatter.

Current methods for evaluating the S status of soils include (i) S released during soil incubation, (ii) S extracted from the soil with (hot) water or with a solution containing an 'exchanging' anion, such as bicarbonate, phosphate and chloride, and (iii) soil S fractionation (Eriksen et al., 1998). The S release during incubation is a measure of the S mineralised from organic matter, which is by far the largest pool of S in the soil. Commonly, S mineralisation co-varies with N and C mineralisation (Verlinden, 2002). The mineralisation strongly depends on temperature and on soil moisture content and recently added organic residues, which complicate the transfer of results from the laboratory to field conditions. Extractions with a solution containing an exchanging anion remove SO_4^{2-}-S from the soil solution and some of the adsorbed SO_4^{2-}-S, depending on the properties of the anion. Soil S fractionation methods are based on size and density fractionations methods, on the differential reactivity of soil organic S fractions to reducing extractants, and/or on a combination of physical-chemical fractionation. These S fractionation procedures are laborious and usually not applied in routine soil test, though much insight can be obtained in soil S availability, as shown recently by Ribeiro et al. (2001) and Kishchuk and Brockley (2002).

In general, soil S tests alone have proven to be of variable help for evaluating the need for additional S fertilisation under field conditions. The variability of the success indicate that soil tests alone do not properly reflect plant available S in the topsoil and/or that plants acquire S also from other S sources (e.g. atmosphere, subsoil, groundwater, irrigation water, pesticides). This is supported by the fact that soil S tests appear to show a better correlation with the amount of S in the plants and the response of the plant to S addition in pot experiments under controlled conditions than in field experiments (Jones, 1986; Verlinden, 2002). As a consequence, results of soil S tests can not be directly transformed into fertiliser requirements, unlike for example the results of P tests. However, results of soil tests do provide essential information about the S supplying capacity of the soil it self (Verlinden, 2002), and therefore can be used as input data for the simple budget modelling approach as described in section 3.2 and Fig. 5).

3.4 Sulphur Balances

Sulphur balances summarise inputs and outputs of S in a defined system over a defined period of time, and as such provide essential information for evaluation of S management. The balances provide insight into the size of the various inputs and outputs, and about the balance, i.e. is there a net deficit or surplus? As such, S balances provide essential information for S management. Mass balances are widely used now for routine S fertiliser recommendations in New Zealand, where S deficiency is widespread when no S fertiliser is applied (Nguyen and Goh, 1993; Goh and Nguyen, 1997). Also in some European countries, S balances are increasingly used to guide decision making in S fertiliser application. Sulphur balances supplement or sometimes replace soil (and plant) tests in decision making. The information from S balances is robust and direct. The increasing popularity of S balances is partly explained also by the fact that soil tests and also plant tests are not always diagnostic because of site specific interfering factors (as discussed before).

Balances can be made at different scale and integration level. Common scales and integration levels for examining nutrient balances of agroecosystems include, 'soil/field', 'farm' and 'region/country/continent' depending on the purpose of the study. Balances at field and farm levels are important for S management in agroecosystems, while balances at catchment level and regional and national scales are useful for policy makers and for managers of, for example, nature conservation areas and surface waters. The data in nutrient budgets can have different origin and as a consequence, the accuracy of the budget can vary. Smaling and Fresco (1993) and Smaling and Oenema (1997) indicated that the data serving as input for nutrient budgets can be classified according to (i) type (i.e. primary data, estimates and assumptions guesstimates), (ii) source (i.e. measurements in the field and in the laboratory, observations, and statistics, computations), and (iii) frequency of measurements and data collection (i.e. continuously, seasonally, and annually). The combination of type-source-frequency defines the data acquisition strategy, and this combination depends on the purpose of the study and the nutrient budgeting approach. Currently, there is still little uniformity in constructing nutrients balances, which complicates comparison of balances (Oenema et al., 2003).

Nutrient balances differ between different types of agroecosystems, i.e. specialised crop production systems, mixed systems with crop and animal production and specialised animal production systems (Fig. 7). Both, input and output entries of the balance differ between different systems. Also input-output ratios of nutrients are fundamentally different between these systems. Below, we discuss balances of specialised crop production systems first and than grassland systems used for animal production (milk, beef, wool).

3.4.1 Sulphur balances of crop production systems

Atmospheric deposition, net S mineralisation of soil organic matter and crop residues, fertilisers, composts and animal manure are commonly the dominant sources of S in crop production systems. Removal of S via harvested crops and leaching of SO_4^{2-}-S are the dominant sinks of S. Table 4 shows how the input and output of S can vary with changes in fertilisation and manuring. Overall, S input by fertiliser and manure additions and output via SO_4^{2-}-S leaching were the dominant balance entries. Losses via leaching exceeded the output via harvested crop by a factor of 2 to 10. Note that leaching losses were the lowest in the treatment that received relatively little S input but sufficiently N fertiliser to increase crop yield and S removal via harvested crop. Storage of S in soils only occurred when the soil was amended with sufficient organic amendments, and not with inorganic fertilisers.

In lowlands with shallow groundwater level, significant amounts of S can be supplied to the soil and the crop via capillary rise of SO_4^{2-}-S containing groundwater, as illustrated by data presented in Table 5. Note that the balance in Table 5 deals with the input and output of SO_4^{2-}-S in the soil solution of the topsoil during the growing season only. Capillary rise of SO_4^{2-}-S containing groundwater during the growing season,

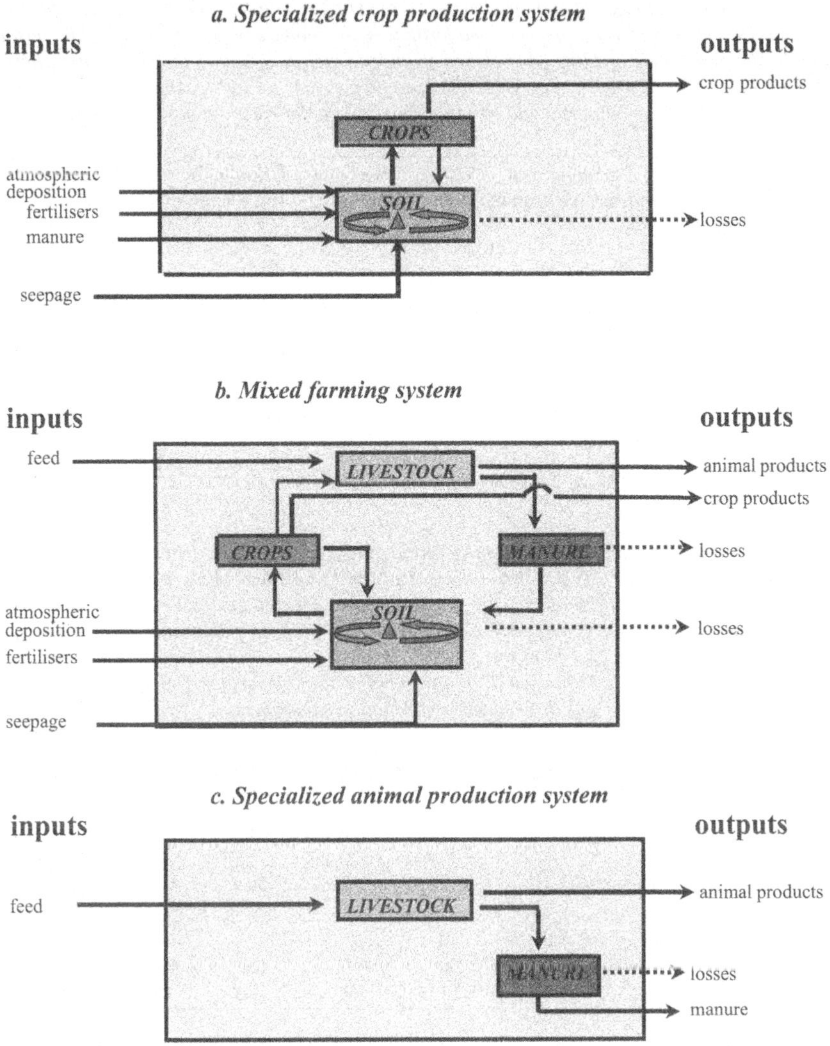

Figure 7. Conceptual diagram of (a) specialised crop production systems, (b) mixed farming systems, and (c) specialised animal production systems. Inputs of S sources are plotted left, outputs are plotted on the right-hand side. Source: Oenema and Pietrzak (2002)

when evapotranspiration exceeds precipitation, was by far the largest source of S in the soil solution of the topsoil. The SO_4^{2-}-S in the groundwater may be of marine origin, because these soils are marine deposits. Alternatively, the origin is industrial, because in previous decades, atmospheric S deposition was very high in industrialised countries in Western Europe (e.g. Fig. 3). The net mineralisation of S from organically bound S

Table 4. Total S balances of eight treatments of the Ultuan long-term-field experiment for the period 1956-1991 (35 yr). Inputs of S via fertilisers and manure included the input of a basal application each year of 27 kg S per ha via ordinary superphosphate. S input via atmospheric deposition and seeds was on average 10.1 kg per ha per year for all treatments. Crop rotation was barley, oat, winter wheat and spring wheat and fodder rape. All data in kg S ha^{-1} yr^{-1}

Treatment	Fertiliser and Manure application	Crop removal	Change in soil S content	Leaching loss
No N, fallow	27	0	-6	43
No N	27	8	-4	33
$Ca(NO_3)_2$	27	15	-2	24
$(NH_4)_2SO_4$	118	17	-2	113
Green manure	38	14	0	34
Animal manure	45	12	+6	38
Sewage sludge	94	21	+13	70
Peat	32	8	-2	36

Source: Kirchmann et al. (1996)

Table 5. Sulphate-S balances of the soil solution of topsoils (0-30 cm) from Belgium (soils A and B; Verlinden, 2002), The Netherlands (soils C and D; Postma et al., 1999) and Germany (soils E and F; Bloem, 1998) during the growing season (March to October). Soils were used for intensively managed grasslands (A-D) and arable crops (E-F). Groundwater level was shallow, ranging from 60-160 in soil A, 40-80 in soil B, 80-160 in soil C, 40-120 in soil D, >300 in soil E and 180-250 in soil F, all in cm below the soil surface. Net S mineralization was determined in incubation experiments.

Balance entries	Soil A	Soil B	Soil C	Soil D	Soil E	Soil F
Inputs						
Atmospheric deposition	10	12	16	16	10	21
Net mineralization	11	26	12	50	9	29
Capillary rise	23	77	0	25	0	60
Animal manure	0	0	5	5	0	0
Fertiliser	0	0	10	10	0	24
Total input	44	115	43	106	19	134
Outputs						
Crop removal	30	41	36	36	18	13
Leaching	12	48	42	21	40	101
Balance	+2	+26	-35	+49	-39	+20

during the growing season, estimated via incubation experiments, was on average a larger source than S input via atmospheric deposition. Differences between soils in net S mineralisation were up to a factor 5, emphasising the importance of soil and site-specific conditions in S management. Removals of S via the harvested crops differed by a factor of 3, and S losses via leaching by a factor of 8. The net S balance indicated that the pool of SO_4^{2-}-S was depleted by 30 to 40 kg per ha in soils C and E, and enriched by

20 to 50 kg per ha in soils B, D and F (Table 5). Based on the information presented by the S balances, it was concluded that additional S fertiliser input was only needed in situations with little net S mineralisation and no S input via capillary rise and/or irrigation. This was confirmed in part by the results of field experiments (Bloem, 1998; Postma et al., 1999; Verlinden, 2002).

3.4.2 Sulphur balances of grazed pastoral systems for animal production

Large areas in the world are managed pastures, used for the production of milk, beef and wool. The productivity and sustainability of these systems depends on the management and on the nutrient cycling and transformation within the soil-plant-animal system. Common to all pastoral systems is the large internal nutrient cycling via animal excrements (e.g. Haynes and Williams, 1993; Goh and Williams, 1997). Only a small fraction of the nutrients contained in the forage ingested by the ruminants is converted to animal products. The remainder returns to the pasture soil via dung and urine and can be utilised again within the system. In sheep-grazed pastures, between 10 and 20% of the S in herbage is transferred into animal products, i.e. wool and lamb. The other 80 to 90% is returned to the pasture via urine and dung depending on the S content of the ingested herbage. Table 6 indicates that S output via animal products more than doubled following fertiliser S and P applications to sheep-grazed pasture. However, S leaching losses increased much stronger (factor of 4 to 7). Evidently, S leaching losses increased much stronger than animal productivity following S and P fertiliser applications. Though herbage growth strongly responded to fertiliser application, leaching losses were larger than the S input via fertilisers. Note that leaching losses were estimated by difference, and that changes in soil S content were not recorded. Interestingly, a significant fraction (16-24%) of the excretal returns were deposited in areas (stock camps, off-farm sites) outside the productive pasture area and, therefore, did not contribute to the recycling of S within the pasture. This amount was about equal to the amount removed via animal products (Nguyen and Goh, 1992).

Annual removal of S via animal products may easily differ by a factor 20 between grassland systems (Table 7). Little S is removed from extensively managed pastures with beef cattle, as only 5% of the ingested herbage S ends up in animal products. On a grass-clover sward with improved management and increased livestock density, the intake of S via herbage is 5 times larger than on the extensively managed grassland, but the conversion of S in plant protein to S in animal protein remains at about 5%. In contrast, dairy cows transform about 10% of the S in herbage to S in animal products (milk and meat). Yet, the removal of S via animal product remains low, i.e. a factor 10 lower than in crop production systems (Table 5, 6 and 7). Even highly intensive dairy farms producing 20,000 kg of milk ha^{-1} $year^{-1}$ remove only 6 kg S ha^{-1} $year^{-1}$ with the sell of milk. Table 7 shows that large amount of S are recycled within the pasture via herbage – animal – manure – soil – herbage. These flow are as large as or larger than the total S inputs and outputs, the latter including SO_4^{2-}–S leaching. Looking only at the total input and output of S may suggest that there is plenty of S in the grassland systems. Yet, there are often local and temporary imbalances between S supply and S demand because of the uneven distribution of excretal returns and because of poor synchronisation of S demand by the growing crop and S supply via mineralisation, atmospheric deposition, and fertiliser application. Simple simulation models as discussed further below can explore such possible imbalances.

Table 6. Sulphur balances (kg S ha^{-1} yr^{-1}) of sheep-grazed pastures in New Zealand, with 3 treatments receiving annual applications of super phosphate fertiliser over a 35 years period

	Superphosphate applications (kg ha^{-1} $year^{-1}$)		
Balance entries	**0**	**188**	**376**
Inputs			
Rainfall	3	3	3
Irrigation	6	6	6
Fertiliser	0	21	42
Total input	9	30	51
Outputs			
Animal S removal	1.4	2.9	3.4
Leaching	6.7	23.9	43.2
Excretal S in stock camps	0.8	2.7	3.6
Excretal S in off-farm sites	0.1	0.5	0.7
Total output	9.0	30.0	51.0

Source: Nguyen and Goh (1992)

Table 7. Sulphur balances (kg S^{-1} ha^{-1} yr^{-1}) of three grassland systems grazed by dairy cows or beef cattle. Systems differ in management (intensive versus extensive management

Balance entries	Intensive Dairy cows	Intensive Beef cattle	Extensive Beef cattle
Inputs			
Atmospheric deposition	20	20	15
Fertiliser	20	0	0
Total input	*40*	*20*	*15*
Outputs			
Animal S removal	3	0.8	0.16
Leaching	30	16	8
Excretal S in off-farm sites	4	0	0
Storage in soil	3	4	6
Total output	*40*	*20*	*15*
Internal recycling			
Uptake in herbage	40	25	6
Intake of herbage by animals	32	16	3
Dead herbage and roots to soil	23	22	8
Excretal returns to grazed area	25	15	3

Source: Whitehead (2000)

3.5 Synchronising S Supply to S Demand Using Simple Models

The nutrient balances discussed in the previous section provide information about the matching of S supply to S demand over a whole growing season (Table 5), a one-year period (Table 7) or over a long-term period of 35 years (Tables 4 and 6). Though S balances over a whole growing period suggest that S supply does balance S demand, there may be imbalances between supply and demand during short periods within a growing season. Such temporary imbalances may greatly affect crop yield and quality.

Figure 8 shows the results of a simulation for winter wheat on clay soil with 3% organic matter (Postma et al., 1999). There was no additional S fertilisation, the only S sources were atmospheric deposition (17 kg per ha per year) and net S mineralisation (15 kg per ha per year). Total crop uptake was almost 22 kg per ha per year, and was not limited by shortage of SO_4^{2-}-S in the soil solution. The stock of S in the soil solution of the top soil (0-60 cm) decreased from an initial value of 10 kg SO_4^{2-}-S per ha per year to a minimum of 2 kg in decade 16 to19, mainly because of uptake of S by the winter wheat. From decade 19 onwards, the stock of SO_4^{2-}-S in the soil solution increased again till late autumn when losses via leaching exceeded the input via atmospheric deposition and net S mineralisation.

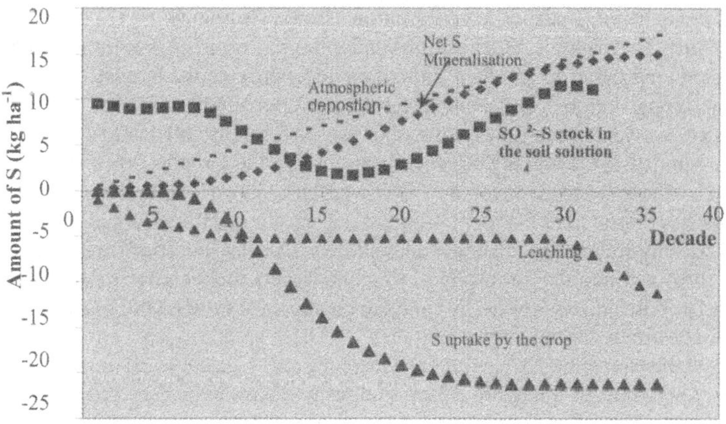

Figure 8. Simulated changes in the amounts of SO_4^{2-}-S in the soil solution of the topsoil (black squares), in response to S additions via atmospheric deposition and net S mineralisation, and S withdrawals via crop offtake and leaching (indicated by negative values). The S additions and withdrawls are plotted cumulative. Results deal with the growth of winter wheat on clay soil with 3% organic matter in The Netherlands (see text)

Results of a simulation for rapeseed grown under the same condition as the winter wheat crop in Figure 8, revealed that the S supply was insufficient to meet the S demand by rapeseed (data not shown). The stock of S in the soil solution decreased from an initial value of 10 kg SO_4^{2-}-S per ha per year to nearly zero in decades 11 to 19. The total S demand was 42 kg per ha, but the actual uptake was only 21 kg per ha per year. The actual S uptake by rapeseed was even lower than the uptake by wheat, because of the much shorter growing season of the rapeseed. Evidently, growing winter wheat did not require additional S fertilisation, but rapeseed did.

The importance of such simulation models offers an opportunity to identify combinations of sites, soil types and crops where S may limit crop yield and crop quality. Even in a small country like The Netherlands, S inputs via atmospheric deposition varies regionally by more than a factor of 2. The S requirement of the major crops also differs by more than a factor 2. Further, there are large differences between soils in net S mineralisation, in the potential for S leaching, and in the supply of S via groundwater seepage. Finally, the S supply from crop residues, animal manure, sludge, compost and fertilisers depends on its composition and the environmental conditions. All these potential variability can be assessed by using simple simulation models and, therefore, are of great value in managing S in agroecosystems.

4. CONCLUSIONS

The management of S in agroecosystems focuses on matching the S supply to the S demand by crops and animals, i.e. the right amount of S in the right form at the right place and at the right time. All these management activities have to be considered in a whole farm perspective, to be able to achieve consistency in the general farm strategy and also to achieve a balanced supply for all essential nutrients.

Globally, atmospheric S depositions and internal S sources (net mineralisation of soil organic matter, crop residues, animal manure) provide the largest supply of S to growing crops. Crop offtake and leaching are the largest sinks of S. However, differences between farms, sites, soils, and crops in S supply and demand are very large which complicates the making of generalisations. Unlike for most other essential plant nutrients, atmospheric S deposition is by far the largest external nutrient source. Atmospheric S originates from three sources, i.e. oceans and micro-organisms in marine and anoxic environments, volcanoes and industry. Oceans and micro-organisms emit a steady flow of S into the atmosphere, which is largely independent of anthropogenic influences. Volcanic emissions, by contrast, greatly vary in time (and space). Industrial S emissions are by far the largest source of S. These anthropogenic emissions started roughly at the end of the 19th century and are believed to continue till the end of the 21st century when most of the hydrocarbon sources will have been depleted.

The management of S in agroecosystems of industrial countries depends on the success of the management by industry and government to decrease industrial S emissions. The management of industrial S emissions has been highly successful in some industrial countries, with the result that atmospheric S depositions decreased by a factor of 2 and more, and that the incidence of S deficiency in crop production increased in these countries.

Net mineralisation of S from soil organic matter, crop residues and animal manure is an important S source in some agroecosystems, but its size and timing are difficult to predict. It is well recognised now that soil tests alone can not accurately predict the need for additional fertiliser S to sustain crop production, but soil test can provide essential information needed by decision support models.

Simple budget models can greatly facilitate the decision-making activities in S management. Such models calculate the input of plant available SO_4^{2-}-S in the soil solution from all sources and the withdrawal of SO_4^{2-}-S from the soil solution by crop uptake, leaching and erosion on a daily or weekly basis. Such models provide insight into the match of S supply to S demand at all critical stages during the growing period.

They are capable of predicting the amount, form and time of S fertilizer application.

Input-output balances for S over a whole growing season or whole year can provide essential information with which to evaluate and improve the model predictions and the S management in general. These input-output balances together with the decision support budget model form essential tools for S management in agroecosystems.

Inputs of S via fertilisers are still a small fraction of the total S input in agroecosystems. About 10 Tg S has been applied via fertilisers to agroecosystems in 2000, most of it as impurity in ordinary superphosphate fertilisers, which is less than 10% of the total S supply. Forecasts indicate that the supply by fertilisers will be doubled during the next two decades in response to the increasing quest for food by the growing global population and in response to the decreasing industrial S emissions.

5. REFERENCES

Anderson JW (1978) Sulphur in Biology (Studies in Biology, No 101). Arnold, London.

Barrow NJ (1987) Returns of nutrients by animals. In: Snayon RW (ed) Managed Grasslands; Analytical Studies, pp 181—186. Elsevier Science Publisher, Amsterdam

Beegle DB, Carton OT and Baily JB (2000) Nutrient management planning: justification, theory, practice. *J Environ Qual* **29:** 72—79

Bloem EM (1998) Sulphur balances of agro-ecosystems with special regard to site hydrology and physical soil characteristics. Ph.D. Dissertation, University of Braunschweig. Landbauforschung Volkenrode, Sonderheft 192

Bowen HJM (1979) Environmental Chemistry of Elements. Academic Press, New York.

Brimblecombe P, Hammer C, Rohde H, Ryaboshapko A and Boutron CF (1989) Evolution of the sulphur cycle over recent millennia. In: Brimblecombe P and Lein AY (eds) Evolution of the Biogeochemical Sulphur Cycle, SCOPE 39, pp 77—123. John Wiley & Sons, New York

Castellano SD and Dick RP (1991) Modelling sulfur transformations in soil. *Soil Sci* **152:** 448—454

David FR (2001) Strategic Management: Concepts. 5th Edition. Prentice Hall Upper Saddle River, NJ

Delagdo JA and Amacher MC (1997) Modelling the uptake of sulfur by crops on three alluvial soils of Lousiana. *Comm Soil Sci Plant Nut* **28:** 225—236

Eriksen J, Murphy MD and Schnug E (1998) The soil sulphur cycle, pp 39—73. In: Schnug E (ed) Sulphur in Agroecosystems. Kluwer Academic Publishers, Dordrecht

Follett RF and Wilkenson SR (1995) Nutrient management of forages. In: Barnes RF, Miller DA and Nelson CJ (eds) Forages: The Science of Grassland Agriculture, Vol II, pp 55—82. IA State Univ Press, Ames

Goh KM and Nguyen ML (1997) Estimating net annual soil sulfur mineralisation in New Zeland grazed pastures using mass balance models. *New Zealand J Agric Res* **48:** 477—484

Goh KM and Williams PH (1997) Comparative nutrient budgets of temperate grazed pastoral systems. In: Smaling EMA, Oenema O and Fresco LO (eds) Nutrient Disequilibria in Agroecosystems: Concepts and Case-studies, pp 265-293. CAB International, Wallingford

Haynes RJ and Williams PH (1993) Nutrient cycling and soil fertility in the grazed pasture ecosystems. *Adv Agron* **49:**119—199

Hoeksma P (1987) Composition of animal slurry transported to arable farms. IMAG, Wageningen University

Houba VJG, Temminghoff EJM, Gaikhorst, GA and van Vark W (2000) Soil analysis procedures using 0.01 M calcium chloride as extraction reagent. *Commun Soil Sci Plant Anal* **31:** 1299—1396

Howarth RW, Stewart JWB, Ivanov MV (eds) (1992) Sulphur Cycling on the Continents: Wetlands, Terrestrial Ecosystems and Associated Water Bodies, SCOPE 48. John Wiley & Sons, Chichester

Ivanov MV and Freney JR (eds) (1983) The Global Biogeochemical Sulfur Cycle, SCOPE 19. John Wiley & Sons, Chichester

Jones MB (1986) Sulfur availability indexes. In: Tabatabai MA (ed) Sulfur in Agriculture, pp 549—566. ASA/CSSA/SSSA, Madison, Wisconsin

Kellog WW, Caddle RD, Allen ER, Lazarus AL and Martell EA (1972) The sulfur cycle. *Sci* **175:** 587—596

Kirchmann H, Pichlmayer F and Gerzabek MH (1996) Sulfur balances and sulfur-34 abundance in a long-term fertiliser experiment. *Soil Sci Soc Am J* **59:** 174—178

Kishchuk BE and Brockley RP (2002) Sulfur availability on lodgepole pine sites in British Columbia. *Soil Sci Soc Am J* **66:** 1325—1333
Kropff M (1989) Quantification of SO_2 effects on physiological processes, plant growth and crop production. Ph.D. thesis Wageningen University
Marschner H (1995) Mineral Nutrition of Higher Plants, 2nd edition. Academic Press, London
Maynard DG (ed) (1998) Sulfur in the Environment. Marcel Dekker, New York
McCaskill MR and Blair GJ (1990) A model of S,P and N uptake by a perennial pasture. *Fert Res* **22:**161-172
Nguyen ML and Goh KM (1992) Nutrient cycling and losses based on a mass-balance model in grazed pastures receiving long-term superphosphate applications in New Zealand. *J Agric Sci* **119:** 107—122
Nguyen ML and Goh KM (1993) An overview of the New Zealand sulphur cycling model for recommending sulphur requirements in grazed pastures. *New Zealand J Agric Res* **36:** 475—491
Oenema O and Pietrzak S (2002) Nutrient management in food production: achieving agronomic and environmental targets. *Ambio* **31:** 159—168
Oenema O, Kros H and De Vries W (2002) Approaches and uncertainties in nutrient budgets; implications for nutrient management and environmental policies. *Euro J Agron* (in press).
Parton WJ, Stewart JWB and Cole CV (1988) Dynamics of C, N, P and S in grassland soils: a model. *Biogeochemistry* **5:** 109—131
Postma R, Van Erp PJ and Saanen R (1999) Quantifying the sulphur supply to agricultural crops (in Dutch) *Meststoffen* **1999:** 28—35
Reuter DJ and Robinson JB (1997) Plant analysis, an interpretation manual. CSIRO, Australia
RIVM (2001) Environmental Balance 2001; Explaining the environment in The Netherlands (In Dutch). Kluwer, Alphen aan de Rijn, The Netherlands
Ribeiro ES, Dias LE, Alvarez VH, Mello JWV and DanielsWL (2001) Dynamics of sulfur fractions in Brazilian soils submitted to consecurtive harvests of sorghum. *Soil Sci Soc Am J* **65:** 789—794
Saunders PWJ (1985) Regulations and research on SO_2 and its effect on plants in the European Community. In: Winner WE, Mooney HA and Goldstein RA (eds) Sulfur Dioxide and Vegetation, Physiology, Ecology and Policy Issues, pp 37—55. Stanford University Press, Stanford, California
Schlesinger WH (1997) Biogeochemistry: An Analysis of Global Change. Academic Press, New York.
Sinclair AG (1986) Sulphur fertiliser recommendations scheme in New Zealand. *Sulphur Agric* **10:** 12—13
Smaling EMA and Fresco LO (1993) A decision support model for monitoring nutrient balances under agricultural land use (NUTMON). *Geoderma* **60:** 235—256
Smaling EMA and Oenema O (1997) Estimating nutrient balances in agroecosystems at different spatial scales. In: Lal R, Blume WH, Valentine C and Stewart BA (eds) Methods for Assessment of Soil Degradation. Advances in Soil Science, pp 229—252. CRC Press, Ames, Iowa
Smil V (1995) Cycles of Life. Civilization and the Biosphere. Scientific American Library, New York
Stevenson FJ and MA Cole (1999) Cycles of Soil: Carbon, Nitrogen, Phosphorus, Sulphur, Micronutrients. John Wiley & Sons, New York
Syers JK, Curtin D and Skinner RJ (1987) Soil and Fertiliser Sulphur in the UK Agriculture. Proceedings 264 The Fertiliser Society, pp 33
Tabatabai MA (ed) (1986) Sulfur in Agriculture. American Society of Agronomy, Madison, WI.
Tilman D, Fargione J, Wolff B, D'Antonio C, Dobson A, Howarth RW, Schindler D, Schlesinger WH, Simberloff D, and Swackhamer D (2001) Forecasting agriculturally driven global environmental change. *Science* **292:** 281—284
Underwood EJ and Suttle NF (1999) The Mineral Nutrition of Livestock. CAB International, Wallingford
Verlinden G (2002) Sulfur Dynamics in Belgian Agricultural Soils. Dissertationes de Agricultura, University of Leuven, Belgium
Walker JCG (1977) Evolution of the Atmosphere. Macmillan, New York
Whitehead DC (2000) Nutrient Elements in Grassland. Soil Plant Animal Relationships. CABI Publishing,
Widdel F (1988) Microbiology and ecology of sulfate- and sulfur-reducing bacteria. In: Zehnder AJB (ed) Biology of Anaerobic Microorganisms, pp 469—585. John Wiley & S Sons, Chichester

6. AFFILIATIONS

Wageningen University and Research Center, Alterra, P.O.Box 47, NL-6700 AA Wageningen (OO) and Nutrient Management Institute NMI, Haagsteeg 2b, NL-6708 PM Wageningen (RP), The Netherlands. For correspondence - *O. Oenema*: O.Oenema@alterra.wag-ur.nl

M.J. HAWKESFORD, P. BUCHNER, L. HOPKINS
AND J.R. HOWARTH

SULPHATE UPTAKE AND TRANSPORT

Abstract. This chapter describes the physiological and biochemical background of sulphate transport in relation to current molecular approaches. Following identification and isolation of the first plant gene encoding H^+/sulphate co-transporter, a large gene family has now been identified. Phylogenetic analysis of this gene family suggests a specialisation of function of subgroups defined by the patterns of gene expression and tissue, cellular and sub-cellular localisation. Functional characteristics of the cloned transporters have been determined by expression in a heterologous yeast expression system, utilising sulphate transporter deficient mutants. A range of affinity constants for sulphate has been determined. An analysis of conserved amino acids together with site-directed mutagenesis indicates residues of probable functional importance.

The control of sulphate uptake and assimilation can be described by a model of a 'highly regulated circuit', mediated by feedback loops involving key metabolites of cysteine biosynthesis. In this model, expression of genes involved in uptake and assimilation are under positive regulation by *O*-acetylserine, which accumulates when insufficient sulphide is available to utilise the *O*-acetylserine for cysteine synthesis. This modulates the activity of the serine acetyltransferase/*O*-acetylserine-thiol-lyase complex. A negative feedback mechanism, mediated by reduced S-compounds may also act at the level of gene expression.

1. INTRODUCTION

Sulphur is mostly absorbed as sulphate from the soil solution through the root system. Once absorbed through the roots, the sulphate is transferred to the site of reduction or of storage via the xylem. Sulphate is then transferred between tissues (Gede et al., 1995) via the phloem. In all cases, transport across the plasma membranes of cells is required. Intracellular transport processes are also involved: reduction occurs in the plastid, requiring specific transport inward across the plastid membrane, and as the vacuole is the major storage pool, both inward and outward transport across the tonoplast is required. Local conditions and requirements for transport will depend upon many factors, including sulphate concentration gradients and potential driving forces (Hawkesford, 2000; Hawkesford and Wray, 2000). It is clear that in higher plants there exists a large gene family encoding related transporters, with multiple members in any one species. However, it is not as yet unequivocally demonstrated that this one gene family is responsible for all of the instances of trans-membrane sulphate transport, which occur in plants. Internal cycling of sulphate is supplemented by transport of various reduced-S compounds such as glutathione (Rennenberg et al., 1979) or S-methyl methionine (Bourgis et al., 1999).

2. SPECIFIC CARRIERS OF SULPHATE

Transport across plasma membranes is via a specific carrier (Leggett and Epstein, 1956). It is generally assumed that sulphate transport is driven by a proton gradient and

Yash P. Abrol and Altaf Ahmad (eds.), Sulphur in Plants, 71—86.

is a $3H^+/SO_4^{2-}$ co-transport mechanism. The requirement for a pH gradient to drive uptake was also shown in higher plants using cultured tobacco cells (Smith, 1975; Jones and Smith, 1981) and in an isolated vesicle system (Hawkesford et al., 1993).

Leggett and Epstein (1956) described uptake kinetics for barley roots showing classical saturation kinetics with a K_m of around 19 μM. Uptake into cultured tobacco cells had a K_m of 20 μM and was stimulated by calcium (Jones and Smith, 1981). Some authors have claimed the existence of a multiphasic uptake system consisting of a single transporter able to exist in multiple states (Nissen, 1971; Vange et al., 1974), however, other explanations may invoke multiple transporters with different affinities. In *Lemna paucicostata*, both a high affinity (K_m 11 μM) and a low affinity uptake were distinguished (Datko and Mudd, 1984). In *Chlorella*, triphasic kinetics were observed (K_ms of 13, 39, 280 μM) (Biedlingmaier and Schmidt, 1989). In the only study of transport into leaf cells, K_ms of 43 and 32 μM were obtained for mesophyll and bundle sheath cells of maize, respectively (Passera and Ferretti, 1988).

Analysis of the first cloned members of a higher plant sulphate transporter gene family in a yeast expression system enabled the determination of K_ms of 10 and 100 μM (Smith et al., 1995a). The recent cloning of members of *Arabidopsis thaliana* multi-gene family and the determination of K_ms, when expressed in yeast (Table 1), may go some way to resolving the variation in K_ms observed in whole tissues or in isolated plant cells. Additionally, such studies will enable the clarification of any role of conformational state transformations on apparent K_ms. Within the restricted range of sulphate concentrations employed, multiphasic kinetics were not observed in the isolated vesicle system (Hawkesford et al., 1993) or when genes for a single plant sulphate transporter were expressed in a sulphate transporter deficient yeast mutant (Smith et al., 1995a; 1997).

3. THE PLANT H^+/SULPHATE CO-TRANSPORTER GENE FAMILY

Since the isolation of the first plant sulphate transporter genes from the tropic legume *Stylosanthes hamata*, using a yeast complementation system (Smith et al. 1995a), a large number of genes and cDNAs encoding sulphate transporters have been isolated from different plant species. In the genome of *A. thaliana*, 12 H^+/sulphate co-transporter like genes have been identified, indicating that a large gene family is responsible for sulphate transport. In most of the plant H^+/sulphate co-transporter protein sequences, the sulphate transporter signature [PAV]-x-Y-[GS]L-Y-[STAG](2)-x(4)-[LIVFYA]-[LIVST]-[YI]-x(3)-[GA]-[GST]-S-[KR] (Pro-site accession number PS01130) is found. This signature is also characteristic for mammalian, yeast and bacterial examples of this family of sulphate transporters. The sulphate transporter superfamily is also characterised by the 'Protein families' database of alignments and HMMs (Hidden Markov Models) consensus accession number PF00916. The C-terminal region of most sulphate transporters contains a STAS domain (Sulphate Transporter and AntiSigma factor antagonist) (Aravind and Koonin, 2000). It has been suggested that this domain may have a general NTP binding function.

The first indication that the plant sulphate transporter gene family may be divided into different subgroups was suggested by transport activity measurements using yeast complementation. With the large number of sequences now available, phylogenetic analysis shows subdivision into 4 distinct groups (Fig. 1). Group 1 is the best-characterised. Members of group 1, which have been characterised by transport activity measurements, have K_ms of less than 12 μM, which indicates a high substrate affinity (Smith et al., 1995a, 1997; Takahashi et al., 2000; Vidmar et al., 2000; Yoshimoto et al., 2002; Table 1).

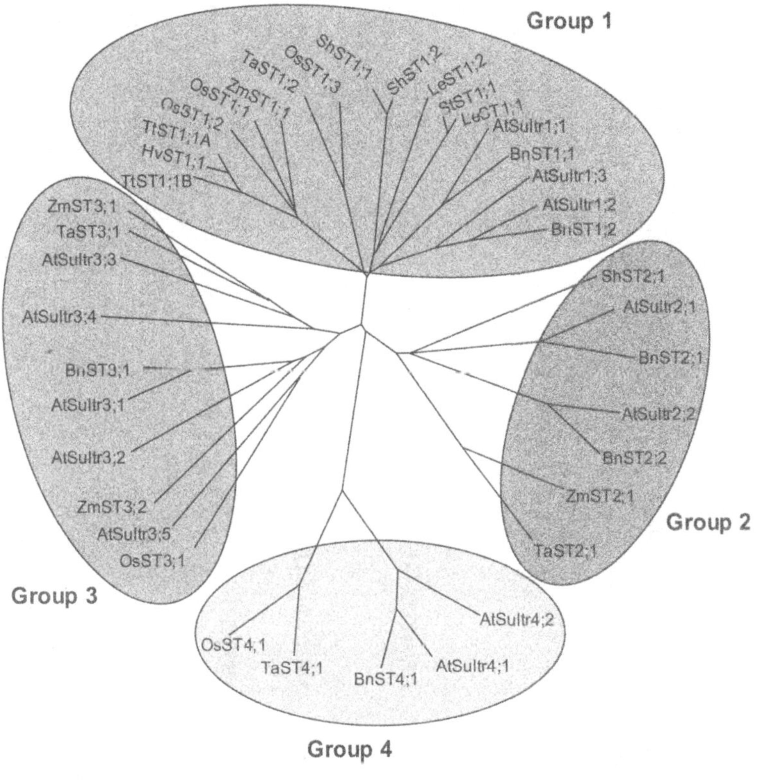

Figure 1. Unrooted dendrogram of selected plant H^+/sulphate transporter amino acid sequences. Accession numbers: for Arabidopsis, see Table 1; for wheat: TtST1;1A and B, AJ238244 and AJ238245, respectively; TaST 3;1 unpublished, and for all other wheat ST genes, see DuPont patent application No. WO 00/04154; for Hordeum vulgare: HvST1;1, X96431; for Brassica napus/juncea: BnST1;1, AJ416460; BnST1;2, AJ311388; BnST2;1, unpublished; BnST2;2, AJ223495; BnST3;1, unpublished; BnST4;1, AJ416461; for Zea mays: ZmST1;1, AF355602; all other maize STs, see DuPont patent application No. WO 00/04154; for S. hamata: ShST1;1, X82255; ShST1;2, X82256; ShST2;1, X82454; for Lycopersicon esculentum: LeST1;1, AF347613; LeST1;2, AF347614; for Solanum tuberosum: StST1;1, AF309643; for Oryza sativa: OsST1;1, AF493790, OsST4;1, AF493791; OsST3;1, BAB68064.1; OsST1;2 and 1;3 from Genome Database of Chinese Super Hybrid Rice. Alignments were performed using CLUSTAL W program (Thompson et al., 1994) version 1.7 and the tree was drawn using the Treeview32 program (Page, 1996)

Table 1. The Arabidopsis gene family for sulphate transporters

Group	Gene designation (alternative names)	Accession numbers: Genomic/ Chromosome	Accession numbers: cDNA	Accession numbers: Protein	K_m	Up-regulation by S-starvation	Expression pattern/ cellular localisation	References
1	AtSultr1;1 (ast101)	AC002983 Chr4	AB018695	O22277	1.5 µM[1]; 3.6±0.6 µM[2]	Yes	Root - root cap; root hairs; epidermis; cortex Leaves under S	Vidmar et al., 2000; Takahashi et al., 2000
	AtSultr1;2	AC009243 Chr1	AB042322	Q9MAX3	6.9±1.0 µM	Yes	Root - root hairs; epidermis; cortex; root tip; Leaves - guard cells	Yoshimoto et al., 2002; Shibagaki et al., 2002
	AtSultr1;3	AC069252 Chr1	AB049624	Q9FEP7	nd	nd	nd	
2	AtSultr2;1 (ast68)	AB003590 Chr5	AB003591	O04722	5 µM[1] 0.41±0.007 mM[2]	Yes (only in roots) negative -S effect in leaves	Root - root cap; pericycles; xylem parenchyma Shoot/leaves xylem prenchyma	Vidmar et al., 2000; Takahashi et al., 1997; 2000
	AtSultr2;2 (ast56, atd14)	AB012047 Chr1	D85416	O64434	1.2 mM	Yes (only in leaves)	Root - phloem Leaves - vascular bundle sheath cells	Takahashi et al., 1997; 2000

3	AtSultr3;1 (ast12, atst1, atd631)	AB012048 Chr3	D89631	CAB41310	nd	Negative -S effect	Leaves	Takahashi et al., 1999b; 2000
	AtSultr3;2 (ast77, at4060)	AC002330 Chr4	AB004060	O04289	nd	Negative -S effect	Leaves	Takahashi et al., 1999b
	AtSultr3;3 (ast91)	AC002311 Chr1	AB023423	O49307	nd	No effect	nd	
	AtSultr3;4	AB012247 Chr3	AB054645	Q9LW86	nd	nd	nd	
	AtSultr3;5	NM_121965 Chr5	AB061739	Q94LW6	nd	nd	nd	
4	AtSultr4;1 (ab8782, ast82)	AB006704 Chr5	AB008782	O22123	nd	Yes (only in leaves)	Vascular tissue; putative protein in plastids.	Takahashi et al., 1999a; 2000
	AtSultr4;2	AC069474 Chr3	AB052775	Q9LHF7	nd	nd	nd	

nd = not determined

Detailed analysis of group 1, particularly the *A. thaliana* and *Hordeum vulgare* examples, suggests that these sulphate transporters are mainly responsible for sulphate uptake by the root (Smith et al., 1997; Takahashi et al., 2000; Vidmar et al., 2000; Yoshimoto et al., 2002). In situ analysis using promoter GFP fusion constructs in transgenic *Arabidopsis* localised expression of the AtSultr1;1 and AtSultr1;2 transporter to the root cap, root hairs, epidermis and cortex, leaf guard cells, hydathodes of cotyledons and in auxiliary buds (Takahashi et al., 2000; Shibagaki et al., 2002; Yoshimoto et al., 2002). In general, expression of group 1 sulphate transporters is induced or up-regulated by sulphate deprivation (Smith et al., 1997; Takahashi et al., 1997, 2000; Vidmar et al., 1999, 2000; Yoshimoto et al., 2002; see also Table 1).

In *S. hamata* and *A. thaliana*, sulphate transporters have been described which have a low affinity for sulphate, with K_m values of 0.41 mM and more (Smith et al., 1995a; Takahashi et al., 1997, 2000). Phylogenetic analysis places these sulphate transporters in group 2. Group 2 sulphate transporters are thought to be responsible for vascular sulphate transport. In situ hybridisation and promoter-GFP fusions of AtSultr2;1 and 2;2 localised the transcript and promoter activity to vascular tissues of root and leaves, respectively (Takahashi et al., 1997, 2000). Both transporters are differentially regulated under sulphate starvation. Whilst the AtSultr2;1 gene expression is upregulated under sulphate deprivation in roots, an opposite effect is found in leaves where sulphate starvation leads to a reduction of the transcript level. A similar situation was seen in *S. hamata* for ShST2;1 (SHST3) (Smith et al., 1995a). The AtSultr2;2 transcript level, in contrast, is higher in leaves under sulphate starvation whereas in roots, no change of the transcript level is found (Takahashi et al., 1997, 2000).

Group 3 contains 5 *Arabidopsis* sulphate transporter genes but these are less well characterised (Takahashi et al., 1999b). PCR-based expression studies indicate that the main expression of AtSultr3;1, AtSultr3;2 and AtSultr3;3 is in leaf tissue, and that their expression does not appear to be significantly modulated by the sulphur status of the plant. There is no information on the cellular localisation of group 3 sulphate transporters.

The sulphate transporters of the group 4 are characterised by a C-terminal plastidal transit peptide. The putative transit peptide region (N-terminal 99 amino acids) of the *Arabidopsis* AtSultr4;1 fused with GFP was able to target the GFP to the chloroplast in a transient assay. Surprisingly, plastidal targeting of the mature AtSultr4;1 protein to chloroplasts was only possible if the C-terminal was deleted by 13 amino acids (Takahashi et al., 1999a).

4. SULPHATE TRANSPORTER: STRUCTURE AND FUNCTION

4.1 Characterisation of Sulphate Transporter Function

In recent years, there has been a substantial increase in the number of known sulphate transporter sequences in the public domain, largely due to the application of RT-PCR cloning and genome sequencing projects. Whilst a limited number of functional studies have identified these genes as sulphate transporters, nucleotide sequence homology has generally been used to assign the assumed function. Initial characterisation of plant sulphate transporters, however, relied on the functional complementation of yeast

mutants lacking sulphate uptake capacity, with plant cDNA libraries (Smith et al., 1995a, 1997).
The molecular mechanism of sulphate uptake in yeast (*Saccharomyces cerevisiae*) was initially predicted by a series of uptake studies. These indicated that sulphate was transported into yeast cells via a system involving two permeases, one with a high and the other a low affinity for sulphate and its toxic homologues chromate and selenate (Breton and Surdin-Kerjan, 1977). The first cloning of a sulphate transporter gene together with confirmation of function was reported when a yeast cDNA library was transformed into a mutagenised yeast strain and selected as sulphate transport-deficient by its resistance to selenate and chromate toxicity. The clone, SUL1, was characterised as a high affinity transporter and was 'knocked-out' to produce a mutant yeast strain, YSD1, deficient in high affinity uptake (Smith et al., 1995b). The YSD1 mutant was a fundamental tool for the cloning of the first plant sulphate transporters and subsequently for the determination of their uptake kinetics in a heterologous expression system. The first plant sulphate transporters to be cloned were from the tropical legume, *S. hamata*. A cDNA library from sulphur-starved *S. hamata* roots was transformed into YSD1, and from colonies that were able to grow on media containing sulphate as the sole S source, three different sulphate transporters, SHST1, SHST2 and SHST3 (ShST1;1, ShST1;2 and ShST2;1, respectively in Fig. 1), were isolated (Smith et al., 1995a). Similarly, in barley the first sulphate transporter from a cereal crop, HVST1 was cloned (HvST1;1) (Smith et al., 1997). These genes were expressed in the YSD1 mutant and their affinities for uptake of radiolabelled sulphate were measured. As had been suggested by the early yeast studies (Breton and Surdin-Kerjan, 1977), the plant transporters were found to fall into groups of high and low affinity for sulphate. ShST1;1 (SHST1), ShST1;2 (SHST2) and HVST1 all had an apparent K_m for sulphate of around 10μM. ShST2;1 (SHST3) exhibited a much lower affinity for sulphate with a K_m of 99μM. Further insight into the function of the newly cloned transporters came from uptake experiments in media with increasing pH. Uptake was found to decrease when the media pH was raised from 5.6 to 7.6, confirming the hypothesis that a proton/sulphate co-transport mechamism was in operation (Smith et al., 1997). To date, sulphate transporter genes have been identified in a wide variety of plant species including *Brassica napus* (unpublished data), *Brassica juncea* (Heiss et al., 1999), *Oryza sativa* (see accession Q942e2 and D25000), *Solanum tuberosum* (see accession AF 309643), *Lycopersicon esculentum* (see accessions AF 347613 and AF347614), *Triticum tauschii* (see accessions AJ238244 and AJ238245) and *Sporobolus stapfiana* (Ng et al., 1996). Completion of *A. thaliana* genome sequencing project has also allowed a full assessment of the gene-family from that organism. Table 1 summarizes data on the 12 *A. thaliana* genes including the K_m data for transporters were determined. It is generally agreed that the affinity of sulphate transporters is related to specific functions of the isoforms in vivo. Northern analysis of the *S. hamata* sulphate transporters localised the expression of the high affinity transporters, ShST1;1 and 1;2 (SHST1 and SHST2), to the root, and the low affinity, ShST2;1 (SHST3), to the leaf tissues (Smith et al., 1997). Similarly the high affinity AtSultr1;1 gene from *A. thaliana* is localised in the root, whereas the lower affinity AtSultr2;1 and AtSultr2;2 genes were specifically expressed in the vascular tissues (Takahashi et al., 2000). The consensus of opinion is that a high affinity mechanism is in operation to transport sulphate from the external environment

and the root apoplast and that systems of lower affinity transporters are responsible for the distribution of sulphate around the plant towards and between sink tissues.

4.2 Sulphate Transporter Structure in Relation to Function

Protein sequence alignments of the rapidly increasing numbers of plant sulphate transporters show a relatively high number of conserved amino acids. In common with several other cation/solute symporters (Reizer et al., 1994), the plant sulphate transporter family has been predicted to consist of membrane bound proteins with 12 distinct membrane-spanning domains (MSDs) (Smith et al., 1997). This was calculated by profiling the hydropathy and relative positioning of predicted α-helical structures along the consensus transporter sequence. The positions of the predicted hydrophobic MSDs agreed closely with those previously proposed for sulphate transporters from rat and human (Hästbacka et al., 1994). The predicted topology model for sulphate transporter structure is shown in figure 2 using ShST1;1 as an example. The 12 MSDs, bordered by long non-membrane associated terminal domains, are thought to be intracellular. There are also clusters of positively charged residues in several of the

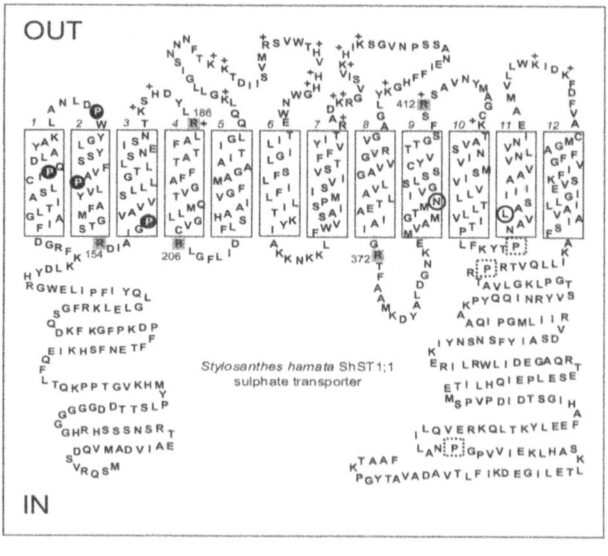

Figure 2. Topology diagram of the amino acid sequence of sulphate transporter ShST1;1 from Stylosanthes hamata showing the 12 membrane spanning domains (numbered). Also highlighted are the conserved arginine (R) residues (grey boxes), proline (P) residues (closed circles) investigated by site-directed mutagenesis (Shelden et al., 2001) and conserved asparagine (N) and leucine (L) residues (open circles) found to be essential for sulphate uptake by Khuarma et al. (2000). Extracellular positively charged basic residues are marked with a +. Uninvestigated conserved proline residues are boxed (dotted lines).

extracellular loops of all the plant sulphate transporters (Fig. 2), a feature which is generally atypical of membrane proteins (von Heinje, 1992). It was suggested that these basic residues and the charged residues buried in MSDs (one positive and one negative residue are highly conserved in MSD 8) may be functionally involved in the binding or channelling of anions (Smith et al., 1997). Other residues postulated to be involved in the functioning of the plant sulphate transporter include a number of highly conserved arginine residues. Clarkson and co-workers found that the uptake of sulphate in barley roots was strongly inhibited by the application of the arginine-binding reagent, phenylglyoxal (PhG) (Clarkson et al., 1992). When applied to barley roots, sulphate uptake was drastically reduced at PhG concentrations 20 times lower than that required to affect phosphate uptake in the same plants. This suggests a key role of arginine residues in sulphate transporter function. Currently, 42 putative sulphate transporter sequences are available in the various public databases, but these are not full length through the coding sequence. An alignment of the deduced amino acid sequences of these proteins allows potentially important functional residues to be identified. Five arginine residues were found to be highly conserved among the known plant transporters. R_{206}, lying between MSDs 2 and 3, is conserved in every known plant sulphate transporter, and R_{412} is present in 37 out of the 38 sequences available, which span this region. These residues may be involved in sulphate transport as previously hypothesised (Clarkson et al., 1992). Three more arginine residues are also highly conserved. R_{154} is present in 29 of the 33 sequences available, R_{186} is present in all group 1 high affinity sulphate transporters (Fig. 1) but not in any group 2, 3 or 4 transporters and R_{372} is conserved in all Type 1, 2 and 3 transporters but not in group 4. These residues may have structural properties; important to the role of specific groups of sulphate transporters rather than being essential to function. Amino acid residues with functional importance in plant sulphate transport have been identified as a consequence of human genetic research (Khurana et al., 2000). Mutations in human sulphate transporter genes have been shown to cause a variety of chondrodysplasias, which result in skeletal deformities (Hästbacka et al., 1994).

Homologous mutations in the S. hamata transporter ShST1;1 were also shown to affect sulphate transport. Asparagine residue N_{395} and leucine L_{453} were both targeted for site directed mutagenesis in ShST1;1. In the human gene DTDST, mutations at these sites disrupts sulphate transport causing achondrogenesis Type B. When ShST1;1 N_{395} (in MSD9) was replaced by aspartic acid, sulphate transport in a yeast uptake system was completely eliminated and was drastically reduced when mutated to alanine. L_{453} (MSD11) when replaced by proline was found to reduce uptake activity to 15% of wild type rates. This indicates the importance of these two residues in sulphate transporter function, probably in the folding of their respective MSDs (Khurana et al., 2000). N_{395} and L_{453} are both conserved in all known 38 plant sulphate transporter sequences indicating their importance in the function of the transporter protein.

Site-directed mutagenesis has also been used to assess the importance of a selection of conserved proline residues in the first 3 MSDs of plant sulphate transporters, again using ShST1;1 as a model (Shelden et al., 2001). MSDs 1 to 3 are tightly coupled with short inter-MSD loops when compared to MSDs 4 to 12. Proline residues are often involved in the function of membrane proteins (Brandl and Deber, 1986). In the region of MSDs 1 to 3 in plant sulphate transporters there are 4 highly conserved proline residues. Sheldon and co-workers studied the effects of replacing these residues in ShST1;1 with alanine and leucine using a yeast expression system. P_{144} was found to be essential for sulphate transport and is present in all plant sulphate transporters except those of group 4 which have leucine, isoleucine or valine in position 144. When SHST1;1 P_{144} was mutated to a leucine residue no sulphate uptake or growth of yeast cells expressing the gene could be detected in 100μM sulphate (Shelden et al., 2001). Transporter activity data for group 4 sulphate transporters is not currently available although their function is thought to be in the plastid membrane rather than the plasma membranes (Fig. 1). For this reason different structural arrangements may be required of group 4 transporters which do not require the conservation of P_{144}. In the same experiments P_{120}, P_{133} and P_{160} were found not to be essential for sulphate uptake but were involved in function, most likely having a structural role (Shelden et al., 2001).

Several other proline residues are also highly conserved among the plant sulphate transporters, mainly around the C-terminus and MSD12 regions (Fig. 2). No functional studies have been carried out on these to date, although the identification of essential protein residues is extremely important to our understanding of ion transport. It is also clear from alignments of all known plant sulphate transporters that the region around amino acids 365 to 341 of MSDs 8 and 9 is also highly conserved. Future research into the function of this domain may provide further interesting insights into the transport of sulphate in plants.

5. REGULATION

5.1 Regulation by Sulphur Availability

A basic characteristic of sulphate uptake is its regulation by S-supply. Under S-limiting conditions S-uptake is increased and reduced following S re-supply (Smith et al., 1997; Bolchi et al., 1999; Lappartient et al., 1999). This corresponds to observed changes in abundance of specific sulphate transporter mRNA, with a de-repression of sulphate transporter expression with S-starvation (Smith et al., 1995, 1997; Takahashi et al., 2000).

A number of high affinity (Smith et al., 1997; Takahashi et al., 2000), low affinity (Smith et al., 1995; Takahashi et al., 2000) and organelle specific sulphate-transporters (Takahashi et al., 1999) have been cloned (Fig. 1). The availability of these clones in conjunction with defined external S-supply provides an experimental system for investigating the underlying mechanisms regulating the response(s) to S-supply. Typically increased S-transporter activity occurs within a few days of exhaustion of external S-supply, however, repression can occur within hours of S re-supply (Smith et al., 1997). It is evident from experimental data that the regulation of S-uptake and assimilation in plants may be explained by a range of mechanisms, including substrate availability, allosteric regulation of enzyme activity and regulation of gene expression. A number of different models of regulation have been proposed.

5.1.1 De-repression model

The classic model of regulation of expression of sulphate transporters and/or the S-assimilatory pathway is the de-repression mediated by feedback loops involving end products of S-assimilation. When plants are deprived of an external source of sulphur, their capacity for uptake increases (Smith, 1975; Lee, 1982; Clarkson et al., 1983). In tobacco, de-repression of S-uptake and transport to the shoots was repressed by glutathione and L-cysteine (Herschbach and Rennenberg, 1994), whereas in S-deficient maize seedlings S-transporter expression and ATP-sulphurylase were both down regulated only by L-cysteine and not glutathione (Bolchi et al., 1999). In young barley seedlings, grown hydroponically, sulphate transporter expression increases substantially after only 2 days of S-starvation, in parallel with observed depletion of intercellular S-pools. Following re-supply of external S, a decrease in mRNA pools and transporter activity occurs within just a few hours. In both cases, activity of sulphate transport parallels mRNA pools (Smith et al., 1997) and sulphate transporter protein (Hawkesford and Wray, 2000). These results support the idea of a de-repression model in which under S-sufficient conditions, the accumulation of S-assimilation end products (i.e. GSH and cysteine) act as repressors of S-uptake at the level of gene expression. Under S-limitation, a decrease in concentration of these compounds removes repression increasing transporter activity and maximising sulphate-uptake.

5.1.2 O-acetylserine (OAS) model

An additional mode of regulation in plants, in which OAS is a positive regulator (i.e. derepressor or inducer) of sulphate uptake and assimilation has been proposed (Hawkesford and Smith, 1997; Smith et al., 1997). This is based on observations described for prokaryotes (Kredich, 1992, 1993). These suggest that the metabolite signal linking the expression of genes for sulphate uptake and assimilation to nutritional S-status is N-acetylserine, which is formed non-enzymatically from OAS. The addition of exogenous OAS has been shown to increase ATP-sulphurylase and APS-reductase activity in *Lemna minor* (Neuenschwander et al., 1991). In a similar experiment with barley seedling roots, increased sulphate transporter mRNA pools and transporter activity together with increased cysteine and glutathione content was observed (Smith et al., 1997). From such observations a model of a 'regulatory circuit' has been proposed (Hawkesford and Smith, 1997, see Fig. 3). In this model, expression of genes involved in uptake and assimilation are under positive regulation by OAS, which accumulates when insufficient sulphide is available to utilise the OAS for cysteine synthesis. A negative feedback mechanism, mediated by reduced S-compounds, such as described for prokaryotes (Kredich, 1993), may also act at the level of gene expression although the identity of the S-pool responsible for this control remains to be verified.

5.1.3 Demand-driven model

The demand-driven regulation of uptake rate is suggested to be mediated via phloem translocated signals. In the case of S-uptake, reduced-S compounds (such as glutathione) have been suggested to be the 'demand-driven' regulatory signals (Lappartient and Touraine, 1996; Lappartient et al., 1999). In 'split-root' experiments using *Brassica napus*, in which one portion of the root was deprived of sulphate while the other portion of the root was fed with adequate sulphate, the GSH and cysteine levels decreased in the sulphate-fed portion of the root and the activity of the transporter

(AtSultr2;1) (AST68) and ATP-sulphurylase increased. The observed increase in sulphate uptake and ATP-sulphurylase activity could be prevented by feeding with GSH and cysteine instead of sulphate (Lappartient et al., 1999). The negative effect of cysteine on the sulphate transporter (AtSultr2;1) and ATP-sulphurylase expression in *Arabidopsis* roots was blocked by buthionine sulphoximine, an inhibitor of glutathione synthesis (Lappartient et al., 1999). These data suggest that glutathione, not cysteine, is

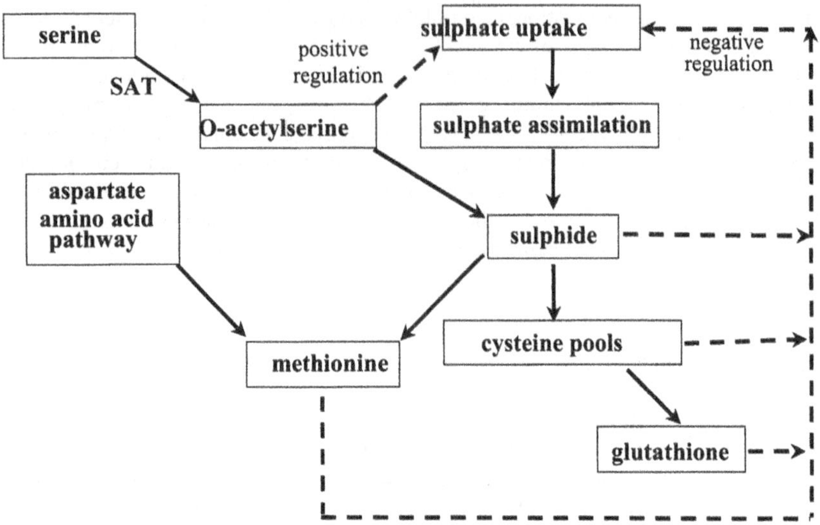

Figure 3. Feedback regulation of sulphate transporter expression. Major metabolites are shown in boxes with metabolic pathways shown as solid arrows. Broken arrows indicate regulatory loops. OAS is suggested to have a positive regulatory influence on gene expression of S-transporters, whilst reduced S metabolites may have a negative effect on gene expression. SAT: serine acetyltransferase

the signal that repress S-transporter and ATP-sulphurylase gene expression. Therefore, glutathione is suggested to be the likely phloem-translocated molecule regulating S-uptake and long-distance signalling of S-nutritional status. In contrast, feeding experiments on S-starved maize seedlings (Bolchi et al., 1999) showed only L-cysteine was able to down-regulate S-transporter and ATP-sulphurylase expression under conditions where glutathione synthesis was blocked by the addition of buthionine sulphoximine. This suggests that cysteine rather than glutathione acts as the repressor signal in this species.

5.2 Signal-Transduction Mechanism

Components involved in S-sensing have been identified by mutant analysis of the unicellular green-alga *Chlamydomonas reinhardtii*, and may provide a working model to start such an analysis in higher plants. *C. reinhardtii* responds to S-limitation in a similar manner to higher plants with increased efficiency for the uptake of sulphate, when sulphate supplies are limiting. The increase in S-uptake during S-limitation was prevented by the addition of cyclohexamide, but not chloramphenicol, suggesting the requirement of de novo protein synthesis on the 80S ribosome of a new high affinity S-transport system (Yildiz et al., 1994). The isolation and characterisation of sac (sulphur acclimation) mutants of *C. reinhardtii*, exhibiting altered responses to sulphur-limitation (Davies et al., 1994, 1996, 1999; Yildiz et al., 1996) suggests that a signal perception and transduction mechanism is involved in regulating the response to S-limitation (Davies et al., 1999). Sac3 positively regulates the high-affinity sulphate transport system, and the sac3 gene product is thought to be required for the full induction/activation of the sulphate transport system during S-starvation (Davies et al., 1999). High homology between sac3 and serine/threonine kinases, which are suggested to be involved in environmental sensing in higher plants, supports the role of these sac genes/proteins in signal transduction mechanisms linking environmental status to physiological processes in the cell.

5.3 Post-Translational Modification

The regulation of sulphate transporters is inter-related with parallel changes in sulphate transport activity, mRNA pools (Smith et al., 1997) and protein (Hawkesford and Wray, 2000). Furthermore, these changes in mRNA pools and detectable transporter protein occurred very rapidly. This would suggest that there is no mechanism operating in these specific cases, which would involve control by post-translational modification. However, given the complexity of the number of transporters expressed in a plant, these mechanisms can not be completely ruled out.

In yeast, there is evidence for a protein:protein interaction between the sulphate transporter and ATP sulphurylase (Logan et al., 1996) which is required for optimal sulphate transporter expression. Transgenic approaches in plants, which examined the effect of over-expressing ATP-sulphurylase on S-assimilation suggested levels of regulation in addition to the transcription response to sulphate deprivation (Hatzfeld et al., 1998). The *Arabidopsis ASA1* cDNA encoding a putative chloroplastic ATP-sulphurylase (Logan et al., 1996) was constitutively overexpressed in Bright Yellow 2 tobacco cells (Hatzfeld et al., 1998). Overexpression under S-sufficient conditions had no effect on S-influx and S-content, although ATP sulphurylase activity and protein was significantly higher in transgenic cells compared to WT (Hatzfeld et al., 1998). Under S-limitation, both tobacco and *Arabidopsis* ATP-sulphurylase protein levels increased. From this observation it is suggested that some types of post-transcriptional modification to enhance ATP-sulphurylase activity are operating, as it is unlikely that the constitutive 35S CaMV promotor is susceptible to S-limitation (Hatzfeld et al., 1998). One suggestion made by these authors is that there is an interaction between the cytosolic ATP-sulphurylase and the plasma membrane bound sulphate transporter (Logan et al., 1996). Under S-limitation, the higher levels of sulphate transporters will interact with more ATP-sulphurylase, irrespective of origin (endogenousor transgene), and results in activation of the ATP-sulphurylase pool. No evidence for an effect on sulphate transport by protein interaction was presented.

6. PROSPECTS

Substantial progress has been made on the understanding of sulphate transport in higher plants. This has been facilitated by the cloning of a large gene family with multiple genes involved, apparently, in sulphate transport. It seems that within the gene family there is a specialisation of function and perhaps some redundancy. Equivalent large gene families in species other than *Arabidopsis* are to be expected but have not yet been resolved. Fine localisation of expression is beginning to show an unexpected sophistication for efficient sulphate management within the whole plant. Many details remain to be elucidated. Structural and mutagenesis experiments will give indications of functional mechanisms. The existence of functional sub-types, and particularly families of isoforms with distinct but defined affinities for sulphate, provide a valuable resource for such mechanistic studies. The analysis of the regulation of gene expression of the transporters in terms of the co-ordination with plant N and S nutrition has important agronomic consequences. The study of the pathway involving OAS, the key metabolite linking N and S metabolism, helps to explain the molecular basis of this interaction. Finally, there are prospects for improving S-nutrient acquisition and utilisation by genetic manipulation. Improved and more efficient resource capture and resource allocation are the clearest targets. Efficient storage, mobilisation, and finally delivery to seed tissues with the consequent enhancement of the content of methionine represents a target for improvement through the manipulation of expression of sulphate transporters.

7. REFERENCES

Aravind L and Koonin EV (2000) The STAS domain – a link between anion transporters and antisigma-factor antagonists. *Curr Biol* **10:** R53—55

Biedlingmaier S and Schmidt A (1989) Sulfate transport in normal and S-deprived *Chlorella fusca. Z Naturforsch* **44c:** 495—503

Bolchi A, Petrucco S, Tenca PL, Foroni C and Ottonello S (1999) Coordinate modulation of maize permease and ATP sulpfurylase mRNAs in response to variations in sulfur nutritional status: stereospecific down-regulation by L-cysteine. *Plant Mol Biol* **39:** 527—537

Bourgis F, Roje S, Nuccio ML, Fisher DB, Tarczynski MC, Li CJ, Herschbach C, Rennenberg H, Pimenta MJ, Shen TL, Gage DA and Hanson AD (1999) S-methylmethionine plays a major role in phloem sulfur transport and is synthesized by a novel type of methyltransferase. *Plant Cell* **11:** 1485—1497

Brandl CJ and Deber CM (1986) Hypothesis about the function of membrane-buried proline residues in transport proteins. *Proc Nat Acad Sci USA* **83:** 917—921

Breton A and Surdin-Kerjan Y (1977) Sulfate uptake in *Saccharomyces cerevisiae*: biochemical and genetic study. *J Bacteriol* **132:** 224—232

Clarkson DT, Smith FW and Vandenberg PJ (1983) Regulation of sulfate transport in a tropical legume, *Macroptilium atropurpureum* cv Sirato. *J Exp Bot* **34:** 1463—1483

Clarkson DT, Hawkesford MJ, Davidian J-C and Grignon C (1992) Contrasting responses of sulfate and phosphate-transport in barley (*Hordeum vulgare* L) roots to protein-modifying reagents and inhibition of protein-synthesis. *Planta* **187:** 306—314

Datko AH and Mudd SH (1984) Sulfate uptake and its regulation in *Lemna paucicostata* Hegelm.6746. *Plant Physiol* **75:** 466—473

Davies JP, Yildiz F and Grossman AR (1994) Mutants of *Chlamydomonas* with aberrant responses to sulfur deprivation. *Plant Cell* **6:** 53—63

Davies JP, Yildiz FH and Grossman A (1996) Sac1, a putative regulator that is critical for survival of *Chlamydomonas reinhardtii* during sulfur deprivation. *EMBO J* **15:** 2150—2159

Davies JP, Yildiz FH and Grossman AR (1999) Sac3, an Snf1-like serine/threonine kinase that positively and negatively regulates the responses of *Chlamydomonas* to sulfur limitation. *Plant Cell* **11:** 1179—1190

Gede I, Adiputra K and Anderson JW (1995) Effect of sulphur nutrition on redistribution of sulphur in vegetative barley. *Physiol Plant* **95:** 643—650

Hästbacka J, de la Chapelle A, Mahtani MM, Clines G, Reeve-Daly MP, Daly M, Hamilton BA, Kusumi K, Trivedi B, Weaver A, Coloma A, Lovett M, Buckler A, Kaitila I and Lander ES (1994) The diastrophic dysplasia gene encodes a novel sulfate transporter: positional cloning by fine-structure linkage disequilibrium mapping. *Cell* **78:** 1073—1087

Hatzfeld Y, Cathala N, Grignon C and Davidian J-C (1998) Effect of ATP sulfurylase overexpression in Bright Yellow 2 tobacco cells. Regulation of ATP sulfurylase and sulphate transport activities. *Plant Physiol* **116:** 1307—1313

Hawkesford MJ (2000) Plant responses to sulphur deficiency and the genetic manipulation of sulphate transporters to improve S-utilisation efficiency. *J Exp Bot* **51:** 131—138

Hawkesford MJ and Smith FW (1997) Molecular biology of higher plant sulphate transporters. In: Cram WJ, De Kok LJ, Stulen I, Brunold C and Rennenberg H (eds) Sulphur Metabolism in Higher Plants, pp 13—25. Backhuys Publishers, Leiden

Hawkesford MJ and Wray JL (2000) Molecular genetics of sulphur assimilation. *Adv Bot Res* **33:** 159—223

Hawkesford MJ, Davidian J-C and Grignon C (1993) Sulfate proton cotransport in plasma membrane vesicles isolated from roots of *Brassica napus* L. Increased transport in membranes isolated from sulfur-starved plants. *Planta* **190:** 297—304

Heiss S, Schäfer HJ, Haag-Kerwer A and Rausch T (1999) Cloning sulfur assimilation genes of *Brassica juncea* L.: cadmium differentially affects the expression of a putative low-affinity sulfate transporter and isoforms of ATP sulfurylase and APS reductase. *Plant Mol Biol* **39:** 847—857

Herschbach C and Rennenberg H (1994) Influence of glutathione (GSH) on net uptake of sulfate and sulfate transport in tobacco plants. *J Exp Bot* **45:** 1069—1076

Jones SL and Smith IK (1981) Sulfate transport in cultured tobacco cells. Effects of calcium and sulfate concentration. *Plant Physiol.* **67:** 445—448

Khurana OK, Coupland LA, Shelden MC and Howitt SM (2000) Homologous mutations in two diverse sulphate transporters have similar effects. *FEBS Lett* **477:** 118—122

Kredich NM (1992) The molecular basis for positive regulation of *cys* promoters in *Salmonella typhimurium* and *Escherichia coli*. *Mol Microbiol* **6:** 2747—2753

Kredich NM (1993) Gene regulation of sulfur assimilation. In: De Kok LJ, Stulen I, Rennenberg H, Brunold C and Rauser WE (eds) Sulfur Nutrition and Assimilation in Higher Plants, pp 37—47. SPB Academic Publishers, The Hague

Lappartient AG and Touraine B (1996) Demand-driven control of root ATP sulfurylase activity and sulfate uptake in intact canola. *Plant Physiol* **111:** 147—157

Lappartient AG, Vidmar JJ, Leustek T, Glass ADM and Touraine B (1999) Inter-organ signaling in plants: regulation of ATP sulfurylase and sulfate transporter genes expression in roots mediated by phloem-translocated compound. *Plant J* **18:** 89—95

Lee RB (1982) Selectivity and kinetics of ion uptake by barley plants following nutrient deficiency. *Ann Bot* **50:** 429—449

Leggett JE and Epstein E (1956) Kinetics of sulfate absorption by barley roots. *Plant Physiol* **31:** 222—226

Logan HM, Cathala N, Grignon C and Davidian J-C (1996) Cloning of a cDNA encoded by a member of the *Arabidopsis thaliana* ATP sulfurylase multigene family. Expression studies in yeast and in relation to plant sulfur nutrition. *J Biol Chem* **271:** 12227—12233

Neuenschwander U, Suter M and Brunold C (1991) Regulation of sulfate assimilation by light and *O*-acetyl-L-serine in *Lemna minor* L. *Plant Physiol* **97:** 253—258

Ng AYN, Blomstedt CK, Gianello R, Hamill JD, Neale AD and Gaff DF (1996) Isolation and characterisation of a lowly expressed cDNA from the resurrection grass *Sporobolus stapfianus* with homology to eukaryotic sulfate transporter proteins (Acc. No. X96761) (PGR96-032). *Plant Physiol* **111:** 651

Nissen P (1971) Uptake of sulfate by roots and leaf slices of barley: mediated by single, multiphasic mechanisms. *Physiol Plant* **24:** 315—324

Passera C and Ferretti M (1988) Sulphate uptake by leaf mesophyll and bundle sheath cells of maize plants. *Biol Plant* **30:** 451—456

Reizer J, Reizer A and Saier MH (1994) A functional superfamily of sodium/solute symporters. *Biochim Biophys Acta* **1197:** 133—166

Rennenberg H, Schmitz K and Bergmann L (1979) Long-distance transport of sulfur in *Nicotiana tabacum*. *Planta* **147:** 57—62

Shelden MC, Loughlin P, Tierney ML and Howitt SM (2001) Proline residues in two tightly coupled helices of the sulphate transporter, SHST1, are important for sulphate transport. *Biochem J* **356:** 589—594

Shibagaki N, Rose A, McDermott JP, Fujiwara T, Hayashi H, Yomeyama T and Davis JP (2002) Selenate-resistant mutants of *Arabidopsis thaliana* identify *Sultr1;2*, a sulfate transporter required for efficient transport of sulfate into roots. *Plant J* **29:** 475—486

Smith FW, Ealing, PM, Hawkesford MJ and Clarkson DT (1995a) Plant members of a family of sulfate transporters reveal functional subtypes. *Proc Nat Acad Sci USA* **92:** 9373—9377

Smith FW, Hawkesford MJ, Prosser IM and Clarkson DT (1995b) Isolation of a cDNA from *Saccharomyces cerevisiae* that encodes a high affinity sulphate transporter at the plasma membrane. *Mol Gen Genet* **247:** 709—715

Smith FW, Hawkesford MJ, Ealing PM, Clarkson, DT, Vandenberg PJ, Belcher A and Warrilow AGS (1997) Regulation of expression of a cDNA from barley roots encoding a high affinity sulphate transporter. *Plant J* **12:** 875—884

Smith IK (1975) Sulfate transport in cultured tobacco cells. *Plant Physiol* **55:** 303—307

Takahashi H, Sasakura N, Noji M and Saito K (1996) Isolation and characterization of a cDNA encoding a sulfate transporter from *Arabidopsis thaliana. FEBS Lett* **392:** 95—99

Takahashi H, Yamazaki M, Sasakura N, Watanabe A, Leustek T, de Almeida Engler J, Engler G, van Montagu M and Saito K (1997) Regulation of sulfur assimilation in higher plants: A sulfate transporter induced in sulfate-starved roots plays a central role in *Arabidopsis thaliana. Proc Natl Acad Sci USA* **94:** 11102—11107

Takahashi H, Asanuma W and Saito K (1999a) Cloning of an *Arabidopsis* cDNA encoding a chloroplast localizing sulphate transporter isoform. *J Exp Bot* **50:** 1713—1714

Takahashi H, Sasakura N, Kimura A, Watanabe A and Saito K (1999b) Identification of two leaf-specific sulfate transporters in *Arabidopsis thaliana* (Acc. No. AB012048 and AB004060) (PGR99-154). *Plant Physiol* **121:** 686

Takahashi H, Watanabe-Takahashi A, Smith FW, Blake-Kalf M, Hawkesford MJ and Saito K (2000) The roles of three functional sulphate transporters involved in uptake and translocation of sulphate in *Arabidopsis thaliana. Plant J* **23:** 171—182

Vange MS, Holmern K and Nissen P (1974) Mutiphasic uptake of sulfate by barley roots. I. Effects of analogues, phosphate, and pH. *Physiol Plant* **31:** 292—301

Vidmar JJ, Schjoerring JK, Touraine B and Glass ADM (1999) Regulation of the *hvst1* gene encoding a high-affinity sulfate transporter from *Hordeum vulgare. Plant Mol Biol* **40:** 883—892

Vidmar JJ, Tagmount A, Cathala N, Touraine B and Davidian J-C (2000) Cloning and characterization of a root specific high-affinity sulfate transporter from *Arabidopsis thaliana. FEBS Lett* **475:** 65—69

von Heijne G (1992) Membrane-protein structure prediction- hydrophobicity analysis and the positive-inside rule. *J Mol Biol* **225:** 487—494

Yildiz FH, Davies JP and Grossman A (1996) Sulfur availability and the *SAC1* gene control adenosine triphosphate sulfurylase gene expression in *Chlamydomonas reinhardtii. Plant Physiol* **112:** 669—675

Yoshimoto N, Takahashi H, Smith FW, Yamaya T and Saito K (2002) Two distinct high-affinity sulfate transporters with different inducibilities mediate uptake of sulfate in *Arabidopsis* roots. *Plant J* **29:** 465—473

8. ACKNOWLEDGEMENTS

Work in the laboratory is sponsored by grants from the BBSRC, DEFRA, Home-Grown Cereals Authority and by Framework V of the EU (QLRT-2000-00103). IACR receives grant-aided support from the Biotechnology and Biological Sciences Research Council of the UK.

9. AFFILIATIONS

IACR Rothamsted, Agriculture and Environment Division, Harpenden, Hertfordshire, AL5 2JQ, UK.

For Correspondence - *MJ Hawkesford*: malcolm.hawkesford@bbsrc.ac.uk

S. KOPRIVA AND A. KOPRIVOVA

SULPHATE ASSIMILATION: A PATHWAY WHICH LIKES TO SURPRISE

Abstract. Sulphate assimilation is a pathway used by prokaryotes, fungi and photosynthetic organisms to convert inorganic sulphates to sulphide, which is incorporated into carbon skeletons of amino acids to form cysteine or homocysteine. Our knowledge about the molecular mechanisms and regulation of sulphate assimilation lags behind that of assimilation of carbon and nitrogen. Nevertheless, during the last few years considerable progress has been made in understanding its assimilation. Many of the new findings, however, are rather unexpected and do not fit into the previously established view of this pathway; (**a**) A detailed biochemical analysis seems to resolve the old question of reduction via bound or free sulphite and the controversy about the reaction mechanism of the enzyme catalyzing APS reduction. The surprising finding of an iron-sulphur cluster bound to the enzyme clearly supports a reductase mechanism and, thus, APS reductase is the correct name for this enzyme. Correspondingly, free sulphite is the reaction product and the bound intermediates are most probably the result of chemical reaction of sulphite with oxidized thiols; (**b**) Attempts to solve the significance of bundle sheath cell specific distribution of sulphate assimilation in C_4 plants failed since the pathway was detected in all cell types in *Flaveria trinervia*, a C_4 dicot plant. (**c**) A novel type of APS reductase was found in various heterotrophic bacteria, which were believed to require the two step activation of sulphate to PAPS for sulphate reduction. More surprising and unpredicted results can be expected in the near future. This represents quite a challenge and makes the sulphur metabolism an exceptionally exciting field for further investigations.

1. INTRODUCTION

During the last few years a considerable progress in understanding the pathway of sulphur assimilation in plants was achieved due to combination of physiological and molecular methods. This is reflected in the recent reviews (Leustek and Saito, 1999; Hawkesford and Wray, 2000; Leustek et al., 2000; Hofgen et al., 2001). Many new findings were quite surprising and did not match with the expectations. In this chapter, we concentrate on the newly obtained information for the current view on the pathway of assimilatory sulphate reduction. A brief historical account of investigations of sulphate assimilation is followed by biochemical and molecular characterisation of the enzymes and their regulation. Finally, a summary of the remaining open questions and directions for future research is given.

2. HISTORICAL OVERVIEW

The sulphate assimilation pathway was first resolved in enteric bacteria, *Escherichia coli* and *Salmonella typhimurium* (Fig. 1). In these organisms, sulphate is first activated by adenylation to adenosine 5'-phosphosulphate (APS) catalyzed by ATP sulphurylase (ATPS). In the second activation step, APS is further phosphorylated

Yash P. Abrol and Altaf Ahmad (eds.), Sulphur in Plants, 87—112.

by APS kinase to form adenosine 3'-phosphate 5'-phosphosulphate (PAPS). The activation of sulphate, thus, consumes two molecules of ATP. PAPS is reduced in a thioredoxin dependent reaction by PAPS reductase to sulphite. In the following step, sulphite is reduced by a NADPH dependent sulphite reductase (SiR) to sulphide, which is finally incorporated into the amino acid skeleton of *O*-acetyl-L-serine (OAS) forming cysteine (Jones-Mortimer, 1968; Kredich, 1971).

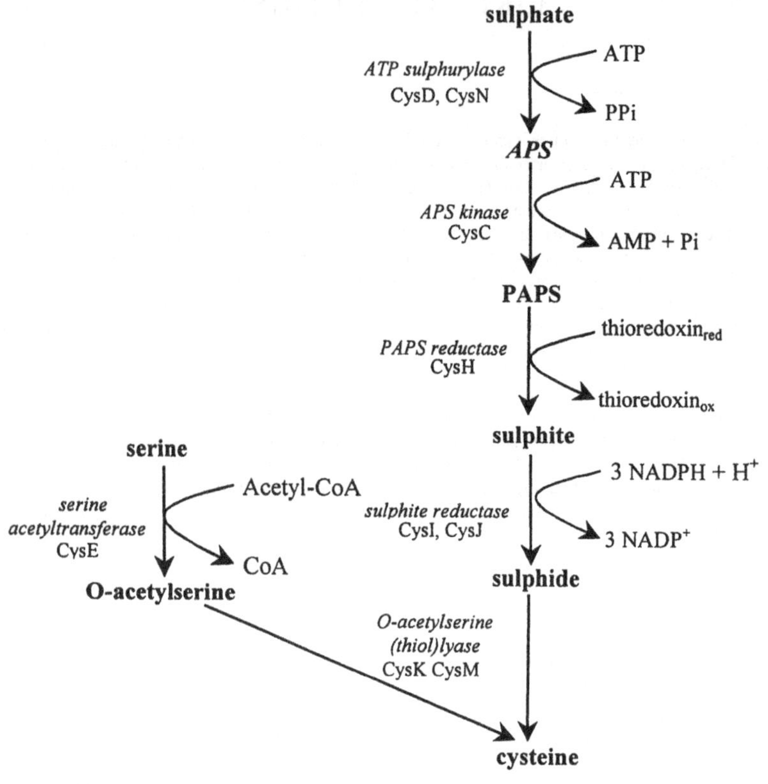

Figure 1. Assimilatory sulphate reduction pathway in Escherichia coli

The observation that isolated spinach chloroplasts were able to reduce sulphate to sulphide was the starting point of investigations of sulphate assimilation in plants (Schmidt and Trebst, 1969). The first experiment suggested an only slightly modified sequence of reactions compared to microorganisms: ATP dependent sulphate activation, reduction to sulphite by a sulphydryl compound, and reduction of sulphite to sulphide by a ferredoxin-dependent sulphite reductase. Studies of photosynthetic sulphate reduction in the green alga, *Chlorella,* revealed that APS rather than PAPS was the sulphonucleotide reduced form (Tsang et al., 1971; Schmidt, 1972a). Another difference compared to bacteria was an identification of sulphite bound onto a thiol carrier, replaceable by glutathione (GSH) in vitro, among the products of APS reduction (Schiff and Hodson, 1970). The bound sulphite could be further reduced by a ferredoxin dependent thiosulphonate reductase to bound sulphide, which reacted with OAS to form cysteine (Fig. 2) (Schmidt, 1973). The enzyme catalyzing the transfer of sulphate to the thiol carrier was named APS sulphotransferase (APSST) (Schmidt, 1972b). APSST activity was then measured in a variety of higher plants. The enzyme was found to be extensively regulated and, therefore, considered to be the major sulphate reducing enzyme in photosynthetic eukaryotes (Brunold, 1990; Schmidt and Jäger, 1992). Nevertheless, a PAPS dependent pathway of sulphate reduction could not be excluded, especially when the purification of PAPS reductase from spinach had been reported (Schwenn, 1989).

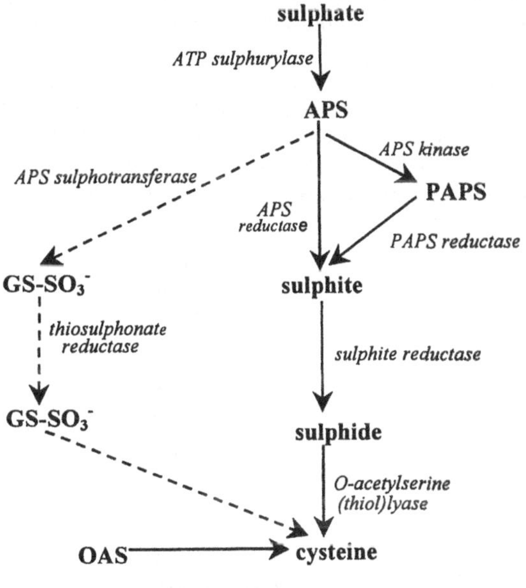

Figure 2. Possible pathways of plant assimilatory sulphate reduction

In attempts to clone (P)APS reducing enzyme(s) from plants by complementation of *E. coli* mutants lacking PAPS reductase, three classes of cDNA clones were obtained from *Arabidopsis thaliana* (Gutierrez-Marcos et al., 1996; Setya et al., 1996). These cDNA clones coded for different isoforms of a novel enzyme with a N-terminal organelle targeting peptide, a central part homologous to *E. coli* PAPS reductase, and a C-terminal extension similar to thioredoxin. The cDNAs were able to functionally complement *E. coli* mutant strains defective in APS kinase and thioredoxin. Since the enzyme produced acid volatile sulphite from APS and not PAPS, it was named APS reductase (APR) (Setya et al., 1996). In the next few years in several reviews, the role of APR in sulphate assimilation was discussed with contradictory results (Hell, 1997; Wray et al., 1998; Leustek and Saito, 1999). The three possible pathways of sulphate assimilation are presented in Fig. 2. Whereas Wray et al. (1998) speculated that APS is directly reduced to free sulphite and further to free sulphide by sulphite reductase, Leustek and Saito (1999) revived the bound sulphite pathway. According to the latter authors, APR acts as a sulphotransferase forming S-sulphoglutathione as the reaction product. After a non-enzymatical reaction of S-sulphoglutathione with reduced glutathione, free sulphite can be released and further reduced to sulphide by sulphite reductase.

The controversy about the enzyme catalyzing the reduction of APS seems to be resolved by the recent report of Suter et al. (2000). The APSST from *Lemna minor* was isolated, the corresponding cDNA was cloned, and the deduced amino acid sequence was found to be very similar to that of APR from *A. thaliana*. Free sulphite was the only reaction product detected under non-oxidizing conditions, while S-sulphoglutathione was only formed when oxidized glutathione was present in the enzyme assay (Suter et al., 2000). From these experiments the authors concluded that (i) APS sulphotransferase and APS reductase were identical enzymes, (ii) the bound sulphite intermediates were only due to presence of oxidized thiols in the reaction assay, and (iii) the enzyme should, therefore, be named APS reductase. These conclusions were corroborated by further biochemical characterization of APR (Weber et al., 2000; Kopriva et al., 2001a). In Figure 3 the new findings are incorporated and the current view on sulphate assimilation pathway is presented.

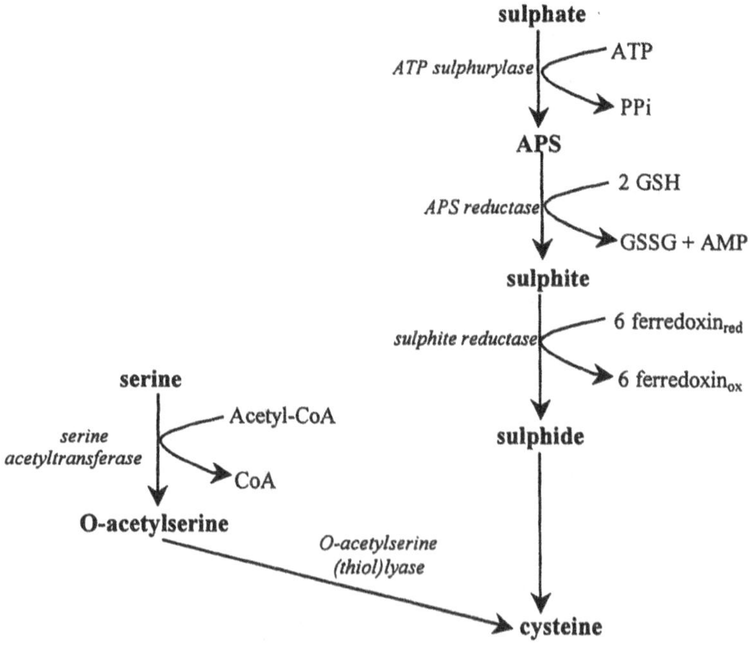

Figure 3. Assimilatory sulphate reduction pathway in higher plants, current view

3. ENZYMES INVOLVED

3.1 ATP Sulphurylase

3.1.1 Biochemical characterization

ATP sulphurylase (EC 2.7.7.4, ATPS) catalyzes the first step in sulphate assimilation, the adenylation of sulphate to APS. The formation of APS is an energetically unfavorable process, which is driven forwards by the consumption of APS by subsequent reactions, reduction to sulphite by APS reductase or phosphorylation to PAPS by APS kinase (APK). Because of the thermodynamic balance, ATPS enzyme activity is routinely measured either in the back reaction by measurement of ATP synthesized from APS and pyrophosphate or indirectly by the molybdolysis assay which determines the rate of AMP production (Segel et al., 1987). Since PAPS is also a sulphonate donor for various sulphotransferases (Varin et al., 1997), ATPS is found also in organisms that do not reduce sulphate, such as human, rat, or *Urechis campo*. In these species, ATPS and APK reside on a single bifunctional protein, the PAPS synthetase (Rosenthal and Leustek, 1995; Venkatachalam et al., 1998). This organization allows an intermediate channeling which further increases the efficiency of APS removal and pushes forward its synthesis. ATPS perform complete different function in anaerobic bacteria where where APS serves as terminal electron acceptor for respiration. Several chemolitotrophic bacteria use ATPS to synthesize ATP from APS and pyrophosphate (Kappler and Dahl, 2001). ATPS from different sources have very different molecular structure. Plant ATPS is a homotetramer of 52-54 kDa polypeptides (Murillo and Leustek, 1995), bacterial ATPS is a completely unrelated tetramer of heterodimers composed of 35 kDa CysD and 53 kDa CysN subunits (Leyh et al., 1988), animal PAPS synthetase functions as a single protein of 56 kDa (Lyle et al., 1994), the ATPS from yeast and fungi is a homohexamer of 59 to 64 kDa subunits, and in sulphate reducing bacteria ATPS is an homo-oligomer, similar to plants (Sperling et al., 1998). The fungal ATPS contains a APS kinase-like domain, which is, however, not functional and is the site of allosteric regulation by PAPS (MacRae et al., 2001; Ullrich et al., 2001).

3.1.2 Localization

In plants, ATPS activity was detected in chloroplasts and in the cytosol of spinach leaves (Lunn et al., 1990; Renosto et al., 1993) and in proplastids of pea roots (Brunold and Suter, 1989). In *Euglena gracilis,* the enzyme was also found in two compartments, in the cytosol and in the mitochondria (Li et al., 1991). The mRNA for ATPS was detected in potato leaves, stems, and roots, but not in tubers (Klonus et al. 1994). Low ATPS activity was measured in etiolated pea seedlings, which increased after transfer into the light, but decreased again in the leaves during further incubation (von Arb and Brunold, 1986). In poplars, ATPS activity diminished slowly with the leaf age (Hartmann et al., 2000). In *Arabidopsis,* the foliar ATPS activity continually declined during the plant growth. During this time the more abundant chloroplastic ATPS activity was decreasing, the cytosolic activity, however, increased (Rotte and Leustek, 2000). This observation indicates different functions of ATPS in the two compartments, sulphate reduction in the plastids and activation of sulphate for synthesis of sulphonated compounds in the cytosol (Rotte and Leustek, 2000).

Several groups reported that 75-100 % of total leaf ATPS activity in maize was confined to bundle sheath cells (Gerwick et al., 1980; Passera and Ghisi, 1982; Schmutz and Brunold, 1984). These findings were extended to 17 other C_4 species, where 95-100 % of total leaf ATPS activity was found in chloroplasts of bundle sheath cells. Accordingly, mRNA of ATPS was detected exclusively in RNA isolated from bundle-sheath cells of maize, revealing that in maize, the intercellular distribution of ATPS is regulated on the transcriptional level (Kopriva et al., 2001b). Surprisingly however, in the C_4 dicot *Flaveria trinervia*, ATPS mRNA was present in both bundle-sheath and mesophyll cells (Koprivova et al., 2001) (see Section 3.2).

3.1.3 Genomic organization

ATP sulphurylase is encoded by small multigene families in all plant species analyzed to date. cDNAs encoding chloroplastic and cytosolic isoforms of ATPS were isolated from potato (Klonus et al., 1994). On the other hand, four isoforms of ATPS were isolated from *Arabidopsis*, all of them containing a chloroplast transit peptide (Murillo and Leustek, 1995; Logan et al., 1996; Hatzfeld et al., 2000a). Based on sequence analysis, the APS2 isoform is a good candidate to encode the cytosolic isoform (Hatzfeld et al., 2000a). All four ATPS genes contain 5 exons and 4 introns and are localized on different chromosomes. Also from *Brassica juncea* two isoforms of ATPS were cloned both containing an organelle-targeting peptide (Heiss et al., 1999). The role of the different chloroplastic isoforms of ATPS, however, still remains to be elucidated.

3.2 APS Reductase

3.2.1 Biochemical characterization

Adenosine 5'-phosphosulphate reductase (EC 1.8.99.2, APR) catalyzes the thiol dependent two-electron reduction of adenosine 5'-phosphosulphate to sulphite and AMP. The enzyme is identical with the formerly described APS sulphotransferase (see Section 3.1). The APR activity is measured as acid volatile radioactivity after incubation with [^{35}S]APS and thiols (Brunold and Suter, 1990). APR is a homodimer of 43 kDa subunits, held together by a disulphide bridge. Although thiols are the electron donors for APS reduction, a pre-incubation with thiols (dithiothreitol, glutathione) when APS or AMP are not present, leads to inactivation of the enzyme (Bick et al., 2001; Kopriva S., unpublished). The mature APR protein consists of two domains, a PAPS reductase-like and a thioredoxin-like. Beside thioredoxin and thioredoxin reductase, an alternative reduction system exists consisting of glutathione and glutaredoxin, a small protein which belongs to the class of thiol:disulphide oxidoreductases containing the CXXC active site motif

(Washburn and Wells, 1999). Although, the sequence of C-terminal domain of APR is more similar to thioredoxin than to glutaredoxin, it possesses the enzymatic activity of glutaredoxin (Bick et al., 1998; Prior et al., 1999). Surprisingly, thioredoxin but not glutaredoxin was able to replace the C-domain in the APS reductase of *Catharanthus roseus* (Prior et al., 1999). The reaction of APR can be divided into two independent steps: a reductive transfer of sulphate to the active cysteine residue, conserved between APS and PAPS reductases, and the release of the sulphite by the C-terminal domain (Weber et al., 2000). APS reductases from *A. thaliana* and *C. roseus* which had been over-expressed in *E. coli*, were described as proteins without prosthetic groups or cofactors (Gutierrez-Marcos et al., 1996; Setya et al., 1996; Bick et al., 1998; Prior et al., 1999). Surprisingly however, the enzyme was purified from *Lemna minor* as a yellow-brown protein (Suter et al., 2000) indicating the presence of a cofactor, possibly FAD or/and iron-sulphur cluster. UV/visible spectra of recombinant APR indicated the presence of iron-sulphur centers and, indeed, iron and acid-labile sulphide were found to bind to the APR protein (Kopriva et al., 2001a). Electron paramagnetic resonance and Mössbauer spectroscopy then confirmed the presence of a diamagnetic $[4Fe\text{-}4S]^{2+}$ cluster as cofactor of plant APS reductase. This cluster is unusual since only three of the iron sites exhibited the same Mössbauer parameters. Although, there is no signature for binding an FeS cluster in the sequence of plant APR, the N-terminal part of APR differs from PAPS reductase by presence of two additional cysteine pairs in the plant enzyme. Only three from these four additional cysteine residues can bind the FeS cluster explaining its extraordinary characteristics (Kopriva et al., 2001a). Although possessing the same cofactor and catalyzing the same reaction, plant APR is completely unrelated to the APS reductase found in sulphate-reducing and sulphide-oxidizing bacteria and *Archaea* (Fritz et al., 2000; Kappler and Dahl, 2001).

3.2.2 Localization

Isolated chloroplasts are capable of reducing sulphate (Schmidt and Trebst, 1969) and, correspondingly, the APR activity was localized in spinach chloroplasts (Schmidt, 1976; Fankhauser and Brunold, 1978). No APR activity was measured in peroxisomes, mitochondria, or cytosol (Fankhauser and Brunold, 1978). APR also appears to be exclusively localized to proplastids of root cells (Brunold and Suter, 1989). Indeed, by western analysis of pea chloroplast fractions, APR protein was detected in stroma but not in any of the membrane fractions (Prior et al., 1999). In accordance, immunogold labeling revealed the presence of APR protein in chloroplasts of three *Flaveria* species of different type of photosynthesis (Koprivova et al., 2001). There the gold particles were predominantly associated with thylakoid membranes.

APR activity changes during development of the plant. Similarly to ATPS, it increased when etiolated pea seedlings were transferred to the light, but decreased with the leaf age (von Arb and Brunold, 1986). In *Arabidopsis*, the enzyme declined during leaf development (Rotte and Leustek, 2000). In contrast, the maximum activity is found in mature leaves of poplars (Hartmann et al., 2000). In roots, APR activity is the highest in the root tip, decreasing continuously with the distance from the root tip (Brunold and Suter, 1989; Kopriva et al., 2001b). Accordingly, by in situ RNA hybridization APR transcript was found predominantly in the root tips with a significantly lower signal in the mature roots, where it was localized predominantly in the exodermis (Kopriva et al., 2001b).

Like ATPS, APR was also found exclusively or almost exclusively in bundle sheath cells of maize (Passera and Ghisi, 1982; Burnell, 1984; Schmutz and Brunold, 1984). Correspondingly, in maize the mRNAs for APR, ATPS, and sulphite reductase accumulated in bundle-sheath only, whereas the mRNA for O-acetylserine-(thiol)lyase was also detected in mesophyll cells (Kopriva et al., 2001b). To answer the question whether the exclusive or almost exclusive localization of APR was a pre-requisite or a consequence of C_4 photosynthesis, the distribution of APR was studied in *Flaveria* species with C_3, C_3-C_4, C_4-like, and C_4 photosynthesis (Koprivova et al., 2001). The genus *Flaveria* (*Flaveriinae)* is an excellent model to study the evolution of C_4 photosynthesis because, beside C_3 and C_4 species, a relatively large number of C_3-C_4 intermediates were found in this genus (Ku et al., 1991) and a continual gradation exists among *Flaveria* species both in the physiology and biochemistry of photosynthesis (Monson and Moore, 1989).

Measurements of APR activity, mRNA level, and protein accumulation in 6 *Flaveria* species revealed that APR activity, cysteine and glutathione levels were significantly higher in C_4-like and C_4 species than in C_3 and C_3-C_4 species. However, surprisingly, by northern analysis of cell-specific RNA and in situ hybridization ATPS and APR mRNA were present at comparable levels in both mesophyll and bundle sheath cells of the C_4 species, *Flaveria trinervia*. Also immunogold electron microscopy confirmed the presence of APR protein in chloroplasts of both cell types (Koprivova et al., 2001). The discrepancy with the older results may be explained by the fact that *Flaveria* species are dicots, whereas in all previous reports only monocot species were analyzed. It seems, therefore, that the localization of assimilatory sulphate reduction in the bundle sheath cells is not ubiquitous among C_4 plants and is, therefore, neither a pre-requisite nor a consequence of C_4 photosynthesis. The functional significance of the compartmentation of sulphate assimilation in bundle-sheath cells of maize and other C_4 monocot species is not clear yet. The fact, that in *Flaveria* sulphate assimilation is not differentially distributed, however, brings serious doubts whether the compartmentation in maize is functionally important at all.

3.2.3 Genomic organization

APS reductase cDNA was cloned from several plant species: *A. thaliana* (Gutierrez-Marcos et al., 1996; Setya et al., 1996), *Catharanthus roseus* (Prior et al., 1999), *Brassica juncea* (Heiss et al., 1999), *Lemna minor* (Suter et al., 2000), *Allium cepa*, *Zea mays*, *Populus tremula x P. alba*, and the green alga *Enteromorpha intestinalis* (Gao et al., 2000). The PAPS reductase-like domains are 22-27% identical to the PAPS reductase from *E. coli*; all APS and PAPS reductases contain an (KRT)ECG(LI)H motif as a sequence signature. APR is encoded by multigene families in most species analyzed. Beside *Arabidopsis* where 3 isoforms exist (Gutierrez-Marcos et al., 1996; Setya et al., 1996), two cDNAs were obtained from

B. juncea (Heiss et al., 1999) and southern analysis revealed at least two APR genes in *L. minor* (Suter et al., 2000), poplar, and several species of the genus *Flaveria* (Koprivova et al., 2001). In *A. thaliana* the corresponding genes were cloned and sequenced (Chen and Leustek, 1998). The gene for APR2 isoform is found on chromosome 1 whereas the genes for APR1 and APR3 are found on chromosome 4. APR1 and APR3 are more similar with 78% identical nucleotides, APR1 and APR2 possess only 68% identity. On the other hand, the genes of APR2 and APR3 have the same exon/intron organization with 3 introns, APR1 gene lacks the intron separating exon 2 and 3 in APR2 and APR3. Surprisingly, the two domains of APR, PAPS reductase-like and thioredoxin-like, are not encoded by separate exons, as would be expected if the *apr* gene was formed by fusion of the corresponding genes. However, in the APR1 gene sequence duplication is found at the 5'end of intron 2 and just before the thioredoxin active site in exon 3, which might possibly represent a remnant of an original intron separating the two domain-coding exons (Chen and Leustek, 1998).

3.3 Sulphite Reductase

3.3.1 Biochemical characterization

Sulphite reductase (SiR) catalyzes the six electron reduction of sulphite to sulphide. The electron donor for the reduction is ferredoxin (Fd) in plants (EC 1.8.7.1) or NADPH in bacteria (EC 1.8.1.2). SiR activity can be measured in a coupled assay as a production of cysteine from sulphite and OAS in the presence of cysteine synthase by determination of rate of methylviologen oxidation, or by coupling the oxidation of Fd to oxidation of NADPH via Fd-$NADP^+$ reductase (Krueger and Siegel, 1982; Yonekura-Sakakibara et al., 2000). SiR is a homodimer of 65 kDa polypeptide and contains one siroheme and one [4Fe-4S] cluster per monomer as cofactors (Krueger and Siegel, 1982). Such protein structure is very similar to that of Fd-nitrite reductase, which catalyzes a six electron reduction of nitrite to ammonia and, indeed, both enzymes can reduce both substrates, sulphite and nitrite, albeit with different efficiency (Krueger and Siegel, 1982). In contrast, the bacterial NADPH-SiR is an oligomer of eight 66 kDa flavoprotein subunits (CysJ) and four 64 kDa hemoproteins (CysI), binding a siroheme and a [4Fe-4S] cluster, analogous to plant Fd-SiR (Crane et al., 1995).

3.3.2 Localization

SiR activity was localized exclusively to plastids both in photosynthetic and non-photosynthetic organs (Brunold and Suter, 1989). In maize, SiR interacts with different isoforms of Fd and Fd-$NADP^+$ reductase in leaves and roots (Yonekura-Sakakibara et al., 2000). The combination of isoproteins present in roots is more efficient in NADPH dependent sulphite reduction than those present in leaves, in accordance with the physiological reducing power present in the respective organs (Yonekura-Sakakibara et al., 2000). During plant development, SiR is regulated in the same manner as ATPS and APR. The activity of SiR increases after transfer of etiolated pea seedlings into light and decreases afterwards (von Arb and Brunold, 1986) and in poplar it diminishes with leaf age (Hartmann et al., 2000).

In maize SiR activity was detected both in bundle-sheath and mesophyll cells (Schmutz and Brunold, 1984, 1985). However, the mRNA for SiR was detected in bundle-sheath cells and not in mesophyll in maize growing at 25°C (Kopriva et al., 2001b). The function of SiR in mesophyll cells is not very clear, since the two preceding enzymes of sulphate assimilation pathway, ATPS and APR, are present exclusively in bundle-sheath cells. A recent report by Sato et al. (2001) has shown that SiR is abundant in the nucleoids of pea chloroplasts and is able to compact chloroplast DNA. It might represent an interesting explanation.

3.3.3 Genomic organization

A cDNA and a single copy gene encoding SiR were isolated from *Arabidopsis* (Brühl et al., 1996; Bork et al., 1998). The presence of a putative transit peptide supported the plastidic localization of SiR. The gene consists of 8 exons, with the first intron localized behind the first amino acid of the mature protein, and is localized on chromosome 5. In contrast, two SiR isoforms are present in leaves and non-photosynthetic organs of *Brassica rapa* and tobacco (Takahashi et al., 1997; Yonekura-Sakakibara et al., 1998). One cDNA and corresponding gene encoding SiR were isolated from tobacco, the gene possessed the same exon/intron organization as the gene from *Arabidopsis* (Yonekura-Sakakibara et al., 1998). Additional full-length cDNA clones for SiR were obtained from *Glycine max* and maize. SiR contains 19 % identical amino acids with nitrite reductase indicating that these genes may have the same evolutionary origin.

3.4 APS Kinase

3.4.1 Biochemical characterization

APS kinase (EC 2.7.1.25, APK) catalyzes the transfer of phosphate from ATP to APS to form PAPS. PAPS is an important metabolite not only as a form of activated sulphate for reduction in fungi and some heterotrophic bacteria but as a substrate also for sulphonation of various molecules, such as glucosinolates, flavonoids, hormones, proteoglycans etc. Deficiency in PAPS synthesis results in brachymorphism in mice and spondyloepimetaphyseal dysplasia in humans (Kurima et al., 1998; ul Haque et al. 1998). Although APK does not participate in the pathway of sulphate reduction in plants, it strongly interacts with the pathway by competing for APS with the APR. APK activity is determined by coupling the ADP production to NADH oxidation (Renosto et al., 1984). The enzyme has a very high affinity to APS; for the enzymes from *Arabidopsis* and *E. coli* K_m values are 3.6 μM and 1.4 μM APS, respectively, and is inhibited by concentrations of APS above 10 μM (Satishchandran and Markham, 1989; Lee and Leustek, 1998). APK from *Penicillium chrysogenum* was crystallized as homodimer of 24 kDa subunits (MacRae et al., 2000). Also the plant enzyme is most probably a dimer, however, since multiple isoforms of APK occur at least in *A. thaliana* chloroplasts, the dimers may be formed also from different monomers (Schiffmann and Schwenn, 1998).

3.4.2 Localization and genomic organization

APS kinase is localized in chloroplast and cytosol. The mRNA was detected both in leaves and roots of *A. thaliana* (Lee and Leustek, 1998). Four genes encoding isoforms of APK are found in *Arabidopsis* genome: all localized on different chromosomes. Three of the *A. thaliana* APKs possess a chloroplast transit peptide, the fourth one, localized on chromosome 3, thus, most probably represents the cytosolic isoform. Until now little is known about plant APK, only two isoforms from *Arabidopsis* and one from *Catharanthus roseus* were cloned and characterized.

3.5 Serine Acetyltransferase

Serine acetyltransferase (EC 2.3.1.30, SAT) catalyzes the formation of *O*-acetyl-serine from L-serine and acetyl-CoA. It represents the first step in the synthesis of cysteine. The enzymes involved in cysteine synthesis are described in detail by Noji and Saito (this book). Only a short overview will be given here and in the next section 3.6. SAT was purified and cloned from several plant species. The enzyme is localized in plastids, cytosol, and mitochondria. It is, therefore, not surpri- sing that the enzyme is encoded by multigene families of 3-6 members. SAT and *O*-acetyl-serine (thiol)lyase (OAS-TL) form a multienzyme complex in bacteria (Kredich and Tomkins, 1966) and in plants (Bogdanova and Hell, 1997; Droux et al., 1998; Wirtz et al., 2001). Interaction with OAS-TL is a pre-requisite for SAT activity whereas OAS-TL is active as a free dimer (Droux et al., 1998; Wirtz et al., 2001). SAT undergoes a feedback regulation by cysteine; in plants both cysteine sensitive and insensitive isoforms were described (Saito et al., 1995; Noji et al., 1998).

3.6 O-acetylserine (thiol)lyase

O-acetylserine (thiol)lyase (EC 4.2.99.8, OAS-TL) catalyzes the synthesis of cysteine from OAS and sulphide. The enzyme belongs to the family of β-replacement enzymes utilizing pyridoxal-5'-phosphate as a cofactor. OAS-TL is a homodimer of 35 kDa subunits found inactive in the complex with SAT or, in the active form, as a free enzyme (Droux et al., 1998; Wirtz et al., 2001). Isoforms of OAS-TL, as in case of SAT, are localized in cytosol, mitochondria and plastids in all plant organs, and interesting is its high expression in root trichomes. Accordingly, at least three genes are present in genomes of higher plants (Jost et al., 2000). OAS-TL also produces sulphide either by simultaneous formation of β-cyanoalanine from cysteine and cyanide or by degradation of cysteine into sulphide, ammonia and pyruvate or into sulphide and alanine (Burandt et al., 2001b). Since sulphate assimilation is localized in plastids, the mitochondrial and cytosolic SAT and OAS-TL might have alternative functions. Consequently, the β-cyanoalanine synthetase is considered to be the primary function of the mitochondrial OAS-TL (Hatzfeld et al., 2000b; Maruyama et al., 2000; Warrilow and Hawkesford, 2000).

4. REGULATION

Sulphur containing compounds play essential role in many metabolic processes. Among these compounds, the tripeptide glutathione (GSH) is of major importance (Rennenberg and Brunold, 1994; May et al., 1998; Noctor et al., 1998). It is, therefore, not surprising that sulphate assimilation is thoroughly regulated in order to keep the concentration of glutathione and other reduced sulphur compounds stable. The first reaction steps catalyzed by ATPS and APR are most susceptible to regulation. From these two enzymes, APR seems to possess higher control over the sulphate assimilation (Brunold, 1990, 1993; Leustek et al., 2000). Cysteine synthesis is, however, regulated not only by the supply of sulphide but also by its amino acid acceptor. Consequently, SAT is a good candidate for control of cysteine synthesis. Indeed, overexpression of SAT in tobacco and potato led to increased cysteine and GSH content (Blaszczyk et al., 1999; Harms et al., 2000).

4.1 Regulation by Sulphur Compounds and Reaction Intermediates

Sulphate assimilation is controlled by the sulphur status of the plant and seems to be repressed under normal levels of external sulphate. Accordingly, the limitation of external sulphate supply results in strong increase of sulphate uptake and activity of ATPS and APR (Reuveny et al., 1980; Smith, 1980; Brunold et al., 1987). This increase is regulated at the level of transcription; the levels of mRNAs for high affinity sulphate transporter, ATPS, APR, and the chloroplastic isoform of serine acetyltransferase were increased in *A. thaliana* after 2 days of sulphur starvation both in leaves and roots (Takahashi et al., 1997; Yamaguchi et al., 1999). When the plants are transferred into a nutrient solution with normal sulphate concentration, APR activity returned quickly to the normal levels. A ten times increase in sulphate concentration, however, also reduces the extractable APR activity (Brunold et al., 1987). OAS-TL activity was reported either to be reduced by sulphur deficiency in spinach (Warrilow and Hawkesford, 1998) or not to be affected (Takahashi and Saito, 1996).

On the other hand, sulphate uptake and assimilation are decreased when plants were fed with reduced forms of sulphur, such as H_2S, cysteine or GSH. H_2S exposure results in accumulation of thiols, reduction in sulphate uptake, and reduction in activity of APR but not ATPS, SAT and OAS-TL in the shoots (Westerman et al., 2001). By the investigation of the feedback regulation of sulphate assimilation by thiols, attention was first paid to ATPS. In *A. thaliana* ATPS activity and mRNA levels of APS1 isoform were decreased by GSH treatment (Lappartient et al., 1999). These results led to the conclusion of demand-driven control of sulphate uptake and assimilation (Lappartient and Touraine, 1996; Lappartient et al., 1999). However, a major contribution of ATPS for regulation of sulphate assimilation was questioned by Koprivova et al. (2000) and Westerman et al. (2001) who showed that ATPS was less susceptible to regulatory signals than APR. Indeed, in *A. thaliana* root cultures APR activity and transcript levels were decreased after feeding with cysteine and GSH (Vauclare et al., 2002). Because external GSH supply results in increased accumulation of cysteine and GSH, both compounds might be responsible for the control of SO_4^{2-} uptake and assimilation. Phloem sap analysis indicated, however, that GSH rather than cysteine was the signal acting in *Brassica napus* (Lappartient and Touraine, 1996). This hypothesis was corroborated by demonstrating that the repression of APS1 mRNA accumulation by external cysteine was relieved by L-buthionine [S, R] sulphoximine, an inhibitor of GSH formation (Lappartient et al., 1999). In contrast, in maize cysteine was able to regulate the level of ATPS mRNA without the need for conversion to GSH (Bolchi et al., 1999). Cysteine regulates both branches leading to its synthesis because it also exerts a feedback regulation on SAT. In *A. thaliana*, only one SAT isoform which was localized to cytosol, was susceptible to this regulation (Noji et al., 1998). However, in pea the chloroplast located SAT activity was inhibited by cysteine (Brunold and Suter, 1982). In addition H_2S has an effect on SAT, since it stabilises the complex with OAS-TL and thus increases the activity (Droux et al., 1998).

Additionally, partially reduced sulphur participates in the feedback regulation of sulphate assimilation. Fumigation with SO_2 resulted in decrease of APR activity but not of ATPS activity. Thus, sulphate uptake was not affected and increase in sulphate content of the leaves was observed (Wyss and Brunold, 1980; Brunold et al., 1983).

4.2 Regulation by Nitrogen

Several studies have established regulatory interactions between assimilatory sulphate and nitrate reduction in plants (Brunold, 1993; Yamaguchi et al., 1999; Koprivova et al., 2000). The two assimilatory pathways are very similar and well co-ordinated; deficiency for one element represses the other pathway. The activities of ATPS, APR and OAS-TL decreased under nitrogen deficient conditions in *Lemna minor* L. and cultured tobacco cells (Reuveny et al., 1980; Smith, 1980; Brunold and Suter, 1984). At the same time, addition of nitrate or ammonia to the N-deficient medium quickly restored the activity of these enzymes. On the other hand, addition of ammonia or amino acids (Arg, Asn, Gln) to nutrient solution caused an 50-110% increase in extractable APR activity in *Lemna* but did not affect ATPS and OAS-TL (Brunold and Suter, 1984; Suter et al., 1986). Addition of ammonia increased the flux through the sulphate assimilation, measured as incorporation of ^{35}S in proteins after feeding with [^{35}S]sulphate (Brunold and Suter, 1984). In *A. thaliana* deprivation of nitrogen source for 3 days led to 30% and 50% decrease of APR activity in leaves and roots, respectively, while cysteine and glutathione content not affected (Koprivova et al., 2000). The decrease of APR activity correlated with decreased mRNA and enzyme levels. $^{35}SO_4^{2-}$ feeding showed that after addition of NH_4^+, Gln, or OAS to N-starved plants, incorporation of ^{35}S into proteins significantly increased, however GSH and cysteine labelling was higher only with Gln or OAS (Koprivova et al., 2000). In an expression analysis of OAS-TL, only the mRNA for mitochondrial isoform was upregulated in spinach plants deprived of nitrogen source (Takahashi and Saito, 1996).

OAS, a precursor of cysteine, plays an important role in regulation of sulphate uptake and assimilation. OAS seems to be limiting for cysteine synthesis in presence of excess sulphate (Rennenberg, 1983). Indeed, overexpression of serine acetyltransferase, the enzyme synthesizing OAS led to increased cysteine and GSH concentrations in transgenic potato and tobacco (Blaszczyk et al., 1999; Harms et al., 2000). OAS accumulates during sulphur starvation and may thus act as a signal of sulphur status (Kim et al., 1999). Addition of OAS increases sulphate uptake and APR activity at normal sulphate levels (Neuenschwander et al., 1991; Smith et al., 1997). OAS acts most probably as a transcriptional regulator, since its addition strongly increased mRNA levels of all the three APR isoforms in roots but also those of SiR, chloroplastic OAS-TL and cytosolic SAT. In addition, [^{35}S]sulphate feeding experiments demonstrated that the closer the N-sources were metabolically related to OAS, the higher was their impact on $^{35}SO_4^{2-}$ incorporation (Koprivova et al., 2000). It seems that sulphate reduction is regulated by nitrogen nutrition at transcriptional level and that OAS plays a major role in this regulation.

4.3 Regulation by Light and Carbohydrates

Sulphate assimilation is influenced by light as demonstrated by the light dependency of sulphate reduction by broken chloroplasts (Schmidt and Trebst, 1969). ATPS activity increased with irradiation in oat, barley and maize, and decreased when inhibitors of photosynthetic electron transport were added to the nutrient solution (Passera et al., 1989). The activity of APR was light induced in *Lemna minor* (Neuenschwander et al., 1991). The activities of ATPS, APR, SiR, and OAS-TL increased when etiolated pea seedlings were exposed to light (von Arb and Brunold, 1986). Similarly, in *A. thaliana* the mRNA levels of APK, SiR, OAS-TL, and SAT were several times higher in green leaves than in etiolated tissues (Hell et al., 1997). A diurnal rhythm of APR with maximal activity during the light period was observed in maize (Kocsy et al., 1997). Also in *A. thaliana* APR activity undergoes a diurnal rhythm, reaching maximum values after 4 hours of light onset in both shoots and roots and is minimal at the beginning of the night (Kopriva et al., 1999). However, since during prolonged light the activity remained stable and low in shoots, the regulation by endogenous rhythm is not involved. The activity decreases rapidly to undetectable levels within 24 hours of prolonged darkness, both in shoots and roots, and is restored by subsequent re-illumination. The changes in APR activity correspond to changes in mRNA levels of all 3 APR isoforms and APR protein accumulation. Also the mRNA levels for SiR and chloroplastic isoform of OAS-TL in the leaves displayed similar day/night rhythm. $^{35}SO_4^{2-}$ feeding experiments showed that the incorporation of ^{35}S into reduced sulphur compounds in vivo was significantly higher in light than in the dark. However, since the accumulation of mRNA and protein, as well as enzyme activity increased during the last 4 hours of dark period, light cannot be the only factor involved in APR regulation. Indeed, OAS mimics the effect of light on dark treated plants (Neuenschwander et al., 1991), but other compounds are also effective. Thus, addition of 0.5% sucrose to the nutrient solution after 38 hours of darkness led to a seven-fold increase of root APR activity during 6 hours. The sucrose-induced increase of APR activity was even quicker than that measured in shoots after re-illumination of dark-adapted plants (Kopriva et al., 1999). The corresponding effect of externally applied sucrose on the activity and mRNA accumulation of APR and

NR, key enzymes of assimilatory sulphate and nitrate reduction, adds an interesting facet to the co-ordinated regulation of these two pathways.

4.4 Regulation by Oxidative Stress

Oxidative stress triggered by active oxygen species results from different environmental stresses, e.g. light, ozone, chilling, drought, nutrient deficiency, wounding, pathogens, herbicides, and sulphur dioxide. The main detoxification pathways for these compounds are the radical scavengers such as ascorbate, GSH, carotenoids, or tocopherol, superoxide dismutase reaction, catalase, and various peroxidases. The most abundant scavenger and peroxidase substrate is ascorbate. Ascorbate can be regenerated from its oxidized form, dehydroascorbate, by reduction with glutathione (May et al., 1998; Noctor et al., 1998). In addition to its role in defense against active oxygen species, GSH is also involved in detoxification of xenobiotics, which are conjugated with GSH by glutathione S-transferase with subsequent transport of the conjugate into vacuole (Edwards et al., 2000). Exposure to heavy metals induces synthesis of phytochelatins, small peptides enzymatically synthesized from GSH that chelate the metals into a form transportable to the vacuoles (Rauser, 1999). Clearly, oxidative stress results in higher need of glutathione and, in concordance with the demand driven control of sulphate assimilation, in the upregulation of sulphate assimilation.

Surprisingly, very little is known about direct effects of active oxygen species on sulphate assimilation. Jecklin (1997) observed increased APR activity in maize roots exposed to H_2O_2. Fumigation of *Brassica juncea* with ozone also led to a massive transient increase in APR activity accompanied by a rise in relative amount of oxidized glutathione; the protein accumulation was, however, not affected, revealing a post-translational regulation (Bick et al., 2001). Transcriptional regulation is involved in modulation of APR activity by oxidative stress, since microarray analysis of *Arabidopsis* transcriptome after treatment with 20 mM H_2O_2 revealed a two-fold induction for APR2 mRNA (Desikan et al., 2001). Also in maize roots, treatment with 2 mM H_2O_2 leads to induction of APR and ATPS activities and mRNA levels (Flückiger et al., 2000).

Chilling temperatures result in oxidative stress due to production of reactive oxygen and, accordingly, in increased GSH levels in different plant species (Walker and McKersie, 1993; Brunner et al., 1995). Chilling of maize at 12°C resulted in a prominent increase of mRNA and activity of APR in bundle-sheath cells of the leaves and in the roots (Kopriva et al., 2001b). Surprisingly, after chilling, mRNAs for APR and SiR were detected also in mesophyll cells. These results demonstrate that chilling stress affected not only the levels but also the intercellular distribution of mRNAs for enzymes of sulphate assimilation. Furthermore, in maize, chilling eliminated diurnal changes in the activities of nitrate reductase and APR (Kocsy et al., 1997). Except for glucosinolates, the role of sulphur compounds in defense aga inst pathogens is not fully understood. Correlation analysis indicated a sulphur-induced resistance, the mechanism is, however, not clear yet (Burandt et al., 2001a). The possible involvement of reduced sulphur is strengthened by the observation that ATPS, APR, and APK mRNA were induced after treatment with methyljasmonate, a well characterized signal in plant response to pathogens and wounding (Harada et al., 2000).

The main pathway of detoxification of xenobiotics, e.g. herbicides, goes through the conjugation with glutathione. It is, thus, not surprising that growth of transgenic poplars with enhanced GSH synthesis was less reduced upon treatment with chloroacetanilide herbicides than that of the wild type (Gullner et al., 2001). Increase of GSH concentration is also the mechanism of protection of crop plants from injury by herbicides due to treatment with safeners (Farago et al., 1994). Again, in accordance with the demand-driven control of sulphate assimilation, treatment with safeners increases the flux through the pathway and extractable activities of ATPS and APR (Farago and Brunold, 1990). This increase is based on transcriptional regulation, since the levels of mRNAs for these enzymes were increased upon safener treatment (Spielhofer et al., 2000).

Exposure of plants to heavy metals is connected to higher demand on GSH synthesis due to the induction of phytochelatin synthesis (Rauser, 1999). Therefore, the flux through the sulphate reduction must be increased, which is again achieved by increase of ATPS and APR activities. Historically, most experiments were performed with cadmium, because Cd is an excellent inducer of phytochelatins. The upregulation of APR and ATPS by Cd was documented in various plant species, including maize (Nussbaum et al., 1988), pea (Rüegsegger et al., 1990) and *Brassica juncea* (Heiss et al., 1999; Lee and Leustek, 1999). In *B. juncea* the increase in enzymatic activity was accompanied by an increase in the corresponding mRNA levels indicating a transcriptional regulation (Heiss et al., 1999; Lee and Leustek, 1999). Also the activity and mRNA level of cytosolic isoform of OAS-TL were induced after treatment of *Arabidopsis* with Cd and overexpression of this enzyme led to increased Cd tolerance (Dominguez-Solis et al., 2001). The detoxification of heavy metals might be the reason for high expression of OAS-TL and other enzymes of GSH synthesis in *Arabidopsis* root trichomes (Gutierrez-Alcala et al., 2000).

Different approaches were utilized to increase plant tolerance to various forms of oxidative stress; most of them aimed at increasing the GSH or cysteine synthesis. Such experiments are, however, not trivial and often end up with unexpected results. In several lines of poplars, overexpressing bacterial enzymes of GSH synthesis, GSH content was increased to 2 to 4-times. Despite this, these transgenic lines were not substantially more tolerant to oxidative stress or treatment with heavy metals. On the other hand, overexpression of SAT, resulting in approx. 3-fold increase in GSH content, led to significantly reduced damage by H_2O_2 (Blaszczyk et al., 1999). The overexpression of OAS-TL in tobacco led to increased tolerance to oxidative stress (Noji et al., 2001; Youssefian et al., 2001). Increasing flux through sulphate assimilation is certainly a promising approach to obtain stress tolerant plants which can be cultivated at extreme climatic conditions or utilized for phytoremediation of contaminated soils. Unfortunately, we are still far from understanding fully the molecular mechanisms of regulation of assimilatory sulphate reduction.

5. EVOLUTION

Sulphate assimilation is present in plants, fungi and yeast, a wide range of eubacteria, but is missing in metazoa. Only very little is known about *Archae*, but since their ecological habitat is usually rich in reduced forms of sulphur or they possess the enzymes for dissimilatory sulphate reduction, sulphate assimilation may be missing in these species as well. Among the 35 species of the *Eubacteria* superkingdom, whose genomes have been sequenced to date, only 15 have the capacity for sulphate reduction. As described in section 2, two pathways of sulphate assimilation exist, using APS or PAPS for reduction to sulfite. It was generally accepted that chemotrophic bacteria and fungi utilize PAPS in a reaction catalyzed by a thioredoxin dependent PAPS reductase, whereas photosynthesising organisms reduce APS directly (Tsang and Schiff, 1975; Schmidt, 1977; Schmidt and Trüper, 1977; Brunold, 1990). The boundary-line between APS and PAPS utilizing organisms could not be sharply defined, as among phototrophic bacteria and cyanobacteria both APS and PAPS reducing species were described (Schmidt, 1977; Schmidt and Trüper, 1977). Sequence analysis revealed that in some chemotrophic bacteria, the PAPS reductase homologues were more similar to plant APR than to PAPS reductase from *E. coli* (Fig. 4).

Figure 4 shows a neighbor-joining tree of APS and PAPS reductase related sequences, retrieved from the GenBank database. The phylogenetic tree is divided into two major branches. The first branch contains a cluster of APS reductases from plants and algae together with many bacterial enzymes, the other one is subdivided into clusters comprising fungal PAPS reductases and well-characterized PAPS reductases from enteric bacteria and cyanobacteria. The clear separation of these two clusters raised the question whether the sulphate reduction in the species clustering with plants is really PAPS dependent. Indeed, many bacteria, such as *Pseudomonas, Rhizobium*, *Ralstonia*, and *Burkholderia*, were able to produce sulphite from APS but not from PAPS (Abola et al., 1999; Bick et al., 2000). It seems, therefore, that the node separating the two major clusters of the phylogenetic tree represents the border between organisms utilizing APS and PAPS dependent pathways.

The novel bacterial assimilatory APR, although similar to the plant APR, is lacking its C-terminal domain and requires thioredoxin as an electron donor. The major difference between these proteins and PAPS reductase from *E. coli* or *S. typhimurium* is the presence of the two additional cysteine pairs conserved in the plant enzyme. These cysteines participate in binding of the iron-sulfur cluster to the plant APR (see Section 3.2.1). Also the APS reductases from *S. meliloti* and *P. aeruginosa* bind an FeS center, which has the same characteristics as the FeS cluster from *Lemna minor* APR (Kopriva et al., 2002). APS reduction thus seems to be dependent on the FeS cluster as a cofactor and the two conserved cysteine pairs represent a marker to distinguish between APS and PAPS dependent sulphate assimilation. APS reduction is thus much more widely distributed than it was believed. Since the dissimilatory sulphate reducing bacteria and bacteria oxidizing sulphur, such as green sulphur bacteria, possess APS reductase and the activation to APS requires less energy than PAPS synthesis, it is plausible to deduce that APS reductase is the ancient enzyme. The bacterial counterparts of plant APR require thioredoxin as a reductant, therefore, it is reasonable to conclude that the plant *apr* gene originated from a fusion of genes for bacterial APS reductase and thioredoxin. Since, all APRs isolated or cloned from higher plants, and as the enzyme from a

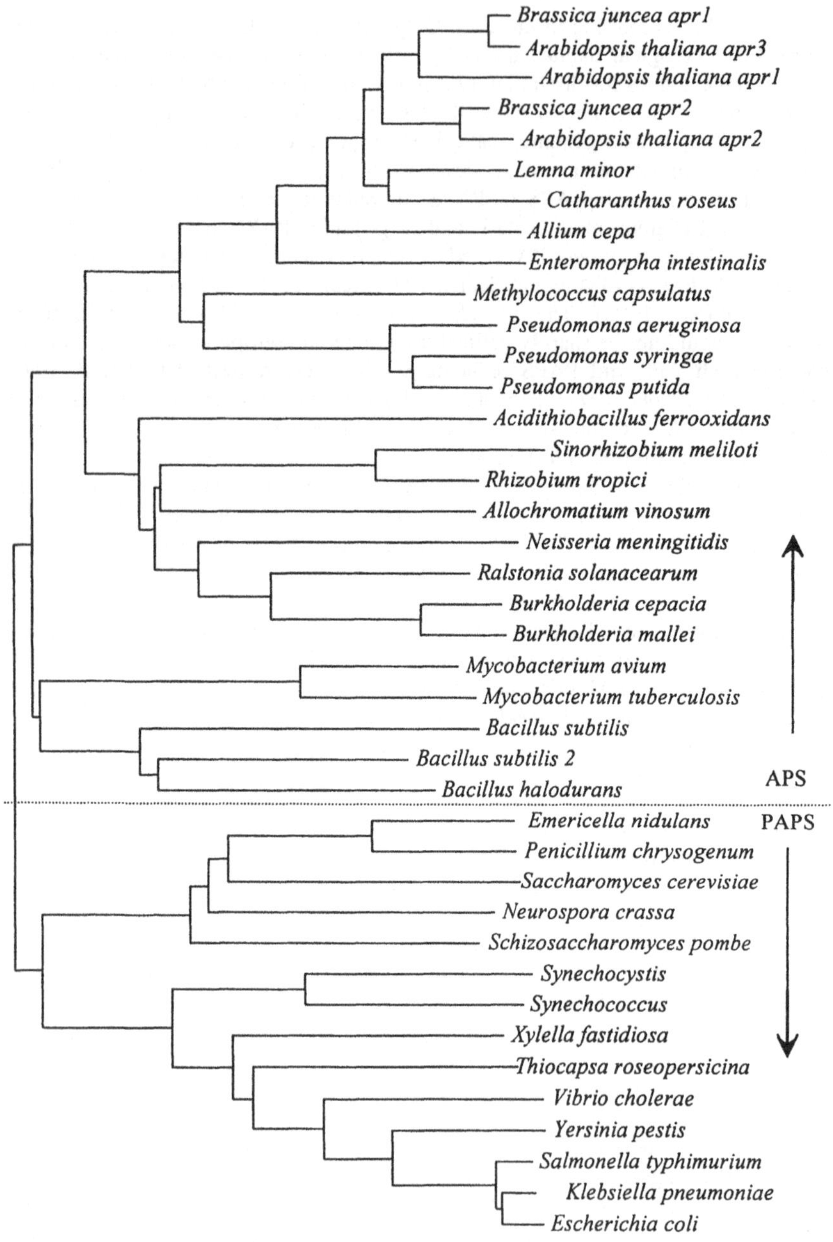

Figure 4. The neighbor-joining tree of APS and PAPS reductase protein sequences. The sequences were retrieved from GenBank, aligned with Clustal and the tree was constructed with the Phylip software.

green algae *Enteromorpha intestinalis* (Gao et al., 2000), have the same structure, this fusion must have occurred early in the evolution of plants.

Little is known about evolution and origin of plant genes for the enzymes of the sulphate assimilation pathway. The structures of plant enzymes differ often very substantially from their bacterial equivalents (see section 3). Phylogenetic reconstructions are difficult since the genes are missing in many organisms, their loss being probably an adaptation of a parasitic way of life. Remarkably, sulphur containing amino acids are massively underrepresented in the enzymes of sulphate assimilation (Baudouin-Cornu et al., 2001). This feature might be especially important for habitats with limiting supply of sulphur, allowing fast and fine regulation of flux through the pathway. Figure 5 shows a simplified phylogenetic tree of ATPS. Plant enzyme is completely unrelated to ATPS from sulphate assimilating bacteria and is most similar to ATPS from other eukaryotes, metazoa and fungi. Surprisingly, plant ATPS are homologous to ATPS from organisms utilizing dissimilatory sulphate reduction for energy conservation. Although the assimilatory and dissimilatory sulphate reduction are catalyzed by the same set of enzymes, APR and SiR are completely unrelated. The genes for ATPS from plants as well as from other Eukaryotes, thus most probably are derived from archaebacterial dissimilatory ATPS. The same conclusion can be made for the evolution of APK. On the other hand, the genes for APR and SiR are most probably of cyanobacterial origin.

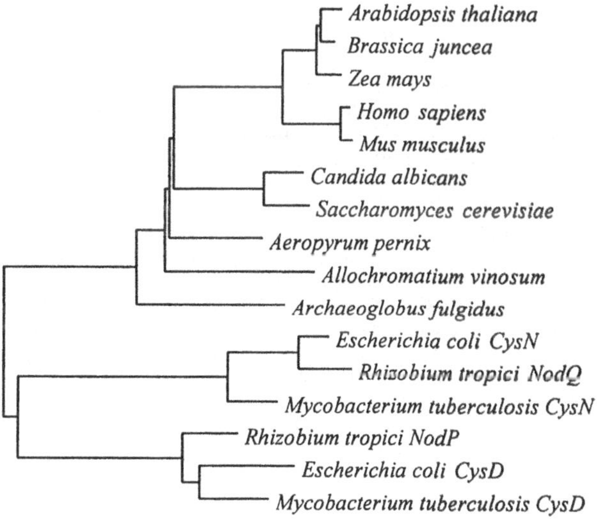

Figure 5. Simplified neighbor-joining tree of ATP sulphurylase protein sequences. The sequences were retrieved from GenBank, aligned with Clustal, and the tree was constructed with the Phylip software.

6. CONCLUSIONS

The recent investigations of the sulphate assimilation pathway in plants have delivered several surprising results, that correct our view of the pathway. The old controversy about the "free sulphite" and "bound sulphite" routes seems to be resolved and the enzyme responsible identified. Most of the results point to APS reductase as the enzyme possessing control over the flux through sulphate assimilation. However, several questions still need to be resolved. In the first place, the existence of a PAPS dependent sulphate reduction is still neither proven nor excluded. For this, a functional genomics approach would be necessary. Secondly, the presence of the FeS cluster in APR stimulates questions on the reaction mechanism of this enzyme. Also the functional role of the cell-specific distribution of sulphate assimilation in maize is still an open question. The different functions (if any) and regulations of the multiple isoforms of enzymes involved in sulphate assimilation remain to be elucidated, most importantly the contribution of cytosolic and mitochondrial SAT and OAS-TL isoforms to total cysteine synthesis. Identification and analysis of mutants and gene knockouts will thus be of great importance. Finally, the molecular mechanisms of regulation of the pathway are not known. Therefore, a careful analysis of the promoters might help to understand these processes. Clearly, although our knowledge of sulphate assimilation and its regulation substantially is improved, there still remains a lot for further investigations. Without doubt, unexpected results will appear in the course of further research to remind us that sulphate assimilation is a pathway that likes to surprise.

7. REFERENCES

Abola AP, Willits MG, Wang RC and Long SR (1999) Reduction of adenosine-5 '-phosphosulfate instead of 3 '-phosphoadenosine-5 '-phosphosulfate in cysteine biosynthesis by *Rhizobium meliloti* and other members of the family *Rhizobiaceae*. *J Bacteriol* **181**: 5280—5287

Baudouin-Cornu P, Surdin-Kerjan Y, Marliere P and Thomas D (2001) Molecular evolution of protein atomic composition. *Science* **293**: 297—300

Bick JA, Aslund F, Chen Y and Leustek T (1998) Glutaredoxin function for the carboxyl-terminal domain of the plant-type 5'-adenylylsulfate reductase. *Proc Natl Acad Sci USA* **95**: 8404—8409

Bick JA, Dennis JJ, Zylstra GJ, Nowack J and Leustek T (2000) Identification of a new class of 5 '-adenylylsulfate (APS) reductases from sulfate-assimilating bacteria. *J Bacteriol* **182**: 135—142

Bick JA, Setterdahl AT, Knaff DB, Chen Y, Pitcher LH, Zilinskas BA and Leustek T (2001) Regulation of the plant-type 5'-adenylyl sulfate reductase by oxidative stress. *Biochemistry* **40**: 9040—9048

Blaszczyk A, Brodzik R and Sirko A (1999) Increased resistance to oxidative stress in transgenic tobacco plants overexpressing bacterial serine acetyltransferase. *Plant J* **20**: 237—243

Bogdanova N and Hell R (1997) Cysteine synthesis in plants: Protein-protein interactions of serine acetyltransferase from *Arabidopsis thaliana*. *Plant J* **11**: 251—262

Bolchi A, Petrucco S, Tenca PL, Foroni C and Ottonello S (1999) Coordinate modulation of maize sulfate permease and ATP sulfurylase mRNAs in response to variations in sulfur nutritional status: stereospecific down-regulation by L-cysteine. *Plant Mol Biol* **39**: 527—537

Bork C, Schwenn JD and Hell R (1998) Isolation and characterization of a gene for assimilatory sulfite reductase from *Arabidopsis thaliana*. *Gene* **212**: 147—153

Brühl A, Haverkamp T, Gisselmann G and Schwenn JD (1996) A cDNA clone from *Arabidopsis thaliana* encoding plastidic ferredoxin:sulfite reductase. *Biochim Biophys Acta* **1295**: 119—124

Brunner M, Kocsy G, Rüegsegger A, Schmutz D and Brunold C (1995) Effect of chilling on assimilatory sulfate reduction and glutathione synthesis in maize. *J Plant Physiol* **146**: 743—747

Brunold C (1990) Reduction of sulfate to sulfide. In: Rennenberg H, Brunold C, De Kok LJ, Stulen I (eds) Sulphur Nutrition and Sulphur Assimilation in Higher Plants, pp 13—31. SPB Academic Publishing, The Hague

Brunold C (1993) Regulatory interactions between sulfate and nitrate assimilation. In: De Kok LJ, Stulen I, Rennenberg H, Brunold C and Rauser WE (eds) Sulfur Nutrition and Sulfur Assimilation in Higher Plants, pp 61—75. SPB Academic Publishing, The Hague

Brunold C and Suter M (1982) Intracellular localization of serine acetyltransferase in spinach leaves. *Planta* **155:** 321—327

Brunold C and Suter M (1984) Regulation of sulfate assimilation by nitrogen nutrition in the duckweed *Lemna minor* L. *Plant Physiol* **76:** 579—583

Brunold C and Suter M (1989) Localization of enzymes of assimilatory sulfate reduction in pea roots. *Planta* **179:** 228—234

Brunold C and Suter M (1990) Adenosine 5'-phosphosulfate sulfotransferase. In: Lea P (ed), Methods in Plant Biochemistry, Vol.3, pp339—343. Academic Press, London

Brunold C, Landolt W and Lavanchy P (1983) SO_2 and assimilatory sulfate reduction in beech leaves. *Physiol Plant* **59:** 313—318

Brunold C, Suter M and Lavanchy P (1987) Effect of high and low sulphate concentrations on adenosine 5'-phosphosulphate sulfotransferase activity from *Lemna minor* L. *Physiol Plant* **70:** 168—174

Burandt P, Papenbrock J, Schmidt A, Bloem E, Haneklaus S and Schnug E (2001a) Genotypical differences in total sulfur contents and cysteine desulfhydrase activities in *Brassica napus* L. *Phyton* **41:** 75—85

Burandt P, Schmidt A and Papenbrock J (2001b) Cysteine synthesis and cysteine desulfuration in *Arabidopsis* plants at different developmental stages and light conditions. *Plant Physiol Biochem* **39:** 861—870

Burnell JN (1984) Sulfate assimilation in C_4 plants. *Plant Physiol* **75:** 873—875

Chen Y and Leustek T (1998) Three genomic clones from *Arabidopsis thaliana* encoding 5'-adeylylsulfate reductase (Acc. Nos. AF016282, AF016283, and AF016284). *Plant Physiol* **116:** 869

Crane BR, Siegel LM and Getzoff ED (1995) Sulfite reductase structure at 1.6 A: evolution and catalysis for reduction of inorganic anions. *Science* **270:** 59—67

Desikan R, Mackerness SA-H, Hancock JT and Neill SJ (2001) Regulation of the Arabidopsis transcriptome by oxidative stress. *Plant Physiol* **127:** 159—172

Dominguez-Solis JR, Gutierrez-Alcala G, Vega JM, Romero LC and Gotor C (2001) The cytosolic O-acetylserine(thiol)lyase gene is regulated by heavy metals and can function in cadmium tolerance. *J Biol Chem* **276:** 9297—9302

Droux M, Ruffet ML, Douce R and Job D (1998) Interactions between serine acetyltransferase and O-acetylserine (thiol)lyase in higher plants - structural and kinetic properties of the free and bound enzymes. *Eur J Biochem* **255:** 235—245

Edwards R, Dixon DP and Walbot V (2000) Plant glutathione S-transferases: enzymes with multiple functions in sickness and in health. *Trends Plant Sci* **5:** 193—198

Fankhauser H and Brunold C (1978) Localization of adenosine 5'-phosphosulfate sulfotransferase in spinach leaves. *Planta* **143:** 285—289

Farago S, Brunold C and Kreuz K (1994) Herbicide safeners and glutathione metabolism. *Physiol Plant* **91:** 537—542

Farago S and Brunold C (1990) Regulation of assimilatory sulfate reduction by herbicide safeners in *Zea mays* L. *Plant Physiol* **94:** 1808—1812

Flückiger H, Jecklin S, von Ballmoos P, Suter M and Brunold C (2000) Effect of hydrogen peroxide induced oxidative stress in maize roots. In: Brunold C, Rennenberg H, De Kok LJ, Stulen I, Davidian JC (eds) Sulphur Nutrition and Sulphur Assimilation in Higher Plants: Molecular, Biochemical and Physiological Aspects, pp 401—403. Paul Haupt, Bern

Fritz G, Büchert T, Huber H, Stetter KO and Kroneck PM (2000) Adenylylsulfate reductases from archaea and bacteria are 1 : 1 αß-heterodimeric iron-sulfur flavoenzymes - high similarity of molecular properties emphasizes their central role in sulfur metabolism. *FEBS Lett* **473:** 63—66

Gao Y, Schofield OM and Leustek T (2000) Characterization of sulfate assimilation in marine algae focusing on the enzyme 5 '-adenylylsulfate reductase. *Plant Physiol* **123:** 1087—1096

Gerwick BC, Ku SB and Black CC (1980) Initiation of sulfate activation: a variation in C_4 photosynthesis plants. *Science* **209:** 513—515

Gullner G, Kömives T and Rennenberg H (2001) Enhanced tolerance of transgenic poplar plants overexpressing γ-glutamylcysteine synthetase towards chloroacetanilide herbicides. *J Exp Bot* **52:** 971—979

Gutierrez-Alcala G, Gotor C, Meyer AJ, Fricker M, Vega JM and Romero LC (2000) Glutathione biosynthesis in *Arabidopsis* trichome cells. *Proc Natl Acad Sci USA* **97:** 11108—11113

Gutierrez-Marcos JF, Roberts MA, Campbell EI and Wray JL (1996) Three members of a novel small gene-family from *Arabidopsis thaliana* able to complement functionally an *Escherichia coli* mutant defective in PAPS reductase activity encode proteins with a thioredoxin-like domain and "APS reductase" activity. *Proc Natl Acad Sci USA* **93:** 13377—13382

Harada E, Kusano T and Sano H (2000) Differential expression of genes encoding enzymes involved in sulfur assimilation pathways in response to wounding and jasmonate in *Arabidopsis thaliana. J Plant Physiol* **156:** 272—276

Harms K, von Ballmoos P, Brunold C, Hofgen R and Hesse H (2000) Expression of a bacterial serine acetyltransferase in transgenic potato plants leads to increased levels of cysteine and glutathione. *Plant J* **22:** 335—343

Hartmann T, Mult S, Suter M, Rennenberg H and Herschbach C (2000) Leaf age-dependent differences in sulphur assimilation and allocation in poplar (*Populus tremula* x *P. alba*) leaves. *J Exp Bot* **51:** 1077—1088

Hatzfeld Y, Lee S, Lee M, Leustek T and Saito K (2000a) Functional characterization of a gene encoding a fourth ATP sulfurylase isoform from *Arabidopsis thaliana. Gene* **248:** 51—58

Hatzfeld Y, Maruyama A, Schmidt A, Noji M, Ishizawa K and Saito K, (2000b) beta-cyanoalanine synthase is a mitochondrial cysteine synthase-like protein in spinach and Arabidopsis. *Plant Physiol* **123:** 1163—1171

Hawkesford MJ and Wray JL (2000) Molecular genetics of sulphate assimilation.*Adv Bot Res* **33:** 159—223

Heiss S, Schäfer HJ, Haag-Kerwer A and Rausch T (1999) Cloning sulfur assimilation genes of *Brassica juncea* L.: cadmium differentially affects the expression of a putative low-affinity sulfate transporter and isoforms of ATP sulfurylase and APS reductase. *Plant Mol Biol* **39:** 847—857

Hell R (1997) Molecular physiology of plant sulfur metabolism. *Planta* **202:** 138—148

Hell R, Schwenn JD and Bork C (1997) Light and sulfur sources modulate mRNA levels of several genes of sulfate assimilation. In: Cram WJ, De Kok LJ, Stulen I, Brunold C and Rennenberg H (eds) Sulfur Metabolism in Higher Plants, pp 181—185. Backhuys Publishers, Leiden

Höfgen R, Kreft O, Willmitzer L and Hesse H (2001) Manipulation of thiol contents in plants. *Amino Acids* **20:** 291—299

Jecklin S (1997) Effect of oxidative stress on assimilatory sulfate reduction in maize (*Zea mays* L.) roots. Diploma work, University of Bern

Jones-Mortimer MC (1968) Positive control of sulphate reduction in *Escherichia coli. Biochem J* **110:** 589—595

Jost R, Berkowitz O, Wirtz M, Hopkins L, Hawkesford MJ and Hell R (2000) Genomic and functional characterization of the oas gene family encoding O-acetylserine (thiol) lyases, enzymes catalyzing the final step in cysteine biosynthesis in *Arabidopsis thaliana. Gene* **253:** 237—247

Kappler U and Dahl C (2001) Enzymology and molecular biology of prokaryotic sulfite oxidation. *FEMS Microbiol Lett* **203:** 1—9

Kim H, Hirai MY, Hayashi H, Chino M, Naito S and Fujiwara T (1999) Role of *O*-acetyl-l-serine in the coordinated regulation of the expression of a soybean seed storage-protein gene by sulfur and nitrogen nutrition. *Planta* **209:** 282—289

Klonus D, Hofgen R, Willmitzer L and Riesmeier JW (1994) Isolation and characterization of two cDNA clones encoding ATP-sulfurylases from potato by complementation of a yeast mutant. *Plant J* **6:** 105—112

Kocsy G, Owttrim G, Brander K and Brunold C (1997) Effect of chilling on the diurnal rhythm of enzymes involved in protection against oxidative stress in a chilling-tolerant and a chilling-sensitive maize genotype. *Physiol Plant* **99:** 249—254

Kopriva S, Büchert T, Fritz G, Suter M, Weber M, Benda R, Schaller J, Feller U, Schürmann P, Schünemann V, Trautwein AX, Kroneck PM and Brunold C (2001a) Plant adenosine 5'-phosphosulfate reductase is a novel iron-sulfur protein. *J Biol Chem* **276:** 42881—42886

Kopriva S, Jones S, Koprivova A, Suter M, von Ballmoos P, Brander K, Flückiger J and Brunold C (2001b) Influence of chilling stress on the intercellular distribution of assimilatory sulfate reduction and thiols in *Zea mays. Plant Biol* **3:** 24—31

Kopriva S, Muheim R, Koprivova A, Trachsler N, Catalano C, Suter M and Brunold C (1999) Light regulation of assimilatory sulfate reduction in *Arabidopsis thaliana. Plant J* **20:** 37—44

Kopriva S, Büchert T, Fritz G, Suter M, Benda R, Schünemann V, Koprivova A, Schürmann P, Trautwein AX, Kroneck PMH and Brunold C (2002) The presence of an iron-sulfur cluster in adenosine 5'-phosphosulfate reductase separates organisms utilizing adenosine 5'-phosphosulfate and phosphoadenosine 5'-phosphosulfate for sulfate assimilation. *J Biol Chem* **277:** 21786—21791

Koprivova A, Suter M, Op den Camp R, Brunold C and Kopriva S (2000) Regulation of sulfate assimilation by nitrogen in Arabidopsis. *Plant Physiol* **122:** 737—746

Koprivova A, Melzer M, von Ballmoos P, Mandel T, Brunold C and Kopriva S (2001) Assimilatory sulfate reduction in C_3, C_3-C_4, and C_4 species of *Flaveria*. *Plant Physiol* **127:** 543—550

Kredich NM (1971) Regulation of L-cysteine biosynthesis in *Salmonella typhimurium*. *J Biol Chem* **246:** 3474—3484

Kredich NM and Tomkins GM (1966) The enzymatic synthesis of L-cysteine in *Escherichia coli* and *Salmonella typhimurium*. *J Biol Chem* **244**: 4955—4965

Krueger RJ and Siegel LM (1982) Spinach siroheme enzymes: Isolation and characterization of ferredoxin-sulfite reductase and comparison of properties with ferredoxin-nitrite reductase. *Biochemistry* **21**: 2892—2904

Ku MSB, Wu JR, Dai ZY, Scott RA, Chu C and Edwards GE (1991) Photosynthetic and photorespiratory characteristics of *Flaveria* species. *Plant Physiol* **96:** 518—528

Kurima K, Warman ML, Krishnan S, Domowicz M, Krueger Jr RC, Deyrup A and Schwartz NB (1998) A member of a family of sulfate-activating enzymes causes murine brachymorphism. *Proc Natl Acad Sci USA* **95:** 8681—8685

Lappartient AG and Touraine B (1996) Demand-driven control of root ATP sulphurylase activity and SO_4^{2-} uptake in intact canola. Role of phloem-translocated glutathione. *Plant Physiol* **111:** 147—157

Lappartient AG, Vidmar JJ, Leustek T, Glass ADM and Touraine B (1999) Inter-organ signaling in plants: regulation of ATP sulphurylase and sulphate transporter genes expression in roots mediated by phloem-translocated compound. *Plant J* **18**: 89—95

Lee S and Leustek T (1998) APS kinase from *Arabidopsis thaliana*: Genomic organization, expression, and kinetic analysis of the recombinant enzyme. *Biochem Biophys Res Commun* **247:** 171—175

Lee S and Leustek T (1999) The effect of cadmium on sulfate assimilation enzymes in *Brassica juncea*. *Plant Sci* **141:** 201—207

Leustek T and Saito K (1999) Sulfate transport and assimilation in plants. *Plant Physiol* **120:** 637—644

Leustek T, Martin MN, Bick JA and Davies JP (2000) Pathways and regulation of sulfur metabolism revealed through molecular and genetic studies. *Annu Rev Plant Physiol Plant Mol Bio* **51:** 141—165

Leyh TS, Taylor JC and Markham GD (1988) The sulfate activation locus of *Escherichia coli* K12: cloning, genetic, and enzymatic characterization. *J Biol Chem* **263:** 2409—2416

Li JJ, Saidha T and Schiff JA (1991) Purification and properties of two forms of ATP sulfurylase from Euglena. *Biochim Biophys Acta* **1078:** 68—76

Logan HM, Cathala N, Grignon C and Davidian JC (1996) Cloning of a cDNA encoded by a member of the *Arabidopsis thaliana* ATP sulfurylase multigene family. Expression studies in yeast and in relation to plant sulfur nutrition. *J Biol Chem* **271:** 12227—12233

Lunn J, Droux M, Martin J and Douce R (1990) Localization of ATP sulfurylase and O-acetylserine (thiol)lyase in spinach leaves. *Plant Physiol* **94:** 1345—1352

Lyle S, Ozeran JD, Stanczak J, Westley J and Schwartz NB (1994) Intermediate channeling between ATP sulfurylase and adenosine 5'-phosphosulfate kinase from rat chondrosarcoma. *Biochemistry* **33:** 6822—6827

MacRae IJ, Segel IH and Fisher AJ (2000) Crystal structure of adenosine 5'-phosphosulfate kinase from *Penicillium chrysogenum*. *Biochemistry* **39:** 1613—1621

MacRae IJ, Segel IH and Fisher AJ (2001) Crystal structure of ATP sulfurylase from *Penicillium chrysogenum*: insights into the allosteric regulation of sulfate assimilation. *Biochem* **40:** 6795—6804

Maruyama A, Ishizawa K and Takagi T (2000) Purification and characterization of beta-cyanoalanine synthase and cysteine synthases from potato tubers: Are beta-cyanoalanine synthase and mitochondrial cysteine synthase same enzyme? *Plant Cell Physiol* **41:** 200—208

May MJ, Vernoux T, Leaver C, van Montagu M and Inzé D (1998) Glutathione homeostasis in plants: implications for environmental sensing and plant development. *J Exp Bot* **49:** 649—667

Monson RK and Moore BD (1989) On the significance of C_3-C_4 intermediate photosynthesis to the evolution of C_4 photosynthesis. *Plant Cell Environ* **12:** 689—699

Murillo M and Leustek T (1995) Adenosine-5'-triphosphate-sulfurylase from *Arabidopsis thaliana* and *Escherichia coli* are functionally equivalent but structurally and kinetically divergent: nucleotide sequence of two adenosine-5'-triphosphate-sulfurylase cDNAs from *Arabidopsis thaliana* and analysis of a recombinant enzyme. *Arch Biochem Biophys* **323:** 195—204

Neuenschwander U, Suter M and Brunold C (1991) Regulation of sulfate assimilation by light and O-acetyl-L-serine in *Lemna minor* L. *Plant Physiol* **97:** 253—258

Noctor G, Arisi A-CM, Jouanin L, Kunert KJ, Rennenberg H and Foyer CH (1998) Glutathione: biosynthesis, metabolism and relationship to stress tolerance explored in transformed plants. *J Exp Bot* **49:** 623—647

Noji M, Inoue K, Kimura N, Gouda A and Saito K (1998) Isoform-dependent differences in feedback regulation and subcellular localization of serine acetyltransferase involved in cysteine biosynthesis from *Arabidopsis thaliana*. *J Biol Chem* 273: 32739—32745

Noji M, Saito M, Nakamura M, Aono M, Saji H and Saito K (2001) Cysteine synthase overexpression in tobacco confers tolerance to sulfur containing environmental pollutants. *Plant Physiol* **126:**973—980

Nussbaum S, Schmutz K and Brunold C (1988) Regulation of assimilatory sulfate reduction by cadmium in *Zea mays* L. *Plant Physiol* **88:** 1407—1410

Passera C and Ghisi R (1982) ATP sulphurylase and O-acetylserine sulphydrylase in isolated mesophyll protoplasts and bundle sheath strands of S-deprived maize leaves. *J Exp Bot* **33:** 432—438

Passera C, Ghisi R and Ferretti M (1989) Light-activation of ATP-Sulfurylase in leaves and chloroplasts of *Zea mays*. *Photosynthetica* **23:** 166—172

Prior A, Uhrig JF, Heins L, Wiesmann A, Lillig CH, Stoltze C, Soll J and Schwenn JD (1999) Structural and kinetic properties of adenylyl sulfate reductase from *Catharanthus roseus* cell cultures. *Biochim Biophys Acta* **1430:** 25—38

Rauser WE (1999) Structure and function of metal chelators produced by plants: the case for organic acids, amino acids, phytin, and metallothioneins. *Cell Biochem Biophys* **31:** 19—48

Rennenberg H (1983) Role of O-acetylserine in hydrogen sulfide emission from pumpkin leaves in response to sulfate. *Plant Physiol* **73:** 560—565

Rennenberg H and Brunold C (1994) Significance of glutathione metabolism in plants under stress. *Prog Bot* **55:** 142—156

Renosto F, Patel HC, Martin RL, Thomassian C, Zimmerman G and Segel IH (1993) ATP sulfurylase from higher plants: kinetic and structural characterization of the chloroplast and cytosol enzymes from spinach leaf. *Arch Biochem Biophys* **307:** 272—285

Renosto F, Seubert PA and Segel IH. (1984) Adenosine 5'-phosphosulfate kinase from *Penicillium chrysogenum*. Purification and kinetic characterization. *J Biol Chem* **259:** 2113—2123

Reuveny Z, Dougall DK and Trinity PM (1980) Regulatory coupling of nitrate and sulfate assimilation pathways in cultured tobacco cells. *Proc Natl Acad Sci USA* **77:** 6670—6672

Rosenthal E and Leustek T (1995) A multifunctional *Urechis caupo* protein, PAPS synthetase, has both ATP sulfurylase and APS kinase activities. *Gene* **165:** 243—248

Rotte C and Leustek T (2000) Differential subcellular localization and expression of ATP sulfurylase and 5 '-adenylylsulfate reductase during ontogenesis of *Arabidopsis* leaves indicates that cytosolic and plastid forms of ATP sulfurylase may have specialized functions. *Plant Physiol* **124:** 715—724

Rüegsegger A, Schmutz D and Brunold C (1990) Regulation of glutathione synthesis by cadmium in *Pisum sativum* L. *Plant. Physiol* **93:** 1579—1584

Saito K, Yokoyama H, Noji M and Murakoshi I (1995) Molecular cloning and characterization of a plant serine acetyltransferase playing a regulatory role in cysteine biosynthesis from watermelon. *J Biol Chem* **270:** 16321—16326

Satishchandran C and Markham GD (1989) Adenosine-5'-phosphosulfate kinase from *Escherichia coli* K12. Purification, characterization, and identification of a phosphorylated enzyme intermediate. *J Biol Chem* **264:** 15012—15021

Sato N, Nakayama M and Hase T (2001) The 70-kDa major DNA-compacting protein of the chloroplast nucleoid is sulfite reductase. *FEBS Lett* **487:** 347—350

Schiff JA and Hodson RC (1970) Pathway of sulfate reduction in algae. *Ann NY Acad Sci* **175:** 555--576

Schiffmann S and Schwenn JD (1998) Isolation of cDNA Clones Encoding Adenosine-5'-phosphosulfate (APS)-kinase EC: 2.7.1.25 from *Catharanthus roseus* (Accession No. AF044285) and an isoform (akn2) from *Arabidopsis thaliana* (Accession No. AF043351). *Plant Physiol* **117:** 1125

Schmidt A (1972a) Über Teilreaktionen der photosynthetischen Sulfatreduktion in zellfreien Systemen aus Spinatchloroplasten und *Chlorella*. *Z Naturforsch* **27b:** 183—192

Schmidt A (1972b) On the mechanism of photosynthetic sulfate reduction. An APS-sulfotransferase from *Chlorella*. *Arch Microbiol* **84:** 77—86

Schmidt A (1973) Sulfate reduction in a cell-free system of *Chlorella*. The ferredoxin dependent reduction of a protein-bound intermediate by a thiosulfonate reductase. *Arch Microbiol* **93**: 29—52
Schmidt A (1976) The adenosine-5'-phosphosulfate sulfotransferase from spinach (*Spinacea oleracea* L.). Stabilization, partial purification and properties. *Planta* **130**: 257—263
Schmidt A (1977) Assimilatory sulfate reduction via 3'-phosphoadenosine-5'-phosphosulfate (PAPS) and adenosine-5'-phosphosulfate (APS) in blue-green algae. *FEMS Microbiol Lett* **1**: 137—140
Schmidt A and Jäger K (1992) Open questions about sulfur metabolism in plants. *Annu Rev Plant Physiol Plant Mol Biol* **43**: 325—349
Schmidt A and Trebst A (1969) The mechanism of photosynthetic sulfate reduction by isolated chloroplasts. *Biochim Biophys Acta* **180**: 529—535
Schmidt A and Trüper HG (1977) Reduction of adenylylsulfate and 3'-phosphoadenylylsulfate in phototrophic bacteria. *Experientia* **33**: 1008—1009
Schmutz D and Brunold C (1984) Intercellular localization of assimilatory sulfate reduction in leaves of *Zea mays* and *Triticum aestivum*. *Plant Physiol* **74**: 866—870
Schmutz D and Brunold C (1985) Localization of nitrite and sulfite reductase in bundle sheath and mesophyll cells of maize leaves. *Physiol Plant* **64**: 523—528
Schwenn JD (1989) Sulfate assimilation in higher plants: a thioredoxin-dependent PAPS-reductase from spinach leaves. *Z Naturforsch* **44c**: 504—508
Segel IH, Renosto F and Seubert PA (1987) Sulfate-activating enzymes. *Methods Enzymol* **143**:334—349
Setya A, Murillo M and Leustek T (1996) Sulfate reduction in higher plants: Molecular evidence for a novel 5'-adenylsulfate reductase. *Proc Natl Acad Sci USA* **93**: 13383—13388
Smith IK (1980) Regulation of sulfate assimilation in tobacco cells: effect of nitrogen and sulfur nutrition on sulfate permease and *O*-acetylserine sulfhydrylase. *Plant Physiol* **66**: 877—883
Smith FW, Hawkesford MJ, Ealing PM, Clarkson DT, van den Berg PJ, Belcher AR and Warrilow AG (1997) Regulation of expression of a cDNA from barley roots encoding a high affinity sulphate transporter. *Plant J* **12**: 875—884
Sperling D, Kappler U, Wynen A, Dahl C and Truper HG (1998) Dissimilatory ATP sulfurylase from the hyperthermophilic sulfate reducer *Archaeoglobus fulgidus* belongs to the group of homo-oligomeric ATP sulfurylases. *FEMS Microbiol Lett* **162**: 257—264
Spielhofer P, Kopriva S, Täschler S, Oberer H, Suter M and Brunold C (2000) Stimulation of key enzymes in sulfate assimilation by safener CGA 173072 may induce higher protection against heavy metal stress in Zea mays L. In: Brunold C, Rennenberg H, De Kok LJ, Stulen I, Davidian JC (eds) Sulphur Nutrition and Sulphur Assimilation in Higher Plants: molecular, biochemical and physiological aspects, pp 387—388. Paul Haupt, Bern
Suter M, Lavanchy P, von Arb C and Brunold C (1986) Regulation of sulfate assimilation by amino acids in *Lemna minor* L. *Plant Sci* **44**: 125—132
Suter M, von Ballmoos P, Kopriva S, Op den Camp R, Schaller J, Kuhlemeier C, Schürmann P and Brunold C (2000) Adenosine 5'-phosphosulfate sulfotransferase and adenosine 5'-phosphosulfate reductase are identical enzymes. *J Biol Chem* **275**: 930—936
Takahashi H and Saito K (1996) Subcellular localization of spinach cysteine synthase isoforms and regulation of their gene expression by nitrogen and sulfur. *Plant Physiol* **112**: 273—280
Takahashi H, Yamazaki M, Sasakura N, Watanabe A, Leustek T, de Almeida Engler J, Engler G, Van Montagu M and Saito K (1997) Regulation of sulfur assimilation in higher plants: A sulfate transporter induced in sulfate-starved roots plays a central role in *Arabidopsis thaliana*. *Proc Natl Acad Sci USA* **94**: 11102—11107
Takahashi S, Yip WC and Tamura G. (1997) Purification and characterization of ferredoxin-sulfite reductase from turnip (*Brassica rapa*) leaves and comparison of properties with ferredoxin-sulfite reductase from turnip roots. *Biosci Biotechnol Biochem* **61**: 1486—1490
Tsang ML-S and Schiff JA (1975) Studies of sulfate utilization by algae 14: Distribution of adenosine 5'-phosphosulfate (APS) and adenosine 3'-phosphate 5'-phosphosulfate (PAPS) sulfotransferase in assimilatory sulfate reducers. *Plant Sci. Lett* **4**: 301—307
Tsang ML, Goldschmidt EE and Schiff JA (1971) Adenosine-5'-phosphosulfate (APS^{35}) as an intermediate in the conversion of adenosine-3'-phosphate-5'-phosphosulfate ($PAPS^{35}$) to acid volatile radioactivity. *Plant Physiol* **47**: Suppl 20
ul Haque MF, King LM, Krakow D, Cantor RM, Rusiniak ME, Swank RT, Superti-Furga A, Haque S, Abbas H, Ahmad W, Ahmad M and Cohn DH. (1998) Mutations in orthologous genes in human spondyloepimetaphyseal dysplasia and the brachymorphic mouse. *Nat Genet* **20**: 157—162

Ullrich TC, Blaesse M and Huber R (2001) Crystal structure of ATP sulfurylase from *Saccharomyces cerevisiae*, a key enzyme in sulfate activation. *EMBO J* **20:** 316—329

Vauclare P, Kopriva S, Fell D, Suter M, Sticher L, von Ballmoos P, Krähenbühl U, Op den Camp R and Brunold C (2002) Flux control of sulphate assimilation in Arabidopsis thaliana: Adenosine 5'-phosphosulphate reductase is more susceptible to negative control by thiols than ATP sulphurylase. *Plant J* (in press)

Varin L, Marsolais F, Richard M and Rouleau M (1997) Sulfation and sulfotransferases 6: Biochemistry and molecular biology of plant sulfotransferases. *FASEB J* **11:** 517—525

Venkatachalam KV, Akita H and Strott CA (1998) Molecular cloning, expression, and characterization of human bifunctional 3'-phosphoadenosine 5'-phosphosulfate synthase and its functional domains. *J Biol Chem* **273:** 19311—19320

von Arb C and Brunold C (1986) Enzymes of assimilatory sulfate reduction in leaves of *Pisum sativum*: activity changes during ontogeny and in vivo regulation by H_2S and cysteine. *Physiol Plant* **67:**81—86

Walker MA and McKersie BD (1993) Role of the ascorbate-glutathione antioxidant system in chilling resistance of tomato. *J Plant Physiol* **141:** 234—239

Warrilow AGS and Hawkesford MJ (1998) Separation, subcellular location and influence of sulphur nutrition on isoforms of cysteine synthase in spinach. *J Exp Bot* **49:** 1625—1636

Warrilow AGS and Hawkesford MJ (2000) Cysteine synthase (O-acetylserine (thiol) lyase) substrate specificities classify the mitochondrial isoform as a cyanoalanine synthase. *J Exp Bot* **51:** 985—993

Washburn MP and Wells WW (1999) The catalytic mechanism of the glutathione-dependent dehydroascorbate reductase activity of thioltransferase (glutaredoxin) *Biochemistry* **38:** 268—274

Weber M, Suter M, Brunold C and Kopriva S (2000) Sulfate assimilation in higher plants: Characterization of a stable intermediate in the adenosine 5'-phosphosulfate reductase reaction. *Eur J Biochem* **267:** 3647—3653

Westerman S, Stulen I, Suter M, Brunold C and De Kok LJ (2001) Atmospheric H_2S as sulphur source for Brassica oleracea: Consequences for the activity of the enzymes of the assimilatory sulphate reduction pathway. *Plant Physiol Biochem* **39**: 425—432

Wirtz M, Berkowitz O, Droux M and Hell R (2001) The cysteine synthase complex from plants. Mitochondrial serine acetyltransferase from *Arabidopsis thaliana* carries a bifunctional domain for catalysis and protein-protein interaction. *Eur J Biochem* **268**: 686—693

Wray JL, Campbell EI, Roberts MA and Gutierrez-Marcos JF (1998) Redefining reductive sulfate assimilation in higher plants: a role for APS reductase, a new member of the thioredoxin superfamily? *Chem Biol Interact* **109**: 153—167

Wyss H-R and Brunold C (1980) Regulation of adenosine 5′-phosphosulfate sulfotransferase by sulfur dioxide in primary leaves of beans (*Phaseolus vulgaris*). *Physiol Plant* **50:** 161—165

Yamaguchi Y, Nakamura T, Harada E, Koizumi N and Sano H (1999) Differential accumulation of transcripts encoding sulfur assimilation enzymes upon sulfur and/or nitrogen deprivation in *Arabidopsis thaliana. Biosci Biotechnol Biochem* **63:** 762—766

Yonekura-Sakakibara K, Ashikari T, Tanaka Y, Kusumi T and Hase T. (1998) Molecular characterization of tobacco sulfite reductase: enzyme purification, gene cloning, and gene expression analysis. *J Biochem* **124:** 615—621

Yonekura-Sakakibara K, Onda Y, Ashikari T, Tanaka Y, Kusumi T and Hase T (2000) Analysis of reductant supply systems for ferredoxin-dependent sulfite reductase in photosynthetic and nonphotosynthetic organs of maize. *Plant Physiol* **122:** 887—894

Youssefian S, Nakamura M, Orudgev E and Kondo N (2001) Increased cysteine biosynthesis capacity of transgenic tobacco overexpressing an O-acetylserine(thiol) lyase modifies plant responses to oxidative stress. *Plant Physiol* **126:** 1001—1011

8. AFFILIATIONS

Institute of Forest Botany and Tree Physiology, Georges-Köhler-Allee 053, D-79085 Freiburg, Germany (S.K.) and Department of Plant Biotechnology, Faculty of Biology, Schänzlestr. 1, D-79104 Freiburg, Germany (A.K.)

For correspondence – *S. Kopriva*: Stanislav.Kopriva@ctp.uni-freiburg.de

J.W. ANDERSON AND M.A. FITZGERALD

SULPHUR DISTRIBUTION AND REDISTRIBUTION IN VEGETATIVE AND GENERATIVE PLANTS

Abstract. Plants normally acquire sulphur (S) from soil as SO_4^{2-}. It is then loaded into the xylem, and directed in the transpiration stream into the shoot. S is not distributed in proportion to leaf area but is preferentially directed into leaves that are 60-80% expanded. This implies that the short-term delivery of SO_4^{2-} involves extensive xylem/phloem transfer. In S-adequate plants, most of the SO_4^{2-} that is initially delivered to the leaves is gradually redistributed, mostly to younger leaves where it is assimilated into S-amino acids. This results in the formation of long term products, the most important being protein-S, which is not readily re-exported, and GSH. Redistribution of S between leaves is achieved by loading S, principally SO_4^{2-}, into the phloem. This implies the expression of SO_4^{2-} transporters which support net export of S when the leaf is about 70-80% expanded. Organic S is redistributed principally as GSH or SMM in the phloem, and the amount of each depends on the tissue, species and possibly growth conditions. Phloem transport provides organic S to the root system. In S-inadequate plants, newly acquired S is retained in the root and S in existing leaves is not redistributed in response to the demand of young leaves.

Redistribution of endogenous S occurs during various phases of plant development. Germinating seeds redistribute S from storage proteins to the growing seedling in the absence of exogenous S. Recent studies with germinating legumes indicate that sulphate, rather than GSH/hGSH, is the principal form in which protein-S is transported to the seedling, implying that S-amino acids, from the hydrolysis of storage proteins, are catabolised to SO_4^{2-}. Developing seeds require S for the synthesis of storage proteins. The S demand during early seed development in soybean is largely supplied by redistributing pools of SO_4^{2-} and hGSH that accumulated previously in the pod, but later most of the demand is supplied by exogenous S. Developing cotyledons contain both SO_4^{2-} and hGSH. It is yet to be established whether exogenous SO_4^{2-} delivered to the pods is metabolised to hGSH in the pod or in developing cotyledons. In wheat, seed growth normally takes place when water, and hence uptake of exogenous S, is restricted. Under these circumstances, S for grain growth is normally recruited from sources within the plant. Plants which receive adequate S during vegetative growth recruit S principally as SO_4^{2-} from the roots and, to a lesser extent, as GSH from the leaves. SMM has also been found in the phloem sap travelling to wheat ears. Developing grains have active mechanisms for the assimilation of SO_4^{2-} into Cys and Met and catabolism of GSH and presumably SMM. Wheat plants that receive inadequate S during vegetative growth do not contain significant pools of soluble S to support grain growth. Instead S is recruited from protein-S in the leaves, especially the flag leaf. Since GSH accounts for most of the soluble S in both the endosperm and the endosperm cavity it is concluded that protein-S is metabolised to GSH in leaves and transported in the phloem to the endosperm cavity where it is recruited into the endosperm and serves as the source of S for the production of S-amino acids for incorporation into grain proteins.

1. INTRODUCTION

1.1 Overview of Distribution and Redistribution of Sulphur

Plants require sulphur (S) for the production of essential S-containing compounds, the most important being the two protein S-amino acids, cysteine (Cys) and methionine (Met). This necessitates that plants have mechanisms for acquiring S from the environment and delivering it to sites of demand. In vegetative plants, the principal short-term sinks for newly acquired S are the growing tissues at the shoot and root

Yash P. Abrol and Altaf Ahmad (eds.), Sulphur in Plants, 113—134.

apices. In generative plants, S is used principally for reproductive tissues and, most importantly, developing seeds. The importance of short-term delivery of exogenous S to generative tissues varies between species, the availability of water and exogenous S. Some plants, under S adequate conditions, acquire exogenous S during generative growth to satisfy the S demand of generative tissues and developing seeds (e.g. soybean). However, not all plants are able to recruit sufficient exogenous S during generative growth to fulfil the demand of the generative tissues in which case other sources of S must be used.

Internal redistribution of endogenous S within the plant assumes great importance if the supply of exogenous S is unavailable (e.g. during seed germination), or restricted due to low water availability or low availability of S in the soil. Redistribution of S occurs at various stages of the plant's life cycle. For example, germinating seeds grow and produce new tissues in the absence of exogenous S. In this case, the demand for S in the growing seedling is sustained by mobilising protein-S from storage tissues in the seed; this explains the large demand of developing seeds for S to synthesise and accumulate those storage proteins. The seeds of most dryland cereals are usually formed when the availability of water is limited with the result that uptake of exogenous S from the soil is correspondingly limited. Under these conditions, redistribution of S plays a key role in supplying the S required for the synthesis of seed storage proteins. Strictly speaking, the initial short-term delivery of S during normal vegetative growth also involves redistribution. In this instance, newly acquired S is initially transported to the leaf canopy in the transpiration stream and then promptly reloaded into the phloem for redirecting to leaves that are about 60-80% expanded.

1.2 Influence of Sulphur Distribution and Redistribution on the Production of S-rich and S-poor Proteins in Developing Seeds

The amount of protein-S in seeds varies between species. Any one species synthesises a number of seed storage proteins each of which has a characteristic amino acid composition. Within a species, the S content of individual storage proteins varies greatly. For example, in wheat, the first three proteins listed in Table 1 have a high proportion of S-amino acid residues per mol and are regarded as S-rich proteins. On the other hand, high molecular weight glutenin contains relatively few S-amino acids (low-S protein). The seed storage proteins of legumes are collectively deficient in S-amino acids. In soybean seeds, glycinin and β-conglycinin collectively account for about 70% of the total seed protein. The data in Table 1 show that these two proteins are especially low in S relative to N and that β-conglycinin contains no Met.

Plants are unable to vary the S content of specific storage proteins as this is determined by the amino acid composition, which in turn is determined by the gene of the relevant storage protein. However, for any one species, the amount of protein-S in a seed is strongly influenced by the level of S nutrition of the maternal plant. This is because generative plants respond to low S by synthesising more of the low-S seed proteins relative to the high-S proteins during seed development. Since the storage proteins dominate the amount of total protein in seeds, over-expression of the low-S proteins results in seeds with a low S content (Blaygrove et al., 1976; Zhao et al., 1999). Conversely, at high S, developing seeds synthesise more of the high-S proteins relative to the low-S proteins. The nutritional quality of seeds with low protein-S is limited by the level of S-amino acids relative to the other amino acids. Seeds with a high N:S ratio (>32 mol/mol or >14 wt/wt) have a low nutritional rating. Thus, the acquisition of exogenous S and the redistribution of internal S to developing seeds are matters of considerable economic importance. The production of S-rich plant material is important with respect to wool production by sheep since the quality of wool is determined by S-rich proteins (Pickering and Reiss, 1993). Moreover, the S content of cereal seeds affects the processing quality of seeds and flour. In particular, the bread-making quality of flour from wheat is extremely dependent on the level of protein-S as the S-rich proteins determine the extensibility and strength of the dough and thereby the propensity of the dough to rise without collapsing (Zhao et al., 1999). These considerations show that distribution/redistribution of S is not only an important physiological process but is also relevant to the production of various commercially important commodities.

Table 1. Sulphur and sulphur-amino acid composition of some storage proteins in seeds of wheat and soybean

Species	Protein	Amino acid residues			N:S ratio	
		Total	Cys	Met	Mol/mol	Wt/wt
Wheat[1]	LMW-Glutenin	354	8	6	39.3	17.2
	α/β-Gliadin	243	5	2	47.0	20.6
	γ-Gliadin	292	9	6	27.3	11.9
	HMW-Glutenin	842	4	3	170.1	74.4
Soybean[2]	Glycinin	2534	16	25	88	38.5
	β-Conglycinin	1292	5	0	136	59.5

Sources: [1]Anderson and Fitzgerald (2001); [2]Moreira et al. (1979); [2]Thanh and Shibasaki (1977)

2. SHORT AND LONG DISTANCE PATHWAYS FOR THE INITIAL DISTRIBUTION OF EXOGENOUS SULPHUR

Plants acquire S from soil as SO_4^{2-} via SO_4^{2-} uptake mechanisms (Hawkesford et al., this book). SO_4^{2-} is not normally subjected to extensive metabolism in the root but is directed into the xylem and thence conducted into the leaf canopy in the transpiration stream (Clarkson et al., 1993). Most species of plants examined contain several SO_4^{2-} transporters in the root. They have high affinity for SO_4^{2-} (K_mc. 10 μM) and are strongly expressed if the level of S nutrition is low (Hawkesford and Smith, 1997; Hawkesford, 2000). The role of the various SO_4^{2-} transporters in the root is yet to be established (Hawkesford, 2000) but the acquisition of exogenous SO_4^{2-} by the root cortex and loading of SO_4^{2-} into the xylem are just two sites in the root that could involve separate SO_4^{2-} transporters.

Theory predicts that SO_4^{2-} is delivered in the transpiration stream to the various organs of the shoot in proportion to the rate of evaporation of water. However, when $^{35}SO_4^{2-}$ is supplied to the root system of vegetative plants grown in nutrient culture the amount of ^{35}S-label found in the various structures in the shoot is not related to the surface area. Rather, most of the label is associated with leaves that are approaching full expansion (Smith and Lang, 1988; Adiputra and Anderson, 1992; Sunarpi and Anderson, 1996a). This implies that SO_4^{2-} in transit in the xylem must be reloaded into the phloem to reach the expanding leaves rather than leaves which are already fully expanded. Consistent with this, Bonas et al. (1982) found that 60-80% of the ^{35}S-label supplied to mature leaves as $^{35}SO_4^{2-}$ was transported basipetally as SO_4^{2-}. Thus, the initial delivery of SO_4^{2-} to sites of demand must involve both xylem and phloem transport.

The sites and mechanisms involved in reloading SO_4^{2-} from the xylem into the phloem have not been identified. Most species examined contain several SO_4^{2-} transporters and, as already noted, high affinity transporters are strongly expressed in the root. Other SO_4^{2-} transporters with a lower affinity (K_m c. 100 μM) are expressed in shoots e.g. SHST3 in *Stylosanthes hamata* (Hawkesford and Smith, 1997; Hawkesford, 2000). Transporters such as SHST3 are prime candidates for roles in xylem/phloem SO_4^{2-} transfer in shoots and await evaluation in this regard. In view of the well established role of transfer cells at the base of the leaf trace in various xylem/phloem transfer processes (Gunning, 1977), these cells would seem to be a likely site for SO_4^{2-} transfer.

3. REDISTRIBUTION OF S FROM LEAVES OF VEGETATIVE PLANTS

3.1 Pulse/chase Labelling Studies with $^{35}SO_4^{2-}$

Much can be learnt about the fluxes of S into and out of leaves by pulsing hydroponically-grown plants with $^{35}SO_4^{2-}$ at various stages of development and observing the long-term fate of the ^{35}S-label within the plant after terminating the pulse. In barley, relatively little label was delivered to mature leaves during a 48h pulse period (Adiputra and Anderson, 1992). Leaves, which were about 70-80% expanded were the strongest net importers during the pulse period, presumably as a consequence of xylem/phloem transfer at the base of fully expanded leaves. During the ensuing chase period, these heavily labelled leaves lost as much as 75% of their label over the life of the leaf. Leaves in the very early stages of expansion also contained a disproportionately large amount of label at the end of the pulse but unlike leaves that are 60-80% expanded they experienced relatively little loss of label during the chase.

Similar results to those described for barley were obtained with labelling studies in soybean (Sunarpi and Anderson, 1996a). Importantly, a balance analysis of the ^{35}S-label showed that the increase in label in the young leaves during the chase could be ascribed mostly to the import of label from leaves which acquired their label during the pulse. As in barley, the label imported by young leaves during the chase was less readily redistributed when the young leaves eventually approached full expansion. Collectively, the studies on barley (Adiputra and Anderson, 1992) and soybean (Sunarpi and Anderson, 1996a) suggest that, for plants supplied with adequate S, SO_4^{2-} is preferentially delivered to leaves that are 60-80% expanded. The S acquired by these leaves does not enter into the long-term pool of S within the leaf and is slowly redistributed to other organs in the shoot. The data also suggest that whereas the S reaching the younger leaves follows a more indirect route and that the S acquired during early leaf development is much less readily redistributed. Presumably this is because the S reaching the leaves at this time is built into long-term products, like leaf proteins, which persist for the life of the leaf whereas this is not the case for S acquired in the later stages of leaf expansion.

3.2 Net Gains and Losses of S in Leaves in Relation to Leaf Growth

Pulse/chase labelling studies provide information about the short-term delivery of S at a particular time and the movement of S to other parts of the plant after that time but they do not provide information on the S status of individual plant parts during plant growth. ICP-OES techniques have been used to make detailed studies of the gains and losses of S in the various plant organs as they grow and mature. In vegetative soybean maintained at a constant level of S nutrition, the S content of a leaf increases rapidly until the leaf attains 40% expansion. Thereafter, the total S in the leaf continues to increase slowly until the leaf is 70% expanded, after which the total S commences to decline, falling by about 50% as leaves reach full expansion (Sunarpi and Anderson, 1996a, 1996b). Thus, the time of the greatest direct short-term delivery of exogenous S as determined by labelling studies, coincides with the onset of net S export, a process which is initiated well before the leaf is fully expanded.

Two most important sources of S in leaves are protein-S and SO_4^{2-} within the vacuole. Protein-S is required for leaf function and can only be withdrawn pursuant to protein hydrolysis, a process that would impair leaf function. In soybean plants grown with adequate S and pulsed with $^{35}SO_4^{2-}$, virtually all of the ^{35}S exported from leaves is derived from the soluble fraction and all of the label in the soluble fraction is eventually exported (Sunarpi and Anderson, 1996b). Since SO_4^{2-} accounts for about 80% of the soluble S and it too was quantitatively exported it was concluded that the SO_4^{2-} delivered to leaves, when they are 60-80% expanded, is stored in the vacuole and serves as a slow-leaking, long-term source of S for redistribution to developing leaves. Bell et al. (1990, 1994) reached similar conclusions with respect to the slow redistribution of vacuolar SO_4^{2-} in leaves of *Macroptilium*. This theory is supported by measurements of a low rate of efflux of SO_4^{2-} across the vacuolar membrane (Cram, 1983).

The studies described above lead to the general conclusion that leaves that are 60-80% expanded are very strong sinks for the short term delivery of S from the nutrient solution but very little of the newly imported S enters the metabolically active pool of S within the leaf and is most likely directed, as SO_4^{2-}, into the vacuole. At the very time that import of S is the greatest (leaves 60-80% expanded), the S content of leaves begins to decline implying net export of S and strong expression of one or more SO_4^{2-} exporters in the leaf at this time. The exported S is most likely directed in the phloem to immature leaves (less than 30% expanded).

The studies of Datko et al. (1978) (see Giovanelli et al., 1980 for a review) provide a rationale for many of the observations noted above. They showed that SO_4^{2-} only accumulates in plant tissues after satisfying the demand for the production of essential organic metabolites, protein-S in particular. Thus, with respect to leaf development, the first phase presumably involves the importation of S in the phloem, mostly as SO_4^{2-} but also as GSH, from more mature leaves. This is used for producing organic S compounds that are essential for growth (mostly protein-S). As the leaf expands, it commences to import exogenous SO_4^{2-}, some of which is promptly exported to younger leaves. If the S supply to the plant is adequate, some of the newly imported SO_4^{2-} in the leaf is loaded into vacuoles. By the time the leaf is about 80% expanded, the rate of export exceeds the rate of import and loading of SO_4^{2-} into vacuoles begins to decline, ceasing shortly after the leaf attains full expansion. Thereafter, any vacuolar SO_4^{2-} remaining in the leaf is gradually exported over the life of the leaf. In conditions of long-term S deficiency, loading of vacuoles with SO_4^{2-} does not occur, and perhaps also, the younger leaves (30% expanded) are the first sink in the shoot for newly acquired SO_4^{2-}. The response of leaves to the sudden imposition of S stress is discussed in Section 4.1.

3.3 Forms of S Transported in Plants

The foregoing discussion indicates that SO_4^{2-} is quantitatively the most important form of S transported in plants, both for the initial distribution in the xylem and subsequent redistribution in the phloem. Protein-S in the storage tissues of legume seeds is also redistributed principally as SO_4^{2-} when the seeds germinate (see Section 5.1). Further, SO_4^{2-} is found in the endosperm cavity of developing wheat grains (Fitzgerald et al., 2001), an apoplastic compartment which is serviced only by phloem. As already noted, one or more of the SO_4^{2-} transporters that occur within a species could be involved in phloem loading at various sites within the plant.

Transport of organic forms of S in the phloem has received a lot of attention. These have mostly involved flap feeding [^{35}S]Cys and $^{35}SO_4^{2-}$ to leaves and identifying the form of the ^{35}S-labelled compound(s) at another site in the transport pathway. Using this approach Rennenberg et al. (1979) found that 70% of the ^{35}S-label in organic compounds in the stem and petiole occurred as GSH. Met was also labelled but the amount of label associated with Cys was negligible. Similarly, when $^{35}SO_4^{2-}$ was supplied to mungbean most of the organic ^{35}S-label occurred as hGSH (MacNicol and Bergman, 1984). When [^{35}S]Cys is supplied, very little label remains associated with Cys and GSH/hGSH is the main organic S product (Herschbach and Rennenberg, 1995). As a result of studies such as these and the fact that Cys accounts for an insignificant amount of S in phloem, there is consensus that Cys is not a significant transport compound, even in germinating seeds where large amounts of free Cys are produced by hydrolysis of seed storage proteins (see Section 5.1). GSH/hGSH and Met are invariably detected in phloem and are presumed to be physiologically important. These metabolites also occur in tissues serviced by phloem such as the vascular crease of the developing seeds of cereals (Fitzgerald et al., 2001). Generally speaking, GSH/hGSH and Met are normally quantitatively less important for long distance transport of S compared to SO_4^{2-} but the transport of GSH to developing grains in S-deficient wheat (section 6.2.4) and to roots (section 3.4) are important exceptions.

SMM, has recently been identified in many flowering plants and in many economically and nutritionally important crops. In generative wheat, SMM accounted for more organic S in phloem than GSH (Bourgis et al., 1999). Furthermore, when [^{35}S]Met was supplied, the label was detected predominantly in SMM. Bourgis et al. (1999) conclude that, for technical reasons, SMM is a previously unrecognised transport form of S that contributes significantly to the S budget of plants. Considering evidences indicating that GSH and SMM are major forms of organic S that are transported in the phloem, and the dynamics of source, supply and sink demand govern phloem transport, it is attractive to hypothesise that GSH is the transport mechanism for exporting Cys, and SMM is a mechanism to export not just methyl groups, but Met. An ideal system to test this would be in germinating seeds where the only source of S is from Cys and Met from the hydrolysis of storage proteins. Pimenta et al. (1998) measured the level of the gene for MMT, the enzyme that synthesises SMM, in germinating barley from day 0 until day 7. They found that expression of the gene increased significantly over that time, as did the amount of SMM. Dethier et al. (1991) conducted a similar experiment in germinating wheat and found a similar result for SMM content. Both those studies approached SMM metabolism from the perspective of methyl group metabolism, so it is not yet known whether SMM is a regulated transport form of S.

3.4 Shoot/root Recycling of S

Roots have a considerable demand for assimilated S especially if they are challenged with a heavy metal which induces the synthesis of S-rich PCs (McMahon and Anderson, 1998). Roots exhibit some SO_4^{2-} assimilation activity (Bell et al., 1994, 1995a) but leaves are regarded as the primary site of SO_4^{2-} assimilation (Anderson, 1990). Accordingly, Cooper and Clarkson (1989) proposed that the demand of the roots for organic S could be met by importing it in the phloem from the leaves. Studies of shoot to root movement of endogenous S (Herschbach and Rennenberg, 1994, 1995; Bell et al., 1995a) support the role of the shoot in S recycling. To test this idea directly, Larsson et al. (1991) supplied $^{35}SO_4^{2-}$ to a small proportion of the root system of wheat plants grown in nutrient culture. They found that ^{35}S-label gradually accumulated in the untreated portion of the root system. Sunarpi and Anderson (1998) provided further support for the recycling hypothesis. They pulsed part of the root system of soybean to label the leaves and then removed the treated part of the root system to prevent root to root movement of label. In the ensuing chase, the remainder of the root system became labelled, a process that was inhibited by removing some of the leaves. GSH would seem a likely candidate as an internal signal in view of the many demonstrations of GSH transport in phloem (Rennenberg, 1987). Indeed, Lappartient et al. (1999) have shown that GSH content signals to the root the S status of the shoot in *Brassica napus*. In *Zea mays* this role is played by Cys (Bolchi et al., 1999). In barley, OAS applied to S-sufficient and S-deficient roots caused a rapid increase in the SO_4^{2-} transport activity and amount of transcript for the high affinity transporter (Smith et al., 1997). Also in barley, when S-deprived plants were exposed to either Cys or GSH, GSH was much more effective in decreasing the amount of transcript for the SO_4^{2-} transporter hvst1 (Vidmar et al., 1999). In leaves, OAS accumulates under S deficiency and GSH accumulates in S adequacy; evidence suggests that both can perform positive and negative signalling functions to the root that regulate transcription for S transporters.

4. EFFECT OF S AND N STRESS ON THE DISTRIBUTION AND REDISTRIBUTION OF S IN VEGETATIVE PLANTS

4.1 Effect of S Stress

As noted in the foregoing sections, the major pools of insoluble-S and vacuolar SO_4^{2-} within the leaf of plants grown with adequate S are not readily redistributed. This raises the question whether the imposition of S stress enhances or accelerates the redistribution of S from these pools and whether a decrease in the availability of exogenous SO_4^{2-} influences the short term distribution of S to the various organs of the plant. Using the labelling procedures described in section 3.1, Adiputra and Anderson (1995) concluded that the export of S from leaves is determined by the S status of the donor leaf and not by the demand of S sinks. Consistent with this, some of the label acquired by very young leaves of S adequate plants was re-exported but this was not the case for S-inadequate plants. In S-adequate plants, SO_4^{2-} would occur in the vacuoles of leaves approaching full expansion and in the root system (Giovanelli et al. 1980). Mobilisation of vacuolar SO_4^{2-} is generally slow from root vacuoles (Bell et al., 1994, 1995a), slower from leaf vacuoles (Bell et al., 1995b) and particularly slow in *Brassica napus* (Blake-Kalff et al., 1998). Prosser et al. (2001) induced S stress in spinach plants that were previously not S stressed. In those plants, growth in the absence of exogenous SO_4^{2-} was supported by internal pools of SO_4^{2-} first from roots, then from young leaves and after some delay, from mature leaves. Collectively, these studies show that in most cases plants do not respond to the demand for S for the production of new vegetative organs in the shoot by withdrawing S from pre-existing organs. Mobilisation of protein-S in the cotyledons of germinating legumes is also not regulated by the S status of the plant (see also section 5.1).

The level of S nutrition has a strong influence on the retention of S in the root. When plants are grown hydroponically with adequate S and pulsed for 48 h with $^{35}SO_4^{2}$, a large amount of label is retained in the root at the end of the pulse (Adiputra and Anderson, 1992; Sunarpi and Anderson, 1996a). Typically about 40% of this label moves out of the root over the ensuing 10 days and the amount remaining in the root then stabilises. However, the root seems to exercise considerable control over the amount of S that is retained and subsequently made available for export. When plants, which have been grown at very low concentrations of SO_4^{2-}, are pulsed for 48 h with $^{35}SO_4^{2-}$, most of the label is retained in the root, moving into the insoluble fraction (Sunarpi and Anderson, 1996b). However, if the plants are supplied with an adequate amount of S after the pulse, virtually all of the label in the soluble fraction of the root is exported and virtually none is incorporated into the insoluble fraction within the root (Sunarpi and Anderson, 1996b). This demonstrates that export of S from the root is influenced by the level of S nutrition.

4.2 Effect of N Stress

Although S stress does not initiate remobilization of protein-S in the leaves of vegetative plants, it is well known that N stress initiates mobilisation of protein-N in mature leaves and that the resulting products are transported to sites of N demand, young expanding leaves in particular (Pate, 1980; Guitman et al., 1991; Hocking and Steer, 1995). Since proteins contain S as well as N, this raises the question whether N stress (but not S stress) initiates mobilisation of protein-S as well as protein-N. Sunarpi and Anderson (1997b, 1997c) found that the S content of leaves was strongly influenced by N nutrition. When plants were initially grown with adequate N and a marginally adequate level of S, the loss of total S from mature leaves, as well as the loss of total N, was greatly enhanced by subsequently lowering the level of N nutrition. This is consistent with the known coordination between N and S metabolism (Dijkshoorn and van Wijk, 1967). When the supply of N exceeds the supply of S to the degree that it exceeds the ratio of N and S in leaf protein, nitrate accumulates and is transported into vacuoles. Decreasing N nutrition would establish the conditions for export of N from mature leaves to meet the requirement of younger leaves. Protein-N would be hydrolysed to meet the demand of young leaves (releasing S) since transport of nitrate across the tonoplast in slow. Sunarpi and Anderson (1997b, 1997c) found that the amount of total S in the youngest leaf was greatly enhanced by lowering the level of N nutrition. This would restore a balance of N and S in an actively metabolising leaf to a ratio consistent with that of leaf protein (Dijkshoorn and van Wijk, 1967). It would appear therefore that vegetative plants remobilize protein-S as a passive consequence of N stress rather than a regulated response to S stress.

Limiting N nutrition to S deficient plants has recently been shown to have an interesting effect on S metabolism, and consequently opportunities for S redistribution. When plants are grown in S deficient conditions, ATP sulphurylase and APS reductase activity increases (Brunold, 1993) correlated with an increase in their transcripts (Yamaguchi et al., 1999) and an accumulation of OAS. Limiting the N nutrition blocked the accumulation of those transcripts and decreased the amount of OAS (Yamaguchi et al., 1999), thereby removing the positive response to S deficiency. Leustek et al. (2000) suggest that the accumulation of OAS could act as a signal for insufficient SO_4^{2-} reduction; indeed application of OAS to *Lemna minor* caused an increase in APS reductase activity and the same response was detected in barley. Collectively, these studies indicate that a nitrogen-containing compound, likely to be OAS, is involved in de-repression of sulphur transport and metabolism.

5. MOBILISATION OF S IN GERMINATING SEEDS

5.1 Germinating Legume Seeds

Many plants raised from seed can grow for several weeks in the absence of exogenous sulphur. Soybean seedlings raised from seed in the absence of exogenous S provide an excellent system for studying S redistribution since they can be grown to the second trifoliate leaf stage with no net gain or loss of total S (Sunarpi and Anderson, 1995). As the plant grows, the total amount of S in the cotyledons declines and this loss is accompanied by a quantitative gain of S in the developing seedling, thereby demonstrating net redistribution of S from the cotyledon to the seedling. In most plant species, nearly all of the S in unimbibed seeds occurs as protein-S. For example, in soybean 91% of the total S in the seeds occurs as Cys and Met residues in the ethanol insoluble fraction (Crouch, 2002). In the initial stages of germination, the amount of insoluble S in the cotyledons declines rapidly. This is accompanied by an increase in soluble S in the cotyledons. However, after a few days the amount of soluble S in the cotyledons also decreases (Sunarpi and Anderson, 1995). These data imply that protein-S in the cotyledons is hydrolysed and metabolised to soluble S-transport compounds which are conducted into the seedling and, used for the synthesis of S-containing compounds essential for seedling growth.

Recently, Herschbach et al. (2002) reported that the cotyledons of 8-day old pea seedlings contain extremely high levels of Cys relative to other organs (Table 2). Normally, the concentration of free Cys in plant tissues is very low as shown by the data in Table 2 for the first developed leaf of pea seedlings. This is because the SO_4^{2-} assimilation pathway, which leads to the synthesis of Cys is tightly regulated (Brunold, 1993; Hawkesford, 2000). However, the very high levels of free Cys in the cotyledons of germinating peas would not be from synthesis, but the result of a largely unregulated proteolysis of seed storage proteins. As discussed below, free Cys is not an important S-transport compound in germinating seeds/seedlings.

Until recently, it has been commonly assumed that GSH and/or one or more of its homologues (e.g. hGSH) is the principal transport compound in germinating seeds since GSH has long been recognised as a significant S-transport compound in mature plants (Rennenberg, 1987). Moreover, metabolism of protein-S to GSH (and/or its homologues) would maintain the S in an organic form, easily metabolised in sink tissues. However, several recent studies suggest that protein-S is mostly redistributed as SO_4^{2-} or SMM. Pimenta et al. (1998) propose that SMM is a transport form of methyl moieties in germinating seeds, suggesting that it is involved in S transport by default. Herschbach et al. (2002) found that 50% of the exogenous [^{35}S]Cys supplied to the cotyledons of 8-day old pea seedlings was metabolised to SO_4^{2-} and was largely transported to the developing seedling in this form. When exogenous $^{35}SO_4^{2-}$ was supplied in lieu of [^{35}S]Cys, very little label was recovered from the cotyledons as organic S although a large amount of label was detected in the various tissues of the seedling confirming the role of SO_4^{2-} transport in this system. Crouch (2002) examined changes in endogenous S during germination of soybean and found that SO_4^{2-} accounted for most of the increase in the large transient rise in soluble S in the cotyledons of germinating soybeans as protein-S was metabolised. Collectively these studies suggest that the storage tissues of germinating seedlings have mechanisms for the catabolism of both endogenous and exogenous Cys to SO_4^{2-} which was partially inhibited by aminooxyacetate, an inhibitor of Cys desulphhydrase suggesting a possible role for this enzyme (Rennenberg, 1983).

The data in Table 2 indicate that although pea cotyledons contain exceptionally high concentrations of Cys, GSH is nonetheless even more abundant. Further, Herschbach et al. (2002) found that while most of the exogenous [^{35}S]Cys was metabolised to SO_4^{2}, and was found mostly in the tissues of the seedling in this form, about 15% of the ^{35}S-label was detected in both the cotyledons and tissues of the seedling as GSH. Given that Cys is a major metabolite in the cotyledon, the labelling data imply that a significant amount of protein-S is transported as GSH from cotyledons to the developing tissues in germinating peas. Considering that SMM assumes a role in S transport in germinating barley seeds, either active or passive (Pimenta et al., 1998), it would be useful to extend the work of Herschbach et al. (2002) and study the fate of exogenous [^{35}S]Met in

Table 2. Concentrations of Cys, GSH and SO_4^{2-} in the cotyledons and in the first developed leaf of 8-day old pea seedlings

Tissue	Cys (nmol g^{-1} FW)	GSH (nmol g^{-1} FW)	SO_4^{2-} (µmol g^{-1} FW)
First leaf	10.1 +/-2.3	992.3 +/- 170.5	17.18 +/- 2.06
Cotyledons	174.4 +/- 22.1	312.0 +/- 76.2	8.40 +/- 2.03

Source: Herschbach et al. (2002)

germinating legume seeds.

In soybean, hGSH serves in place of GSH and the hGSH accounts for most of the soluble organic S in the cotyledons of unimbibed seeds. The amount of this metabolite in the cotyledons falls progressively to zero as the seeds germinate (Crouch, 2002). This is matched by a corresponding rise and fall in hGSH of seedling, consistent with the possibility that hGSH acts as a transport compound. BSO, an inhibitor of hGSH synthetase, enhanced the rate of decline in the steady state pool of hGSH in the cotyledons implying turnover of hGSH as hGSH declined. However, BSO had no effect on the total redistribution of S from the cotyledons to the developing seedling or on seedling growth (Crouch, 2002). This suggests that hGSH does not act as an important transport compound for the redistribution of protein-S from soybean cotyledons.

Sunarpi and Anderson (1995) reported that the total S content of soybean cotyledons decreased by about 92% regardless of whether the seeds/seedlings were grown with or without SO_4^{2-} even though the total S content of the seedlings that received SO_4^{2-} was about 100% greater. This suggests that the mobilisation of protein-S in germinating seeds is not regulated by S demand.

5.2 Germinating Cereal Seeds

Hydrolysis of the reserves in the cereal endosperm is initiated by secretion of hydrolytic enzymes from the aleurone and the epithelial layer of the scutellum. The hydrolysis products are then absorbed via the scutellum and conducted in the phloem to the developing seedling. Moussavi-Nik et al. (1998) reported that the S content of endosperm declined by 85% within 7 d of imbibition and this decrease was matched by a similar increase in the seedling. Rauser et al. (1991) studied the transfer of S from the endosperm to the seedling in maize by injecting [^{35}S]Cys into the endosperm. They found label associated with GSH and SO_4^{2-} in the scutellum and with GSH, but not $^{35}SO_4^{2-}$ in the root. In the shoots, some ^{35}S-label was associated with SO_4^{2-} but GSH was the major labelled product. SMM is another transport form of organic S, albeit by default, since it is probably subject to regulation by requirements and metabolism of methyl groups (Pimenta et al., 1998). In that work, expression of the gene for MMT which catalyses the synthesis of SMM, increased up to the 7th day after germination. Most activity was in the root and shoot, not in the endosperm (Pimenta et al., 1998). It appears as though the strategy of exporting S as GSH and SMM requires the spatial separation of protein hydrolysis and the production of free S-amino acids (in the endosperm) from the anabolic process of synthesising GSH and SMM from the products of protein hydrolysis. Whilst enzymes for the production of GSH from free Cys have been identified in plants (Noctor et al., 1998), the mechanism for the production of Cys from Met which involves the enzyme γ-cystathionase, is speculative. The synthesis of SMM has been determined and established in plants (Giovanelli et al., 1980), but whether it contributes actively or by default to S transport remains to be established.

6. SOURCES OF SULPHUR FOR SEED DEVELOPMENT

Legumes and cereals show important differences in the production of generative tissues in relation to the formation of leaves. Whereas cereals flower and commence grain filling after the production of leaves is complete, legumes flower and commence grain filling in the pods at the lower axils while continuing to grow and form new leaves and additional reproductive structures in the upper leaf axils. This results in very different patterns of S demand and S redistribution in relation to plant development.

6.1 Sources of S for Seed Development in Legumes

6.1.1 Acquisition of exogenous S in relation to generative growth

Most of the S acquired by generative soybean plants takes place after the onset of generative growth. The amount of S in the root system and stem increases throughout generative growth but the developing seeds account for most of the net gain of S at this time (Sunarpi and Anderson, 1997a). Thus, most of the S acquired (as SO_4^{2-}) during generative growth is allocated, directly or indirectly, to seed growth and to the production of seed storage proteins. One important consequence of this is that soybeans and presumably other legumes have an ongoing demand for exogenous S throughout generative growth, and an interruption to the S supply will have deleterious effects on the production of protein-S in the seeds. The leaves account for most of the S at the beginning of generative growth. As generative growth begins, the S content of the leaves increases slightly but then declines as the rate of importation of S by developing seeds also decreases. However, the amount of S lost from the leaves collectively could account for no more that 15% of the gain of S during the final phase of seed growth. The pods exhibited net loss of S during early seed development, sufficient to account for some of the increase in the seeds (see below).

6.1.2 Origin and forms of S imported into developing seeds

Although the increase in S in the seeds is not balanced by net loss of S from any other part of the plant for most of the duration of generative growth, the events at an individual axil are of some interest. The onset of seed development is accompanied by a rapid net loss of soluble S, but not insoluble S from both the pod and the leaf at the same axis (Sunarpi and Anderson, 1997a). These losses amount to as much as 40% of the gain of S in the seeds at the same axil during the first 15-20 days of seed growth suggesting that the leaf and pod at the same axil form important short term reservoirs of

S for the early phase of seed development. Most of the loss of soluble S from the leaf involved the quantitative net loss of SO_4^{2-}. Conversely, the steady state pool of hGSH underwent little change, though at the onset of generative growth, the amount of hGSH in the leaves was typically about 7-10% of the initial amount of SO_4^{2-}. The pods initially contained approximately equal amounts of SO_4^{2-} and hGSH. The amount of SO_4^{2-} declined rapidly to undetectable levels. The amount of hGSH in the pod declined by about 60% during seed formation but it was still present in the pod in the final stages of seed development (Sunarpi and Anderson, 1997a). These data do not establish whether SO_4^{2-} is imported directly from the pod into the seed or whether it is first metabolised to hGSH in the pod. The amount of hGSH in developing grains is typically 7-10 times the amount of SO_4^{2-} for most of the period of grain development. This could signify that hGSH is the major form in which S is imported into the developing seed or, alternatively, that hGSH is essential for some other function (e.g. redox poise). During the later stages of generative growth, the most likely scenario is that newly acquired SO_4^{2-} is transported from the root in the xylem and selectively unloaded at the appropriate leaf axil into a phloem trace that leads to the pod. Some or all of the SO_4^{2-} is reductively assimilated into Cys and thence into hGSH in the pod which is then delivered in the phloem into the seed coat where it is recruited from the embryo sac by the developing cotyledons. However, the possibility that S is also imported mostly as SO_4^{2-} can not be ruled out particularly since SO_4^{2-} as well as Met serves as a S source for the synthesis of seed proteins in soybean cotyledons cultured in vitro (Holowach et al., 1984). Nonetheless, SO_4^{2-} is less effective than Met for the synthesis of seed protein in vitro (Holowach et al., 1984) consistent with the possibillity that organic metabolites such as hGSH serve as the principal form of S imported from the pod. In lupins, SO_4^{2-} is main source of S for seed growth and SO_4^{2-} is a major contributor of S for grain protein (Tabe and Droux, 2001).

6.1.3 GSH uptake mechanisms

If GSH/hGSH acts as an important S source for developing legume seeds, they must have a mechanism for recruiting GSH/hGSH from the embryo sac. Little is known about transmembrane transport of GSH and a consensus has yet to be reached concerning the properties of GSH transporters, let alone their role in the uptake of GSH by developing seeds. Schneider et al. (1992) have reported mechanisms with high and low affinity for GSH in cultured tobacco cells and Jamai et al. (1996) have reported a GSSG proton-cotransport mechanism in bean mesophyll cells which also shows activity towards GSH. Since these mechanisms are associated with the plasma membrane, they may have some relevance to the uptake of GSH/hGSH by developing legume seeds.

6.2 Sources of S for Grain Filling in Cereals

6.2.1 Importance of endogenous S with respect to generative growth

Cereals, unlike soybean, have a terminal inflorescence which forms subsequent to the production of the leaves. Although cereals can continue to recruit exogenous S and other nutrients during generative growth (Smith and Whitfield, 1990; Larsson et al.,

1991; Fitzgerald et al., 1999a), this is usually restricted to irrigated crops and rarely occurs under field conditions. Most economically important cereal crops, wheat in particular, are grown under dryland conditions and commonly set and develop seed under conditions of diminishing rainfall. Thus, uptake of exogenous S becomes severely restricted at the very time that S is required to support the synthesis of grain storage proteins. Under these conditions, S for grain production must be derived largely from sources acquired during vegetative growth. This raises various difficulties about the redistribution of S, since the S associated with protein-S and vacuolar SO_4^{2-}, the two major pools of S in vegetative plants, do not readily respond to a demand for S by metabolically active vegetative sinks (section 4.1). Some of the issues outlined above can be investigated experimentally by manipulating the amount and availability of S nutrition during vegetative and generative growth (Fitzgerald et al., 1999a, b, 2001). It would be expected that the smaller the nutrient supply and the sooner it is terminated the smaller the pool of internal S available for redistribution during generative growth. Changes in the pools of S in both the vegetative and generative tissues can then be measured during grain-filling.

6.2.2 Influence of S nutrition and the time of terminating the exogenous S supply on the recruitment of S from vegetative tissues for grain growth

The vegetative tissues of wheat plants supplied with adequate S during vegetative growth as described in section 6.2.1 (see also legend to Table 3) experienced much larger losses of S during generative growth than plants that received inadequate S (Table 3). In general, for S-adequate plants, the later the S supply was terminated, the larger the total net loss of S from the vegetative tissues, reflecting the greater time for

Table 3. Losses of soluble and insoluble S from the vegetative tissues and gains of soluble and insoluble S in the grains of S-adequate and S-inadequate wheat plants between 13 and 25 days after anthesis (dpa). Plants were grown throughout vegetative growth in nutrient solution containing 20 µM SO_4^{2-} (S-inadequate) or100 µM SO_4^{2-} S-adequate). During generative growth the nutrient supply was discontinued at the times of physiological development indicated in the table.

Form of S	**Time nutrients discontinued**	**Loss of S from vegetative tissues (µmol plant^{-1})**		**Gain of S in grains (µmol plant^{-1})**	
		Adequate S	Inadequate S	Adequate S	Inadequate S
Soluble	Booting	3	2		
	Emergence	26	3		
	Anthesis	45	1		
	8 dpa	41	7		
Insoluble	Booting	27	24	28	25
	Emergence	8	14	42	20
	Anthesis	1	16	52	20
	8 dpa	7	11	50	22

Source: Anderson and Fitzgerald (2001)

the plant to acquire S. These losses were approximately balanced by gains of S in the grains, indicative of internal redistribution of S. The data in Table 3 show that most of the redistributed S in S-adequate plants was derived from soluble S in the vegetative tissues except when the S supply was terminated during early generative growth (booting) when most of the redistributed S was derived from insoluble S. Thus, wheat plants which receive adequate S up to anthesis can recruit all of the S required for grain growth by redistributing S from soluble pools within the vegetative tissues of the plant.

6.2.3 Sources of soluble S imported into developing grains in S-adequate plants

A balance analysis of the various soluble metabolites in each of the tissues in S-sufficient plants shows that 50-70% of the soluble S acquired by grains is derived from the root, almost entirely as SO_4^{2-}, and the balance from the leaves, mostly as GSH (see Anderson and Fitzgerald, 2001, for an overview). This suggests, therefore that the demand for S in the grains promotes the redistribution of SO_4^{2-} from the roots and of GSH from the leaves. The relative contributions of SO_4^{2-} and GSH in the roots and the leaves towards the redistribution of S to the grains is of some interest. Whereas SO_4^{2-} accounted for most of the S in the roots and this was almost entirely redistributed to the grains, SO_4^{2-} accounted for about 60-70% of the soluble S in the leaves but, unlike the root, virtually none of the leaf SO_4^{2-} was redistributed. Other data also hint at this possibility. Although the form of S was not characterised, the study of Adiputra and Anderson (1995) also showed no evidence of movement of ^{35}S-label from leaves in response to S demand in young leaves. Also, in vegetative plants, the alleviation of S insufficiency by adding unlabelled SO_4^{2-} to plants previously supplied with $^{35}SO_4^{2-}$, initiated immediate redistribution of ^{35}S-label from the root (Sunarpi and Anderson, 1996b), most likely as SO_4^{2-}. It therefore appears that whereas SO_4^{2} in the vacuole of mesophyll cells is slowly made available, consistent with the low permeability of the tonoplast (Cram, 1983), SO_4^{2} redistribution from the roots appears to be much more responsive to the prevailing circumstances, be it the S demand of generative tissues or the alleviation of S stress in vegetative plants. The vegetative tissues of S-inadequate plants contain very small pools of soluble S with the result that the loss of soluble S from vegetative tissues was very small except in those plants, which continued to receive inadequate amounts of S until 8 days after anthesis.

6.2.4 Redistribution of insoluble S from vegetative tissues to developing grains

The data in Table 3 show that most of the limited amount of S imported by developing grains of S-inadequate plants is derived from the insoluble fraction (i.e. from protein-S). This implies that vegetative tissues, in the absence of a sufficient pool of soluble S to meet the S demand of developing grains, respond by mobilising protein-S and redistributing it to the grains. If the supply of S to S-adequate plants is terminated at booting, the plants respond the same way as S-inadequate plants (Table 3), reflecting the shorter time for the plants to acquire reserves of soluble S. An analysis of the net changes of the insoluble fraction of the leaves shows that the amount of protein-S in S-adequate plants remains constant or even increases in those plants that continue to receive S until well into grain formation. However, in S-inadequate plants, the flag leaf was an important contributor of protein-S for grain growth (Fitzgerald et al., 1999a). The flag leaf was also an important contributor of protein-S in S-adequate plants when the S supply was terminated at booting. The other leaves exhibited small net losses in both S-adequate and S-inadequate plants. The selective withdrawal of protein-S from the flag leaf could be interpreted as a strategy to maintain the carbohydrate/protein balance of grains in S-stressed plants. The flag leaf is normally an important source of sucrose for starch synthesis in the developing grain and mobilising protein-S from the flag leaf, a process which would involve protein hydrolysis, would presumably impair CO_2 assimilation and sucrose production. The mobilisation of protein-S in response to S demand in S-inadequate generative wheat plants (Table 3) contrasts with the absence of a response in vegetative plants (section 4.1) where no such response is evident except as a consequence of N deficiency (section 4.2). One interpretation is that S insufficiency elicits different responses in vegetative plants and generative plants. As noted in Section 4.1, an important outcome when vegetative plants are stressed for S is that export of S from the root system virtually ceases as the plant diverts S into root growth as a strategy to procure more S to enable the plant to survive whereas generative plants are committed to the production of seed to ensure survival of the species. Another interpretation is that the nutritional conditions for growing the S-inadequate plants could have induced physiological N deficiency as the amount of S supplied was insufficient to balance the amount of N available (Dijkshoorn and van Wijk, 1967; Brunold, 1993). In support of this possibility, the leaves of S-inadequate plants accumulated very high levels of nitrate in the leaves (Fitzgerald et al., 1999b). As noted in Section 4.2, N deficiency promotes protein hydrolysis in leaves with consequent mobilisation of protein-S.

6.2.5 Forms of S imported into developing grains

Developing cereal grains recruit metabolites from the endosperm cavity and the embryo sac. These apoplastic compartments are serviced by the phloem but not by the xylem (Wang et al., 1994). Thus, control over the composition of these compartments is exercised by the loading of metabolites into the phloem during generative growth. The forms of S transported, and any metabolic changes that occur in the transport pathway, can be ascertained by analysing the soluble S at various sites in the vascular pathway leading from the vegetative tissues to the developing endosperm. This involves, in sequence, the rachis, grain crease vascular bundle, endosperm cavity (from which the S-transport compounds are finally recruited) and endosperm. SO_4^{2-} accounts for most of the soluble S in the endosperm cavity in S-adequate plants, consistent with the conclusion that SO_4^{2-} from the root system is the main source of S for developing grains in these plants with a smaller contribution from GSH and a minor contribution from Met (Table 4). In S-insufficient plants, however, SO_4^{2-} was not detected in the endosperm cavity. In the absence of SO_4^{2-}, GSH was by far the most important source of S in the cavity and the contribution from Met was correspondingly more important. SMM has also been detected at the base of the ear, in similar proportions to GSH (Bourgis et al., 1999) though Fitzgerald et al. (2001) did not detect it in the endosperm cavity of any plants suggesting that they missed it or that SMM did not reach the endosperm cavity. SO_4^{2-} was the main form of S in the endosperm of S-adequate plants.

In S-inadequate plants, GSH was the main form of S in the endosperm, with significant amounts of Met in the later stages of seed development but SO_4^{2-} was not detected (Table 4). These data concur with earlier reports that the demand for Cys and Met for the synthesis of wheat seed proteins is unlikely to be supplied by recruitment of these amino acids from the endosperm cavity (e.g. Ugalde and Jenner, 1990).

Table 4. Soluble S-metabolites in the endosperm and endosperm cavity of S-adequate and S-inadequate wheat plants at 13 and 25 days after anthesis (dpa). The growing conditions were as described in Table 3. The values shown are for plants that did not receive nutrients after anthesis. Cys was not detected.

S nutrition	**Tissue**	**SO_4^{2-} (nmol tissue^{-1})**		**GSH (nmol tissue^{-1})**		**Met (nmol tissue^{-1})**	
		13 dpa	25 dpa	13 dpa	25 dpa	13 dpa	25 dpa
Adequate	Endsp cavity	5.1	5.3	1.4	1.5	0.4	0.3
	Endosperm	4.2	48.1	6.5	11.2	0.2	0.2
Inadequate	Endsp cavity	nd	nd	1.7	2.8	0.3	0.4
	Endosperm	nd	nd	6.3	14.7	0.2	3.2

Endsp = endosperm; nd = not detected
Source: Anderson and Fitzgerald (2001)

In S-adequate plants, SO_4^{2-} in the endosperm cavity must be derived from the root (Section 6.2.3) whereas GSH must be derived from the pool of soluble GSH in the leaves. However, the vegetative tissues of S-inadequate plants lack significant pools of soluble S (Section 6.2.3). Therefore, the GSH and the Met in the endosperm cavity of S-inadequate plants must have a different origin. They must be derived from the insoluble-S in the leaves, especially the flag leaf (section 6.2.4). Thus, the leaves, particularly the flag leaf of S-inadequate generative plants must have a mechanism for synthesising GSH from Cys and other products of protein hydrolysis. It is unlikely that Met, formed by hydrolysis of leaf protein, is also metabolised to GSH since Met would first need to be converted to Cys catalysed by γ-cystathionase. However, this enzyme has not been demonstrated in plants (Giovanelli and Mudd, 1971; Crouch, 2002). Transport of Met as SMM is a more attractive idea. However, it is unclear whether the synthesis and transport of SMM is regulated by sink demand for methyl groups or for S. The transfer of methyl groups is a common biochemical reaction that modifies the biological activity of a compound (Kuchel and Ralston, 1988). Perhaps the charge resulting from the methylation of Met increases its mobility in the phloem.

MMT, the enzyme that catalyses the synthesis of SMM, has been cloned from germinating barley (Pimenta et al. 1998) and various other species. SMM occurs at high concentrations in germinating barley but, importantly, MMT activity was not found in endosperm (Pimenta et al., 1998), nor was SMM found in dry and germinating wheat seeds (Dethier et al., 1991). Therefore, if SMM has a role in Met transport in germinating seeds, Met from protein hydrolysis must be methylated at a site(s) in the seedling after it has been absorbed from endosperm. The absence of a mechanism in endosperm for synthesising SMM raises the question whether developing endosperm lacks a mechanism for demethylating SMM. In this event, SMM would possibly need to be demethylated before entering the endosperm for it to contribute to the S budget. This idea would be consistent with the observations of Fitzgerald et al. (2001) who did not detect SMM in the cavity or endosperm. Lemmas and glumes are richly vascularised photosynthetic tissues and their role in receiving, interconverting and redistributing N to endosperms is well known (e.g. Simpson et al., 1983). Perhaps SMM is delivered to lemmas and glumes where it is used as a methyl-group donor for various metabolic processes. Met could then travel in the phloem of the vascular bundle in the crease region to the endosperm cavity.

6.2.6 S uptake mechanisms

The concentration of soluble S metabolites in the endosperm cavity and in the endosperm can be calculated from the data in Fitzgerald et al. (2001). These data show that both GSH and SO_4^{2-} are taken up against concentration gradients. This raises the question of transporters for the uptake of these metabolites. The characteristics of GSH transporters are discussed in Section 6.1.3 though nothing is known about their occurrence in cereal endosperm. Similarly, although plants contain several SO_4^{2-} transporters (section 2), they have yet to be identified in cereal endosperm.

6.2.7 Metabolism of S-transport compounds in developing grains

Almost all of the S in cereal grains occurs as protein-S. Therefore, SO_4^{2-}, GSH and Met imported into grains must be assimilated/metabolised to Cys and Met for incorporation into grain proteins. Theory predicts that SO_4^{2-} would be reductively assimilated into OAS to form Cys, some of which would be metabolised via the transsulphurylation pathway to Met. On the other hand, GSH would presumably be metabolised to Cys and Met. S imported as Met could be incorporated directly. Collectively, the rates of these processes need to account for the observed rate of incorporation of S into protein during seed development. Fitzgerald et al. (2001) examined the activities of various enzymes of S metabolism in extracts of endosperm prepared from developing grains of S-adequate and S-inadequate wheat plants. Extracts from all plants supported ATP sulphurylase, cystathionine synthase, β-cystathionase and Cys synthase activities indicating that the endosperms had active pathways for SO_4^{2-} assimilation (Cys synthesis) and Met biosynthesis at rates that were well in excess of the rate of incorporation of S into protein-S in vivo. The activity of ATP sulphurylase was less than 5% of the activities of the other enzymes and the activity of ATP sulphurylase was a further 4-fold less in the endosperms of S-inadequate plants (which lack SO_4^{2-}) compared to the endosperms of S-adequate plants which contain SO_4^{2-} (Table 4). Whilst the relevance of in vitro rates to in vivo rates is uncertain, the data nevertheless raise the possibility that SO_4^{2-} might be an important effector in enhancing the rate limiting step of SO_4^{2-} assimilation in endosperms that import their S, principally as SO_4^{2-}.

Crude extracts of endosperms from both S-adequate and S-inadequate plants supported the hydrolysis of GSH to γGluCys at similar rates (Fitzgerald et al., 2001). They also catalysed the hydrolysis of γGluCys to glutamate and Cys at about 70-80% of the rate of GSH hydrolysis but they did not support the hydrolysis of cysteinylglycine. As noted above, GSH, formed from protein-S in leaves, accounts for most of the S recruited by the endosperms of S-inadequate plants. Importantly, the in vitro rates of hydrolysis of GSH and γGluCys were more than sufficient to account for the rate of GSH consumed by endosperms of S-inadequate plants.

7. REFERENCES

Adiputra IGK and Anderson JW (1992) Distribution and redistribution of sulphur taken up from nutrient solution during vegetative growth in barley. *Physiol Plant* **85:** 453—460

Adiputra IGK and Anderson JW (1995) Effect of sulphur nutrition on redistribution of sulphur in vegetative barley. *Physiol Plant* **95:** 643—650

Anderson JW (1990) Sulfur metabolism in plants. In: Miflin BJ and Lea PJ (eds) The Biochemistry of Plants, Vol 16, pp 327—381. Academic Press, San Diego

Anderson JW and Fitzgerald MA (2001) Physiological and metabolic origin of sulphur for the synthesis of seed storage proteins. *J Plant Physiol* **158:** 447—456

Bell CI, Clarkson DT and Cram WJ (1990) Turnover of sulfate in leaf vacuoles limits retranslocation under sulfur stress. In: de Kok LJ, Rennenberg H, Brunold C and Rauser WE (eds) Sulfur Nutrition and Assimilation in Higher Plants, pp 3—19. SPB Academic Publishing, The Hague

Bell CI, Clarkson DT and Cram WJ (1994) Compartmental analysis of $^{35}SO_4^{2-}$ exchange kinetics in roots and leaves of a tropical legume *Macroptilium atropurpureum* cv. Siratro. *J Exp Bot* **45:** 879—886

Bell CI, Clarkson DT and Cram WJ (1995a) Sulphate supply and its regulation of transport in roots of a tropical legume *Macroptilium atropurpureum* cv. Siratro. *J Exp Bot* **46:** 65—71

Bell CI, Clarkson DT and Cram WJ (1995b) Partitioning and redistribution of sulphur during S-stress in *Macroptilium atropurpureum* cv. Siratro. *J Exp Bot* **46:** 73—81

Blake-Kalff MMA, Harrison KR, Hawkesford MJ, Zhao FJ and McGrath SP (1998) Allocation of sulfur within oilseed rape (*Brassica napus* L.) leaves in response to sulfur-deficiency. *Physiol Plant* **118:** 1337—1344

Blaygrove RJ, Gillespie JM and Randall PJ (1976) Effect of sulphur supply on the seed globulin composition of *Lupinus angustifolius. Aust J Plant Physiol* **3:** 173—184

Bolchi A, Petrucco S, Tenca PL, Foroni C and Ottonello S (1999) Coordinate modulation of maize sulfate permease and ATP sulfurylase mRNAs in response to variations in sulfur nutritional status: stereospecific down-regulation by L-cysteine. *Plant Mol Biol* **39:** 527—537

Bonas U, Scmitz K, Rennenberg H and Bergmann L (1982) Phloem transport of sulphur in *Rincinus. Planta* **155:** 82—88

Bourgis F, Roje S, Nuccio M, Fisher DB, Tarczynski MC, Li C, Herschbach C, Rennenberg H, Pimenta MJ, Shen TL, Gage DA and Hanson AD (1999) *S*-Methylmethionine plays as major role in phloem transport and is synthesized by a novel type of methyltransferase. *Plant Cell* **11:** 1485—1497

Brunold (1993) Regulatory interactions between sulfate and nitrate assimilation. In: de Kok LJ, Stulen I, Rennenberg H, Brunold CH and Rauser WE (eds) Sulphur Nutrition and Assimilation in Higher Plants, pp 61—77. SPB Academic Publishing, The Hague

Clarkson DT, Hawkesford MJ and Davidian, J-C (1993) Membrane and long-distance transport of sulfate. In: de Kok LJ, Rennenberg H, Brunold C and Rauser WE (eds) Sulfur Nutrition and Assimilation in Higher Plants, pp 3—19. SPB Academic Publishing, The Hague

Cooper D and Clarkson DT (1989) Cycling of amino nitrogen and other nutrients between shoots and roots in cereals: a possible mechanism integrating shoot and root in the regulation of nutrient uptake. *J Exp Bot* **40:** 753—762

Cram WJ (1983) Characteristics of sulfate transport across the plasmalemma and the tonoplast of carrot root cells. *Plant Physiol* **72:** 204—211

Crouch PJ (2002) The reallocation of sulphur from source tissues of germinating and vegetative soybean and the acquisition of sulphur by developing soybean seeds. PhD thesis, LaTrobe University

Datko AH, Mudd SH, Giovanelli J and MacNicol PK (1978) Sulfur-containing compounds in *Lemna perpusilla* 6746 grown at a range of sulfate concentrations. *Plant Physiol* **62:** 629—635

Dethier M, Jaeger BD, Barszczak E and Dufour JP (1991) In vivo and in vitro investigations of the synthesis of methylmethionine during barley germination. *J Am Soc Brew Chem* **49:** 31—37

Dijkshoorn W and van Wijk AL (1967) The S requirement of plants as evidenced by the sulphur-nitrogen ratio in the organic matter: a review of published data. *Plant Soil* **26:** 129—157

Fitzgerald MA, Ugalde TD and Anderson JW (1999a) Sulphur nutrition changes the sources of S in vegetative tissues of wheat during generative growth. *J Exp Bot* **50:** 499—508

Fitzgerald MA, Ugalde TD and Anderson JW (1999b) S nutrition affects the pools of S available to developing grains of wheat. *J Exp Bot* **50:** 1587—1592

Fitzgerald MA, Ugalde TD and Anderson JW (2001) Sulphur nutrition affects delivery and metabolism of S in developing endosperms of wheat. *J Exp Bot* **52:** 1519—1526

Giovanelli J and Mudd H (1971) Transsulfuration in higher plants: Partial purification and properties of β-cystathionase of spinach. *Biochim Biophys Acta* **227:** 654—670

Giovanelli J, Mudd SH and Datko AH (1980) Sulfur amino acids in plants. In: Miflin BJ and Lea PJ (eds) The Biochemistry of Plants, Vol 5, pp 453—505. Academic Press, New York

Guitman MR, Arnozis PA and Barneix AJ (1991) Effect of source-sink relations and nitrogen nutrition on senescence and N remobilization in the flag leaf of wheat. *Physiol Plant* **82:** 278—94

Gunning BES (1977) Transfer cells and their roles in transport of solutes in plants. *Sci Prog Oxf* **64:** 539—568

Hawkesford MJ (2000) Plant responses to sulphur deficiency and the genetic manipulation of sulphate transporters to improve S-utilization efficiency. *J Exp Bot* **51:** 131—138

Hawkesford MJ and Smith FW (1997) Molecular biology of higher plant sulphate transporters. In: Cram WJ, DeKok LJ, Stulen I, Brunold C and Rennenberg H (eds) Sulphur Assimilation in Higher Plants: Molecular, Ecophysiological and Nutritional Aspects, pp 13—25. Backhuys Publishers, Leiden

Herschbach C and Rennenberg H (1994) Influence of glutathione (GSH) on net uptake of sulfate and sulfate transport in tobacco plants. *J Exp Bot* **45:** 1069—1076

Herschbach C and Rennenberg H (1995) Long distance transport of ^{35}S-sulphur in 3-year old beech trees (*Fagus sylvatica*). *Physiol Plant* **95:** 379—386

Herschbach C, Pilch B, Tausz M, Rennenberg H and Grill D (2002) Metabolism of reduced and inorganic sulphur in pea cotyledons and distribution into developing seedlings. *New Pytol* **153:** 73—80

Hocking PJ and Steer BT (1995) Effects of timing and supply of nitrogen on nitrogen remobilization from vegetative organs and redistribution to developing seeds of sunflower. *Plant Soil* **170:** 359—370

Holowach LP, Thompson JF and Madison JT (1984) Storage protein composition of soybean cotyledons grown in vitro in media of various sulfate concentrations in the presence and absence of exogenous methionine. *Plant Physiol* **74:** 584—589

Jamai AR, Tommasini R, Martinola E and Delrot S (1996) Characterization of glutathione uptake in broad bean leaf protoplasts. *Plant Physiol* **111:** 1145—1152

Kuchel PW and Ralston GB (1988) Schaum's Outline of Theory and Problems of Biochemistry. McGraw Hill NY.

Larsson CM, Larsson M, Purves JV and Clarkson DT (1991) Translocation and cycling through roots of recently absorbed nitrogen and sulphur in wheat (*Triticum aestivum*) during vegetative and generative growth. *Physiol Plant* **82:** 345—352

Lappartient A, Vidmar JJ, Leustek T, Glass ADM and Touraine B (1999) Inter-organ signaling in plants: regulation of ATP sulfurylase and sulfate transporter genes expressed in roots mediated by phloem translocated compounds. *Plant J* **18:** 89—9

Leustek T, Martin MN, Bick J and Davies JP (2000) Pathways and regulation of sulfur metabolism revealed through molecular and genetic studies. *Annu Rev Physiol Plant Mol Biol* **51**: 141—165

MacNicol PK and Bergmann L (1984) A role for homoglutathione in organic sulphur transport to the developing mungbean seed. *Plant Sci Lett* **36:** 219—223

McMahon PJ and Anderson JW (1998) Preferential allocation of sulphur into γ-glutamylcysteinyl peptides in wheat plants grown at low sulphur nutrition in the presence of cadmium. *Physiol Plant* **104:** 440--448

Moreira MA, Hermodson MA, Larkins BA and Nielsen NC (1979) Partial characterization of the acidic and basic polypeptides of glycinin. *J Biol Chem* **254:** 9921—9926

Moussavi-Nik M, Pearson JN, Hollamby GJ and Graham RD (1998) Dynamics of nutrient remobilization during germination and early seedling development in wheat. *J Plant Nutr* **21:** 421—434

Noctor G, Arisi ACM, Jouanin L, Kunert KJ, Rennenberg H and Foyer CH (1998) Glutathione: biosynthesis, metabolism and relationship to stress tolerance explored in transformed plants. *J Exp Bot* **49:** 623—47

Pate JS (1980) Transport and partitioning of nitrogenous solutes. *Annu Rev Plant Physiol* **31:** 313—340

Pickering FS and Reiss PJ (1993) Effects of abomasal supplements of methionine on wool growth of grazing sheep. *Aust J Exp Agric* **33:** 7—12

Pimenta, MJ, Kaneta T, Larondelle Y, Dohmae N and Kamiya Y (1998) *S*-Adenosyl-L-methionine:L-methionine *S*-methyltransferase from germinating barley: Purification and localization. *Plant Physiol* **118:**4 31—438

Prosser IM, Purves JV, Saker LR and Clarkson DT (2001) Rapid disruption of nitrogen metabolism and nitrate transport in spinach plants deprived of sulphate. *J Exp Bot* **52:** 113—121

Rauser WE, Schupp R and Rennenberg H (1991) Cysteine, γ-glutamylcsteine and glutathione levels in maize seedlings. *Plant Physiol* **97:** 128—138

Rennenberg H (1983) Cysteine desulfhydrase activity in cucurbit plants: stimulation by preincubation with L- or D-cysteine. *Phytochemistry* **22:** 1557—1560

Rennenberg H (1987) Aspects of glutathione function and metabolism in plants. In: VonWettstein D and Chua N-H (eds) Plant Molecular Biology, pp 279—292. Plenun Press, New York

Rennenberg H, Schmitz K and Bergmann L (1979) Long distance transport of sulphur in *Nictoiana tabacum*. *Planta* **147:** 52—62

Schneider AN, Martini N and Rennenberg H (1992) Reduced glutathione (GSH) transport in cultured tobacco cells. *Plant Physiol Biochem* **30:** 29—38

Simpson RJ, Lambers H and Dalling MJ (1983) Nitrogen redistribution during grain growth in wheat (*Triticum aestivum* L.). IV Development of a quantitative model of the translocation of nitrogen to the grain. *Plant Physiol* **71:** 7—14

Smith IK and Lang AL (1988) Translocation of sulfate in soybean (*Glycine max* L. Merr). *Plant Physiol* **86:** 798—802

Smith CJ and Whitfield DM (1990) Nitrogen accumulation and redistribution of late applications of ^{15}N-labelled fertiliser by wheat. *Field Crops Res* **24:** 211—226

Smith FW, Hawkesford MJ, Ealing PM, Clarkson DT and VandenBerg PJ. (1997) Regulation of expression of a cDNA from barley roots encoding a high affinity sulphate transporter. *Plant J* **12:** 875—84

Sunarpi and Anderson JW (1995) Mobilization of sulphur in soybean cotyledons during germination. *Physiol Plant* **94:** 143—150

Sunarpi and Anderson JW (1996a) Distribution and redistribution of sulfur supplied as [^{35}S]sulfate to roots during vegetative growth of soybean. *Plant Physiol* **110:** 1151—1157

Sunarpi and Anderson JW (1996b) Effect of sulfur nutrition on the redistribution of sulfur in vegetative soybean plants. *Plant Physiol* **112:** 623—631

Sunarpi and Anderson JW (1997a) Allocation of S in generative growth of soybean. *Plant Physiol* **114:** 687—693

Sunarpi and Anderson JW (1997b) Effect of nitrogen nutrition on the export of sulphur from leaves in soybean. *Plant Soil* **188:** 177—187

Sunarpi and Anderson JW (1997c) Effect of nitrogen nutrition on remobilization of protein sulfur in the leaves of vegetative soybean and associated changes in soluble sulfur metabolites. *Plant Physiol* **115:** 1671—1680

Sunarpi and Anderson JW (1998) Direct evidence for the involvement of the root in the redistribution of sulphur between leaves. *J Plant Nutr* **21:** 1273—1286

Tabe L and Droux M (2001) Sulfur assimilation in developing lupin cotyledons could contribute significantly to the accumulation of organic sulfur reserves in seed. *Plant Physiol* **126:** 176—187

Thanh VH and Shibasaki K (1977) β-Conglycinin from soybean proteins: Isolation and immunological and physicochemical properties of the monomeric forms. *Biochim Biophys Acta* **490**: 370—384

Ugalde TD and Jenner CF (1990) Substrate gradients and regional patterns of dry matter deposition within developing wheat endosperm. II. Amino acids and protein. *Aust J Plant Physiol* **17:** 377—394

Vidmar JJ, Schoerring JK, Touraine B and Glass ADM (1999). Regulation of the hvst1 gene encoding a high affinity sulfate transporter from *Hordeum vulgare*. *Plant Mol Biol* **40:** 855—892

Wang HL, Offler CE, Patrick JW and Ugalde TD (1994) The cellular pathway of photosynthate transfer in the developing wheat grain. I. Delineation of a potential transfer pathway using fluorescent dyes. *Plant Cell Environ* **17:** 257—266

Yamaguchi Y, Nakamura T, Harada E, Koizumi N and Sano H. (1999) Differential accumulation of transcripts encoding sulphur assimilation enzymes upon sulfur and /or nitrogen deprivation in *Arabidopsis thaliana*. *Biosci Biotech Biochem* **63:** 762—766

Zhao FJ, Hawkesford MJ and McGrath SP (1999) Sulphur assimilation and effects on yield and quality of wheat. *J Cereal Sci* **30:** 1—17

8. ACKNOWLEDGEMENT

We would like to thank Dr PJ Crouch for allowing us to incorporate data from his PhD thesis in this article.

9. AFFILIATIONS

Botany Department, La Trobe University, Bundoora, Victoria 3083, Australia (JWA); Yanco Agricultural Institute, NSW Agriculture, Private Mail Bag, Yanco, NSW 2703, Australia (MAF)

For correspondence - *JW Anderson*: john.anderson.@latrobe.edu.au

M. NOJI AND K. SAITO

SULPHUR AMINO ACIDS: BIOSYNTHESIS OF CYSTEINE AND METHIONINE

Abstract. Cysteine and methionine are sulphur-containing amino acids in the twenty protein amino acids. The inorganic sulphur, which is an essential nutrient for plants, is first fixed into cysteine by the cysteine biosynthetic pathway. Cysteine is then incorporated into proteins and glutathione or serves as the sulphur donor of methionine and sulphur-containing secondary products in plants. The cysteine biosynthetic pathway involves several enzymatic reactions. The final step of cysteine biosynthesis is the reaction incorporating sulphide moiety into β-position of alanine. The amino acid moiety is derived from serine through *O*-acetyl-L-serine. Serine acetyltransferase and cysteine synthase (*O*-acetylserine (thiol)-lyase) are committed in this final step of cysteine biosynthesis. The feedback inhibition for serine acetyltransferase activity plays one of the major roles in regulating cysteine synthesis. Methionine is synthesized from cysteine and *O*-phosphohomoserine, which is derived from aspartic acid through a pathway involving trans-sulphuration reaction. Three enzymes, cystathionine γ-synthase, cystathionine β-lyase and methionine synthase, are involved in the methionine biosynthesis from cysteine. The methionine biosynthesis is controlled by supply of O-phosphohomoserine and the enzyme, cystathionine γ -synthase.

1. INTRODUCTION

L-Cysteine (Cys) is one of the sulphur-containing amino acids in the twenty standard amino acids found in proteins. The inorganic sulphur, which is an essential nutrition for growth of plants, is first fixed into Cys by the Cys biosynthetic pathway (Saito, 1999). Cys is then incorporated into proteins and glutathione (GSH), or serves as the sulphur donor of methionine (Met) and sulphur-containing secondary products in plants (Fig. 1). Since thiol group of Cys is highly reactive, Cys residue in active centre plays critical roles in the catalytic function of some proteins, so-called SH proteins. Moreover, two thiol groups can form a covalent disulphide bond by oxidation. Disulphide bonds are most important for the establishment of tertiary and in some cases, quaternary protein structures. Thus, Cys residues in some proteins are necessary to retain the protein structure by disulphide bond. Interconversion of sulphide and disulphide residues by oxidation-reduction also play a critical role in redox cycle in cells.

L-Methionine (Met), another sulphur-containing amino acid, is synthesized from cysteine and *O*-phosphohomoserine, which is derived from aspartic acid, through a pathway involving transsulphuration reaction (Fig. 1). Met is then incorporated into proteins, or converted into *S*-adenosylmethionine (SAM). SAM serves as a methyl donor of DNA and RNA modification and the synthesis of plant structural components, e.g. lignin precursors, choline and its derivatives, and pectin. Met is also an intermediate in the biosynthesis of the polyamines, and plant ripening hormone ethylene. Since the carbon skeleton of Met is derived from aspartate in

Yash P. Abrol and Altaf Ahmad (eds.), Sulphur in Plants, 135—144.

plants, Met is classified into aspartate-derived amino acid (i.e. Met, lysine and threonine). Aspartate-derived amino acids are required in the diets of non-ruminant animals, including humans.

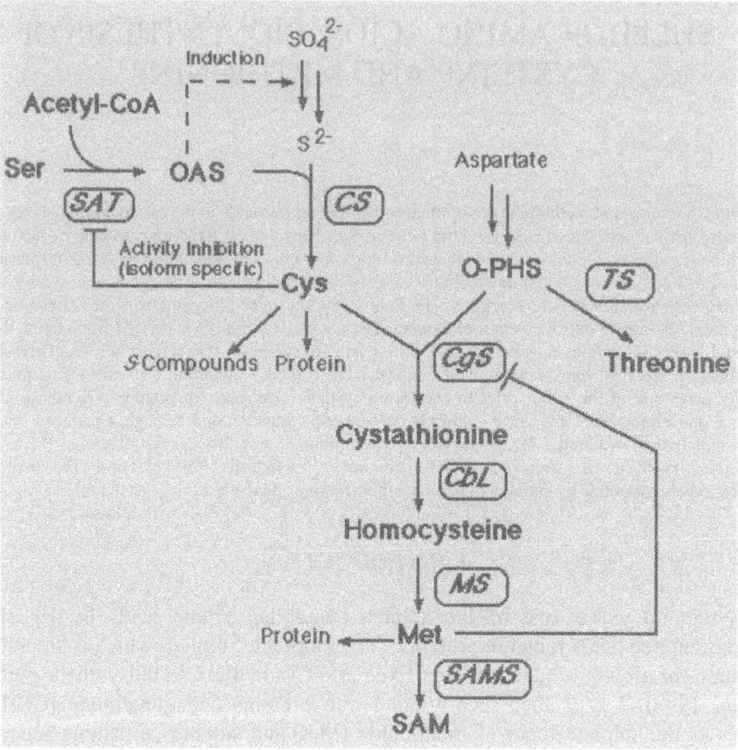

Figure 1. Biosynthetic pathway for the sulphur-containing amino acids, Cys and Met. The enzymes involved in the pathway are: SAT = Ser acetyltransferase; CS = Cys synthase; CgS = Cystathionine g-synthase; CbL = Cystathionine β-lyase; MS = Met synthase; SAM = S-adenosylhomocysteine. Dotted arrow indicates induction (derepression). Bar indicates inhibition.

2. CYSTEINE

2.1 Biosynthesis

In plants, the biosynthetic pathway of Cys involves several enzymatic reactions (Brunold and Rennenberg, 1997; Hell, 1997; Leustek and Saito, 1999; Saito, 1999; Hawkesford and Wray, 2000; Leustek et al., 2000; Saito, 2000). The Cys biosynthesis in plants consists of two biochemical pathways: the pathway of transport, activation and reduction sulphate into sulphide, and the pathway supplying amino acid moiety, which is derived from serine (Ser) through *O*-acetyl-L-serine (OAS), and then yielding Cys by the reaction of incorporating sulphide moiety into β-position of alanine. Two enzymes, Ser acetyltransferase and Cys synthase (OAS (thiol)-lyase) are committed in the final step of this pathway.

Serine acetyltransferase catalyzes the formation of OAS from serine (Ser) and acetyl-CoA (Fig. 2). This enzyme is responsible for the entry step from Ser metabolism to Cys biosynthesis; therefore, it would be logical to postulate the existence of regulation at this step.

$$\text{L-Serine} + \text{Acetyl-CoA} \longrightarrow O\text{-acetyl-L-serine (OAS)} + \text{CoA-SH}$$

Cys synthase catalyzes the formation of Cys from OAS and hydrogen sulphide, with the release of acetic acid at the final step of the Cys biosynthetic pathway (Fig. 2). This enzyme requires pyridoxal 5'-phosphate (PLP) as a cofactor for the catalytic reaction. PLP binds with the OAS substrate as the amino acrylate intermediate. The kinetic investigation revealed that Cys synthase has two catalytically nonequivalent active sites (Kuske et al., 1994; Rolland et al., 1996).

$$\text{OAS} + H_2S \longrightarrow \text{Cysteine} + \text{acetic acid}$$

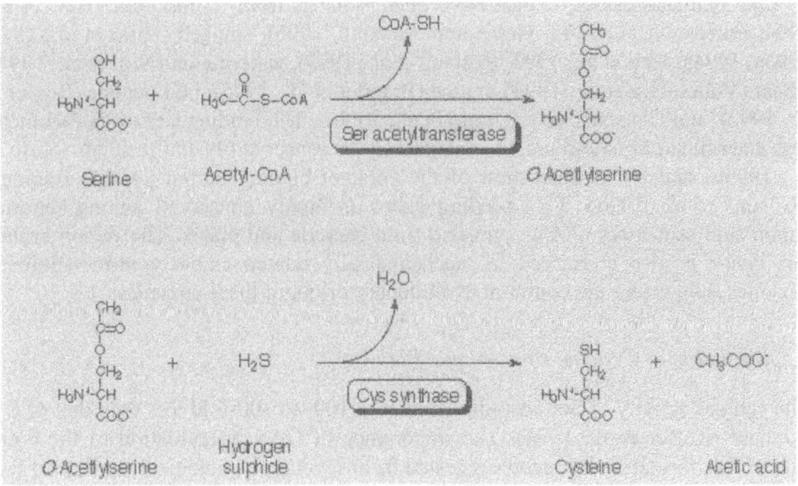

Figure 2. Sequential reactions from serine to cysteine

It is known that Ser acetyltransferase can be multimeric, and is able to form a complex in association with Cys synthase (Nakamura et al., 1988; Nakamura and Tamura, 1990; Ruffet et al., 1994; Bogdanova and Hell, 1997; Droux et al., 1998).

Cys synthase concentrations in spinach chloroplasts are, however, far in excess of Ser acetyltransferase concentrations indicating that only a fraction of Cys synthase associates with Ser acetyltransferase (Ruffet et al., 1994). A large amount of Cys synthase is, therefore, present in a free form (Droux et al., 1998).

Reduction of sulphate into sulphide takes place almost exclusively in chloroplasts. However, Ser acetyltransferase and Cys synthase are localized in three major compartments of plant cells viz., cytosol, chloroplasts and mitochondria (Lunn et al., 1990: Ruffet et al., 1995; Takahashi and Saito, 1996; Noji et al., 1998; Hesse et al., 1999). In mitochondria, there is an additional β-cyanoalanine synthase that exhibits similar catalytic activity to that of Cys synthase (Warrilow and Hawkesford, 1998; Hatzfeld et al., 2000). Cytosolic β-cyanoalanine synthase activity is subscribed to side activity of Cys synthase (Maruyama et al., 1998). Mitochondrial β-cyanoalanine synthase was previously identified as the mitochondrial form of Cys synthase (Saito et al., 1994a; Takahashi and Saito, 1996; Hatzfeld et al., 2000).

cDNA clones encoding Ser acetyltransferase have been isolated from watermelon (Saito et al., 1995), spinach (Noji et al., 2001), *Arabidopsis thaliana* (Bogdanova et al., 1995; Hell and Bogdanova, 1995; Ruffet et al., 1995; Roberts and Wray, 1996; Howarth et al., 1997) and *Allium tuberosum* (Chinese chive) (Urano et al., 2000). In particular, cDNAs of three Ser acetyltransferase isoforms that exhibit different subcellular localization, SAT-c (cytosolic isoform), SAT-p (plastidic isoform) and SAT-m (mitochondrial isoform) have been cloned from *A. thaliana* (Noji et al., 1998).

Cys synthase cDNA clones have been isolated from *A. thaliana* (Hell et al., 1994; Barroso et al., 1995; Hesse and Altmann, 1995), spinach (Saito et al., 1992, 1993a, 1994a; Hell et al., 1993; Rolland et al., 1993), watermelon (Noji et al., 1994), wheat (Youssefian et al., 1993), maize (Brander et al., 1995), bell pepper (Römer et al., 1992), and Chinese chive (Urano et al., 2000). The binding site of PLP cofactor was determined by site-directed mutagenesis of conserved lysine residues (Saito et al., 1993b) and by identification of the cofactor-binding partial peptide fragment (Rolland et al., 1996). This binding lysine is highly conserved among reported amino acid sequences of Cys synthase from bacteria and plants. The region around this lysine is also conserved in mechanistically related amino acid-metabolizing enzymes, suggesting the common evolutionary origin of these enzymes.

2.2 Regulation of Cysteine Biosynthetic Pathway

The cellular activity of Ser acetyltransferase is 100- to 300-fold less than that of Cys synthase (Ruffet et al., 1994). The importance of OAS in regulation of the entire flux of Cys formation has been suggested from several physiological studies. In fact, the availability of OAS has been found to be rate-limiting for Cys biosynthesis in feeding experiments (Neuenschwander et al., 1991; Saito et al., 1994b). Furthermore, addition of OAS to plants led to increase in mRNA accumulation of the sulphate transporter and the rate of sulphate uptake in barley roots (Smith et al., 1997). The activities of ATP sulphurylase and APS sulphotransferase (APS reductase) in *Lemna minor* were also enhanced by supplement of OAS (Neuenschwander et al., 1991). These studies suggest that Ser acetyltransferase and

OAS are the major regulatory factors in the biosynthesis of Cys in plants.

In bacteria, the feedback inhibition of Ser acetyltransferase activity by micromolar concentrations of Cys plays a key role for the regulation of Cys biosynthesis (Kredich and Tomkins, 1966). OAS serves as an inducer of Cys regulon besides as a substrate for Cys synthesis in bacteria (Kredich, 1996).

In plants, there are two types of Ser acetyltransferase that differ in their sensitivity to the Cys inhibition (Noji et al., 1998). The activities of *Arabidopsis* SAT-c, that is the cytosolic isoform from *A. thaliana* (Howarth et al 1997) and Ser acetyltransferases from watermelon (Saito et al., 1995), spinach (Noji et al., 2001) and Chinese chive (Urano et al., 2000) are inhibited by Cys at physiological concentration in an allosteric manner. In contrast, *Arabidopsis* SAT-p (plastidic isoform, Ruffet et al., 1995), and SAT-m (mitochondrial isoform, Bogdanova et al., 1995; Hell and Bogdanova, 1995; Roberts and Wray, 1996) are insensitive to the feedback inhibition. Difference of sensitivity to Cys means that Ser acetyltransferase have a regulatory role through the feedback inhibition in Cys biosynthesis and it depends on the subcellular compartmentation (Noji et al., 1998; Saito, 2000).

The overexpression of Ser acetyltransferase, of which the activity is not inhibited by Cys in transgenic *A. thaliana* caused the overaccumulation of OAS and Cys, but the overexpression of Ser acetyltransferase from watermelon, which is highly sensitive to the feedback inhibition by Cys, failed the overproduction of OAS and Cys (Noji and Saito, 2002). This in vivo transgenic experiment confirmed that feedback inhibition to the activity of Ser acetyltransferase by Cys is important for the regulation of the Cys levels in plant cells.

Ser acetyltransferase and Cys synthase form a hetero-oligomeric enzyme comp lex that may function to regulate Ser acetyltransferase activity. Ser acetyl-transferase is activated within the complex. The bound form of Cys synthase, which has dramatically reduced catalytic activity, appears to function as a regulatory subunit in the enzyme complex (Droux et al., 1998). Free form of Cys synthase is presumably responsible for the actual catalytic function of this enzyme. Because the enzyme complex formation is promoted by sulphide and is disrupted by OAS, these metabolites seem to regulate the activity of Ser acetyltransferase (Leustek et al., 2000).

mRNAs of cytosolic and chloroplastic isoforms of Cys synthase were slightly induced by sulphate starvation (Hell et al., 1994; Barroso et al., 1995; Takahashi and Saito, 1996). In *A. thaliana*, the expression of cytosolic Cys synthase is induced by exposure to salt and heavy-metal stress via a mechanism that probably involves mediation by abscisic acid (Barroso et al., 1999). Under these stress condition, cytosolic Cys synthase is expressed in the leaf trichome, suggesting a possible relationship between this enzyme and detoxification of heavy metals in trichomes (Gotor et al., 1997).

3. METHIONINE

3.1 Biosynthesis

Met is synthesized in three steps from Cys and *O*-phosphohomoserine that is derived from aspartic acid (Azevedo et al., 1997; Matthews, 1999) (Fig. 3). Three enzymes, cystathionine γ-synthase, cyatathionine β-lyase and Met synthase, are involved in the Met biosynthesis in plant.

Cystathionine γ-synthase catalyzes the first step that involves the β-replacement of the phosphate group of *O*-phosphohomoserine with Cys to form cystathione (Fig. 3). Cystathionine γ-synthase has been shown to be localized in the plastids of spinach leaves (Ravanel et al., 1995a) and *A. thaliana* (Ravanel et al., 1998a). The predicted cystathionine γ-synthase structures deduced from its cDNAs from *A. thaliana* (Kim and Leustek, 1996) and potato (Riedel et al., 1999) also suggest the plastid-localization of this enzyme.

The second step of the Met biosynthesis is the cleavage of the C_3 skeleton of cystathionine. Cystathionine β-lyase cleaves the β-C-S bond of cystathionine to produce homocysteine, ammonia and pyruvate (Fig. 3). There are two distinct forms of cystathionine β-lyase in spinach, one is localized in the plastids, and the other is localized in the cytosol (Droux et al., 1995). The cytosolic isoform of cystathionine β-lyase catalyzes the β-cleavage of cystine more efficiently than that of cystathionine, indicating that the proper substrate for the cytosolic cystathionine β-lyase is cystine rather than cystathionine (Ravanel et al., 1998b). In contrast, the plastidic isoform efficiently catalyzes the β-clevage of cystathionine. These findings were confirmed in *A. thaliana* (Ravanel et al., 1995a) and *Echinochloa colonum* (Turner et al., 1998), in which cystathionine β-lyase is only detected in the plastids. A cDNA encoding cystathionine β-lyase was cloned from *A. thaliana* (Ravanel et al., 1995b). The predicted polypeptide encoded by this cDNA contained a putative transit peptide for chloroplast translocation.

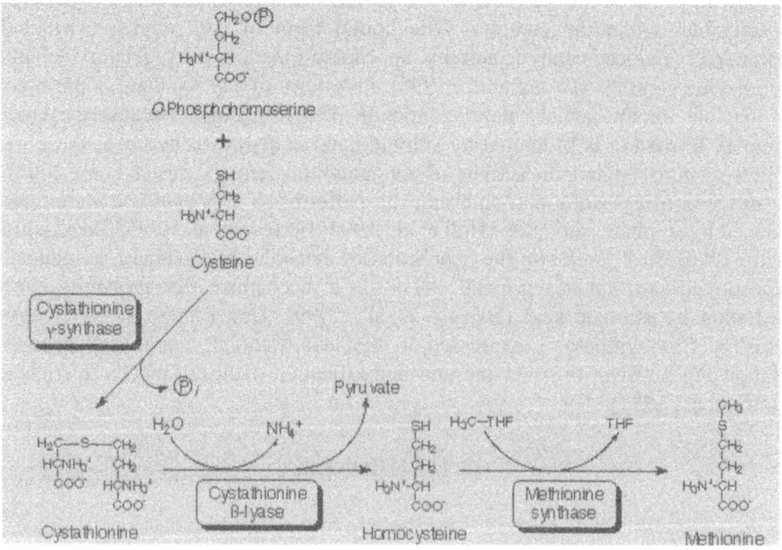

Figure 3. Methionine biosynthetic pathway in plants. THF = tetrahydrofolate

The final step of the Met biosynthesis in plant is catalyzed by Met synthase, which transfers the methyl group from methyltetrahydrofolate to homocysteine (Fig. 3). This reaction is important not only in de novo Met synthesis but also in the recycling of *S*-adenosylmethionine, which is the primary methyl-group donor in a variety of methyl-transfer reaction. Although all of the enzymes involved in the conversion of aspartate to homocysteine are localized in chloroplasts, the steps from homocysteine to Met and the recycling of *S*-adenosylmethionine are localized in the cytosol. The cDNAs of Met synthase have been cloned from several plant species (Eichel et al., 1995; Petersen et al., 1995; Ravanel et al., 1998b).

3.2 Regulation of Methionine Biosynthetic Pathway

Multiple regulatory mechanisms control the synthesis of amino acids from the aspartate-derived amino acids i.e. Met, lysine and threonine. The activity of cystathionine γ-synthase is regulated negatively by Met, and positively by lysine and threonine (Matthews, 1999). A recent analysis of a Met over-producing *A. thaliana* mutant revealed that the stability of the mRNA that encodes cystathionine γ-synthase is negatively auto-regulated by an amino-acid stretch in its own translational product (cystathionine γ-synthase) that takes place when Met or its metabolites are present (Chiba et al., 1999). This amino-acid stretch is highly conserved among several cystathionine γ-synthases from different species, and hence, this unique regulatory mechanism is likely to be functionally conserved in plant cells.

The antisense suppression of the cystathionine γ-synthase encoding gene resulted in severe growth reduction, which is restored by the exogenous addition of Met (Ravanel et al., 1998b; Kim and Leustek, 2000). The overexpression of cystathionine γ-synthase in *A. thaliana* led to increased levels of Met and *S*-methylmethionine (Gakière et al., 2002), whereas the overexpression of cystathionine β-lyase in transgenic potato did not enhance the metabolic flux towards Met biosynthesis (Maimann et al., 2001). These findings suggest an essential role for cystathionine γ-synthase in controlling Met biosynthesis in plants. A mutation of threonine synthase gene in *A. thaliana* caused over-accumulation of Met (Bartlem et al., 2000). Threonine synthase and cystathionine γ-synthase are positioned at a metabolic branchpoint between Met and threonine (Fig. 1). In this mutant plant, *O*-phosphohomoserine, which is the common substrate of both enzymes, is used exclusively for Met synthesis. Over-accumulation of Met resulted also from a mutation of one of SAM synthase genes in *A. thaliana* (Shen et al., 2002). This accumulation of Met observed in mutant plants is likely the result of a decrease in the flux from Met to SAM due to the alteration in one of the SAM synthase genes.

4. REFERENCES

Azevedo RA, Arruda P, Turner WL and Lea PJ (1997) The biosynthesis and metabolism of the aspartate derived amino acids in higher plants. *Phytochemistry* **46:** 395—419

Barroso C, Vega JM and Gotor C (1995) A new member of the cytosolic *O*-acetylserine (thiol)lyase gene family in *Arabidopsis thaliana. FEBS Lett* **363:** 1—5

Barroso C, Romero LC, Cejudo FJ, Vega JM and Gotor C (1999) Salt-specific regulation of the cytosolic *O*-acetylserine(thiol)lyase gene from *Arabidopsis thaliana* is dependent on abscisic acid. *Plant Mol Biol* **40:** 729—736

Bartlem D, Lambein I, Okamoto T, Itaya A, Uda Y, Kijima F, Tamaki Y, Nambara E and Naito S (2000) Mutation in the threonine synthase gene results in an over-accumulation of soluble methionine in *Arabidopsis*. *Plant Physiol* **123:** 101—110

Bogdanova N and Hell R (1997) Cysteine synthesis in plants: Protein-protein interactions of serine acetyltransferase from *Arabidopsis thaliana*. *Plant J* **11:** 251—262

Bogdanova N, Bork C and Hell R (1995) Cysteine biosynthesis in plants: isolation and functional identification of a cDNA encoding a serine acetyltransferase from *Arabidopsis thaliana*. *FEBS Lett* **358:** 43—47

Brander KA, Owttrim GW and Brunold C (1995) Isolation of a cDNA encoding a putative chloroplastic isoform of cysteine synthase from maize. *Plant Physiol* **108:** 1748

Brunold C and Rennenberg H (1997) Regulation of sulfur metabolism in plants: First molecular approaches. *Prog Bot* **58:** 164—186

Chiba Y, Ishikawa M, Kijima F, Tyson RH, Kim J, Yamamoto A, Nambara E, Leustek T, Wallsgrove RM and Naito S (1999) Evidence for autoregulation of cystathionine γ-synthase mRNA stability in *Arabidopsis*. *Science* **286:** 1371—1374

Droux M, Ravanel S and Douce R (1995) Methionine biosynthesis in higher plants. II. Purification and characterization of cystathionine β-lyase from spinach chloroplasts. *Arch Biochem Biophys* **316:** 585—595

Droux M, Ruffet ML, Douce R and Job D (1998) Interactions between serine acetyltransferase and *O*-acetylserine(thiol)lyase in higher plants. Structural and kinetic properties of the free and bound enzymes. *Eur J Biochem* **255:** 235—245

Eichel J, Gonzalez JC, Hotze M, Matthews RG and Schroeder J (1995) Vitamini-B_{12}-independent methionine synthase from higher plant (*Catharanthus roseus*). Molecular characterization, regulation, heterologous expression, and enzyme properties. *Eur J Biochem* **230:** 1053—1058

Gakière B, Denis L, Droux M and Job D (2002) Over-expression of cystathionine γ-synthase in *Arabidopsis thaliana* leads to increased levels of methionine and S-methylmethionine. *Plant Physiol Biochem* **40:** 119—126

Gotor C, Cejudo FJ and Barroso C (1997) Tissue-specific expression of ATCYS-3A, a gene encoding the cytosolic isoform of *O*-acetylserine(thiol)lyase in *Arabidopsis*. *Plant J* **11:** 347—352

Hatzfeld Y, Maruyama A, Schmidt A, Noji M, Ishizawa K and Saito K (2000) β-Cyanoalanine synthase is a mitochondrial cysteine synthase-like proteine in spinach and *Arabidopsis*. *Plant Physiol* **123:** 1163—1171

Hawkesford MJ and Wray JL (2000) Molecular genetics of sulphate assimilation. *Adv Bot Res* **33:** 160—223

Hell R (1997) Molecular physiology of plant sulfur metabolism. *Planta* **202:** 138—148

Hell R and Bogdanova N (1995) Characterization of a full-length cDNA encoding a serine acetyl-transferase from *Arabidopsis thaliana*. *Plant Physiol* **109:** 1498

Hell R, Schuster G and Gruissem W (1993) An *O*-acetylserine (thiol) lyase cDNA from spinach. *Plant Physiol* **102:** 1057—1058

Hell R, Bork C, Bogdanova N, Frolov I and Hauschild R (1994) Isolation and characterization of two cDNAs encoding for compartment specific isoforms of *O*-acetylserine(thiol)lyase from *Arabidopsis thaliana*. *FEBS Lett* **351:** 257—262

Hesse H and Altmann T (1995) Molecular cloning of a cysteine synthase cDNA from *Arabidopsis thaliana*. *Plant Physiol* **108:** 851—852

Hesse H, Lipke J, Altmann T and Hoefgen R (1999) Molecular cloning and expression analyses of mitochondrial and plastidic isoforms of cysteine synthase (*O*-acetylserine(thiol)lyase) from *Arabidopsis thaliana*. *Amino Acids* **16:** 113—131

Howarth JR, Roberts MA and Wray JL (1997) Cysteine biosynthesis in higher plants: A new member of the *Arabidopsis thaliana* serine acetyltransferase small gene family obtained by functional complementation of an *Escherichia coli* cysteine auxotroph. *Biochim Biophys Acta* **1350:** 123—127

Kim J and Leustek T (1996) Cloning and analysis of the gene for cystathionine γ-synthase from *Arabidopsis thaliana*. *Plant Mol Biol* **32:** 1117—1124

Kim J and Leustek T (2000) Repression of cystathionine γ-synthase in *Arabidopsis thaliana* produces partial methionine auxotrophy and developmental abnormalities. *Plant Sci* **151:** 9—18

Kredich NM (1996) Biosynthesis of cysteine. In: Neidhardt FC, Curtiss R III, Ingraham JL, Lin ECC, Low KB, Magasanik B, Renznikoff WS, Riley M, Schaechter M and Umbargar HE (eds) *Escherichia coli* and *Salmonella typhimurium*: Cellular and Molecular Biology, 2nd edn, Vol 1, pp 514—527. American Society for Microbiology, Washington DC

Kredich NM and Tomkins GM (1966) The enzymic synthesis of L-cysteine in *Escherichia coli* and *Salmonella typhimurium*. *J Biol Chem* **241:** 4955—4965

Kuske CR, Ticknor LO, Guzman E, Gurley LR, Valdez, JG, Thompson ME and Jackson PJ (1994) Purification and characterization of *O*-acetylserine sulfhydrylase isoenzymes from *Datura innoxia*. *J Biol Chem* **269:** 6223—6232

Leustek T and Saito K (1999) Sulfate transport and assimilation in plants. *Plant Physiol* **120:** 637—64

Leustek T, Martin MN, Bick JA and Davies JP (2000) Pathways and regulation of sulfur metabolism revealed through molecular and genetic studies. *Annu Rev Plant Physiol Plant Mol Biol* **51:** 141—166

Lunn JE, Droux M, Martin J and Douce R (1990) Localization of ATP sulfurylase and *O*-acetylserine (thiol) lyase in spinach leaves. *Plant Physiol* **94:** 1345—1352

Maimann S, Hoefgen R and Hesse H (2001) Enhanced cystathionine β-lyase activity in transgenic potato plants does not force metabolite flow towards methionine. *Planta* **214:** 163—170

Maruyama A, Ishizawa K, Takagi T and Esashi Y (1998) Cytosolic β-cyanoalanine synthase activity attributed to cysteine synthases in Cocklebur seeds. Purification and characterization of cytosolic cysteine synthases. *Plant Cell Physiol* **39:** 671—680

Matthews BF (1999) Lysine, threonine, and methionine biosynthesis. In: Singh BK (ed) Plant Amino Acids: Biochemistry and Biotechnology, pp 205—225. Marcel Dekker Inc., New York

Nakamura K and Tamura G (1990) Isolation of serine acetyltransferase complexed with cysteine synthase from *Allium tuberosum*. *Agric Biol Chem* **53:** 2537—2538

Nakamura K, Hayama A, Masada M, Fukushima K and Tamura G (1988) Purification and some properties of plant serine acetyltransferase. *Plant Cell Physiol* **29:** 689—693

Neuenschwander U, Suter M and Brunold C (1991) Regulation of sulfate assimilation by light and *O*-acetyl-L-serine in *Lemma minor* L. *Plant Physiol* **97:** 253—258

Noji M and Saito K (2002) Molecular and biochemical analysis of serine acetyltransferase and cysteine synthase towards sulfur metabolic engineering in plants. *Amino Acids* **22:** 231—243

Noji M, Murakoshi I and Saito K (1994) Molecular cloning of a cysteine synthase cDNA from *Citrullus vulgaris* (watermelon) by genetic complementation in an *Escherichia coli* Cys$^-$ auxotroph. *Mol Gen Genet* **244:** 57—66

Noji M, Inoue K, Kimura N, Gouda A and Saito K (1998) Isoform-dependent differences in feedback regulation and subcellular localization of serine acetyltransferase involved in cysteine biosynthesis from *Arabidopsis thaliana*. *J Biol Chem* **273:** 32739—32745

Noji M, Takagi Y, Kimura N, Inoue K, Saito M, Horikoshi M, Saito F, Takahashi H and Saito K (2001) Serine acetyltransferase involved in cysteine biosynthesis from spinach: Molecular cloning, characterization and expression analysis of cDNA encoding a plasticic isoform. *Plant Cell Physiol* **42:** 627—634

Petersen M, Van Der Straeten D and Bauw G (1995) Full-length cDNA clone from *Coleus Blumei* with high similarity to cobalamine independent methionine synthase. *Plant Physiol* **109:** 338

Ravanel S, Droux M and Douce R (1995a) Methionine biosynthesis in higher plants. I. Purification and characterization of cystathionine γ-synthase from spinach chloroplasts. *Arch Biochem Biophys* **316:** 572—584

Ravanel S, Ruffet ML and Douce R (1995b) Cloning of an *Arabidopsis thaliana* cDNA encoding cystathionine β-lyase by functional complementation in *Escherichia coli*. *Plant Mol Biol* **29:** 875--882

Ravanel S, Gakiere B, Job D and Douce R (1998a) Cystathionine γ-synthase from *Arabidopsis thaliana*: purification and biochemical characterization of the recombinant enzymes overexpressed in *Escherichia coli*. *Biochem J* **331:** 639—648

Ravanel S, Gakiere B, Job D and Douce R (1998b) The specific features of methionine biosynthesis and metabolism in plants. *Proc Natl Acad Sci USA* **95:** 7805—7812

Riedel K, Mangelsdorf C, Streber W, Willmitzer L, Hoefgen R and Hesse H (1999) Cloning and characterization of cystathionine γ-synthase from *Solanum tuberosum* L. *Plant Biol* **1:** 638—644

Roberts MA and Wray JL (1996) Cloning and characterization of an *Arabidopsis* cDNA clone encoding an organellar isoform of serine acetyltransferase. *Plant Mol Biol* **30:** 1041—1049

Rolland N, Droux M, Lebrun M and Douce R (1993) *O*-Acetylserine (thiol) lyase from spinach (*Spinacia oleracea* L) leaf: cDNA cloning, characterization, and overexpression in *Escherichia coli* of the chloroplast isoform. *Arch Biochem Biophys* **300:** 213—222

Rolland N, Ruffet ML, Job D, Douce R and Droux M (1996) Spinach chloroplast *O*-acetylserine(thiol)lyase exhibits two catalytically nonequivalent pyridoxal-5'-phosphate-containing active sites. *Euro J Biochem* **236:** 272—282

Römer S, d'Harlingue A, Camara B, Schantz R and Kuntz M (1992) Cysteine synthase from *Capsicum annuum* chromoplasts. *J Biol Chem* **267:** 17966—17970

Ruffet ML, Droux M and Douce R (1994) Purification and kinetic properties of serine acetyltransferase free of *O*-acetylserine (thiol) lyase from spinach chloroplasts. *Plant Physiol* **104:** 597—604

Ruffet ML, Lebrun M, Droux M and Douce R (1995) Subcellular distribution of serine acetyltransferase from *Pisum sativum* and characterization of an *Arabidopsis thaliana* putative cytosolic isoform. *Euro J Biochem* **227:** 500—509

Saito K (1999) Biosynthesis of cysteine. In: Singh BK (ed) Plant Amino Acids: Biochemistry and Biotechnology, pp 267—291. Marcel Dekker Inc., New York

Saito K (2000) Regulation of sulfate transport and synthesis of sulfur-containing amino acids. *Curr Opin Plant Biol* **3:** 188—195

Saito K, Miura N, Yamazaki M, Hirano H and Murakoshi I (1992) Molecular cloning and bacterial expression of cDNA encoding a plant cysteine synthase. *Proc Natl Acad Sci USA* **89:** 8078—8082

Saito K, Tatsuguchi K, Murakoshi I and Hirano H (1993a) cDNA cloning and expression of cysteine synthase B localized in chloroplasts of *Spinacia oleracea*. *FEBS Lett* **324:** 247—252

Saito K, Kurosawa M and Murakoshi I (1993b) Determination of a functional lysine residue of a plant cysteine synthase by site-directed mutagenesis, and molecular evolutionary implication. *FEBS Lett* **328:** 111—114

Saito K, Tatsuguchi K, Takagi Y and Murakoshi I (1994a) Isolation and characterization of cDNA that encodes a putative mitochondrion-localizing isoform of cysteine synthase [O-acetylserine (thiol)-lyase] from Spinacia oleracea. *J Biol Chem* **269:** 28187—28192

Saito K, Kurosawa M, Tatsuguchi K, Takagi Y and Murakoshi I (1994b) Modulation of cysteine biosynthesis in chloroplasts of transgenic tobacco overexpressing cysteine synthase [*O*-acetylserine (thiol)-lyase]. *Plant Physiol* **106:** 887—895

Saito K, Yokoyama H, Noji M and Murakoshi I (1995) Molecular cloning and characterization of a plant serine acetyltransferase playing a regulatory role in cysteine biosynthesis from watermelon. *J Biol Chem* **270:** 16321—16326

Shen B, Li C and Tarczynski MC (2002) High free-methionine and decreased lignin content result from a mutation in the *Arabidopsis* S-adenosyl-L-methionine synthetase 3 gene. *Plant J* **29:** 371—380

Smith FW, Hawkesford MJ, Ealing PM Clarkson DT, Vanden Berg PJ, Belcher AR and Warrilow AGS (1997) Regulation of expression of a cDNA from barley roots encoding a high affinity sulphate transporter. *Plant J* **12:** 875—884

Takahashi H and Saito K (1996) Subcellular localization of spinach cysteine synthase isoforms and regulation of their gene expression by nitrogen and sulfur. *Plant Physiol* **112:** 273—280

Turner WL, Pallett KE and Lea PJ (1998) Cystathionine β-lyase from *Echinochloa colonum* tissue culture. *Phytochemistry* **47:** 189—196

Urano Y, Manabe T, Noji M and Saito K (2000) Molecular cloning and functional characterization of cDNAs encoding cysteine synthase and serine acetyltransferase that may be responsible for high cellular cysteine content in *Allium tuberosum*. *Gene* **257:** 169—277

Warrilow AGS and Hawkesford MJ (1998) Separation, subcellular location and influence of sulphur nutrition on isoforms of cysteine synthase in spinach. *J Exp Bot* **49:** 1625—1636

Youssefian S, Nakamura M and Sano H (1993) Tobacco plants transformed with the *O*-acetylserine (thiol) lyase gene of wheat are resistant to toxic levels of hydrogen sulphide gas. *Plant J* **4:** 759—769

5. AFFILIATIONS

Department of Molecular Biology and Biotechnology, Graduate School of Pharmaceutical Sciences, Chiba University, Yayoi-cho 1-33, Inage-ku, Chiba 263-8522, Japan. For correspondence - *K. Saito:* ksaito@p.chiba-u.ac.jp

E. GLAWISCHNIG, M.D. MIKKELSEN AND B.A. HALKIER

GLUCOSINOLATES: BIOSYNTHESIS AND METABOLISM

Abstract. Glucosinolates are amino acid derived secondary metabolites of the Capparales order, which include the *Brassica* crops and the model plant *Arabidopsis*. Tissue disruption brings glucosinolates into contact with myrosinases, thereby producing numerous biologically active compounds (e.g. isothiocyanates, nitriles, and thiocyanates). Following the completion of the *Arabidopsis* genome, significant progress has been made towards understanding the biosynthetic pathway of glucosinolates as genes involved in amino acid chain-elongation, glucosinolate core structure biosynthesis and side chain modification have been identified and characterized. These results have given new insight into glucosinolate metabolism within the metabolic network of the plant. Furthermore, they have provided important tools for metabolic engineering of glucosinolate profiles to improve nutritional value and pest resistance of crop plants.

1. INTRODUCTION

Glucosinolates are plant secondary metabolites consisting of a thioglucose moiety, a sulphonated aldoxime, and a variable side chain derived from amino acids (Fig. 1). When the plant tissue is disrupted glucosinolates are hydrolyzed by endogenous thioglucosidases, myrosinases to produce a wide range of products namely nitriles,

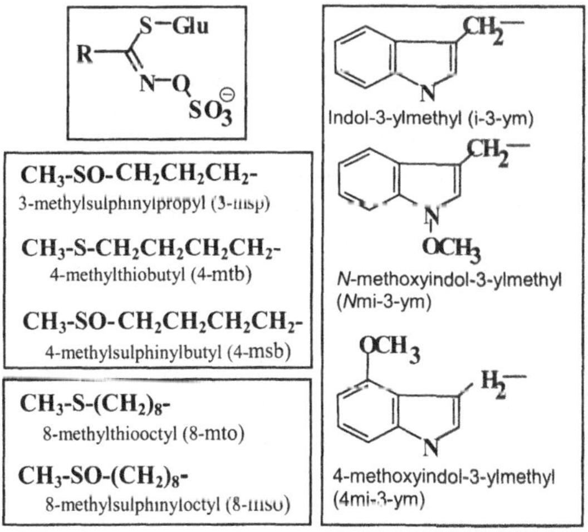

Figure 1. The major glucosinolates in Arabidopsis thaliana ecotype Columbia. Side chains (R) are shown

Yash P. Abrol and Altaf Ahmad (eds.), Sulphur in Plants, 145—162.

thiocyanates, isothiocyanates and oxazolidine-2-thiones with a variety of biological effects (Fig. 2). Many of these are well known as characteristic flavor compounds in *Brassica* vegetables (e.g. cabbage, cauliflower) and condiments such as mustard.

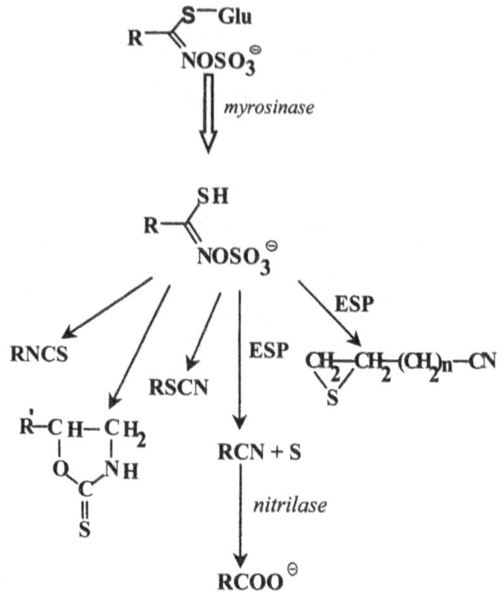

Figure 2. Degradation of glucosinolates. Products of known genes are indicated. Nitrilases that convert nitriles to carboxylic acids have been suggested to be involved in glucosinolate degradation

Isothiocyanates have antibacterial and antifungal activities and can act as insect repellents (Giamoustaris and Mithen, 1995). On the contrary, they can function as attractants for specialist insects (Chew, 1988). A well-studied effect on animals is the growth inhibitory activity of 5-vinyl-oxazolidine, a degradation product of 2-hydroxy-3-butenyl glucosinolate, one of the major glucosinolates in rapeseed. This glucosinolate causes the goitre disease in animals fed with the seed meal, a by-product of rapeseed oil production (Fenwick et al., 1983). As the seed meal is rich in protein and would otherwise be a valuable animal feed, an elimination of 2-hydroxy-3-butenyl glucosinolate in rapeseed is desirable.

Some glucosinolates are beneficial for human health. 4-methylsulphinyl-butyl isothiocyanate (sulphoraphane) is a very potent cancer-preventive phytochemical (Zhang et al., 1992) as it induces a quinone reductase, a phase II enzyme in the detoxification process in mammals. Sulphoraphane is derived from 4-methyl-sulphinylbutyl glucosinolate, which is a major glucosinolate in *Brassica* vegetables, such as broccoli. Control of the composition of glucosinolates will make it possible to improve nutritional qualities and flavor of food crops.

This review focuses on the recent advances in the elucidation of glucosinolate biosynthesis. Several different approaches have led to the identification of biosynthetic genes, which provides important tools for metabolic engineering of the glucosinolate composition. Characterization of many of the biosynthetic genes have revealed new aspects of the interplay within the network of primary and secondary metabolism, as exemplified by the inter-relationship between biosynthetic pathways of indole glucosinolates and the plant hormone auxin.

2. TAXONOMIC DISTRIBUTION

More than 120 different glucosinolates have been identified (see review Fahey et al., 2001). Glucosinolates have only been found in families of dicotyledonous plants belonging to the order Capparales (*Brassicaceae, Capparaceae, Resedaceae, Tovariaceae, Akaniaceae, Bataceae, Bretschneideraceae, Caricaceae, Koeberlinaceae, Limnanthaceae, Morginaceae, Pentadiplandraceae, Salvadoraceae, Tropaeolaceae*) and in the genus *Drypetes* (*Euphorbiaceae*) (Rodman et al., 1998). Sub-classes of glucosinolates are restricted to a small number of families, e.g. indole glucosinolates are found in *Brassicaceae, Capparaceae, Tropaeolaceae, Bataceae* and *Resedaceae* (Griffiths et al., 2001) and chain-elongated glucosinolates are restricted to *Brassicaceae*, *Capparaceae* and *Resedaceae* (Ettlinger and Kjær, 1968). The occurrence of glucosinolates in *Drypetes* is an interesting example of secondary metabolites being synthesized in species distantly related to the families in which these compounds are typically found. A similar phenomenon has also been observed for benzoxazinoids that are important secondary metabolites of the *Gramineae*, but are found in few dicotyledonous species (see review Sicker et al., 2000). Identification of biosynthetic genes in the "outgroup" species will elucidate whether convergent evolution of secondary metabolite biosynthesis also involves convergent evolution of biosynthetic genes or whether unrelated enzyme systems are involved in distantly related species.

3. BIOSYNTHESIS

3.1 Chain Elongation

Many glucosinolates are derived from chain-elongated derivatives of methionine and phenylalanine. The derivatives of methionine (CH_3-S-$(CH_2)_2$-$(CH_2)_n$-$CH(NH_2)$-COOH) with 'n' being as high as 9 have been identified. In *Arabidopsis,* predominantly homo-, dihomo-, and hexahomomethionine-derived glucosinolates are found. On the basis of isotope labeling in *Eruca sativa* (Graser et al., 2000) and *Arabidopsis* (Graser et al., 2001), a three-step chain elongation cycle has been proposed. This cycle begins with condensation of acetyl-CoA and a 2-oxo acid derived from methionine and ends with an oxidative decarboxylation forming a new 2-oxo acid with one additional methylene group.

The ratio between the homomethionine-derived propyl glucosinolate and the dihomomethionine-derived butyl glucosinolate varies dramatically between

Arabidopsis ecotypes and a QTL responsible for this variation has been identified. Recombinant inbred lines of Landsberg *erecta*, which synthesizes propyl glucosinolates, and Colombia, which predominantly synthesizes butyl glucosinolates, were used to map the QTL that enabled identification of two members of an isopropylmalate synthase-like gene family (De Quiros et al., 2000). For one of the genes, *MAM1*, two mutant alleles have been identified that are deficient in C_4-glucosinolates (Kroymann et al., 2001). Recombinant *MAM1* expressed in *E. coli* showed methylthioalkylmalate synthase activity. This provided further evidence that *MAM1* controls the chain-elongation of methionine with one or two methylene groups. The function of *MAM*-like other gene is still unknown, although high homology to *MAM1* suggests a similar function. *Arabidopsis* carrying a homozygous *GSL-ELONG* null allele still produces methionine-derived long chain aliphatic glucosinolates. A separate pathway for the biosynthesis of long-chain aliphatic glucosinolates, which does not allow flux of short chain-elongated methionine derivatives into short chain aliphatic glucosinolates, has been suggested (Kroyman et al., 2001). Interestingly, these two pathways partly merge again at the next step, as CYP79F1 converts both short and long chain-elongated methionine derivatives to the corresponding aldoximes (see 3.2).

3.2 Biosynthesis of the Core Structure of Glucosinolates

3.2.1 Formation of aldoximes from amino acids

Completion of the *Arabidopsis* sequencing project and the availability of extensive mutant collections has led to substantial progress in identification of genes involved in the biosynthesis of the core glucosinolate structure (Fig. 3; Table 1). The conversion of amino acids to aldoximes, a key step in the biosynthesis of glucosinolates, has been shown in several plants to be catalyzed by cytochrome P450 enzymes. Recently, several independent approaches suggested that cytochrome P450 enzymes belongs to the CYP79 family.

CYP79A1 from *Sorghum bicolor* converts tyrosine to *p*-hydroxyphenylacetaldoxime in the biosynthesis of the cyanogenic glucoside, dhurrin (Koch et al., 1995; Sibbesen et al., 1995). It was hypothesized that members of the CYP79 family catalyzed aldoxime formation in glucosinolate biosynthesis. The genome of *Arabidopsis* ecotype Columbia contains seven CYP79 genes: *CYP79A2*, *CYP79B2*, *CYP79B3*, *CYP79C1*, *CYP79C2*, *CYP79F1* and *CYP79F2* (see: http://www.biobase.dk/P450). In a functional genomics approach, most of the CYP79s have been functionally expressed and characterized with relation to their substrate specificities (Table 1; Halkier et al., 2002).

CYP79A2 converts phenylalanine to phenylacetaldoxime, the precursor of benzyl glucosinolate (Wittstock and Halkier, 2000) and both CYP79B2 and CYP79B3 convert tryptophan to IAOx (Hull et al., 2000; Mikkelsen et al., 2000), a precursor of indole glucosinolates and potentially of IAA (see 5.4). *CYP79B2* has been identified in a yeast screen conferring resistance to 5-fluorotryptophan (Hull et al., 2000) as well as in a functional genomics approach (Mikkelsen et al., 2000).

CYP79F1, expressed in *E. coli,* converts chain-elongated methionine derivatives

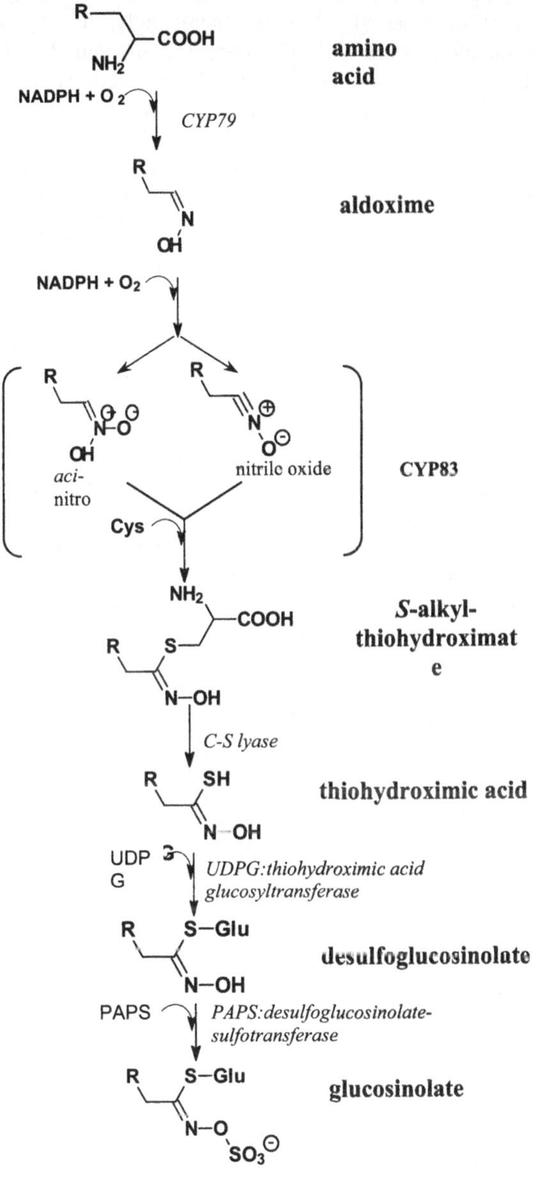

Figure 3. General scheme of the biosynthesis of the core structure of glucosinolates. The cytochrome P450 enzymes catalyzing the first two steps have been cloned in Arabidopsis

to their corresponding aldoximes. The substrate specificity is broad with respect to the chain length as the substrates range from homomethionine to hexahomomethionine (Hansen et al., 2001a). Interestingly, the K_m values for tetra- and pentahomomethionine are high, reflecting the low abundance of the corresponding hexyl and heptyl glucosinolates in the Columbia ecotype. CYP79F2, which is 88% identical to CYP79F1 at the amino acid level, has a similar function, as evidenced by the characterization of recombinant CYP79F2 expressed in *Saccharomyces cerevisiae*. However, CYP79F2 metabolizes only hexahomo- methionine and with lower affinity pentahomomethionine (S Chen, E Glawischnig, P Naur and BA Halkier, unpublished). The expression patterns of both genes differ substantially (Table 1). While CYP79F1 is expressed at high levels in cotyledons, rosette, stem and siliquae, CYP79F2 is highly expressed in hypocotyl and roots. The substrate specificity is reflected in the glucosinolate composition of transposon tagged null mutants. A CYP79F1 mutant completely lacks short-chain aliphatic glucosinolates (Reintanz et al., 2001). In contrast, the levels of long-chain elongated glucosinolates in a CYP79F2 mutant are reduced, while the level of short-chain aliphatic glucosinolates is not affected.

The role of the CYP79C homologues is unclear, as it is not known whether these genes are expressed or they encode functional enzymes. A possible function is the biosynthesis of homophenylalanine-derived glucosinolates, for which no CYP79 has been assigned. CYP79Cs share the highest homology with CYP79Fs, which makes aliphatic substrates seem more likely. Possibly, the CYP79C homologues are involved in the biosynthesis of branched aliphatic glucosinolates, such as methylethyl glucosinolate, found in the seeds of some *Arabidopsis* ecotypes.

Table 1. Substrate specificity and expression of CYP79s in Arabidopsis. Only tissues with high transcript level and high GUS expression, respectively are included

Gene	**Substrates**	**Expression pattern RT-PCR/ Northern Blot**	**Expression pattern GUS-staining**	**Ref.**
CYP79A2	phenylalanine	nd	nd	a)
CYP79B2	tryptophan	rosette leaves, stem, strongest in roots	rosette leaves (mainly vascular tissue), cotyledons, young roots	b) c)
CYP79B3	tryptophan	nd	nd	b)
CYP79F1	homo-, dihomo-, trihomo-, tetrahomo-, pentahomo-, and hexahomomethionine	rosette leaves (all stages), stem and siliques	all upper parts of the plant in roots restricted to meristem and vascular tissue	d); e); f); g)
CYP79F2	penta- and hexahomomethionine	root	root, upper hypocotyl, vascular tissue of leaves	g)

nd = not determined. Sources: (a) Wittstock and Halkier, 2000; (b) Hull et al., 2000; (c) Mikkelsen et al., 2000; (d) Hansen et al., 2001a; (e) Reintanz et al., 2001; (f) Tantikanjana et al., 2001; (g) S Chen, E Glawischnig, P Naur and BA Halkier (unpublished)

3.2.2 Formation of glucosinolates from aldoximes

The next step in glucosinolate biosynthesis is the conversion of aldoximes into thiohydroximates. The S-atom in the thiohydroximate intermediate is likely to be derived from cysteine as evidenced by *in vivo* feeding studies (Wetter and Chisholm, 1968). A proposed reaction mechanism involves the oxidation of the aldoxime to an *aci*-nitro compound or a nitrile oxide, which functions as the acceptor for the thiol donor (Ettlinger and Kjaer, 1968). In *Arabidopsis*, CYP83A1 and CYP83B1 have been shown to catalyze the oxidation of various aldoximes (Bak et al., 2001; Bak and Feyereisen, 2001; Hansen et al., 2001a) to *S*-alkylthio-hydroximates when S-donors are added to the reaction. CYP83A1 and CYP83B1 are non-redundant under physiological conditions, as CYP83B1 has high affinity for tryptophan- and phenylalanine-derived aldoximes while CYP83A1 has high affinity for the aliphatic aldoximes (Bak and Feyereisen, 2001, P Naur and BA Halkier, unpublished). It has been suggested that the formation of thiohydroximic acid from S-alkylthiohydroximate is catalyzed by a C-S lyase (Wallsgrove and Bennett, 1995). No C-S lyase in the glucosinolate pathway has yet been cloned. It remains an open question whether different C-S lyases with different substrate specificities are involved in the pathway or if only a single C-S lyase with broad substrate specificity exists.

The thiohydroximic acid is further S-glucosylated by a soluble UDPG-thiohydroximic acid glucosyltransferase (S-GT), which has been purified from several species (Jain et al., 1990; Guo and Poulton, 1994). The S-GT purified from *Brassica juncea* was specific for the thiohydroximate moiety but accepts variations of the side chain of thiohydroximic acids (GrootWassink et al., 1994). The amino acid sequence information of an S-GT isolated from *B. oleracea ssp. botrytis* was used to clone a potential S-GT gene form *B. napus* (Marillia et al., 2001). The enzyme was expressed in *E. coli* and shown to incorporate ^{14}C from ^{14}C-UDP-glucose when phenylacetothiohydroximic acid was used as a substrate. A candidate S-GT, showing 84% amino acid sequence similarity, has been identified in the *Arabidopsis* genome. PAPS:desulphoglucosinolate-sulphotransferases, catalyzing the final biosynthe-tic step of the glucosinolate core structure, have been purified from *B. juncea* and *Lepidium sativa*. The enzymes showed little substrate specificity with respect to the side chains of the desulphoglucosinolates (Glendening and Poulton, 1988; Jain et al., 1990) In the *Arabidopsis* genome, many putative sulphotransferases have been identified (Marsolais et al., 2000), but none of the enzymes have been assigned a function in glucosinolate biosynthesis.

3.3 Side Chain Modification and Glucosinolate Diversity in Arabidopsis

The great structural diversity of glucosinolates is to a large extent due to secondary modifications, particularly of the side chains of chain elongated methionine-derived glucosinolates. These include oxidation of the thio-sulphur, hydroxylation, methoxylation, desaturation, sulphation and glucosylation. The great variation in the composition of aliphatic glucosinolates among different *Arabidopsis* ecotypes has been used to identify QTLs responsible for this variation (Kliebenstein et al., 2001a). Recently, the genes encoding for *GS-ALK*, *GS-OHP* and *GS-OH* have been identified (Fig. 4). They encode 2-oxoacid dependent dioxygenases. *GS-ALK (AOP2)* and *GS-OHP (AOP3)* are clustered on chromosome IV and responsible for the conversion of methylsulphinylalkyl glucosinolates to alkenyl glucosinolates and hydroxyalkyl glucosinolates, respectively (Kliebenstein et al., 2001b). *AOP2* and *AOP3* have probably derived from gene duplication and evolved to different functions, which appear to be mutually exclusive as no ecotype expressing both genes has been identified.

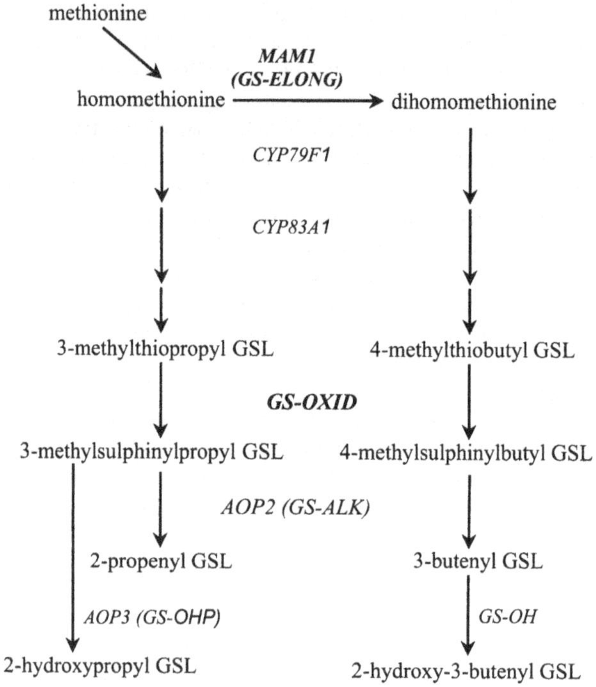

Figure 4. Biosynthesis of short chain aliphatic glucosinolates in Arabidopsis. The products of known genes are indicated. Known QTLs are shown in italics and in brackets when the corresponding gene has been cloned and named differently. Abbreviation: GSL, glucosinolates

The *GS-OH* gene encodes a 2-oxoacid dependent dioxygenase that is only distantly related to AOP2 and AOP3. GS-OH converts but-3-enyl glucosinolate to 2-hydroxybut-3-enyl glucosinolate (Kliebenstein et al., personal communication). Interestingly, a number of ecotypes express a functional *GS-OH* allele, despite the lack of the substrate 3-butenyl glucosinolates. Such hidden enzyme activities should be taken into account in genetic engineering approaches, as they can result in the formation of unexpected compounds (Kliebenstein et al., personal communication).

4. DEGRADATION

Glucosinolates co-occur with endogenous thioglucosidases called myrosinases. Upon tissue disruption, the myrosinases hydrolyze glucosinolates into unstable aglucones, which decompose to a variety of products, dependent on several factors such as e.g. pH, Fe^{2+} concentration and epithiopecifier proteins (Fig. 2). A large number of genes encoding myrosinases have been identified in *Brassica napus*, whereas only 3 myrosinase genes have been identified in the *Arabidopsis* genome (see review Rask et al., 2000). Myrosinases are found in specialized idioblastic myrosin cells (Höglund et al., 1991). In *Arabidopsis* the myrosin cells are exclusively localized in the phloem parenchyma, whereas in *Brassica napus* myrosin cells are detected both in the phloem parenchyma and in the ground tissue (Andreasson et al., 2001). Myrosinases are often found in complexes with myrosinase binding proteins, which have no known function. They may not be required for myrosinase activity as *Brassica napus* MYR1 expressed in *Saccharomyces cerevisiae* was active without the association of myrosinase binding proteins (Chen and Halkier, 1999).

A QTL affecting the composition of glucosinolate breakdown products has been identified in *Arabidopsis* (Jander et al., 2001; Lambrix et al., 2001). Based on this QTL, a gene encoding an epithiospecifier protein (ESP) has been cloned (Lambrix et al., 2001). ESPs have previously been characterized biochemically in different members of the *Brassicaceae* (Tookey, 1973; MacLeod and Rossiter, 1985). They promote the formation of epithionitriles and nitriles instead of isothiocyanates. The generalist insect herbivore, *Trichoplusia ni*, strongly prefers plants that express a functional ESP and, therefore, produce nitriles and epithionitriles upon insect feeding. Expression of an ESP is, therefore, related to reduced resistance against insect feeding. The ecological function of the ESPs remains unclear.

5. GLUCOSINOLATE BIOSYNTHESIS AS PART OF A METABOLIC NETWORK

5.1 Regulation of Glucosinolate Biosynthesis

The biosynthesis of glucosinolates is developmentally regulated and several biosynthetic genes are tissue specifically expressed (for the CYP79s, see table 1). In addition, wounding, insect feeding and infection with fungal and bacterial pathogens can systemically induce glucosinolate biosynthesis. The induction of indole glucosinolates by various stress factors is in accordance with the stress inducibility of tryptophan biosynthetic genes (Zhao et al., 1998) as indole glucosinolates are major tryptophan sinks. A feedback regulation of indole glucosinolate biosynthesis via the Myb factor ATR1 has been proposed (Smolen and Bender, 2002). The biosynthesis of gluconasturtiin (2-phenylethyl glucosinolate) in *B. napus* is specifically induced by salicylic acid (Kiddle et al., 1994), and the induction of indole glucosinolates in *Arabidopsis* is mediated by jasmonic acid (Brader et al., 2001, Mikkelsen et al., 2003). The induction is likely regulated at the rate limiting aldoxime formation step. *CYP79B* transcription is induced by infection with *Pseudomonas syringae* (Hull et al., 2000) and strong GUS expression has been observed in wounded tissue of

CYP79B2::GUS plants (Mikkelsen et al., 2000). Brader et al. (2001) showed that *CYP79B3*, but not *CYP79B2*, was induced by treatment with jasmonic acid as well as a cultural filtrate of *Erwinia carotovora.* Under different experimental conditions, transcription of both *CYP79B2* and *CYP79B3* was induced by methyl jasmonate (Mikkelsen et al., 2003). Whether both genes are expressed in response to wounding and pathogen attack can be investigated using CYP79B knockout mutants. Generally, it appears that indole glucosinolates are induced under various experimental conditions. Further studies are needed to clarify whether the aliphatic glucosinolates are subject also to induction.

5.2 Indole Glucosinolates and Phytoalexins in the Brassicaceae

During the hypersensitive response to infection with *Pseudomonas syringae* strains *Arabidopsis* leaves accumulate the phytoalexin camalexin (3-thiazol-2'-yl-indole; Tsuji et al., 1992). In this case an indole-containing phytoalexin and indole glucosinolates co-exist in *Arabidopsis* leaves. The tryptophan dependence of indole glucosinolate biosynthesis has not been questioned. However, specific incorporation of ^{14}C labeled tryptophan into camalexin was 5 to 20 times lower than incorporation of labeled anthranilate (Tsuji et al, 1993; Zook and Hammerschmidt, 1997). Based on labeling experiments with *Arabidopsis* cell cultures it was proposed that free indole is the precursor of the indole ring of camalexin. Furthermore, it was suggested that an enzyme similar to BX1 (Frey et al., 1997), synthesizing free indole in benzoxazinoid biosynthesis in maize, was involved (Zook, 1998).

Besides camalexin, a great variety of phytoalexins are synthesized in the *Brassicaceae* (see review Pedras et al., 2000). They are composed of an indole moiety that is substituted in the 3-position, often with sulphur and nitrogen containing side chains or ring systems. These structural similarities suggest common biosynthetic precursors and several of these phytoalexins are derived from tryptophan (Monde et al., 1994). Initially, brassinin was thought to be a product of glucobrassicin, but in recent feeding experiments with turnip (*Brassica rapa*) roots, efficient IAOx incorporation into brassinin and brassinin-derived phytoalexins was observed (Pedras et al., 2001). IAOx as metabolic branch point of indole glucosinolates and phytoalexin biosynthesis in *B. rapa* raises the question of whether also camalexin could be synthesized from IAOx. In agreement with this hypothesis, *CYP79B* transcripts are strongly induced in *Arabidopsis* leaves after *Pseudomonas syringae* infection with a kinetic similar to camalexin formation (Hull et al., 2000). Further investigations are necessary to clarify whether this is coincident and there are tryptophan-dependent and tryptophan-independent pathways for the biosynthesis of structurally related phytoalexins in the *Brassicaceae*.

5.3 Evolution of Glucosinolate Biosynthesis

Glucosinolates are related to cyanogenic glucosides as both groups of natural products are derived from amino acids that are converted into aldoximes by cytochrome P450 enzymes belonging to the CYP79 family. Furthermore, the aldoxime-metabolizing enzymes are evolutionarily related, as are the specific degradative enzymes. Theses two pathways are nearly mutually exclusive. Only *Carica papaya* is known to contain both groups of secondary metabolites (Tang, 1971; Spencer and Seigler, 1984; Bennett et al., 1997). Cyanogenic glucosides occur widely throughout the plant kingdom. Glucosinolates, however, are found exclusively among the order Capparales and in the genus *Drypetes* in the evolutionarily distant order Euphorbiales (Rodman et al., 1998). This has led to the hypothesis that the cyanogenic glucoside biosynthesis is an ancient pathway from which glucosinolate biosynthesis has evolved. The first step unique to glucosinolate biosynthesis is the formation of *aci*-nitro compounds or nitrile oxides from the intermediate aldoximes. This step is catalyzed by CYP83, which based on homology is grouped within the CYP71 family. It has been proposed (Hansen et al., 2001a), that this enzyme evolved from a member of the CYP71 family that oxidizes an aldoxime to the corresponding nitrile in the biosynthesis of cyanogenic glucosides. The reactive compound produced by this 'mutated' enzyme was then detoxified by conjugation with a nucleophile (the S-donor), followed by a transglycosylation and sulphation reaction resulting in the production of a glucosinolate (Fig. 5). In agreement with this hypothesis, myrosinases could have evolved from the β-glucosidases responsible for the hydrolysis of cyanogenic glucosides. It has been argued that glucosinolate biosynthesis may not necessarily have evolved from cyanogenic

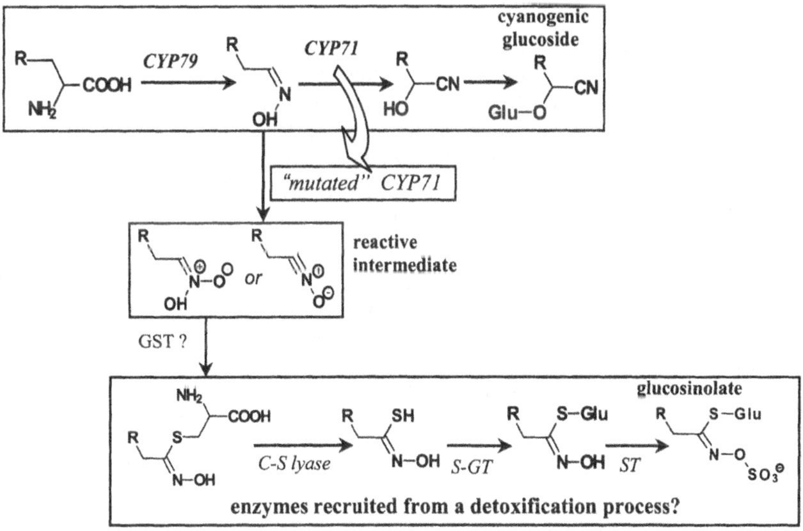

Figure 5. Hypothetical model for the evolution of glucosinolate biosynthesis. An aldoxime-metabolizing enzyme in the cyanogenic pathway was mutated into an enzyme, the ancestor of the CYP83 family, which converted the aldoxime into a toxic compound such as e.g. an aci-*nitro compound or a nitrile oxide, which got detoxified into a glucosinolate. In accordance with this hypothesis, the post-oxime enzymes appear to have been recruited from the detoxification process. Abbreviations: S-GT, UDPG:thiohydroximic acid glucosyltransferase; ST, PAPS:desulphoglucosinolate-sulphotransferase*

glucoside biosynthesis, but alternatively be evolutionarily related to the IAA biosynthetic pathway (Bak et al., 2001).

5.4 Metabolic Engineering of Glucosinolate Biosynthesis

The identification of genes involved in glucosinolate biosynthesis provides important tools for metabolic engineering of the glucosinolate composition (see review Mikkelsen et al., 2002), which will ultimately result in plants with increased resistance against pathogens and improved nutritional value. Overexpression of *CYP79* genes has resulted in plants with dramatically elevated levels of specific glucosinolates derived from protein amino acids. *Arabidopsis* plants expressing CYP79A2 under the control of the 35S promoter accumulate benzyl glucosinolate, which in wild type plants is only observed at very low concentrations (Wittstock and Halkier, 2000). In *Arabidopsis* plants overexpressing CYP79B2, a significant increase in the level of indole glucosinolates is observed (Mikkelsen et al., 2000). This indicates that the downstream enzymes are not rate limiting. In addition, expression of exogenous CYP79 homologues from cyanogenic plants or other glucosinolate-producing species can produce plants accumulating glucosinolates that are not naturally constituents of the given species. As an example, *Arabidopsis* expressing CYP79A1 from sorghum synthesizes large quantities of *p*-hydroxybenzyl glucosinolate (Bak et al., 1999).

Metabolic engineering of the glucosinolate side chain is of particular importance as the physico-chemical properties and the biological activity of the glucosinolate degradation products are to a large extent determined by the structure of the side chain. Rape plants with defective alleles of the *GS-OH* homologues are likely to not accumulate 2-hydroxy-3-butenyl glucosinolate, which causes goitre disease. Similarly oxidation of cancer-preventive 4-methylsulphinylbutyl glucosinolate will be reduced in plants with defective *GS-ALK* homologues. These goals will be major challenges, as modification of the glucosinolate biosynthesis can affect general plant metabolism, as exemplified by *CYP79F1* knockout mutant with altered plant hormone levels (Reintanz et al., 2001; Tantikanjana et al., 2001), resulting in unpredictable outcomes of metabolic engineering (see 5.5). Further studies are needed to unravel the fundamental underlying mechanisms at play in the metabolic network between glucosinolates and primary metabolism.

5.5 Interplay between Glucosinolate Biosynthesis and Hormone Metabolism

Knockout or ectopically overexpression of biosynthetic genes can have dramatic effects on primary plant metabolism. For examples, knockout mutants of the glucosinolate biosynthetic genes *CYP79F1* and *CYP83B1* have altered plant hormone levels and strong morphologic phenotypes. CYP79F1 knockout or co-suppressed plants that are devoid of short-chain aliphatic glucosinolates are characterized by formation of multiple axillary shoots and have, therefore, been named *bus1* (for *bushy*, Reintanz et al., 2001) and *supershoot* (Tantikanjana et al., 2001). Proliferate shoot branching through bud formation as well as bud growth, is connected with a 3 to 9-fold increase of Z-type cytokinin level (Tantikanjana et al.,

2001). Levels of free IAA, IAN and indole glucosinolates were also significantly higher (Reintanz et al., 2001). No function of glucosinolates as plant growth regulators is known. However, co-suppressed 35S:CYP79F1 plants accumulate 50 times more dihomomethionine than wild type, possibly resulting in a general disturbance of the plant's methionine metabolism (Hansen et al., 2001b). This would account for pleiotropic effects, as numerous processes in plants involve methylation by (methionine-derived) *S*-adenosyl-methionine.

Two CYP83B1 knockout mutants, *runt* (*rnt-1*, Winkler et al., 1998) and *superroot2* (Delarue et al, 1998), were identified using a reverse genetics approach. The mutants had elongated hypocotyls and showed strong apical dominance, which are features associated with auxin overproduction. In accordance, increased levels of IAA have been measured in *superroot2* (Barlier et al. 2000). The reverse phenotype has been observed in plants overexpressing CYP83B1 under control of the 35S promoter (Bak et al., 2001). Based on these results, it has been concluded that IAOx is a branchpoint between IAA and indole glucosinolate biosynthesis. The biosynthesis of the major auxin IAA in plants remains unclear, despite considerable efforts to identify biosynthetic intermediates and genes involved. A possible reason for this is the presence of multiple redundant pathways, which make it difficult to identify IAA deficient mutants (see review Normanly and Bartel, 1999; Celenza, 2001).

A proposed pathway involves the conversion of tryptophan to IAOx, which via IAN is converted to IAA (Fig. 6). The IAA pool is small in comparison with the indole glucosinolate pool, e.g. in seedlings of *Arabidopsis,* indole glucosinolates are found in concentrations 10 times higher than IAA conjugates (Müller and Weiler, 2000a). As IAOx and IAN are intermediates of glucosinolate metabolism, it is difficult to specifically address enzymatic reactions with IAOx or IAN as substrates or products in an IAA biosynthetic pathway, in glucosinolate-producing species. This makes *Arabidopsis* a poor model system for studying general IAA biosynthesis. Members of the CYP79B family are to date the only enzymes cloned that are able to convert tryptophan to IAOx (Hull et al., 2000; Mikkelsen et al., 2000; Naur et al., 2003). *Arabidopsis* plants expressing CYP79B2 under the control of the 35S promoter accumulated higher levels of indole glucosinolates. These plants do not show the same severe auxin overproducing phenotype as *CYP83B1* knockouts, probably because excess IAOx is channeled into glucosinolate biosynthesis through the highly specific enzyme CYP83B1 (K_m = 3 μM; Bak and Feyereisen, 2001).

Enzyme activities converting IAOx to IAN or indole-3-acetaldehyde have been

Figure 6. Possible interconnections between glucosinolate metabolism and IAA biosynthesis

found in several plants (Helminger et al., 1985). Plasma membranes isolated from seedlings of *Brassica campestris* metabolized IAOx to IAN with an apparent K_m of 6.3 μM (Ludwig-Müller and Hilgenberg, 1988). The involvement of such an enzyme activity in an IAA biosynthetic pathway via IAOx was questioned by the fact that auxin overproduction in the *cyp83b1* mutant is not suppressed in a *nitrilase1* mutant background (Bak et al., 2001). *Arabidopsis* has 4 genes encoding nitrilases (*NIT1* to *NIT4*), which have been proposed to convert IAN to IAA in vivo (Bartling et al., 1992; Bartel and Fink, 1994). The nitrilases are differentially expressed and NIT1 and NIT2 are strongly induced during infection with *Plasmodiophora brassicae*, which causes clubroot disease. Nitrilase expression has been correlated with IAA production during this disease (Grsic-Rausch et al., 2000). IAN applied exogenously to *Arabidopsis* results in an auxin overproduction phenotype, whereas a *nit1* mutant is resistant to these auxin effects (Normanly et al., 1997). In *Arabidopsis* (Normanly et al., 1997) or tobacco (Schmidt et al., 1996), overexpression of *NIT2* results in auxin effects at low IAN concentrations. Detailed analysis of the substrate specificity of the 4 nitrilases (Piotrowski et al., 2001; Vorwerk et al., 2001) showed that NIT4 is specific for β-cyano-L-alanine and possibly involved in detoxification. NIT1, NIT2, and NIT3 showed very broad substrate specificity. For all the three enzymes, IAN turnover was low and the apparent K_m value was high in comparison to other substrates tested. Glucosinolate breakdown results in formation of a variety of nitriles. The broad substrate specificity of NIT1, NIT2, and NIT3 and the fact that members of the *NIT1/NIT2/NIT3* subfamily of nitrilase genes have only been found in glucosinolate-producing plants suggests a biological role in glucosinolate degradation.

A flavin dependent monooxygenase has been proposed to be involved in IAA biosynthesis (*YUCCA*, Zhao et al., 2001). This enzyme family generally shows broad substrate specificity, but efficient channelling opens the possibility for the involvement of unspecific enzymes, like flavin dependent monooxygenases and nitrilases in biosynthetic pathways. There is immunological evidence that a nitrilase-like protein is involved in an IAA-synthase complex that has been identified in *Arabidopsis* (Müller and Weiler, 2000b). IAA formation by this complex is dependent on molecular oxygen, but cofactor and inhibitor studies could not confirm the involvement of a flavin dependent monooxygenase or a cytochrome P450 enzyme, like CYP79B2 or CYP79B3. The potential role of CYP79B2 or CYP79B3 in IAA biosynthesis will be elucidated by analysis of a double knockout mutant.

6. CONCLUSIONS

Recently, large progress towards understanding of glucosinolate biosynthesis has been made. A number of complementary approaches have been applied, such as e.g. the exploitation of natural variations between ecotypes, functional genomics on the CYP79 family and the analysis of knockout mutants. Future research will focus on the regulatory mechanisms of glucosinolate biosynthesis, the physiological role of glucosinolates in interaction with herbivory insects and microorganisms and on the turnover and transport of glucosinolates. Identification of regulators of glucosinolate biosynthesis will make it possible to metabolically engineer glucosinolate profiles in a tissue specific manner to improve disease resistance and nutritional value. With the rapid progress in glucosinolate research within the last few years, the modification of glucosinolate profile in crop plants has become a realistic possibility.

7. REFERENCES

Andreasson E, Jorgensen LB, Höglund AS, Rask L and Meijer J (2001) Different myrosinase and idioblast distribution in *Arabidopsis* and *Brassica napus*. *Plant Physiol* **127:** 1750—63

Bak S, Olsen CE, Petersen BL, Møller BL and Halkier BA (1999) Metabolic engineering of p-hydroxybenzylglucosinolate in *Arabidopsis* by expression of the cyanogenic CYP79A1 from Sorghum bicolor. *Plant J* **20:** 663—671

Bak S and Feyereisen R (2001) The involvement of two p450 enzymes, CYP83B1 and CYP83A1, in auxin homeostasis and glucosinolate biosynthesis. *Plant Physiol* **127:** 108—118

Bak S, Tax FE, Feldmann KA, Galbraith DA and Feyereisen R (2001) CYP83B1, a cytochrome P450 at the metabolic branchpoint in auxin and indole glucosinolate biosynthesis in *Arabidopsis thaliana*. *Plant Cell* **13:** 101—111

Barlier I, Kowalczyk M, Marchant A, Ljung K, Bhalerao R, Bennett M, Sandberg G and Bellini C (2000) The SUR2 gene of *Arabidopsis thaliana* encodes the cytochrome P450 CYP83B1, a modulator of auxin homeostasis. *Proc Natl Acad Sci USA* **97:** 14819—14824

Bartel B and Fink GR (1994) Differential regulation of an auxin-producing nitrilase gene family in *Arabidopsis thaliana*. *Proc Natl Acad Sci USA* **91:** 6649—6653

Bartling D, Seedorf M, Mithofer A and Weiler EW (1992) Cloning and expression of an *Arabidopsis* nitrilase which can convert indole-3-acetonitrile to the plant hormone, indole-3-acetic acid. *Euro J Biochem* **205:** 417—424

Bennett RN, Kiddle G and Wallsgrove RM (1997) Biosynthesis of benzylglucosinolate, cyanogenic glucosides and phenylpropanoids in *Carica papaya*. *Phytochemistry* **45:** 59—66

Brader G, Tas E and Palva ET (2001) Jasmonate-dependent induction of indole glucosinolates in *Arabidopsis* by culture filtrates of the nonspecific pathogen *Erwinia carotovora*. *Plant Physiol* **126:** 849—860

Celenza JL (2001) Metabolism of tyrosine and tryptophan — new genes for old pathways. *Curr Opin Plant Biol* **4:** 234—240

Chen S and Halkier BA (1999) Functional expression and characterization of the myrosinase MYR1 from *Brassica napus* in *Saccharomyces cerevisiae*. *Protein Exp Purif* **17:** 414—420

Chew FS (1988) Biological effects of glucosinolates. In: Cutler HG (ed) Biologically Active Natural Products Potential Use in Agriculture, pp 155—181. American Chemical Society, Washington DC

Delarue M, Prinsen E, Onckelen HV, Caboche M and Bellini C (1998) Sur2 mutations of *Arabidopsis thaliana* define a new locus involved in the control of auxin homeostatis. *Plant J* **14:** 603—611

De Quiros HC, Magrath R, McCallum D, Kroymann J, Schnabelrauch D, Mitchell-Olds T and Mithen R (2000) α-Keto acid elongation and glucosinolate biosynthesis in *Arabidopsis thaliana*. *Theor Appl Genet* **101:** 429—437

Ettlinger MG and Kjær A (1968) Sulfur compounds in plants. *Rec Adv Phytochem* **1:** 59—144

Fahey JW, Zalemann AT and Talalay P (2001) The chemical diversity and distribution of glucosinolates and isothiocyanates among plants. *Phytochemistry* **56:** 5—51

Fenwick GR, Heaney RK and Mullin WJ (1983) Glucosinolates and their breakdown products in food and food plants. *Crit Rev Food Sci Nutr* **18:** 123—201

Frey M, Chomet P, Glawischnig E, Stettner C, Grün S, Winklmair A, Eisenreich W, Bacher A, Meeley RB, Briggs SP, Simcox K and Gierl A (1997) Analysis of a chemical plant defense mechanism in grasses. *Science* **277:** 696—699

Giamoustaris A and Mithen R (1995) The effect of modifying the glucosinolate content of leaves of oilseed rape (*Brassica napus* ssp. oleifera) on its interaction with specialist and generalist pests. *Ann Appl Biol* **126:** 347—363

Glendening TM and Poulton JE (1988) Glucosinolate biosynthesis. Sulfation of desulfoglucosinolate by cell-free extracts of cress (*Lepidium sativum* L.) seedlings. *Plant Physiol* **86:** 319—321

Graser G, Schneider B, Oldham NJ and Gershenzon J (2000) The methionine chain elongation pathway in the biosynthesis of glucosinolates in *Eruca sativa* (Brassicaceae). *Arch Biochem Biophys* **15:** 411—419

Graser G, Oldham NJ, Brown PD, Temp U and Gershenzon J (2001) The biosynthesis of benzoic acid glucosinolate esters in *Arabidopsis thaliana*. *Phytochemistry* **57:** 23—32

Griffiths DW, Deighton N, Birch AN, Patrian B, Baur R and Stadler E. (2001) Identification of glucosinolates on the leaf surface of plants from the Cruciferae and other closely related species. *Phytochemistry* **57:** 693—700

GrootWassink JWD, Reed DW and Konenovsky AD (1994) Immonupurification and immono-characterization of the glucosinolate biosynthesis enzyme thiohydroximate S-glucosyltransferase. *Plant Physiol* **86:** 319—321

Grsic-Rausch S, Kobelt P, Siemens JM, Bischoff M and Ludwig-Müller J (2000) Expression and localization of nitrilase during symptom development of the clubroot disease in *Arabidopsis*. *Plant Physiol* **122:** 369—378

Guo L and Poulton JE (1994) Partial purification and characterization of *Arabidopsis thaliana* UDPG: thiohydroximate glucosyltransferase. *Phytochemistry* **36:** 1133—1138

Halkier BA, Hansen CH, Naur P, Mikkelsen MD and Wittstock U (2002) The role of cytochromes P450 in biosynthesis and evolution of glucosinolates. In: Romeo JT (ed) Recent Advances in Phytochemistry, Vol. 36, pp 223—248. Elsevier Science Publisher, Amsterdam

Hansen CH, Du L, Naur P, Olsen CE, Axelsen KB, Hick AJ, Pickett JA and Halkier BA (2001a) CYP83B1 is the oxime-metabolizing enzyme in the glucosinolate pathway in *Arabidopsis*. *J Biol Chem* **276:** 24790—24796

Hansen CH, Wittstock U, Olsen CE, Hick AJ, Pickett JA and Halkier BA (2001b) Cytochrome P450 CYP79F1 from *Arabidopsis* catalyzes the conversion of dihomomethionine and trihomomethionine to the corresponding aldoximes in the biosynthesis of aliphatic glucosinolates. *J Biol Chem* **276:** 11078—11085

Helminger J, Rausch T and Hilgenberg W (1985) Metabolism of ^{14}C-indole-3-acetaldoxime by hypocotyls of Chinese cabbage. *Phytochemistry* **24:** 2497—2502

Höglund AS, Lenman M, Falk A and Rask L (1991) Distribution of myrosinase in rapeseed tissues. *Plant Physiol* **95:** 213—221

Hull AK, Vij R and Celenza JL (2000) *Arabidopsis* cytochrome P450s that catalyze the first step of tryptophan-dependent indole-3-acetic acid biosynthesis. *Proc Natl Acad Sci USA* **97:** 2379—2384

Jain JC, GrootWassink, JWD, Kolenovsky AD and Underhill EW (1990) Purification and properties of 3-phosphoadenosine-5-phosphosulphate: desulphoglucosinolate sulphotransferase from *Brassica juncea* cell cultures. *Phytochemistry* **29:** 1425—1428

Jander G, Cui J, Nhan B, Pierce NE and Ausubel FM (2001) The TASTY locus on chromosome 1 of *Arabidopsis* affects feeding of the insect herbivore *Trichoplusia ni*. *Plant Physiol* **126:** 890—898

Kiddle GA, Doughty KJ and Wallsgrove RM (1994) Salicylic acid-induced accumulation of glucosinolates in oilseed rape leaves. *J Exp Biol* **45:** 1343—1346

Kliebenstein DJ, Kroymann J, Brown P, Figuth A, Pedersen D, Gershenzon J and Mitchell-Olds T (2001a) Genetic control of natural variation in *Arabidopsis* glucosinolate accumulation. *Plant Physiol* **126:** 811—825

Kliebenstein DJ, Lambrix VM, Reichelt M, Gershenzon J and Mitchell-Olds T (2001b) Gene duplication in the diversification of secondary metabolism: Tandem 2-oxoglutarate–dependent dioxygenases control glucosinolate biosynthesis in *Arabidopsis*. *Plant Cell* **13:** 681—693

Koch B, Sibbesen O, Svendsen I and Møller BL (1995) The primary sequence of cytochrome P-450tyr, the multifunctional N-hydroxylase catalyzing the conversion of L-tyrosine to p-hydroxy-phenylacetaldehyde oxime in the biosynthesis of the cyanogenic glucoside dhurrin in *Sorghum bicolor* (L.) Moench. *Arch Biochem Biophys* **323:** 177—186

Kroymann J, Textor S, Tokuhisa JG, Falk KL, Bartram S, Gershenzon J and Mitchell-Olds T (2001) A gene controlling variation in *Arabidopsis* glucosinolate composition is part of the methionine chain elongation pathway. *Plant Physiol* **127:** 1077—1088

Lambrix V, Reichelt M, Mitchell-Olds T, Kliebenstein DJ and Gershenzon J (2001) The *Arabidopsis* Epithiospecifier Protein promotes the hydrolysis of glucosinolates to nitriles and influences *Trichoplusia ni* herbivory. *Plant Cell* **13:** 2793—2807

Ludwig-Müller J and Hilgenberg W (1988) A plasma membrane-bound enzyme oxidases L-tryptophan to indole-3-acetaldoxime. *Physiol Plant* **74:** 240—250

MacLeod AJ and Rossiter JT (1985) The occurrence and activity of epithiospecifier protein in some

Cruciferae seeds. *Phytochemistry* **24:**1895—189

Marillia E-F, MacPherson JM, Tsang EWT, Van Audenhove KV, Keller WA and GrootWassink JWD (2001) Molecular cloning of a Brassica napus thiohydroximate S-glucosyltransferase gene and its expression in *Escherichia coli*. *Physiol Plant* **113:** 176—184

Marsolais F, Gidda SK, Boyd J and Varin L (2000) Plant soluble sulfotronsferases: structural and functional similarity with mammalian enzymes. In: Romeo JT, Ibrahim R, Varin L and de Luca V (eds) Evolution of Metabolic Pathways, pp 433—456, Elsevier Science Publisher, Amsterdam

Mikkelsen MD, Hansen CH, Wittstock U and Halkier BA (2000) Cytochrome P450 CYP79B2 from *Arabidopsis* catalyzes the conversion of tryptophan to indole-3-acetaldoxime, a precursor of indole glucosinolates and indole-3-acetic Acid. *J Biol Chem* **275:** 33712—33717

Mikkelsen MD, Petersen BL, Olsen CE and Halkier BA (2002) Biosynthesis and metabolic engineering of glucosinolates. *Amino Acids* **22:** 279—295

Mikkelsen MD, Petersen BL, Glawischnig E, Jensen AB, Anderasson E and Halkier BA (2003) Modulation of *CYP79* gene and glucosinolate profile in *Arabidopsis thaliana* by defense signaling pathways. *Plant Physiol (*in press)

Monde K, Takasugi M and Ohnishi T (1994) Studies on stress metabolites. 19. Biosynthesis of cruciferous phytoalexins. *J Am Chem Soc* **116:** 6650

Müller A and Weiler EW (2000a) Indolic constituents and indole-3-acetic acid biosynthesis in the wild-type and a tryptophan auxotroph mutant of *Arabidopsis thaliana*. *Planta* **211:** 855—863

Müller A and Weiler EW (2000b) IAA-Synthase, an enzyme complex from *Arabidopsis thaliana* catalyzing the formation of indole-3-acetic acid from (S)-tryptophan. *Biol Chem* **381:** 679—686

Naur P, Hansen CH, Bak S, Hansen BG, Jensen NB, Neilsen HL and Halkier BA (2003) *CYP79B1* from *Sinapis albs* converts tryptophan to indole-3-acetaldoxime. *Arch Biochem Biophys* (in press)

Normanly J and Bartel B (1999) Redundancy as a way of life: IAA metabolism. *Curr Opin Plant Biol* **2:** 207—213

Normanly J, Grisafi P, Fink GR and Bartel B (1997) *Arabidopsis* mutants resistant to the auxin effects of indole-3-acetonitrile are defective in the nitrilase encoded by the NIT1 gene. *Plant Cell* **9:** 1781—1790

Pedras MSC, Okanga FI, Zaharia IL and Khan AQ (2000) Phytoalexins from crucifers: synthesis, biosynthesis, and biotransformation. *Phytochemistry* **53:** 161—176

Pedras MSC, Montaut S, Xu Y, Khan AQ and Loukaci A (2001) Assembling the biosynthetic puzzle of crucifer metabolites: indole-3-acetaldoxime is incorporated efficiently into phytoalexins but glucobrassicin is not. *Chem Commun* 1572—1573

Piotrowski M, Schonfelder S and Weiler EW (2001) The *Arabidopsis thaliana* isogene NIT4 and its orthologs in tobacco encode beta-cyano-L-alanine hydratase/nitrilase. *J Biol Chem* **276:** 2616—2621

Rask L, Andreasson E, Ekbom B, Eriksson S, Pontoppidan B and Meijer J (2000) Myrosinase: Gene family evolution and herbivore defense in Brassicaceae. *Plant Mol Biol* **42:** 93—113

Reintanz B, Lehnen M, Reichelt M, Gershenzon J, Kowalczyk M, Sandberg G, Godde M, Uhl R and Palme K (2001) bus, a bushy *Arabidopsis* CYP79F1 knockout mutant with abolished synthesis of short chain aliphatic glucosinolates. *Plant Cell* **13:** 351—367

Rodman JE, Soltis PS, Soltis DE, Sytsma KJ and Karol KG (1998) Parallel evolution of glucosinolate biosynthesis inferred from congruent nuclear and plastid gene phylogenies. *Am J Bot* **85:** 997—1006

Schmidt RC, Müller A, Hain R, Bartling D and Weiler EW (1996) Transgenic tobacco plants expressing the *Arabidopsis thaliana* nitrilase II enzyme. *Plant J* **9:** 683—691

Sibbesen O, Koch B, Halkier BA and Møller BL (1995) Cytochrome $P450_{TYR}$ is a multifunctional heme-thiolate enzyme catalyzing the conversion of L-tyrosine to p-hydroxyphenylacetaldehyde oxime in the biosynthesis of the cyanogene glucoside dhurrin in *Sorghum bicolor* (L.) Moench. *J Biol Chem* **270:** 3506—3511

Sicker D, Frey M, Schulz M and Gierl A (2000) Role of natural benzoxazinones in the survival strategy of plants. *Int Rev Cytol* **198:** 319—346

Smolen G and Bender J (2002) *Arabidopsis* Cytochrome P450 cyp83B1 mutations activate the tryptophan biosynthetic pathway. *Genetics* **160:** 323—332

Spencer KC and Seigler DS (1984) Cyanogenic glucosides of Carica papaya and its phylogenetic position with respect to the violales and capparales. *Am J Bot* **71:** 1444—1447

Tang C-S (1971) Benzyl thioisozyanate of papaya fruit. *Phytochemistry* **15:** 117—131

Tantikanjana T, Yong JW, Letham DS, Griffith M, Hussain M, Ljung K, Sandberg G and Sundaresan V (2001) Control of axillary bud initiation and shoot architecture in *Arabidopsis* through the

SUPERSHOOT gene. *Genes Dev* **15:**1577—1588

Tsuji J, Jackson EP, Gage DA, Hammerschmidt R and Somerville SC (1992) Phytoalexin accumulation in Arabidopsis thaliana during the hypersensitive reaction to *Pseudomonas syringae* pv. syringae. *Plant Physiol* **98:** 1304—1309

Tsuji J, Zook M, Hammerschmidt R, Somerville S and Last R (1993) Evidence that tryptophan is not a direct biosynthetic intermediate of camalexin in *Arabidopsis thaliana. Physiol Mol Plant Pathol* **43:** 221—229

Tookey HL (1973) Crambe thioglucoside glucohydrolase (EC 3.2.3.1): Separation of a protein required for epithiobutane formation. *Can J Biochem* **51:** 1654—1660

Vorwerk S, Biernacki S, Hillebrand, Janzik I, Müller A, Weiler EW and Piotrowski M (2001) Enzymatic characterization of the recombinant *Arabidopsis thaliana* nitrilase subfamily encoded by the NIT2/NIT1/NIT3 gene cluster. *Planta* **212:** 508—516

Wallsgrove RM and Bennett RN (1995) The biosynthesis of glucosinolates in Brassicas. In: Wallsgrove RM (ed) Amino Acids and their Derivatives in Higher Plants, pp 243—259. Cambridge University Press, Cambridge

Wetter LR and Chisholm MD (1968) Sources of sulfur in the thioglucosides of various higher plants. *Can J Biochem* **46:** 931—935

Winkler RG, Frank MR, Galbraith DW, Feyereisen R and Feldmann KA (1998) Systematic reverse genetics of transfer-DNA-tagged lines of *Arabidopsis. Plant Physiol* **118:** 743—750

Wittstock U and Halkier BA (2000) Cytochrome P450 CYP79A2 from *Arabidopsis thaliana* L. Catalyzes the Conversion of L-phenylalanine to phenylacetaldoxime in the biosynthesis of benzylglucosinolate. *J Biol Chem* **275:** 14659—14666

Zhang Y, Talalay P, Cho C-G and Posner GH (1992) A major inducer of anticarcinogenic protective enzymes from broccoli: Isolation and elucidation of structure. *Proc Natl Acad Sci USA* **89:** 2399—2403

Zhao J, Williams CC and Last RL (1998) Induction of *Arabidopsis* tryptophan pathway enzymes and camalexin by amino acid starvation, oxidative stress, and an abiotic elicitor. *Plant Cell* **10:** 359—370

Zhao Y, Christensen SK, Fankhauser C, Cashman JR, Cohen JD, Weigel D and Chory J (2001) A role for flavin monooxygenase-like enzymes in auxin biosynthesis. *Science* **291:** 306—309

Zook M (1998) Biosynthesis of camalexin from tryptophan pathway intermediates in cell-suspension cultures of *Arabidopsis. Plant Physiol* **118:** 1389—1393

Zook M and Hammerschmidt R (1997) Origin of the thiazole ring of camalexin, a phytoalexin from *Arabidopsis thaliana. Plant Physiol* **113:** 463—468

8. AFFILIATIONS

Plant Biochemistry Laboratory, Department of Plant Biology, The Royal Veterinary and Agricultural University, 40, Thorvaldsensvej, DK-1871 Frederiksberg C, Denmark

For Correspondence - *Barbara A Halkier*: bah@kvl.dk

J.A.C. VERKLEIJ, F.E.C. SNELLER AND H. SCHAT

METALLOTHIONEINS AND PHYTOCHELATINS: ECOPHYSIOLOGICAL ASPECTS

Abstract. In plants, various classes of metal-chelating agents are known, among which are metallothioneins (MTs), phytochelatins (PCs), phytins, organic acids and amino-acids. MTs and PCs are the best characterized sulphur-containing metal-binding ligands and are possibly contribute to metal homeostasis and detoxification by buffering cytosolic metal concentrations.

MTs are ubiquitous low molecular weight, cystein rich metal-binding proteins, which are products of mRNA translation and bind metal ions in metal-thiolate ligands. MTs in plants differ considerably from MTs found in animals, micro-organisms and prokaryotes. The functions of MTs in plants are still unclear. These proteins seem to play a role in Cu detoxification and Cu (hyper)tolerance and not for other heavy metals (in contrast to animals and micro-organisms).

PCs form a family of enzymatically synthesized peptides of the general structure $(\gamma\text{-Glu-Cys})_n\text{-Gly}$, where 'n' is normally in the range of 2-5. Cadmium and arsenate are the strongest inducers of PC-synthesis, while zinc and nickel hardly induce PCs. The formation of low molecular weight PC-Cd complexes is well characterised and is significant with respect to Cd detoxification. There is no evidence that PCs play a role in naturally selected heavy metal tolerance. Recent studies, however, suggest that arsenate (hyper)tolerance requires a constitutive PC production. PCs could function as an excellent, specific and highly sensitive biomarker for heavy metal toxicity in plants. Its validation under field conditions, however, showed that the use of PCs as a biomarker for heavy metal toxicity is limited.

Recent advances in our understanding of the functions of MTs and PCs will be discussed. Emphasis will be placed on their role in metal detoxification, metal specific genetically defined tolerance (hypertolerance) and the possible use of PCs as biomarkers of heavy meal toxicity.

1. INTRODUCTION

Upon exposure to toxic levels of heavy metals/metalloids, plants have adopted different strategies to control the uptake and transport of both essential and non-essential heavy metals. For normal growth, the plant is depending on mechanisms that maintain the internal concentrations of essential metals (Zn, Cu, Ni) between limits of deficiency and toxicity and of the non essential metals (Cd, Pb, As) below their toxicity thresholds. One of the strategies is to exclude the metal from the root, which can be achieved by secretion of certain metal chelators or reduced metal influx and/or increased metal efflux. Good examples of these mechanisms are described for Al resistance (secretion of chelators as citrate or malate into the rhizosphere, Kochian, 1993), arsenic-resistance (reduced metal influx through suppression of high affinity phosphate system, Meharg and Macnair, 1992) and for Cu-tolerance (enhanced ATP-dependent higher Cu efflux across the root cell plasma membrane, Van Hoof et al., 2001a). However complete exclusion of metals from the plant tissue is not possible. So plants, in addition, have developed internal detoxification and/or tolerance mechanisms including chelation of heavy metals and compartmentation of the chelated heavy metals into the vacuoles (Schat and Verkleij, 1998).

Yash P. Abrol and Altaf Ahmad (eds.), Sulphur in Plants, 163—176.

Plants produce a number of different ligands viz., organic acids, phytin, amino acid and S-containing compounds, which form complexes with metals (Rauser, 1999). Metallothioneins (MTs) and Phytochelatins (PCs) are the best characterized S-containing metal binding ligands that seem to contribute to metal homeostasis detoxification and/or tolerance by buffering metal concentrations. Recently, the genes encoding the enzyme phytochelatine (PC) synthase have been identified in plants and yeast, and there is now an increasing interest to investigate the precise function of PCs and whether overexpression of the PC synthase gene and other genes encoding enzymes of the PC synthetic pathway could be used to modify plants for phytoremediation of polluted environments. Molecular genetic studies have also greatly enhanced our understanding of the number of MTs in plants, their gene structure and expression mainly due to the completion of the *Arabidopsis* genome sequence. The present knowledge of MTs and PCs, and their possible roles in heavy metal homeostasis, detoxification and tolerance are, therefore, largely based on studies on *Arabidopsis* and (fission) yeast and have been extensively discussed in recent reviews (Cobbett, 2000 a, b; Cobbett and Goldsbrough, 2002; also see Cobbett, 2002 this book).

The response to toxic levels of metals is, however, essentially different among plant species. Metallophytes (plants genetically adapted to metalliferous soils) exhibit high levels of tolerance, which can be defined as the ability to survive in a soil that is toxic to other plants of the same or different species (Macnair et al., 2000). Heavy metal tolerance is not universal phenomenon. Species like *Arabidopsis thaliana,* the best characterized plant species, and *Brassica juncea* have never been found on contaminated soils, but are often referred in literature as species showing metal tolerance (Murphy et al., 1995a,b; Zenk, 1996; Salt, 2001). In order to avoid misunderstanding and to distinguish the physiological and molecular responses to toxic metals of non-adapted plants from those of the metal adapted plants Meharg (1994) introduced the term 'normal constitutive' metal tolerance versus adaptive metal tolerance (also called high level, differential or hyper tolerance, Schat et al., 2000a). Plants have both constitutive (present in most phenotypes) and adaptive (present only in tolerant phenotypes) mechanisms for coping with elevated metal concentrations. The mechanisms of 'normal constitutive' metal tolerance is, therefore, defined as heavy metal detoxification mechanisms operating in plants, not adapted to elevated concentrations of heavy metals in the soil.

This review will cover recent advances in our understanding of the functions of MTs and PCs. Emphasis will be put on their possible role in adaptive, genetically defined metal (hyper)tolerance mechanism and the use of PCs as potential biomarkers of heavy meal toxicity.

2. METALLOTHIONEINS

2.1 Structure

Metallothioneins (MTs) are a group of low molecular weight, cysteine-rich, metal-binding proteins, which provide thiols for metal chelation. Although PCs formally belong to this group of compounds (Class III MTs), the enzymatic synthesis of PCs distinguishes them from MT proteins. Since their discovery as Cd-binding proteins present in equine kidney, MT proteins and genes have been found in animals and eukaryotic micro-organisms, some prokaryotes and plants. The structure of these proteins, MT gene structure and –expression is extensively described and discussed in next chapter and recent reviews (Rauser, 1999, Cobbett and Goldsbrough, 2002).

2.2 Functions

2.2.1 Heavy metal homeostasis/detoxification

Although a number of reports have shown that metals including copper had either no effect or repressed MT gene expression, in some species copper induced expression of a Type I MT gene. Because other stresses like heat shock, aluminium etc. also induced this type of expression, it was suggested that these MTs may be expressed as part of a general stress response (Cobbett and Goldsbrough, 2002). There is, however, some evidence to support the hypothesis that MTs are involved in copper homeostasis and detoxification (Hasegawa et al., 1997). In the non-metallophyte *Arabidopsis thaliana*, a positive correlation was observed between the degrees of copper sensitivity and copper-induced *MT2* transcript accumulation among distinctly copper-sensitive ecotypes (Murphy and Taiz, 1995a; Murphy et al., 1997).

2.2.2 Role of MTs in Cu adaptive (hyper) tolerance

Because the so-called ecotype differentiation in *Arabidopsis thaliana* is rather small with respect to Cu (not more than twice, Murphy and Taiz, 1995b), it would be interesting to know whether MT2 expression could be related to highly copper adaptive tolerance in metallophyte species.

A MT2b-like gene was identified in the metallophyte *Silene vulgaris*, which has evolved populations with extremely high levels of copper tolerance (10-40 order difference in magnitude). When expressed in yeast, this gene (*SvMT2b*), restored cadmium and copper tolerance in different hypersensitive strains. Northern blot analysis and quantitative RT-PCR showed that plants from the copper-tolerant populations had significantly higher transcript levels of *SvMT2b* than plants from the copper-sensitive populations, both in roots and shoots and with and without copper exposure (van Hoof et al., 2001b). The fact that the MT2b overexpression in the tolerant *Silene* plants is found to be constitutive contrasts with the Cu inducible MT2 expression in *Arabidopsis* and points out that patterns are fundamentally different. So there seems to be an essential difference in MT response to excess Cu in the non-metallophyte *Arabidopsis* showing an apparently Cu-detoxifying mechanism with an inducible MT expression and in the metallophyte *Silene* with Cu-tolerant ecotypes possessing a tolerance mechanism in which MTs are constitutively overexpressed. This overexpression of MT2b, possibly due to amplification, did not however produce Cu tolerance, but merely enhanced the degree of tolerance in already tolerant plants. Cu tolerance in *Silene vulgaris* is controlled by two major genes and some hypostatic modifiers (Schat et al., 1993). Studies on isolated plasmamembrane vesicles from roots of Cu-sensitive and -tolerant *Silene vulgaris* ecotypes showed an enhanced ATP dependent copper efflux across the root cell plasmamembrane in Cu-tolerant *Silene vulgaris* (van Hoof et al.,

2001a). These data support the hypothesis that enhanced Cu efflux contributes to the Cu tolerance mechanisms. The function of MT in Cu-detoxification and Cu-tolerance remains, however, still an enigma (Cobbett and Goldsbrough, 2002).

3. PHYTOCHELATINS (PCs)

3.1 Structure and Biosynthesis

Phytochelatins (PCs) form a family of small metal binding peptides that consist of repetition of the γ-Glu-Cys dipeptide, followed by a terminal Gly; the basic structure is $(\gamma\text{-Glu-Cys})_n$ Gly, where 'n' has been reported as high as 11, but generally is found between 2-5 (Grill et al., 1987). The C-terminal amino acid of the peptides can vary. In Fabales, so-called homo-PCs (hPCs, C-terminal glycine is replaced by β-alanine) occur along with glycine-PCs (Klapheck et al., 1995); in Poaceae, with the exception of maize and sorghum, a PC-family with C-terminal serine is detected, namely hydroxymethyl-PC (hm-PC) (Klapheck et al., 1994, 1995). In maize, a PC-family with C-terminal glutamic acid along with glycine-PC and so-called des-gly-PCs with the structure $(\gamma\text{-EC})_n$ have been identified as well (Meuwly et al., 1995).

PCs are produced from glutathione, homoglutathione, hydroxymethyl-gluthathione or γ-glutamylcysteine by a transpeptidase, the constitutive PC-synthase enzyme (Grill et al., 1989; Hayashi et al., 1991; Chen et al., 1997). The gene encoding this enzyme has been identified simultaneously by three research groups in *Arabidopsis thaliana* (Clemens et al., 1999; Ha et al., 1999), wheat (Clemens et al., 1999), and *Schizosaccharomyces pombe* (Clemens et al., 1999; Ha et al., 1999). Induction of PC-synthesis occurs through activation of this enzyme by various metals in decreasing order of strength: Hg >Cd>As>Te>Ag>Cu>Ni>Sb>Au>Sn>Se>Bi >Pb>W>Zn (Ernst, 1996). Induction is instantaneous upon exposure and highly specific for heavy metals, since no other stress factor has been shown to induce PCs. PC-synthesis is spread throughout the plant kingdom (Gekeler et al., 1989); all plants tested so far including the majority of algae and several fungi are capable of PC-synthesis (Mehra et al., 1988, Gekeler et al., 1989; Kneer et al., 1992). Recently, it has been shown that PCs are not only restricted to plants and fungi as was once thought, but also has an essential role in heavy metal detoxification in the nematode *Caenorhabditits elegans* and probably in several other invertebrates (Vatamaniuk et al., 2002). Detail information on biosynthesis are given in next chapter.

3.2 Function

3.2.1 Heavy metal homeostasis

It is suggested that PCs play a role in cellular homeostasis and the mechanism of detoxification of excess concentrations of essential heavy metals (Zn, Cu, Ni), but direct evidence is lacking (Zenk, 1996; Rauser, 1999; Cobbett, 2000a).

Zn and Ni are relatively ineffective activators of PC synthase and exposure levels of these metals to induce PCs are close to or even beyond toxicity thresholds. Treatment with buthionine sulphoximine (BSO) strongly decreased the PC accumulation, but failed to affect root growth response to Zn in Zn-sensitive and -tolerant *Silene vulgaris* (Schat et al., 2000b). Although Cu appeared to be an effective activator of PC synthase in vitro, PCs are induced by Cu in vivo and PC-Cu complexes are demonstrated in *Silene vulgaris* (Verkleij et al., 1989), Cu-chelation by PCs seemed to be not very effective. PC-deficient mutants of *Arabidopsis thaliana* did not show an increased sensitivity to Cu or Zn (Howden et al., 1995) and after a BSO treatment the dose-response relationship for Cu-imposed root growth inhibition in Cu sensitive and -tolerant *Silene vulgaris* ecotypes was not affected (Schat et al., 2000b). For these essential metals alternative and more effective detoxification mechanisms seem to be operating such as MTs for Cu (see section Metallothionein), Histidine for Ni and organic acids (citrate) for Zn (Cobbett, 2000a,b; Cobbett and Goldsbrough, 2002).

However, there is strong evidence that PCs play an important role in detoxification of non-essential metals, such as Cd and As and probably Hg.

3.2.2 Heavy metal detoxification

Although PCs are induced and PC synthase is activated by a wide range of heavy metals, it is not at all evident that PCs play a role in detoxification of all those heavy metals. The most convincing evidence showing PCs are essential in heavy metal detoxification, particularly Cd and As, has been based on:

a) studies with plant and yeast mutants, deficient in PC synthesis. Cd hypersensitive *Arabidopsis* mutants showed an impaired PC synthesis (Howden et al., 1995). Mutants of *Arabidopsis* and fusion yeast deficient in PC synthesis were hypersensitive to Cd and As (Clemens et al., 1999; Ha et al., 1999). In both organisms there was no apparent role in the detoxification of Zn, Ni and Si ions (Cobbett, 2000a, b).

b) inhibition of PC synthesis by treatment with the γ-glutamyl-cysteine synthase (γ ECS) inhibitor, BSO. Inhibition of PC synthesis by treatment with BSO resulted in hypersensitivity to Cd in the sensitive population of *Silene vulgaris* (De Knecht et al., 1992) and enhanced the Cd sensitivity in non-metallicous *Thlaspi caerulescens* (Schat et al., 2002). Similar treatment enhanced Hg sensitivity in *Hydrilla verticulata* and *Vallisneria spiralis* (Gupta et al., 1998) and dramatically increased As sensitivity in cell cultures of tobacco and *Rauvolfia serpentina* (Schmöger et al., 2000) and in non metallicolous populations of *Silene vulgaris*, *Agrostis castellana* and *Holcus lanatus* (Sneller et al., 1999a, Schat et al., 2002; Hartley-Whitaker et al., 2002).

c) overexpression of enzymes involved in PC synthesis i.e. γ ECS and PC synthase. Overexpression of enzymes involved in PC synthesis has been shown to decrease metal sensitivity especially to Cd (Zhu et al., 1999a, b).

BSO treatment did not enhance the sensitivities to Cu, Zn, Ni and Co in nonmetallicous and metallicolous populations of *S. vulgaris, T. caerulescens, A. castellana* and *H. lanatus*. It suggests that PCs are not essential in detoxification and tolerance mechanism for these metals (Schat et al., 2002).

A role for PCs in heavy metal detoxification, at least for Cd, As and Hg is evident. It is likely, however, that the mechanism of heavy metal detoxification is more complex than simply the chelation of the metal ion. After activation of PC synthase by the metal ion and metal chelation by the PCs synthesized, the metal ion complex must presumably be transported to the vacuole and form a more stable complex aggregate in the vacuole with sulphide, organic acids or other complexators. Tonoplast transport and vacuolar compartmentation of PC-Cd complexes has been reported in oat and tobacco (Vögeli-Lange and Wagner, 1989; Salt and Rauser, 1995). PCs can form several complexes with metal ions. After sequestration in the vacuole a low molecular weight (LMW) complex is supposed to be transformed into the high molecular weight (HMW) (Speiser et al., 1992b; Rauser and Meuwly, 1995). Incorporation of sulphide stabilizes the PC-Cd complexes and increases the metal binding capacity as well (Bae and Mehra, 1998). The fact that incorporation of labile S in PC-Cd complexes was observed in *Silene vulgaris* and *B. juncea* could be, therefore, of importance in the general Cd-detoxification mechanism (Speiser et al., 1992a; De Knecht, 1995). Some Cd hypersensitive mutants of the fission yeast *Schizosaccharomyces pombe* with a normal PC synthesis lack an ABC transporter type protein (HMT), which is required for the transport of the PC-Cd complex or are not able to incorporate acid labile S for further stabilization (Juang et al., 1993; Ortiz et al., 1995).

3.2.3 Role of PCs in adaptive metal tolerance

Based on the above mentioned studies as well as on observations of *Arabidopsis* or *S. pombe* mutants that Cd (hyper)sensitivity results from decreased PC biosynthesis or function, it has been suggested that PCs are directly involved in Cd resistance or adaptive tolerance (Zenk, 1996; Salt, 2001). However, there is considerable and increasing evidence to contradict this view and that adaptive Cd tolerance (and also Zn/Cu tolerance) is not caused by an increased ability to produce PCs.

Naturally selected Cd, Zn and Cu-tolerant populations of *Silene vulgaris* contain lower concentrations of PCs than the sensitive populations of the same species (De Vos et al., 1992; Harmens et al., 1993; De Knecht et al., 1994). These results were confirmed by Mehra et al. (1994), who showed that in Cd-resistant *Candida glabrata* strains, PC- and GSH-concentrations were lower than in Cd-sensitive strains. The lower concentration of PCs in *Silene vulgaris* is neither attributable to a decreased capacity to synthesize GSH, nor to a decreased PC-synthase activity in the tolerant plants (De Knecht et al., 1995). The rate of formation of PC-Cd complexes is faster in Cd-sensitive plants. Tolerance can not be explained by a higher binding capacity of the produced PCs, since tolerant plants produce shorter PCs on average. In addition, incorporation of sulphide does not differ between Cd-tolerant and sensitive populations (De Knecht et al., 1994). Degradation of PCs in the tolerant and sensitive populations is not significantly different (De Knecht et al., 1995). In addition, inhibition of PC-synthesis by treating the plants with BSO results in hypersensitivity in the sensitive population, whereas the tolerance of the tolerant population remains the same (De Knecht et al., 1992). Similar results were found in a recent study on non-metallicolous and (hyper)tolerant metallicolous populations of *Silene vulgaris, Thlaspi caerulescencs, Holcus lanatus* and *Agrostis castellana* using BSO (Schat et al., 2002). In accordance with this, Ebbs et al. (2002) concluded that PC synthesis was not responsible for Cd tolerance in another population of *Thlaspi caerulescens* showing a comparable hypertolerance. Furthermore, the PC accumulation was consistently higher in non-metallicolous plants than in the tolerant metallicous plants despite corresponding higher metal concentrations in the tolerant ones (Ebbs et al., 2002; Schat et al., 2002).

All these recent data clearly indicate that adaptive Cd (and also Zn and Cu) tolerances are not dependent on PC-mediated sequestration. This conclusion does not contradict the findings that artificial overexpression of enzymes and transporters involved in the PC-based metal sequestration has been shown to increase Cd tolerance in species like *Arabidopsis* and *B. juncea* (Clemens et al., 1999; Zhu et al., 1999a, b). The increase in tolerance is less than two fold, not comparable with the tolerance levels in metallophytes and interpretable as a more efficient detoxification mechanism. Moreover a PC mediated Cd detoxification might not be the most effective strategy to cope with toxic exposure of this metal for adaptive tolerant metallophytes (Schat et al., 2002).

One exception to this general rule seemed to be the PC mediated response of arsenate (As) tolerant plants to excess As. Arsenate tolerance has been identified in a number of plant species and is achieved by a reduction of arsenate influx via suppression of the high-affinity phosphate/arsenate uptake system (Meharg and Macnair, 1992; Meharg, 1994). In this way the arsenate influx is reduced to a level that can be detoxified, presumably by a constitutive cellular mechanism. PCs are required for As detoxification in non-tolerant plants as has been demonstrated in *Silene vulgaris* (Sneller et al., 1999a), *Rauvolfia serpentina* (Schmöger et al., 2000) and *Rubia tinctorum* (Maitani et al. 1996).

PC accumulation was also observed in tolerant *Holcus lanatus* plants and even at significantly higher concentrations at equivalent levels of stress than the non-tolerant plants (Hartley-Whitaker et al., 2001). Furthermore, BSO treatment produced a high degree of hypersensitivity to arsenate in both non-metallicous and mine plants of *Holcus lanatus, Agrostis castellana* and *Cytisus striatus* (Hartley-Whitaker et al., 2002; Schat et al., 2002) suggesting that PCs are absolutely required for arsenate detoxification in non-tolerant plants as well as in adaptive tolerant ones. It is, however, not clear whether the highly tolerant plants should possess a more effective PC-based arsenate detoxification system as suggested by Hartley-Whitaker et al. (2002). The study of *Cytisus striatus* did not show a higher PC accumulation and a higher PC SH/As ratio in the mine plants compared to the non-metallic plants as was found in *Holcus lanatus*. More research is needed to clarify the precise role of PCs in arsenate tolerance mechanism.

4. PHYTOCHELATINS AS BIOMARKER FOR HEAVY METAL TOXICITY

4.1 Laboratory Studies

Phytochelatins fulfil many of the requirements for a specific biomarker for heavy metal toxicity in plants:

(a) PCs are specifically produced and PC synthesis starts almost instantaneously upon exposure to heavy metals and metalloids (Grill et al., 1987).

(b) PC synthesis is spread throughout the plant kingdom (Gekeler et al., 1989)

As soon as all free metal ions are chelated, PC synthesis is terminated (Loeffler et al., 1989). After arresting the metal exposure, PCs seem to be degraded (De Knecht et al., 1995). PC-synthesis is induced by very low external concentrations (in the pM to nM range) (Ahner and Morel, 1995). A linear relationship between the degree of metal-imposed growth inhibition and the rate of PC accumulation in roots has been shown in short-term experiments with several elements, such as Cu (Schat and Kalff, 1992) and Cd (De Knecht et al., 1995) in *Silene vulgaris* and Cu and Cd in maize (Keltjens and Van Beusichem, 1998a). This relation is valid from very low, non-toxic concentrations to concentrations above the EC50 for growth. At highly toxic Cu exposure PC-levels decrease (De Vos et al., 1992; Schat and Kalff, 1992). Taken together these results indicate that PCs might serve as a specific, early and highly sensitive biomarker for heavy metal toxicity (Sneller, 1999).

The clear relationships beween PC concentrations in plant material and toxicity have been observed in roots and in relatively short term experiments. In chronically exposed *Silene vulgaris* plants, the relationship between Cd toxicity and accumulation of PCs appeared to be linear up to an exposure concentration of 2 μM Cd, which corresponded to the EC50 value for plant growth. Above this concentration, the linearity was lost due to the severe toxicity of the treatments (Sneller et al., 1999b).

While PC concentrations were proven to be linearly related to the short term toxicity of As in *Silene vulgaris*, this relationship could not be assessed in plants that had been chronically exposed to arsenate, since no toxic effects were observed (Sneller et al., 1999b). Due to the very slow turnover rate of PC-As complexes (probably because of their stability at vacuolar pH), plants can accumulate high concentrations of PC-As complexes over time.

Heavy metal contaminants often occur in combination, rather than singly. Therefore, interactions between different metals at the level of binding to soil particles, plant metal transporters, metal-targets or detoxifying metal-binding compounds may be expected to occur in polluted situations. In a study of the effects of mixtures of arsenate and cadmium on *Silene vulgaris* in hydroponic cultures AsO_4 and Cd produced additive effects on accumulation of PCs with respect to both external and root-internal metal concentrations (Sneller et al., 2000). The PC-concentrations in AsO_4/Cd mixtures show a clear quantitative relationship with relative root elongation. In this particular case, PCs can be used as a **biomarker of effect**. On the other hand, in combinations of Cd, Cu and Zn, the PC accumulation in *Silene vulgaris* was additive with respect to the root-internal metal concentrations, but non-correlated to the degree of root growth inhibition (Sharma et al., 1999). Thus, in combinations of Cd, Cu and Zn, PC-accumulation can only be used as a **qualitative biomarker of exposure**. In maize and wheat exposed to combinations of Cd and Cu, however, good correlations between shoot PC-accumulation and shoot biomass were shown by Keltjens and Van Beusichem (1998b). The reaction patterns seem, therefore, to be metal and species specific. Taken together, PCs are useful tools to assess the strain caused by heavy metals (biomarker of effect). In the metals and metalloids studied so far, i.e. Cd, Cu and As, accumulation of PCs is linearly related to single metal/metalloid toxicity up to concentrations severely inhibiting root or whole plant growth. The decrease in PC concentrations at highly toxic doses may not be a problem, since then morphological changes will show the toxic effect of the metals and a biochemical biomarker is no longer needed. For combinations of metals/metalloids, PCs can at least be used as biomarkers of exposure when one of the metals involved is a strong metal inductor.

4.2 Validation under Field Conditions

Phytochelatin accumulation under field conditions have been investigated by several authors. Ahner et al. (1994, 1998) established a negative relationship between PC concentrations in marine phytoplankton and distance to Massachusetts Bay harbour. Gawel et al. (1996) described a positive relationship between PC concentrations in the needles of healthy-looking red spruce, growing in the Adirondack mountains and the number of dead spruce trees in the same transect. However, these authors did not measure metal concentrations in the plants or algae they examined and were therefore unable to explain which metal was responsible for the enhanced PC concentration in the plant or algal material. Grill et al. (1988) measured rather high concentrations of PCs in roots from *Silene vulgaris* and *Acer pseudoplatanus*, growing on a Zn-mine. This coincided with extremely high Zn-concentrations. In the leaves of both species, however, both Zn and PC concentrations were low.

To validate the concept of phytochelatins (PCs) as a biomarker for heavy metal toxicity in plants under (semi)-field conditions, plant materials from an As-contaminated site (Diemen), and a reference site (Eemnes) in The Netherlands were collected, and metal and phytochelatin (PC) concentrations were measured. In the Diemen material, high concentrations of As accumulated in the shoots, coinciding with high PC concentrations. However, the plants were not visibly injured by metal stress. Along a gradient of As-contamination at this site, the PC and As concentrations of both shoots and roots of *Holcus lanatus* decreased in a similar patterns as did the soil As concentration (Table 1). In the Eemnes material, low metal concentrations were measured in soil and plant material. In addition, PCs were not found in plants growing at this site. In another species, *Polygonum hydropiper* growing in the same transect, shoot PC concentrations did not correlate with this gradient. This may be the result of the severe As toxicity shown by the high As concentrations in the leaves and by visual symptoms. At very high metal toxicity PC levels decrease (see 3.1).

To test whether differences in accumulation patterns of metals and PCs found in the field were not just due to microclimatic or genotypic or species differences, a bioassay experiment was carried out under greenhouse conditions. For this purpose, a well-characterised ecotype of *Silene vulgaris* was grown on soils contaminated with different heavy metals. High PC concentrations coincided with either elevated As or Cd levels and confirmed the toxicity of the Diemen soils. In conclusion, for plants grown on soils contaminated with As and Cd (both strong inductors of PC-synthesis), PCs can be used as a biomarker for heavy metal toxicity. However, for Zn, another biomarker will have to be developed.

Accumulation of PCs may be used as an additional tool to assess the effect of exposure to excess amounts of heavy metal. This will particularly be the case for strong PC inductors, such as Cd and As. In addition, the PC accumulation pattern of a well-known plant variety (bioassay) can be used to relate the toxicity of a certain substrate to others.

*Table 1. Metal and PC-SH concentrations in leaves of Holcus lanatus along an As-contaminated transect. Diemen transect, ranging from 1 (at the waterside) to 5 (the middle of the plot). Arsenic concentrations (± S.E.) in soil samples from the rhizosphere along this transect are shown. In addition, metal concentrations in leaf samples and phytochelatin (PC-SH) concentrations of the leaves and roots on June 19, 1998 are given) of Holcus lanatus. (n = 3, * = 2) Standard errors of the metal concentrations were less than 15% of the PC-concentrations less than 30%.*

	As-soil (µmol/g)	**Metal concentration in leaves**			**PC-SH**	
		As (µmol/g)	**Zn (µmol/g)**	**Cd (nmol/g)**	**leaves**	**roots**
1	7.75 (0.80)	0.336*	0.56*	1.16*	23.6	987
2	1.45 (0.02)	0.185	0.62	2.40	14.8	191
3	2.81 (0.31)	0.093*	1.01	2.22	5.4	182
4	0.56 (0.05)	0.008*	0.48*	1.33*	1.4	76
5	0.25 (0.02)	0.064	0.36*	0.98*	1.1	11

Source: Sneller (1999)

5. CONCLUSIONS

The role of PCs in heavy metal homeostasis, detoxification and tolerance is still elusive. There is no direct evidence of PCs functioning in metal homeostasis. The fact that PC-deficient mutants of *Arabidopsis* are hypersensitive to Cd and arsenate seems quite significant for the involvement of PCs in Cd and arsenate detoxification (and hardly to other heavy metals). The importance of PCs for heavy metal detoxification in general must therefore not be overestimated.

There is no hard evidence that PCs play an essential role in adaptive Cd and Cu (or Zn) tolerance; apparently an alternative PC-independent tolerance mechanisms is operating and more attention will be given to that part of the research in view of finding better tools for phytoremediation purposes. An exception are the recent results of a PC-As interaction suggesting that arsenate (hyper)tolerance requires a constitutive PC production. The precise involvement of PCs in the As tolerance mechanism is still unsure.

So MTs in plants indicate that these proteins seem to play a role in Cu detoxification and Cu (hyper)tolerance and not for other heavy metals (in contrast to animals and micro-organisms). In addition, MTs seem to be involved in Zn homeostasis and scavenging of heavy metals during leaf senescence.

The general mechanisms of metal-detoxification and hyper-tolerance is, however, not only chelation of the metal to these ligands, but in most cases includes the subsequent compartmentation of the ligand-metal complex into the vacuoles. The search of genes involved in adaptive heavy metal tolerance has been strongly increased due to the growing interest of scientists in the hyperaccumulator species *Thlaspi caerulescens* and *A. halleri*, closely related species to *A. thaliana* and, therefore, very useful in further characterization of ion transporters involved in the uptake and compartmentation of cadmium, nickel, and zinc, and in the hyperaccumulation of these heavy metals.

It has been suggested that PCs could function as an excellent, specific and highly sensitive biomarker for heavy metal toxicity in plants. Under laboratory conditions, PCs may very well be used as a biomarker of exposure to and/or effect of heavy metal toxicity in plants. Their use is probably restricted to relatively strong inductors such as Cd, As and Cu. Further research could add other metals or metalloids to this list. In the case of mixtures, the PC concentration in plants exposed to mixtures of AsO_4 and Cd still reflect the toxicity. However, for combinations of other metals (i.e. Cd, Cu and Zn), PC concentrations reflect internal metal concentrations, but not toxicity. Under field conditions, the use of PCs as a biomarker will mostly be restricted to metals that are strong inductors, such as As and Cd. Because the collection of sufficient amount of young root material in the field is rather troublesome, PCs in the shoots will mostly be used to assess toxicity of the soil. The PC accumulation pattern of a well-known plant variety (bioassay) can be used to relate the toxicity of a certain substrate to others.

6. REFERENCES

Ahner BA and Morel FMM (1995) Phytochelatin production in marine algae. 2. Induction by various metals. *Limnol Oceanogr* **40:** 658—665

Ahner BA, Price NM and Morel FMM (1994) Phytochelatin production by marine phytoplankton at low free metal ion concentrations: laboratory studies and field data from Massachusetts Bay. *Proc Natl Acad Sci USA* **91:** 8433—8436

Ahner BA, Lee JG, Price NM and Morel FMM (1998) Phytochelatin concentrations in the equatorial pacific. *Deep Sea Res* I **45:** 1179—1796

Bae W and Mehra RK (1998) Properties of glutathione- and phytochelatin-capped CdS bioanocryatallites. *J Inorg Biochem* **69:** 33--43

Chen JJ, Zhou JM and Goldsbrough PB (1997) Characterization of phytochelatin synthase from tomato. *Physiol Plant* **101:** 165—172

Clemens S, Kim EJ, Neumann D and Schroeder JI (1999) Tolerance to toxic metals by a gene family of phytochelatin synthases from plants and yeast. *EMBO J* **18:** 3325—3333

Cobbett CS (2000a) Phytochelatin and their roles in heavy metal detoxification. *Plant Physiol* **123:** 825—832

Cobbett CS (2000b) Phytochelatin biosynthesis and function in heavy metal detoxification. *Curr Op Plant Biol* **3:** 211—216

Cobbett CS and Goldsbrough P (2002) Phytochelatins and metallothioneins: roles in heavy metal detoxification and homeostasis. *Annu Rev Plant Biol* **53:** 159—182

De Knecht JA, Koevoets PLM, Verkleij JAC and Ernst WHO (1992) Evidence against a role for phytochelatins in naturally selected increased cadmium tolerance in *Silene vulgaris* (Moench) Garcke. *New Phytol* **122:** 681—688

De Knecht JA, Van Dillen M, Koevoets PLM, Schat H, Verkleij JAC and Ernst WHO (1994) Phytochelatins in cadmium-sensitive and cadmium-tolerant *Silene vulgaris*. Chain length distribution and sulphide incorporation. *Plant Physiol* **104:** 255—261

De Knecht JA, Van Baren N, ten Bookum WM, Wong Fong Sang HW, Koevoets PLM, Schat H and Verkleij JAC (1995) Synthesis and degradation of phytochelatins in cadmium-sensitive and cadmium-tolerant *Silene vulgaris*. *Plant Sci* **106:** 9—18

De Vos CHR, Vonk MJ, Vooijs R and Schat H (1992) Glutathione depletion due to copper-induced phytochelatin synthesis causes oxidative stress in *Silene cucubalus*. *Plant Physiol* **98:** 853—858

Ebbs S, Lau I, Ahner B and Kochian L (2002) Phytochelatin synthesis is not responsible for Cd tolerance in the Zn/Cd hyperaccumulator *Thlaspi caerulescens* (J & C Presl.). *Planta* **214:** 635—640

Ernst WHO (1996) Schermetalle. In: Brunold CH, Ruegsegger A and Braendle R (eds) Stress bei Pflanzen, pp 191—219. Verlag Paul Haupt, Berlin

Gawel JE, Ahner BA, Friedland AJ and Morel FMM (1996) Role for heavy metals in forest decline indicated by phytochelatin measurements. *Nature* **381:** 64—65

Gekeler W, Grill E, Winnacker E-L and Zenk MH. (1989) Survey of the plant kingdom for the ability to bind heavy metals through phytochelatins. *Z Naturf C* **44:** 361—369

Grill E, Winnacker E-L and Zenk MH (1987) Phytochelatins, a class of heavy-metal-binding peptides from plants are functionally analogous to metallothioneins. *Proc Natl Acad Sci USA* **84:** 439—443

Grill E, Löffler S, Winnacker E-L and Zenk MH (1989) Phytochelatins, the heavy-metal-binding peptides of plants, are synthesized from glutathione by a specific γ-glutamylcysteine dipeptidyl transpeptidase (phytochelatin synthase). *Proc Natl Acad Sci USA* **86:** 6838—6842

Grill E, Winnacker E-L and Zenk MH (1988) Occurrence of heavy metal binding phytochelatins in plants growing in a mining refuse area. *Experientia* **44:** 539—540

Gupta M, Tripathi RD, Rai UN and Chandra P (1998) Role of glutathione and phytochelatin in *Hydrilla verticillata* Royle and *Vallisneria spiralis* L under mercury stress. *Chemosphere* **37:** 785—800

Ha SB, Smith AP, Howden R, Dietrich WM, Bugg S, O'Connell MJ, Goldsbrough PB and Cobbett CS (1999) Phytochelatin synthase genes from *Arabidopsis* and the yeast *Schizosaccharomyces pombe*. *Plant Cell* **11:** 1153—1163

Harmens H, Cornelisse E, Den Hartog PR, Ten Bookum WM and Verkleij JAC (1993) Phytochelatins do not play a key role in naturally selected zinc tolerance in *Silene vulgaris* (Moench) Garcke. *Plant Physiol* **103:** 1305—1309

Hartley-Whitaker J, Ainsworth G, Vooijs R, Ten Bookum WM, Schat H and Meharg AA (2001) Phytochelatins are involved in differential arsenate tolerance in *Holcus lanatus*. *Plant Physiol* **126:** 299—306

Hartley-Whitaker J, Woods C and Meharg AA (2002) Is differential phytochelatin production related to decreased arsenate influx in arsenate tolerant *Holcus lanatus*? *New Phytol* **155:** 219—225

Hasegawa J, Terada E, Sunairi M, Wakita H, Shimachi F, Noguchi A, Nakajima M and Yazaki J (1997) Genetic improvement of heavy metal tolerance in plants by transfer of the yeast metallohionein gene (*CUPI*). *Plant Soil* **196:** 277—281

Hayashi Y, Nakagawa CW and Mutoh N (1991) Two pathways in the biosynthesis of cadystins $(\gamma\text{-EC})_nG$ in the cell-free system of the fission yeast. *Biochem Cell Biol* **69:** 115—121

Howden R and Cobbett CS. (1992) Cadmium-sensitive mutants of *Arabidopsis thaliana*. *Plant Physiol* **99:** 100—107

Howden R, Goldsbrough PB, Andersen CR and Cobbett CS (1995) Cadmium-sensitive, cad1, mutants of *Arabidopsis thaliana* are phytochelatin-deficient. *Plant Physiol* **107:** 1059—1067

Juang RH, McCue and Ow DW (1993) Two purine biosynthetic enzymes that are required for cadmium tolerance in *Schizosaccharomyces pombe* utilize cysteine sulphinate in vitro. *Arch Biochem Biophys* **304:** 392—401

Keltjens WG and Van Beusichem ML (1998a) Phytochelatins as biomarkers for heavy metal toxicity in maize: single metal effects of copper and cadmium. *J Plant Nutr* **21:** 635—648

Keltjens WG and Van Beusichem ML (1998b) Phytochelatins as biomarkers for heavy metal stress in maize (*Zea mays* L.) and wheat (*Triticum aesticum* L.): combined effects of copper and cadmium. *Plant Soil* **203:** 119—126

Klapheck S, Fliegner W and Zimmer I (1994) Hydroxymethyl-phytochelatins [$(\gamma$-glutamylcysteine$)_n$-serine] are metal-induced peptides of the Poaceae. *Plant Physiol* **104:** 325—1332

Klapheck S, Schlunz S and Bergmann L (1995) Synthesis of phytochelatins and homo-phytochelatins in *Pisum sativum* L. *Plant Physiol* **107:** 515—521

Kneer R, Kutchan RM, Hochberger A and Zenk MH (1992) *Saccharomyces cerevisiae* and *Neurospora crassa* contain heavy metal sequestering phytochelatin. *Arch Microbiol* **157:** 305—310

Kochian LV (1993) Cellular mechanisms of aluminum toxicity and resistance in plants. *Annu Rev Plant Physiol Plant Mol Biol* **46:** 237—260

Loeffler S, Hochberger A, Grill E, Winnacker E-L and Zenk M-H (1989) Termination of the phytochelatin synthase reaction through sequestration of heavy metals by the reaction product. *FEBS Lett* **258:** 42—46

Macnair MR, Tilstone GH and Smith SE (2000) The genetics of metal tolerance and accumulation in higher plants. In: Terry N and Bañuelos G (eds) Phytoremediation of Contaminated Soil and Water, pp 235—251. Lewis Publishers, Boca Raton, Florida

Maitani T, Kubota T, Sato K and Yamada T (1996) The composition of metals bound to class III metallothioneins (phytochelatin and its desglycyl peptide) induced by various metals in root cultures of *Rubia tinctorum*. *Plant Physiol* **110:** 1145—1150

Meharg AA (1994) Integrated tolerance mechanisms: constitutive and adaptive plant responses to elevated metal concentrations in the environment. *Plant Cell Environ* **17:** 989—993

Meharg AA and Macnair MR (1992) Suppression of the high-affinity phosphate uptake system: a mechanism of arsenate tolerance in *Holcus lanatus* L. *J Exp Bot* **43:** 519—524

Mehra RK, Tarbet EB, Gray WR and Winge DR (1988) Metal-specific synthesis of 2 metallthioneins and γ-glutamyl-transferase peptides in *Candida glabrata*. *Proc Natl Acad Sci USA* **85:** 8815—8819

Mehra RK, Mulchaldani P and Hunter TS (1994) Role of CdS quantum crystallites in cadmium resistance in *Candida glabrate*. *Biochem Biophy Res Comm* **3:** 1193—1200

Meuwly P, Thibault P, Schwan AL and Rauser WE (1995) Three families of thiol peptides are induced by cadmium in maize. *Plant J* **7:** 391—400

Murphy A and Taiz L (1995a) A new vertical mesh transfer technique for metal tolerance studies in *Arabidopsis*. *Plant Physiol* **108:** 29—38

Murphy A and Taiz L (1995b) Comparison of metallothionein gene expression and non-protein thiols in ten *Arabidopsis* ecotypes. *Plant Physiol* **109:** 945—954

Murphy A, Zhou J, Goldsbrough P and Taiz L (1997) Purification and immunological identification of metallothioneins 1 and 2 from *Arabidopsis thaliana*. *Plant Physiol* **113:** 1293—1301

Ortiz DF, Ruscitti T, McCue KF and Ow DW (1995) Transport of metal-binding peptides by HMT1, a fission yeast ABC-type vacuolar membrane-protein. *J Biol Chem* **270:** 4721—4728

Rauser WE (1995) Phytochelatins and related peptides: structure, biosynthesis, and function. *Plant Physiol* **109:** 1141—1149

Rauser WE (1999) Structure and function of metal chelators produced by plants: the case for organic acids, amino acids, phytin and metallothioneins. *Cell Biochem Biophys* **31:** 19—48

Rauser WE and Meuwly P (1995) Retention of cadmium in roots of maize seedlings. *Plant Physiol* **109:** 195—202

Salt DE (2001) Responses and adaptations of plants to metal stress. In: Hawkesford MJ and Buchner P (eds) Molecular Analysis of Plant Adaptation to the Environment, pp 159—179. Kluwer Academic Publishers, Dordrecht

Salt DE and Rauser WE (1995) MgATP-dependent transport of phytochelatins across the tonoplast of oat roots. *Plant Physiol* **107:** 1293—1301

Schat H and Kalff MMA (1992) Are phytochelatins involved in differential heavy metal tolerance or do they merely reflect metal-imposed strain? *Plant Physiol* **99:** 1475—1480

Schat H and Verkleij JAC (1998) Biological interactions: the role for non-woody plants in phytorestoration: possibilities to exploit adaptive heavy metal tolerance. In: Vangronsveld J and Cunningham SD (eds) Metal-Contaminated Soils: In Situ Inactivation and Phytorestoration, pp 51—65. Springer-Verlag and RG Landes Company, Austin, Texas

Schat H, Kuiper E, Ten Bookum WM and Vooys R (1993) A general model for the genetic control of copper tolerance in *Silene vulgaris*: evidence from crosses between plants from different tolerant populations. *Heredity* **70:** 142—147

Schat H, Llugany M and Bernhard R (2000a) Metal-specific patterns of tolerance, uptake and transport of heavy metals in hyperaccumulating and non hyperaccumulating metallophytes. In: Terry N and Banuelos G (eds) Phytoremediation of Contaminated Soil and Water, pp 171—188. Lewis Publishers, Boca Raton, Florida

Schat H, Van Hoof NALM, Tervahauta A, Hakvoort HWJ, Chardonnens AN, Koevoets PLM, Verkleij JAC and Ernst WHO (2000b) Evolutionary responses to zinc and copper stress in *Bladder Campion, Silene vulgaris* (Moench) Garcke. In: Cherry JH, Locy RD and Richter A (eds) Plant Tolerance to Abiotic Stresses: Role of Genetic Engineering, pp 343—360. Kluwer Academic Publishers, Dordrecht

Schat H, Llugany M, Vooys R, Hartley-Whitaker J and Bleeker PM (2002) The role of phytochelatins in constitutive and adaptive heavy metal tolerances in hyperaccumulator and non-hyperaccumulator metallophytes. *J Exp Bot* (in press).

Schmöger MEV, Oven M and Grill E (2000) Detoxification of arsenic by phytochelatins in plants. *Plant Physiol* **122:** 793—801

Sharma SS, Schat H, Vooijs R and Van Heerwaarden LM (1999) Combination toxicology of copper, zinc and cadmium in binary mixtures: concentration-dependent antagonistic, nonadditive, and synergistic effects on root growth in *Silene vulgaris*. *Environ Toxicol Chem* **18:** 348—355

Sneller FEC (1999) Phytochelatins as a biomarker for heavy metal toxicity in terrestrial plants. Thesis Vrije Universiteit Amsterdam. Print Partners Ispkamp Enschede/Amsterdam

Sneller FEC, Van Heerwaarden LM, Kraaijeveld-Smit FJL, Ten Bookum WM, Koevoets PLM, Schat H and Verkleij JAC (1999a) Toxicity of arsenate in *Silene vulgaris*, accumulation and degradation of arsenate-induced phytochelatins. *New Phytol* **144:** 223—232

Sneller FEC, Ten Bookum WM, Schat H, Bedaux JJM and Verkleij JAC (1999b) Quantitative relationship between phytochelatin accumulation and growth inhibition during prolonged exposure to cadmium in

Silene vulgaris. *Ecotoxicology* **8:** 167—175

Sneller FEC, Van Heerwaarden LM, Schat H and Verkleij JAC (2000) Toxicity, metal uptake and accumulation of phytochelatins in *Silene vulgaris* exposed to mixtures of cadmium and arsenate. *Environ Tox Chem* **19:** 2982—2986

Speiser DM, Abrahamson SL, Banuelos G and Ow DW (1992a) *Brassica juncea* produces a phytochelatin-cadmium-sulfide complex. *Plant Physiol* **99:** 817—821

Speiser DM, Ortiz DF, Kreppel L, Scheel G, McDonald G and Ow DW (1992b) Purine biosynthetic genes are required for cadmium tolerance in *Schizosaccharomyces pombe*. *Mol Cell Biol* **12:** 5301—5310

Van Hoof NALM, Koevoets PLM, Hakvoort HWJ, Ten Bookum WM, Schat H, Verkleij JAC and Ernst WHO (2001a) Enhanced ATP-dependent copper efflux across the root cell plasma membrane in copper-tolerant *Silene vulgaris*. *Physiol Plant* **113:** 225—232

Van Hoof NALM, Hassinen VH, Hakvoort HWJ, Ballintijn KF, Schat H, Verkleij JAC, Ernst WHO, Karenlämpi SO and Tervahauta AI (2001b) Enhanced copper tolerance in *Silene vulgaris* (Moench) Garcke populations from copper mines is associated with increase transcript levels of a 2b-type metallothionein gene. *Plant Physiol* **126:** 1519—1526

Vatamaniuk OK, Bucher EA, Ward JT and Rea PA (2002) Worms take the 'phyto' out of 'pytochelatins'. *Trends Biotech* **20:** 61—64

Verkleij JAC, Koevoets PLM, Van 't Riet J, Van Rossenberg MC, Bank R and Ernst WHO (1989) The role of metal-binding compounds in the copper tolerance mechanism of Silene vulgaris. In: Hamer DH and Winge DR (eds) Metal Ion Homeostasis: Molecular Biology and Chemistry, pp 347—358. Alan R Liss Inc, New York

Vögeli-Lange F and Wagner GJ (1989) Subcellular localization of cadmium and cadmium-binding peptides in tobacco. Implication of a transport function for cadmium-binding peptides. *Plant Physiol* **92:** 1086—1093

Zenk MH (1996) Heavy metal detoxification in higher plants – a review. *Gene* **179:** 21—30

Zhu YL, Pilon-Smits EAH, Jouanin L and Terry N (1999a) Overexpression of glutathione synthetase in Indian mustard enhances cadmium accumulation and tolerance. *Plant Physiol* **119:** 73—79

Zhu YL, Pilon-Smits EAH, Jouanin L and Terry N (1999b) Cadmium tolerance and accumulation in Indian mustard is enhanced by overexpressing γ-glutamylcysteine synthetase. *Plant Physiol* **121:** 1169—1177

7. AFFILIATIONS

Department of Ecology and Physiology of Plants, Vrije Universiteit, Faculty of Earth and Life Sciences, De Boelelaan 1085, 1081 HV Amsterdam, The Netherlands (JACV, HS); Department of Chemistry and Ecotoxicology, National Institute of Inland Water Management and Wastewater Treatment (RIZA), P.O. Box 17, 8200 AA Lelystad, The Netherlands

For Correspondence - *JAC Verkliej*: verkleij@bio.vu.nl

C.S. COBBETT

METALLOTHIONEINS AND PHYTOCHELATINS: MOLECULAR ASPECTS

Abstract. The metabolism and/or detoxification of heavy metals present particular challenges to biological systems. These mechanisms include heavy metal-binding ligands such as the metallothioneins (MTs) and phytochelatins (PCs). MTs and PCs are different classes of cysteine-rich, heavy metal-binding peptide or polypeptide molecules. PCs are enzymatically synthesised peptides while MTs are gene-encoded polypeptides. Recently, advances in understanding their expression, regulation and possible roles in plant metabolism and stress responses have been made, particularly with the use of the model organism, *Arabidopsis thaliana.*

1. INTRODUCTION

A number of sulphur-containing proteins and peptides play important roles in heavy metal detoxification and/or metabolism in plants. These include the metallothioneins (MTs), short, gene-encoded polypeptides and the phytochelatins (PCs), peptides that are enzymatically synthesised from glutathione (GSH). MTs are cysteine-rich, GSH is a cysteine-containing tripeptide and PCs are γ-glutamylcysteine polymers. All form metal-thiolate bonds and metal-bound complexes and have varying roles in metal detoxification and metabolism. MTs were first discovered in animals and initial attempts to identify their equivalent in plants led to the discovery of PCs. The initial apparent absence of PCs from animals and of MTs from plants led to the assumption that PCs were the functional equivalent of animal MTs. However, with the subsequent identification of an extensive and ancient family of MTs in plants and of PC synthase genes in at least some animal species it seems more likely that these two classes of metal binding ligands played different roles in metal detoxification or metabolism. This chapter reviews our current understanding of the biosynthesis, expression, regulation and functions of MTs and PCs, drawn from a range of physiological, biochemical, genetic and molecular biological approaches to their study.

2. METALLOTHIONEINS

2.1 MT Structure and Classification

Metallothioneins are low molecular weight, cysteine-rich, metal-binding proteins (Kägi, 1991, 1993) which bind a variety of heavy metals through thiol bonds with the cysteine (Cys) sulfhydryl groups. MT proteins and/or genes have been found throughout the animal and plant kingdoms as well as in some prokaryotes. They are classified according to the arrangement of Cys residues (Cherian and Chan, 1993). Class I MTs

Yash P. Abrol and Altaf Ahmad (eds.), Sulphur in Plants, 177—188.

are similar to mammalian MTs, generally contain 20 highly conserved Cys residues and are widespread in vertebrates. MTs without this particular arrangement of cysteines are referred to as Class II MTs and include those from plants and fungi as well as non-vertebrate animals. Phytochelatins are also classified as Class III MTs. However, the enzymatic biosynthesis of PCs distinguishes them from MTs.

A MT protein was first identified in wheat (Lane et al., 1987) and subsequently numerous MT genes have been isolated from plants. The plant MT proteins encoded by these genes can be further classified according to amino acid sequence conservation and the arrangement of the Cys residues. Robinson et al. (1993) first proposed a classification system. Alternative systems have been proposed by others (Rauser, 1999; Liu et al., 2000; Binz and Kägi, 2001). The system developed by Robinson et al. (1993) assigns most plant MT genes into four Types based on amino acid sequence and the arrangement of Cys residues in the metal-binding domains.

Type 1 MTs generally contain six Cys-Xaa-Cys motifs distributed equally among two domains separated by approximately 40 amino acids that may include aromatic amino acids. This contrasts with most other MTs in which cysteine-rich domains are separated by a short spacer that does not include aromatic residues. Type 2 MTs have a similar structure. In general, the sequences of the N-terminal domain of Type 2 MTs are highly conserved and contain a Cys-Cys motif in amino acid positions 3 and 4. A Cys-Gly-Gly-Cys motif occurs at the end of the N-terminal cysteine-rich domain and the C-terminal domain contains 3 Cys-Xaa-Cys motifs. By contrast, the spacer region separating these domains in Type 2 MTs is much more variable between species.

Type 3 MTs contain only four Cys residues in the N-terminal domain with the first three arranged as Cys-Xaa-Xaa-Cys-Xaa-Cys. The fourth is separated and generally contained within a highly conserved motif -Gln-Cys-Xaa-Lys-Lys-Gly. The six Cys residues in the C-terminal cysteine-rich domain are arranged in Cys-Xaa-Cys motifs. As with the majority of Type 1 and Type 2 plant MTs, the two domains are separated from each other by approximately 40 amino acid residues. Type 4 MTs differ in having three cysteine-rich domains, each containing 5 or 6 conserved cysteine residues generally arranged as Cys-Xaa-Cys.

2.2 MT Genes and Proteins

Genes encoding each of the four Types of MTs have been identified in individual plant species including both monocots and dicots. Previously, seven functional MT genes (and one pseudogene) including the four Types have been reported in *Arabidopsis* (Zhou and Goldsbrough, 1994, 1995) (although an exhaustive analysis of the genome sequence suggests the presence of at least four additional MT genes (Cobbett and Meagher, 2002)). Rice also has all four MT gene Types indicating that their evolution predates the separation of monocots and dicots and it is likely that the majority of flowering plants also contain the four different MT Types. The diversity and conservation of the plant MT gene family suggests that different MTs may differ in function.

There is now an extensive (and in some cases exhaustive) description of the MT gene families in a number of plant species. There is, however, relatively little data about the MT protein. The few reports of purification and characterisation of plant MTs include the first description of a plant MT, the wheat embryo zinc-binding Ec protein

(Lane et al., 1987) and the purification of several MTs from *Arabidopsis* (Murphy et al., 1997). In the latter case, purification required the use of anaerobic conditions and illustrates that many of the difficulties in purifying MTs in plants are likely due to their instability under aerobic extraction procedures.

An alternative to the purification and analysis of MTs directly from plant tissues has been the expression of plant MT cDNAs in heterologous microbial hosts. This has allowed both the subsequent purification of the MT to examine its metal binding properties and studies of their capacity to confer in vivo metal tolerance and/or accumulation on the heterologous host. For example, *Arabidopsis* MTs have been expressed in MT-deficient strains of yeast (Zhou and Goldsbrough, 1994) and *Synechococcus* (Robinson et al., 1996) and were able to complement Cu and Zn sensitivity phenotypes, respectively. Other studies examining the effects of expression of, for example, animal and microbial MT proteins on metal tolerance in plants have, in some cases, resulted in increased metal tolerance or altered distribution of metals in plants. These studies, however, tell us little about the normal functions of plant MTs.

2.3 MT Gene Expression

Early experiments identifying genes on the basis of differential expression often identified MT cDNAs suggesting a high level of expression of at least some MT genes. This has been confirmed using northern analysis. Other indirect evidence is that among random ESTs derived from various plant cDNA libraries, MT sequences are among the most prevalent (TIGR Gene Indices. 2001. http://www.tigr.org/tdb/tgi.shtml). More direct evidence comes from techniques such as serial analysis of gene expression (SAGE). For example, gene profiling experiments in rice using SAGE (Matsumura et al., 1999) detected four different MT transcripts exceeding 1% of the transcripts of the tissue sampled.

There have been numerous studies of the expression of plant MT genes during development and in response to various environmental factors. Many of the results of these analyses have been summarised elsewhere recently (see Rauser, 1999; Cobbett and Goldsbrough, 2002) and will not be covered in detail here. In general, expression of the different MT Type genes is not highly tissue specific, although, a number of broad generalisations can be made. Type 1 gene expression tends to be higher in roots than shoots while Type 2 expression is higher in shoots than roots. Type 3 MT RNAs are highly expressed in ripening fruits. Type 4 gene expression is confined to developing embryos. A level of cell-specificity of expression has been observed for some genes. For example, in both *Arabidopsis* and *Vicia faba*, MT expression in leaves was especially high in trichomes and in *Arabidopsis* MT gene expression is localized to the phloem in a number of tissues (Foley and Singh, 1994; Garcia-Hernandez et al., 1998; Cobbett and Goldsbrough, 2002). The various observations of tissue or cell-specificity of expression have led to speculation about the function of different MTs. For example, phloem-specific expression suggests a role in metal ion transport; seed-specific expression for the storage of metals essential for germination and seedling development, and trichome expression for metal sequestration and detoxification. MT gene expression is also induced during leaf senescence and this too has suggested a role in both protecting senescing tissues from the effects of free metal ions released from metalloprotein degradation and redistributing essential metals via translocation to other parts of the plant.

Among the various environmental factors that influence MT RNA expression in plants are heavy metal stress, nutrient deprivation and heat shock (see Rauser, 1999; Andrews, 2000; Cobbett and Goldsbrough, 2002). In general, such stimuli have a limited effect, leading to an increase in expression of only a few fold. Thus, although MTs are Cys-rich polypeptides that are, in some cases, expressed at a high level in particular tissues and can be further induced by particular environmental conditions, there is no evidence for, or expectation that such changes in expression lead to any greater demand on the sulphur assimilation pathway than the expression of any other class of proteins in a plant species. From this perspective, unlike the PCs discussed below, MTs are a significant class of cysteine-rich, metal-binding ligand in plant cells and should not be regarded as being a particular sink for sulphur metabolism in a plant.

2.4 MT Function in Plants

In view of the ancient origins of the plant MT families and their conservation through evolution we can presume individual Types of MTs play particular roles in metal metabolism in plants. One such role may be metal detoxification. PCs, as described below, have a clear role in cadmium detoxification in plants. However, PCs have little effect on copper tolerance and a role for MTs in copper tolerance and homeostasis is more likely. The following evidence has been summarised by Cobbett and Goldsbrough (2002): some plant MTs are known to be copper-binding proteins; expression of some MT genes is induced by copper; in some situations MT gene expression is coordinated with other genes involved in copper homeostasis; the level of expression of a particular MT gene correlated closely with copper tolerance in a number of *Arabidopsis* ecotypes (Murphy amd Taiz, 1995); expression of a particular MT is elevated in a copper-sensitive mutant that accumulates copper (van Vliet et al., 1995); copper tolerant natural isolates of *S. vulgaris* have increased gene copy number and expression of a particular MT (van Hoof et al., 2001). There is limited data regarding the possible roles of plant MTs in the metabolism of other essential metals, particularly zinc. Resolving the many questions about MT function in plants will require additional analytical tools including more efficient means of MT purification, Type-specific (or even individual MT-specific) antibodies and a complete panel of MT mutants in *Arabidopsis*. These tools are currently being developed and more definitive answers about MT function are likely in the not-too-distant future.

3. PHYTOCHELATINS

3.1 PC Structure and Occurrence

PCs are a family of related peptides with the structures, (γ-GluCys)$_n$-Gly (where 'n' is generally 2-6). That the Glu and Cys residues in PCs are linked through a γ-carboxylamide bond demonstrates these peptides, unlike the Class I and II MTs, are not encoded directly by genes but are the products of a biosynthetic pathway. These peptides have been referred to as cadystins (from *Schizosaccharomyces pombe*), poly-

$(\gamma\text{-EC})_nG$ peptides, Cd-binding peptides, and phytochelatins and are broadly classified as Class III MTs. Here they will be referred to by the commonly accepted name, phytochelatins (PCs).

A number of structural variants of PCs have been identified in different species. For example, some plants, particularly legumes, contain both GSH and a related compound called homoGSH (γ-GluCys-β-Ala) (Grill et al., 1986). On exposure to Cd these plants synthesise an analogous family of PCs, homoPCs $[(\gamma\text{-GluCys})_n\text{-}\beta\text{-Ala}]$. Similarly, in maize the structure $(\gamma\text{-GluCys})_n$-Glu has been detected (Meuwley et al., 1995) and in other monocots, *Agrostis* and rice, for example, $(\gamma\text{-GluCys})_n$-Ser peptides were identified (Klapheck et al., 1994). The des-Gly variant, $[(\gamma\text{-GluCys})_n]$ has also been found in Cd-binding complexes from maize and some yeasts. These variants were all identified in heavy metal-binding complexes. Presumably, they function in physiologically similar ways.

PCs have been identified in a wide variety of plant species including monocots, dicots, gymnosperms and algae (Gekeler et al., 1989). In addition, PCs have been described in various fungal species including *S. pombe*, *Candida glabrata*, *Saccharomyces cerevisiae* and *Neurospora crassa* and in marine diatoms (see Rauser, 1999). Of these organisms, *S. pombe* and the plant, *Arabidopsis thaliana,* have been used most extensively as models for the molecular genetic characterisation of PC biosynthesis and function. PCs themselves have not been directly identified from an animal species. However, the isolation of PC synthase genes and their identification in some animal species has led to the recent conclusion that PCs are not, as previously thought, essentially a plant-specific metal detoxification mechanism.

3.2 PC Biosynthetic Pathway

3.2.1 Glutathione and PC biosynthesis

Structural analysis of PCs indicates that they are derived from GSH (or, in some cases, a related compound, such as homo-glutathione). This has been confirmed through various physiological, biochemical and genetic studies. A decrease in the level of GSH is coincident with the induction of PCs in plant cell cultures (Grill et al., 1987; Scheller et al., 1987). In addition, the exposure of whole plants or plant cell cultures to an inhibitor of GSH biosynthesis, buthionine sulphoximine (BSO), also inhibits the induction of PCs and/or confers hypersensitivity to metal ions (Reese and Wagner, 1987; Scheller et al., 1987). Furthermore, the effects of BSO on PC biosynthesis in cell cultures can be reversed by the addition of GSH (Mendum et al., 1990). These physiological studies have been confirmed by the study in which mutants of *S. pombe* (Mutoh and Hayashi, 1988; Glaeser et al., 1991) and *A. thaliana* (Cobbett et al., 1998), which have defects in genes encoding GSH biosynthetic enzymes, are deficient in PCs and are hypersensitive to metal ions.

GSH is synthesized from its constituent amino acids in a two-step pathway catalysed successively by the ATP-dependent enzymes, γ-glutamylcysteine synthetase (GCS) and glutathione synthetase (GSH-S). In plants, the GSH biosynthetic enzymes have been purified and some studies have suggested the existence of different isozymes of each in chloroplastic and cytosolic fractions (see May et al., 1998; Noctor et al.,

2002). This has been interpreted as indicating the existence of separate GSH biosynthetic pathways in these two cellular compartments.

Genes encoding these enzyme activities were first cloned from *Arabidopsis* by functional complementation of *E. coli* and yeast GSH-deficient mutants (see May et al., 1998; Noctor et al., 2002). Homologous sequences have been subsequently isolated from various other plants. Many, but not all, of the GCS and GSH-S proteins predicted from cDNAs in the sequence databases include predicted plastid target sequences. For each of GCS and GSH-S in *Arabidopsis*, only a single gene has been identified and both enzymes appear to have chloroplast targeting sequences (May and Leaver, 1994; Skipsey et al., 1999). Not withstanding the apparent occurrence of the enzyme activities in both plastid and cytosolic fractions in other species, this observation leads to questions about the subcellular compartmentalisation of GSH biosynthesis. GSH-deficient mutants have been identified in *Arabidopsis*. The *cad2-1*, Cd-sensitive mutant is partially PC-deficient and contains only 20-40% of wildtype levels of GSH (Cobbett et al., 1998). The mutant has only 40% of wildtype GCS levels and has a mutation in the GCS gene. A more extreme GCS mutant, *rml1* (*rootmeristemless1*) has less than 2% wild-type levels of GSH (Vernoux et al., 2000). This suggests that the *cad2-1* mutant gene product is partially active and, furthermore, confirms there is unlikely to be a second gene encoding GCS activity. The *rml1* mutant fails to develop a root following germination due to an apparent lack of cell division. This implies an unexpected role for GSH in controlling root growth and has led to speculation that GSH may be an important signal by which root growth is modulated in response to environmental stress.

3.2.2 PC synthase

Grill et al. (1989) identified a PC biosynthetic activity from cultured cells of *Silene cucubalis*. This enzyme synthesised PCs from GSH, catalysing the successive appearance of PC_2, PC_3 and PC_4. The reaction was presumed to involve the transpeptidation of the γ–GluCys moiety of GSH onto, initially, a second GSH molecule to form PC_2 or, in later stages of the incubation, onto a PC molecule to produce increasingly larger oligomers. This γ–GluCys dipeptididyl transpeptidase (EC 2.3.2.15) has been named PC synthase.

The PC synthase gene was first identified through the isolation of Cd-sensitive, PC-deficient, *cad1* mutants (Howden et al., 1995). The *cad1* mutants, in contrast to *cad2-1*, have wildtype levels of GSH and, consistent with this, lack PC synthase activity. Thus, *CAD1* was presumed to be the structural gene for PC synthase. The *CAD1* gene has been isolated using a positional cloning strategy (Ha et al., 1999) and in parallel studies an *Arabidopsis* cDNA (*AtPCS1*) (Vatamaniuk et al., 1999), which is identical to *CAD1*, and a similar cDNA from wheat (*TaPCS1*) (Clemens et al., 1999) were identified through their ability to confer Cd-resistance when expressed in yeast. Targeted deletion mutants of a similar sequence identified in *S. pombe* are also Cd-sensitive and PC-deficient, confirming the analogous function of the two genes in the different organisms. Assays of the activity of the *Arabidopsis* and *S. pombe* gene products, purified as epitope-tagged derivatives or expressed in *E. coli*, demonstrated each was sufficient for GSH-dependent PC biosynthesis in vitro.

Similar sequences have now been identifed in a variety of plants. Surprisingly, since

PCs themselves have not been detected in animal species, a functional PC synthase has also been described in the nematode, *C. elegans*, and the slime mould, *D. discoideum* (Clemens et al., 2001; Vatamaniuk et al., 2001; Cobbett and Goldsbrough, 2002). Thus, contrary to early expectations, PC appears to be a metal detoxification mechanism in both plants and at least some animals. Nonetheless, there is, for example, no sequence homologous to a PC synthase in the nearly complete *Drosophila* and human genomes. NOR is there an apparent PC synthase gene in the *S. cerevisiae* genome, suggesting that PCs identified in this organism arise via an alternative mechanism. Hayashi et al. (1991) proposed there are two pathways for PC biosynthesis in *S. pombe*. In addition to the pathway found in plants (transfer of γ–GluCys from GSH to an acceptor), they presented evidence for a pathway in which γ–GluCys was polymerised and then a terminal Gly added by GS. The absence of detectable PCs in both the *cad1-3* mutant and the analogous mutant of *S. pombe* argues against the presence of a second pathway of PC biosynthesis in these organisms.

3.3 Regulation of PC Biosynthesis

The kinetics of induction of PCs have been studied using easily manipulated *S. pombe* and plant cell cultures. In medium containing trace levels of only essential heavy metals, low levels of PCs can be detected (Grill et al., 1988). On exposure to Cd, generally in 1 to 100 μM range, depending on the species or conditions of cultivation, PCs can be detected within minutes with a concomitant decrease in GSH levels and induction is independent of de novo protein synthesis (Grill et al., 1987; Scheller et al., 1987). Continued exposure leads to the accumulation of progressively larger PC oligomers (Grill et al., 1987). A wide range of metal ions induce PC biosynthesis (Grill et al., 1987; Maitani et al., 1996).

These early in vivo studies of PC induction led to the expectation that direct activation of PC synthase would be the primary determinant of PC biosynthesis. Consistent with this, the enzyme appears to be constitutively expressed and has been detected in plant cells grown in culture medium and in plants grown in soil or agar medium in the presence of only trace levels of essential heavy metals (Grill et al., 1989; Klapheck et al., 1995; Chen et al., 1997). In vitro assays using the partially purified enzyme from *S. cucubalis* demonstrated that it was active only in the presence of a range of metal ions, including Cd, Ag, Bi, Pb, Zn, Cu, Hg and Au cations and the anions, AsO_2^- and AsO_4^{-3} (Grill et al., 1989). Broadly, the range of activating metal ions has been confirmed by *in vitro* studies using purified epitope-tagged derivatives of cloned PC synthases expressed and purified from *E. coli* or *S. cereviseae* (Clemens et al., 1999; Vatamianuk et al., 1999, 2000; Oven et al., 2002).

There is debate about the mechanism by which PC synthase is activated by such a wide range of metal ions. While early models for the activation of PC synthase assumed a direct interaction between metal ions and the enzyme, the model by Vatamaniuk et al. (2000) suggested that, in contrast, metal binding to the enzyme *per se* is not responsible for catalytic activation. While PC synthase binds Cd ions at high affinity and capacity, it has a much lower affinity for other metal ions, such as Cu, which are equally effective activators (Vatamaniuk et al., 1999; 2000), suggesting that a direct interaction with the metal ion was not essential for activation. The observation by Vatamaniuk et al. (2000)

that S-alkyl-GSHs can participate in PC biosynthesis in the absence of heavy metals supported this model. However, in contrast, the study by Oven et al. (2002) found S-methyl-GSH activated AtPCS1 to a very limited extent and that this activation was completely inhibited in the presence of Cd ions. This difference was ascribed to the use of different buffers in the reactions. Oven et al. (2002) demonstrated that the presence of thiols was essential for PC synthase activation. Thus, at this point the detailed mechanism by which PC synthase is activated by metal ions remains to be resolved.

Although activation of PC synthase is the primary mechanism regulating PC biosynthesis, regulation at other levels may occur. For example, *TaPCS1* expression in wheat roots is induced on exposure to Cd (Clemens et al., 1999), although, in contrast, *AtPCS1/CAD1* mRNA levels were not influenced by exposure of plants to Cd and other metals thus suggesting an absence of regulation at the level of transcription (Ha et al., 1999; Vatamaniuk et al., 2000).

In view of the demand for cysteine and GSH for PC biosynthesis, observations that genes in both the sulphur assimilation and GSH biosynthetic pathways are regulated in response to Cd exposure are not surprising. Studies in *Brassica juncea* (Heiss et al., 1998; Schafer et al., 1998; Lee and Leustek, 1999) and in *Arabidopsis* (Xiang and Oliver, 1998) have demonstrated the coordinated transcriptional regulation of genes involved in sulphur transport and assimilation and in GSH biosynthesis. In *Arabidopsis* the signal molecule, jasmonate, mediated an effect similar to Cd exposure (Xiang and Oliver, 1998). However, whether or not jasmonate is itself the signal in responses to Cd has not yet been demonstrated. Thus, there is clear evidence that the sulphur assimilation pathway is regulated in response to Cd exposure and the consequent rapid induction of PC synthesis from GSH. Other possible mechanisms of regulation of GSH biosynthesis in response to other environmental signals may act at both the translational and post-translational levels and have been reviewed recently (May et al., 1998; Noctor and Foyer, 1998; Noctor et al., 2002).

3.4 Sulphide and PC Function

Cadmium effective inducer of PCs has been used in many studies of metal detoxification in *S. pombe* and in plants. In both fungi and plants, labile sulphide can also be detected in PC-Cd complexes. Native Cd-binding complexes can be resolved on the basis of size and in some studies two peaks of Cd-PC complexes (referred to as LMW and HMW) have been resolved (see Rauser, 1999; Cobbett, 2000). In fungi and plants, HMW complexes contain both Cd and acid-labile sulphide and in some cases consist of aggregates of 20-A-diameter particles which contain a CdS crystallite core coated with PCs (Dameron et al., 1989; Reese et al., 1992; Mehra et al., 1994). The presence of sulphide in the HMW complexes increases both the stability of the complex and the amount of Cd per molecule (Reese and Winge, 1988).

The characterisation of Cd-sensitive mutants of *S. pombe* and *Candida glabrata* has provided genetic evidence for the importance of sulphide in the function of PCs. These include various single or double mutants deficient in steps of adenine biosynthetic pathway (Speiser et al., 1992). The enzymes in this pathway, in addition to converting aspartate to intermediates in adenine biosynthesis, can also utilise cysteine sulphinate, a sulphur-containing analog of aspartate, to form other sulphur-containing compounds

(Juang et al., 1993). In *S. pombe,* the *hmt2* mutant hyperaccumulates sulphide in both the presence and absence of Cd (Vande-Weghe and Ow, 1999) and is deficient in a mitochondrial sulphide:quinone oxidoreductase which was suggested to function in the detoxification of endogenous sulphide. In *C. glabrate,* the *hem2* mutant is deficient in porphobilinogen synthase involved in siroheme biosynthesis (Hunter and Mehra, 1998). Siroheme is a cofactor for sulphite reductase required for sulphide biosynthesis. Thus, each of these mutants reinforces the importance of sulphide in the capacity of PCs to detoxify Cd. Further studies are needed to determine the details in each case. Although, sulphide is a component of PC-Cd complexes in plants, mutants affecting this function have not yet been identified and the pathways of metabolism have not been clearly defined.

3.5 PC Function in Plants

Earlier studies of PCs suggested a role in metal detoxification. This was clearly demonstrated by PC-deficient, *cad1-3* mutant of *Arabidopsis* and the PC synthase targeted deletion mutant of *S. pombe*. Both mutants are highly sensitive to both Cd and arsenate but with little or no sensitivity to metals such as Cu, Hg and Ni, indicating that PCs are essential for the detoxification of the former but have little involvement in the detoxification of the latter (Ha et al., 1999). For metal ions that do not appear to be detoxified by PCs in *Arabidopsis* there may be alternative, more effective detoxification mechanisms, such as MTs or histidine in the case of Cu and Ni, respectively. Thus, even if PCs could detoxify these metal ions, in the presence of an alternative, more effective mechanism, a deficiency in PCs may have no observable effect on detoxification. Some observations, such as a low constitutive level of PC expression in cells exposed only to trace levels of essential metals and the capacity of PC-Cu and PC-Zn complexes to reactivate the apo forms of the Cu- or Zn-dependent enzymes in vitro, diamino oxidase and the Zn-dependent enzyme carbonic anhydrase, respectively (Thumann et al., 1991) have led to suggestions that PCs have a role in essential metal metabolism (see Rauser, 1995, 1999; Zenk, 1996). Further roles for PCs in Fe or sulphur metabolism have also been proposed (see Zenk, 1996; Sanita di Toppi and Gabrielli, 1999). However, there is currently no direct evidence that PCs have functions other than in metal detoxification.

4. CONCLUSIONS

MTs and PCs are significant cysteine-rich, metal-binding ligands in plants and other organisms. Various physiological, biochemical, genetic and molecular approaches have led to the advances in our understanding of their functions, particularly in recent years. The use of the model organism, *Arabidopsis thaliana*, with its now completed genome sequence and functional genomics approaches has been significant. However, even using *Arabidopsis*, thus far, the suite of MT genes has been identified but no MT-deficient mutants have been reported. Similarly, only mutants and genes in the PC biosynthetic pathway have been isolated. The parallel studies of PCs in *S. pombe* suggest a number of additional functions yet to be discovered in plants. Nonetheless, further studies of mutants in *Arabidopsis* over the coming years are likely to identify additional aspects of PC function. Questions relating to the functions of MTs in vivo will only be fully resolved when the entire set of MT mutants has been isolated. The use of plants for the detoxification or phytoremediation of polluted environments is being increasingly explored. A sound understanding of the roles of MT and PCs in metal detoxification and metabolism will contribute significantly to this endeavour.

5. REFERENCES

Andrews GK (2000) Regulation of metallothionein gene expression by oxidative stress and metal ions. *Biochem Pharmacol* **59:** 95—104

Binz PA and Kägi JHR (2001) *Metallothionein.* http://www.unizh.ch/~mtpage/MT.html

Chen J, Zhou J and Goldsbrough PB (1997) Characterization of phytochelatin synthase from tomato. *Physiol Plant* **101:** 165—172

Cherian GM and Chan HM (1993) Biological functions of metallothioneins - a review. In: Suzuki KT, Imura N and Kimura M (eds) Metallothionein III, pp 87—109. Birkhauser, Basel

Clemens S, Kim EJ, Neumann D and Schroeder JI (1999) Tolerance to toxic metals by a gene family of phytochelatin synthases from plants and yeast. *EMBO J* **18:** 3325—3333

Clemens S, Schroeder JI and Degenkolb T (2001) *Caenorhabditis* expresses a functional phytochelatin synthase. *Euro J Biochem* **268:** 3640—3643

Cobbett CS, May MJ, Howden R and Rolls B (1998) The glutathione-deficient, cadmium-sensitive mutant, *cad2-1*, of *Arabidopsis thaliana* is deficient in γ-glutamylcysteine synthetase. *Plant J* **16:** 73—78

Cobbett CS (2000) Phytochelatins and their role in heavy metal detoxification. *Plant Physiol* **123:** 463—469

Cobbett CS and Goldsbrough PB (2002) Phytochelatins and metallothioneins: roles in heavy metal detoxification and homeostasis. *Annu Rev Plant Biol* **53:** 159—182

Cobbett CS and Meagher RB (2002) Arabidopsis and the genetic potential for the phytoremediation of toxic elemental and organic pollutants. In: Somerville CR and Meyerowitz EM (eds) The *Arabidopsis* Book, doi/10.1199/tab.0032 http://www.aspb.org/publications/arabidopsis/. American Society of Plant Biologists, Rockville, MD

Dameron CT, Reese RN, Mehra RK, Kortan AR, Carroll PJ, Steigerwald ML, Brus LE and Winge DR (1989) Biosynthesis of cadmium sulphide quantum semiconductor crystallites. *Nature* **338:** 596—597

Foley RC and Singh KB (1994) Isolation of a *Vicia faba* metallothionein-like gene: expression in foliar trichomes. *Plant Mol Biol* **26:** 435—444

Garcia-Hernandez M, Murphy A and Taiz L (1998) Metallothioneins 1 and 2 have distinct but overlapping expression patterns in Arabidopsis. *Plant Physiol* **118:** 387—397

Gekeler W, Grill E, Winnacker E-L and Zenk MH (1989) Survey of the plant kingdom for the ability to bind heavy metals through phytochelatins. *Zeit Naturfor Sec C Biosci* **44:** 361—369

Glaeser H, Coblenz A, Kruczek R, Ruttke I, Ebert-Jung A and Wolf K (1991) Glutathione metabolism and heavy metal detoxification in *Schizosaccharomyces pombe.* Isolation and characterization of glutathione-deficient, cadmium-sensitive mutants. *Curr Genet* **19:** 207—213

Grill E, Winnacker E-L and Zenk MH (1986) Homo-phytochelatins are heavy-metal-binding peptides of homo-glutathione containing Fabales. *FEBS Lett* **205:** 47—50

Grill E, Winnacker E-L and Zenk MH (1987) Phytochelatins, a class of heavy-metal-binding peptides from plants are functionally analogous to metallothioneins. *Proc Natl Acad Sci USA* **84:** 439—443

Grill E, Thumann J, Winnacker E-L and Zenk MH (1988) Induction of heavy metal binding phytochelatins by innoculation of cell cultures in standard media. *Plant Cell Rep* **7:** 375—378

Grill E, Loffler S, Winnacker E-L and Zenk MH (1989) Phytochelatins, the heavy-metal-binding peptides of plants, are synthesized from glutathione by a specific γ-glutamylcysteine dipeptidyl transpeptidase (phytochelatin synthase). *Proc Natl Acad Sci USA* **86:** 6838—6842

Ha S-B, Smith AP, Howden R, Dietrich WM, Bugg S, O'Connell MJ, Goldsbrough PB and Cobbett CS (1999) Phytochelatin synthase genes from *Arabidopsis* and the yeast, *Schizosaccharomyces pombe. Plant Cell* **11:** 1153—1164

Hayashi Y, Nakagawa CW, Mutoh N, Isobe M and Goto T (1991) Two pathways in the biosynthesis of cadystins $(\gamma\text{-EC})_n$G in the cell-free system of the fission yeast. *Biochem Cell Biol* **69:** 115—121

Heiss S, Schafer HJ, Haag-Kerwer A and Rausch T (1998) Cloning sulfur assimilation genes of *Brassica juncea* L.: evidence of Cd-induction of a putative mitochondrial gamma-glutamylcysteine synthetase isoform. *Plant Mol Biol* **37:** 87—97

Howden R, Goldsbrough PB, Andersen CR and Cobbett CS (1995) Cadmium-sensitive, *cad1*, mutants of *Arabidopsis thaliana* are phytochelatin deficient. *Plant Physiol* **107:** 1059—1066

Hunter TC and Mehra RK (1998) A role for *HEM2* in cadmium tolerance. *J Inorg Biochem* **69:** 293—303

Juang R-H, MacCue KF and Ow DW (1993) Two purine biosynthetic enzymes that are required for cadmium tolerance in *Schizosaccharomyces pombe* utilize cysteine sulfinate *in vitro*. *Arch Biochem Biophys* **304:** 392—401

Kägi JHR (1991) Overview of metallothionein. *Methods Enzymol* **205:** 613—626

Kägi JHR (1993) Evolution, structure and chemical activity of class I metallothioneins: an overview. In: Suzuki KT, Imura N and Kimura M (eds) Metallothionein III, pp 29—56. Birkhauser, Basel

Klapheck S, Fliegner W and Zimmer I (1994) Hydroxymethyl-phytochelatins [(γ-glutamylcysteine)$_n$-serine] are metal induced peptides of the Poaceae. *Plant Physiol* **104:** 1325—1332

Klapheck S, Schlunz S and Bergmann L (1995) Synthesis of phytochelatins and homo-phytochelatins in *Pisum sativum* L. *Plant Physiol* **107:** 515—521

Lane BG, Kajioka R and Kennedy TD (1987) The wheat germ Ec protein is a zinc-containing metallothionein. *Biochem Cell Biol* **65:** 1001—1005

Lee SM and Leustek T (1999) The effect of cadmium on sulfate assimilation enzymes in *Brassica juncea*. *Plant Sci* **141:** 201—207

Liu JY, Lu T and Zhao NM (2000) Classification and nomenclature of plant metallothionein-like proteins based on their cysteine arrangement patterns. *Acta Bot Sinica* **42:** 649—652

Maitani T, Kubota H, Sato K and Yamada T (1996) The composition of metals bound to class III metallothionein (phytochelatin and its desglycyl peptide) induced by various metals in root cultures of *Rubia tinctorum*. *Plant Physiol* **110:** 1145—1150

Matsumura H, Nirasawa S and Terauchi R (1999) Transcript profiling in rice (*Oryza sativa* L.) seedlings using serial analysis of gene expression (SAGE). *Plant J* **20:** 719—726

May MJ and Leaver CJ (1994) *Arabidopsis thaliana* γ-glutamylcysteine synthetase is structurally unrelated to mammalian, yeast, and *Escherichia coli* homologs. *Proc Natl Acad Sci USA* **91:** 1005—10063

May MJ, Vernoux T, Leaver C, Van Montague M and Inze D (1998) Glutathione homeostasis in plants: implications for environmental sensing and plant development. *J Exp Bot* **49:** 649—667

Mehra RK, Mulchandani P and Hunter TC (1994) Role of CdS quantum crystallites in cadmium resistance in *Candida glabrata*. *Biochem Biophys Res Comm* **200:** 1193—1200

Mendum ML, Gupta SC and Goldsbrough PB (1990) Effect of glutathione on phytochelatin synthesis in tomato cells. *Plant Physiol* **93:** 484—488

Meuwley P, Thibault P, Schwan AL and Rauser WE (1995) Three families of thiol peptides are induced by cadmium in maize. *Plant J* **7:** 391—400

Murphy A and Taiz L (1995) Comparison of metallothionein gene expression and nonprotein thiols in ten *Arabidopsis* ecotypes. *Plant Physiol* **109:** 945—954

Murphy A, Zhou J, Goldsbrough P and Taiz L (1997) Purification and immunological identification of metallothioneins 1 and 2 from *Arabidopsis thaliana*. *Plant Physiol* **113:** 1293—1301

Mutoh N and Hayashi Y (1988) Isolation of mutants of *Schizosaccharomyces pombe* unable to synthesize cadystin, small cadmium-binding peptides. *Biochem Biophys Res Commun* **151:** 32—39

Noctor G and Foyer CH (1998) Ascorbate and glutathione: keeping active oxygen under control. *Annu Rev Plant Physiol Plant Mol Biol* **49:** 249—279

Noctor G, Gomez L, Vanacker H and Foyer CH (2002) Interactions between biosynthesis, compartmentation and transport in the control of glutathione homeostasis and signalling. *J Exp Bot* **53:**1283—1304

Oven M, Page JE, Zenk MH and Kutchan TM (2002) Molecular characterization of the homo-phytochelatin synthase of soybean *Glycine max*. *J Biol Chem* **277:** 4747—4754

Rauser WE (1995) Phytochelatins and related peptides: structure, biosynthesis, and function. *Plant Physiol* **109:** 1141—1149

Rauser WE (1999) Structure and function of metal chelators produced by plants: The case for organic acids, amino acids, phytin and metallothioneins. *Cell Biochem Biophys* **31:** 19—48

Reese RN and Wagner GJ (1987) Effects of buthionine sulfoximine on Cd-binding peptide levels in suspension-cultured tobacco cells treated with Cd, Zn, or Cu. *Plant Physiol* **84:** 574—577

Reese RN and Winge DR (1988) Sulfide stabilization of the cadmium-γ-glutamyl peptide complex of *Schizosaccharomyces pombe*. *J Biol Chem* **263:** 12832—12835

Reese RN, White CA and Winge DR (1992) Cadmium sulfide crystallites in Cd-(γ-EC)$_n$G peptide complexes from tomato. *Plant Physiol* **98:** 225—229

Robinson NJ, Tommey AM, Kuske C and Jackson PJ (1993) Plant metallothioneins. *Biochem J* **295:** 1610

Robinson NJ, Wilson JR and Turner JS (1996) Expression of the type 2 metallothionein-like gene *MT2* from *Arabidopsis thaliana* in Zn^{2+}-metallothionein deficient *Synechococcus* PCC 7942. Putative role for *MT2* in Zn^{2+}-metabolism. *Plant Mol Biol* **30:** 1169—1179

Sanita di Toppi L and Gabbrielli R (1999) Response to cadmium in higher plants. *Env Exp Bot* **41:**105—130

Schafer HJ, Haag-Kerwer A and Rausch T (1998) cDNA cloning and expression analysis of genes encoding GSH synthesis in roots of the heavy metal accumulator *Brassica juncea* L.: evidence or Cd-induction of a putative mitochondrial gamma-glutamylcysteine synthetase isoform. *Plant Mol Biol* **37:** 87—97

Scheller HB, Huang B, Hatch E and Goldsbrough PB (1987) Phytochelatin synthesis and glutathione levels in response to heavy metals in tomato cells. *Plant Physiol* **85:** 1031—1035

Skipsey M, Andrews CJ, Townson J, Jepson I and Edwards R (1999) Isolation of cDNA (AJ243813) and genomic clones (AJ243812) of glutathione synthetase containing plastidic targeting sequences from *Arabidopsis thaliana. Plant Physiol* **121:** 312

Speiser DM, Ortiz DF, Kreppel L and Ow DW (1992) Purine biosynthetic genes are required for cadmium tolerance in *Schizosaccharomyces pombe. Mol Cell Biol* **12:** 5301—5310

Thumann J, Grill E, Winnacker E-L and Zenk MH (1991) Reactivation of metal-requiring apoenzymes by phytochelatin-metal complexes. *FEBS Lett* **284:** 66—69

Vande Weghe JG and Ow DW (1999) A fission yeast gene for mitochondrial sulfide oxidation. *J Biol Chem* **274:** 13250—13257

van Hoof NALM, Hassinen VH, Hakvoort H, Ballintijn KF, Schat H, Verkleij JAC, Ernst WHO, Karenlampi SO and Tervahauta AI (2001) Enhanced copper tolerance in *Silene vulgaris* (Moench) Garcke populations from copper mines is associated with increased transcript levels of a 2b-type metallothionein gene. *Plant Physiol* **126:** 1519—1527

van Vliet C, Anderson CR and Cobbett CS (1995) Copper-sensitive mutant of *Arabidopsis thaliana. Plant Physiol* **109:** 871—878

Vatamaniuk OK, Mari S, Lu Y-P and Rea PA (1999) AtPCS1, a phytochelatin synthase from *Arabidopsis*: isolation and *in vitro* reconstitution. *Proc Natl Acad Sci USA* **96:** 7110—7115

Vatamaniuk OK, Mari S, Lu Y-P and Rea PA (2000) Mechanism of heavy metal ion activation of phytochelatin (PC) synthase - Blocked thiols are sufficient for PC synthase-catalyzed transpeptidation of glutathione and related thiol peptides. *J Biol Chem* **275:** 31451—31459

Vatamaniuk OK, Bucher EA, Ward JT and Rea PA (2001) A new pathway for heavy metal detoxification in animals - Phytochelatin synthase is required for cadmium tolerance in *Caenorhabditis elegans. J Biol Chem* **276:** 20817—20820

Vernoux T, Wilson RC, Seeley KA, Reichheld JP, Muroy S, Brown S, Maughan SC, Cobbett CS, Van Montagu M, Inze D, May MJ and Sung ZR (2000) The *ROOT MERISTEMLESS1/CADMIUM SENSITIVE2* gene defines a glutathione-dependent pathway involved in initiation and maintenance of cell division during postembyronic root development. *Plant Cell* **12:** 97—109

Xiang C and Oliver DJ (1998) Glutathione metabolic genes co-ordinately respond to heavy metals and jasmonic acid in *Arabidopsis. Plant Cell* **10:** 1539—1550

Zenk MH (1996) Heavy metal detoxification in higher plants - a review. *Gene* **179:** 21—30

Zhou J and Goldsbrough PB (1994) Functional homologs of animal and fungal metallothionein genes from Arabidopsis. *Plant Cell* **6:** 875—84

Zhou J and Goldsbrough PB (1995) Structure, organization and expression of the metallothionein gene family in *Arabidopsis. Mol Gen Genet* **248:** 318—328

6. ACKNOWLEDGEMENT

Research in the author's laboratory is supported by the Australian Research Council.

7. AFFILIATION

Department of Genetics, The University of Melbourne, Parkville, Victoria, 3052, Australia
For correspondence: ccobbett@unimelb.edu.au

J.L. HARWOOD AND A.A. OKANENKO

SULPHOQUINOVOSYL DIACYLGLYCEROL (SQDG) - THE SULPHOLIPID OF HIGHER PLANTS

Abstract: Plant sulpholipid, sulphoquinovosyl diacylglycerol (SQDG), has been found in all photosynthetic plants. It appears to be concentrated mainly in the chloroplasts of plants. Higher plant chloroplasts are completely autonomous for the SQDG synthesis. Concerning the pathway for the head group synthesis it was suggested that sulphoquinovose could be formed from a compound related to UDP-6-deoxy-4-ketoglucose which is an intermediate in L-rhamnose synthesis. The final intermediate in the pathway is UDP-sulphoquinovose which is used by the enzyme catalyzing the final stage of SQDG formation. The generally accepted view is that SQDG is hardly metabolized by higher plants apart from initial hydrolysis by acyl hydrolases and glycosidases, and it is considered widely that liberation of sulphate from sulphoquinovose is due to microbial activity. SQDG is associated with purified chloroplast CF_0-CF_1 ATPase and, like MGDG, plays a more specific and intimate role in the catalytic activity of proteins. Two SQDG molecules with predominant palmitic acid were found in the association of the LHC II-apoproteins. Bound SQDG molecules are localized as prosthetic groups at the surface of native D1/D2 heterodimer, holding the dimmer together. But a SQDG-deficient mutants showed the same growth rates under optimal phototrophic growth conditions as the wild type, suggesting that SQDG was not essential for oxygenic photosynthesis. The results of studies, devoted to lipid involvement in adaptation processes, showed that both SQDG quantitative changes and fatty acid composition shifts take place. Besides, SQDG accumulation was observed in wheat plants infected by *Puccinia graminis* and kidney bean plants infected by tobacco mosaic and potato x viruses.

1. INTRODUCTION

Plants, algae, cyanobacteria and some photosynthetic bacteria contain a unique sulpholipid discovered and studied by Benson et al. (1963). Three products resulted from hydrolysis of this sulpholipid are glycerol, a sulphur-containing C6 compound and two fatty acids. The sulphosugar was identified as a sulphonic acid by its resistance to acid hydrolysis, and the hexose thought to be galactose with a β-linkage because the deacylated sulpholipid was hydrolysed by β-galactosidase. But it was found later, in fact, that all the major glycolipids occurring in plants (the sulpholipid and two β-D-galactosyl diacylglycerols) possess a linkage which may be cleaved by β-galactosidase and the hexose itself was identified as sulphoquinovose (Shibuya and Benson, 1961). Finally, the sulpholipid structure was defined as 1,2-di-O-acyl-3-O-(6-deoxy-6-sulpho-α-D-glucopyranosyl)-*sn*-glycerol (SQDG) (Fig. 1). The distinctive feature of this substance is carbon bonded directly to sulphur as $C—SO_3$. Sulphonic acids of this type are chemically stable and strong acids over a wide pH range (Barber and Gounaris, 1986). The deacylation products of sulpholipid, sulphoquinovosyl glycerol, occur in plants at variable concentrations, but little or no sulphoquinovose has been found in plant extracts. SQDG is the most concentrated anionic sugar derivative in plants under most conditions.

Yash P. Abrol and Altaf Ahmad (eds.), Sulphur in Plants, 189—219.

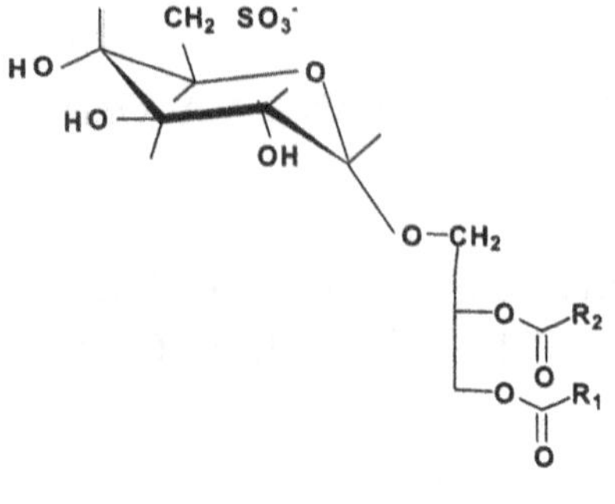

Figure 1. A sulphoquinovose

2. OCCURENCE AND LOCALISATION OF SULPHOLIPID

Plant sulpholipid has been found in all plants, algae, cyanobacteria, purple sulphur and non-sulphur bacteria (Barber and Gounaris, 1986). Early studies suggested that it was a typical chloroplast constituent in eukaryotes e.g. green alga *Chlamydomonas reinhardtii 137*$^+$ about 70% of SQDG localised in the thylakoids (Janero and Barrnett, 1981). Brown algae tend to have rather high levels of the compound, which represents 30% of the total lipids in *Hizikia fusiformis* (Fucales) (Harwood, 1998). The betaine lipid and the glycolipids – monogalactosyldiacylglycerol (MGDG), digalactosyldiacylglycerol (DGDG) and SQDG were major components in the marine brown alga *Dictyopteris membranacea*. MGDG, DGDG and SQDG are the major polar lipids in *Ectocarpus fasciculatus* (Ectocarpales, Phaeophyceae) also. The positional distribution of fatty acids in *Ectocarpus* showed that molecular species of "eukaryotic structure" accounted for 99% in MGDG, 98% in DGDG, 62% in PG but only 23% in SQDG. Thus, in the latter, labelled fatty acids were found in prokaryotic as well as in eukaryotic molecular species (Makewicz et al., 1997). This means, of course, that in the former species the diacylglycerol backbone of the SQDG originates predominately from acylation reactions within the chloroplast (Browse and Somerville, 1991). SQDG has also been reported in non-photosynthetic bacteria (e.g. *Rhizobium meliloti* 2011) (Cedergreen and Hollingsworth, 1994), in non-photosynthetic root nodule-forming bacteria (e.g. *Sinorhizobium meliloti*) (Weissenmayer et al., 2000) and in non-photosynthetic diatoms (e.g. *Navicula alba*) (Harwood and Jones, 1989).

In some lichens, the SQDG content is nearly one third of the total glycolipids, and sometimes surpasses the MGDG content (e.g. by two-fold in *Trebouxia impressa*) and the DGDG (*Trebouxia erici*) levels (Bychek-Gushchina, 1997). Two sulpholipids, SQDG and SQMG (sulphoquinovosyl monoacylglycerol) containing mainly palmitic ($\approx$ 90%) and stearic (3-11%) acid residues with small amounts of heptadecanoic, oleic, arachidonic, uncosanoic and lignoceric acyl residues have been isolated from the basidiolichen *Dictyonema glabratum* (Sassaki et al., 2001). In addition, a number of unusual sulphur-containing lipids have been detected in algae (Haines, 1973; Harwood and Jones, 1989; Kaya et al., 1993), but none of the other sulphur-containing lipids are quantitatively significant when compared to SQDG in mature (Heinz, 1993). In higher plants it can be noted that SQDG comprises up to 24% of the four major chloroplast lipids. This means that SQDG is a very significant contributor to the global sulphur cycle where it provides amounts of sulphur which are comparable to that derived from protein (Harwood and Nicholls, 1979).

SQDG usually contains a substantial quantity of saturated and, hence, high melting point fatty acids (see section 6.1). For example, total saturates can range between 26% in *Triticum aestivum* and 62% in *Carica papaya*. These figures are much higher than for the chloroplast galactosylglycerides, MGDG and DGDG. All of the 16:3 plants (i.e. *Solanaceae* and *Apiaceae*) contained at least 40% saturated fatty acids, although *Passifloraceae* (which are 18:3 plants) also have similar levels of SQDG (Kenrick and Bishop, 1986). SQDG of some plants contains mainly palmitic and linolenic residues, e.g. spinach, maize, rice, lettuce, sponge cucumber (Murata and Hoshi, 1984), wheat (*Triticum durum*) (Quartacci et al., 1995), but others have a significant quantity of linoleate with smaller quantities of oleate, e.g. *Pelargonium* (Shipley et al., 1973) *Atriplex lentiformis* (Pearcy, 1978) and cabbage (Murata and Hoshi, 1984). Other plants have significant amounts of both oleate (*Nerium oleander* L., 12-14%) and linoleate (Caricaceae, Passifloraceae, 10-16%) residues in SQDG (Murata and Hoshi, 1984; Kenrick and Bishop, 1986; Orr and Raison, 1987), which can sometimes surpass the linolenate content e.g. in *Boea hygroscopica* and sugar beet (Kylin et al., 1972). In summary, SQDG contains a saturated fatty acid content which is comparable to the other chloroplast anionic lipid, phosphatidyl-glycerol. It is much higher than that of MGDG and DGDG (Harwood, 1980a).

Part of the reason for the high saturation in SQDG comes from its method of biosynthesis. Thus, whereas MGDG was almost exclusively of the "eukaryotic" type, SQDG was predominantly of the "prokaryotic'" type (Hofmann and Eichenberger, 1997). Plants that contain 18:3 but not 16:3 in MGDG (synthesized via extra-chloroplastic PC) are known as 18:3 plants and such lipids are called 'eukaryotic lipids', whereas in 16:3 plants part of MGDG contains 16:3 esterified at the *sn*-2 position which is synthesized within chloroplasts. These lipids are called 'prokaryotic lipids' because their fatty acid distribution is similar to that in cyanobacteria. The significance of the subcellular locations of the biosynthetic pathways for chloroplast lipids are discussed well in Browse and Somerville (1991). In 16:3 plants a high proportion of SQDG has a 'prokaryotic' structure, whereas 18:3 plants contain SQDG mostly with a 'eukaryotic' fatty acid distribution (Joyard et al., 1998). Because of some detailed difficulties in describing the pathways as 'eukaryotic' or 'prokaryotic', Murata and Siegenthaler (1998) suggested that the terms 'chloroplast' and 'cytosolic' pathways would be better used.

On a per gram fresh weight basis SQDG varies from 0.5 μmol in white clover leaves to 1.3 μmol in perennial rye grass (Roughan and Batt, 1968); on a per gram dry weight basis the values range from 1.55 μmol in red clover in December to 5.24 μmol in rye grass (Russel, 1966); on a protein basis the reported variation is 6-14 μmol mg^{-1} (Heinz, 1993). Although SQDG appears to be concentrated in the chloroplast lamellar membranes it can be found in small amounts in roots and chlorophyll-free leaves (Benson, 1963). Its presence throughout thylakoids could be connected both with its interaction with Photosystem II (PS II), which is located mainly in the appressed region and the coupling factor of ATP synthase (CF_0-CF_1), which is localised exclusively in the non-appressed regions (Barber and Gounaris, 1986).

The sided distribution of SQDG in thylakoid membranes has been studied. Rawyler and Siegenthaler (1985) concluded that SQDG is almost completely localised in the inner leaflet of the bilayer for spinach thylakoids because no significant degradation was observed in thylakoid vesicles treated with *Rhizopus arrhizus* lipase. In a further, more detailed study Rawyler et al. (1987) found that of the SQDG that was accessible to digestion, only 10-12% was present in the outer leaflet. However, it should be noted that nearly 30% of the total SQDG was unavailable for digestion, making accurate conclusions impossible. Nevertheless, it was pointed out by Sundby and Larsson (1985) that, if acidic lipids are located in the inner leaflet of the bilayer, they could act as counter ions for protons accumulated in the lumen, thus serving as proton–conducting pathways. This also supports suggestions by Haines (1983). SQDG is also present in both outer and inner chloroplast envelope membranes at about 6% of total acyl lipids (Douce et al., 1984).

An examination has been made of SQDG distribution in relation to chlorophyll-protein (CP) complexes and the photosystems. It is known that the content of SQDG is much higher in PS II (13.7 mol%) than in photosystem I (PS I) (3.1 mol%). However, the highest amounts were observed in light harvesting complex II (LHC II)-B preparations (23.6 mol%). Molecular ratios of SQDG/chlorophyll-protein (CP) complexes are similar for PS II (26%) and LHC II (25% for LHC II-A and 20% for LHC II-B) whereas for PS I, it is much lower (<4%) (Sato et al., 1995). In a reconstitution study of delipidated LHC II it was found that uptake of MGDG, DGDG, SQDG, and PG was 6200, 2200, 1150 and 950 µg lipid/mg Chl, respectively. This suggests that the maximum amount that can be bound by LHC II is proportional to the relative concentrations of these lipids in thylakoids (Simidjiev et al., 2000).

As mentioned above, much smaller quantities of SQDG are found in plant roots e.g. 0.084-0.188 µmol lipid g^{-1} fresh weight in wheat roots and 0.087-0.113 µmol g^{-1} in oat roots, depending on growing condition. Although early studies (Kuiper, 1972) suggested that SQDG was present in subcellular compartments other than plastids, these data may reflect the difficulty of obtaining uncontaminated fractions from plant root tissues. Nevertheless, the SQDG content was reported to be 12-18% and about 37% of total lipids in bean and sugar beet roots, respectively (Kuiper, 1972). In fact, for sugar beet plants grown in climate chambers this compares to 20% of total lipids for SQDG in leaves and 28% of total lipids in stalks (Kylin et al., 1972).

3. SQDG CODING GENES

Identification of genes for SQDG synthesis was first made with photosynthetic purple bacteria. Four genes involved in SQDG biosynthesis were designated as *sqdA, sqdB, sqdC* and *sqdD* in *Rhodobacter sphaeroides*. Three of them, *sqdA, sqdB, sqdC*,- were organised in two transcription units. The fourth, *sqdD,* was identified by insertional inactivation of the open reading frame between *sqdB* and *sqdC* in the *R. sphaeroides* sulpholipid operon. Results from insertional inactivation of *sqdB* suggested that *sqdB* and *sqdC* were co- transcribed and part an operon for sulpholipid biosynthesis. The N-terminal region of the 46-kDa putative protein encoded by the *sqdB* gene showed some homology to the gene for UDP-glucose epimerase from various organisms (Benning and Somerville, 1992). This fact is of relevance to the biosynthetic pathway (see section 4.1). More recently, a cDNA from *A. thaliana,* designated *SQD1,* which shows high sequence similarity to bacterial *sqdB* genes has been shown to be essential for SQDG biosynthesis (Essigman et al., 1998). Using a *R. sphaeroides* mutant inactivated for *sqdD*, Rossak et al. (1995) reported the accumulation of a water-soluble ^{35}S-labeled compound, identified as UDP-sulphoquinovose. The authors suggested that the inactivation of *sqdD* causes a metabolic block in the last step of sulpholipid biosynthesis. Thus, it was supposed that, most probably, the gene coding for SQDG synthase in *R. sphaeroides* is *sqdD*, and the protein coded had a similar sequence to glycosyl transferases of the glycogenin type. The 30-kDa putative protein showed very striking homology to rabbit muscle glycogenin, a UDP-glucose utilising autoglycosylating glycosyl transferase (Benning and Somerville, 1992). However, enzymatic activity was not observed in vitro (Benning, 1998) perhaps because *R. sphaeroides sqdD* coded only one SQDG synthase subunit whereas, for the fully functional SQDG synthase, auxillary proteins were necessary.

Data obtained with a *R. sphaeroides sqdC* null mutant showed only a 90% reduction in SQDG content and accumulation of a ^{35}S-labeled compound identified as α-D-sulphoquinovosyl-1-O-dihydroxyacetone. This compound was considered not to be a SQDG biosynthesis intermediate but a side product accumulated in the absence of SQDG synthesis. It is not found during SQDG synthesis by *R. sphaeroides* extracts or spinach chloroplasts (Rossak et al., 1997). Moreover, it should be borne in mind that *sqdC* mutants contain significant amounts of SQDG. Thus it was assumed (Benning and Somerville, 1992; Rossak et al., 1997) that the protein coded by *sqdC* can be substituted by another protein or that it plays an auxiliary function in SQDG biosynthesis. Taking these results into account, it was suggested that the *sqdC* gene product mediates the substrate specificity of the sulphoquinovosyl transferase encoded by *sqdD* that catalyses the final reaction of sulpholipid biosynthesis or that it ensures that the enzyme associates with the membrane. In *sqdC* deficient mutants, it was suggested that SQDG synthase could transfer sulphoquinovose to diacylglycerol (DAG) at low levels but, for the most part, sulphoquinovose was transferred to a soluble acceptor, dihydroxyacetone, to form sulphoquinovosyl dihydroxyacetone (Rossak et al., 1997).

Benning (1998) suggested that genes similar to *sqdB* were present in all SQDG-containing organisms. However, a novel gene, *sqdX*, essential for sulpholipid biosynthesis was reported in the cyanobacterium *Synechococcus* sp. strain PCC7942. The derived sequence of its product was similar to glycosyltransferases and contained two hydrophobic domains, indicating that the *sqdX* gene product might be a transmembrane protein. Guler et al. (2000) proposed that *sqdX* encoded the cyanobacterial SQDG synthase which catalysed the last reaction of the pathway. Inactivation of *sqdX* in *Synechococcus* led to a complete SQDG deficiency which could be restored by introducing the *sqdX* genes from either *Synechococcus* or *Synechocystis*. The sequence similarity to glycosyl transferases strongly suggested that *sqdX* encoded the SQDG synthase catalysing the transfer of sulphoquinovose from UDP-sulphoquinovose onto a suitable acceptor, presumably DAG. Thus, taking into account that no sequences with similarity to *sqdA*, *sqdC*,or *sqdD* of *R. sphaeroides* were detected upon examination of the *Synechocystis* genomic sequence, it is possible that different types of glycosyl-transferases may catalyze this reaction in different organisms (Guler et al., 2000). The formation of the sulphonic acid precursor, UDP-sulphoquinovose, in higher plants is proposed to be catalysed by the orthologous plant SQD1 proteins. Like the bacterial *sqdB* gene products, the SQD1 protein has a sequence similar to sugar nucleotide modifying enzymes and it was assumed to be involved in the biosynthesis of UDP-sulphoquinovose from UDP-glucose (Benning and Somerville, 1992; Rossak et al., 1995; Guler et al., 1996). Results of recent work showed that recombinant SQD1 of *A. thaliana* catalysed the formation of UDP-sulphoquinovose, the sulpholipid head group donor from UDP-glucose and sulphite (see section 4.1). The latter could be derived from the APS reductase reaction (Sanda et al., 2001).

4. SQDG METABOLISM

4.1 Biosynthesis

Despite the quantitative importance of SQDG as a component of thylakoid membranes (Harwood, 1980a) and as a major contributor to the global sulphur cycle (Harwood and Nicholls, 1979; Heinz, 1993), attempts to unravel the pathway of its biosynthesis proved fruitless for many years. In fact, at one stage it could be remarked that there were more suggestions as to possible routes for SQDG formation than there was hard experimental evidence (Harwood, 1980b)!

Much of the early work concentrated on the so-called sulphoglycolytic pathway which was first outlined in detail by Davies et al. (1966). Although experiments with *Euglena gracilis* (Davies et al., 1966) and germinating seeds of alfalfa and brussel sprouts (Harwood, 1975) were consistent with the pathway, later results obtained with spinach seedlings (Mudd et al., 1980) or isolated chloroplasts (Kleppinger-Sparace et al., 1985) were not. In fact, as summarised in two reviews, there are a number of theoretical problems with the sulphoglycolytic pathway for the formation of SQDG (Harwood, 1980b; Mudd and Kleppinger-Sparace, 1987).

A route for the chemical synthesis of sulphonic acids involves the addition of sulphite to double bonds (see Haines, 1971) and Lehmann and Benson (1964) proposed that 6-sulphoquinovose could be produced in such a manner in vivo. Evidence for the involvement of sulphite in SQDG formation was obtained from experiments with germinating seeds (Harwood, 1975), with *Chlamydomonas* membranes (Hoppe and Schwenn, 1981) and in *Euglena* (Saidha and Schiff, 1989). If sulphite were involved in SQDG biosynthesis then the simplest substrate for it to react with would be a 6C sugar derivative such as a glucose-5-enide which would be analogous to the reaction used by Lehmann and Benson (1964) to synthesise methyl α–sulphoquinovoside in vitro.

Based on previous evidences, both theoretical and practical, Pugh et al. (1995) proposed a new pathway for SQDG biosynthesis which is shown in Fig. 2. Supporting evidence from the literature included the following observations:

a. It was suggested that, in theory, sulphoquinovose could be formed from a compound related to UDP-6-deoxy-4-ketoglucose which is an intermediate in L-rhamnose synthesis in tobacco (Barber, 1963).
b. The *sqdB* gene from *R.sphaeroides* showed sequence similarity to UDP-glucose epimerase (Benning and Somerville, 1992).

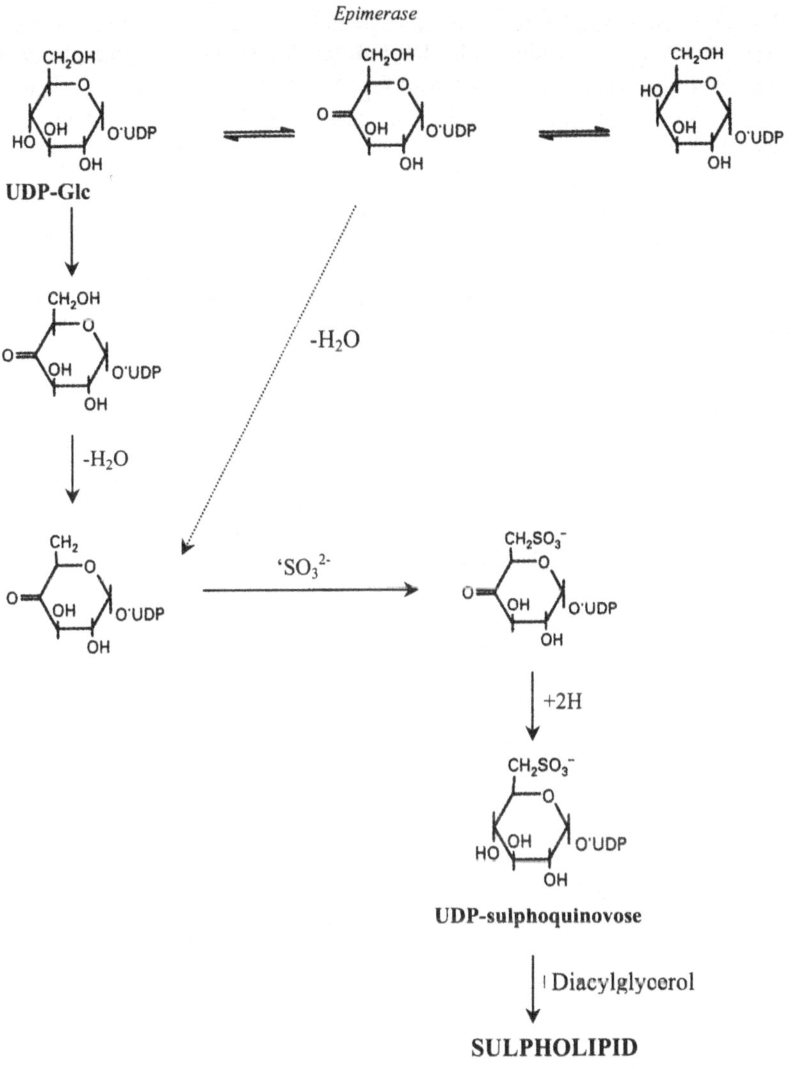

Figure 2. Pathway of the biosynthesis of SQDG (Source: Pugh et al. (1995)

c. Several enzymes of carbohydrate metabolism involve abstraction of an axial hybrid equivalent from C4 (Gabriel, 1973).
d. Formation of a nucleotide derivative of 4-ketoglucose-5-ene had been shown for other sugar synthesis pathways (Snipes et al., 1977).
e. The double bond in glucose-5-enide readily accepts sulphite in vitro to form sulphoquinovoside (Lehmann and Benson, 1964).

f. In fact, the addition of sulphite to hexose-5-enides is a general reaction and would be predicted to occur readily with UDP-6-deoxyglucose-5-ene (Pugh et al., 1995).
g. The final intermediate in the pathway (Fig. 2) is UDP-sulphoquinovose which is used by the enzyme catalysing the final stage of SQDG formation (Seifert and Heinz, 1992).

Evidence for the pathway was then obtained by a number of experiments as performed by Pugh et al. (1995). Although most of these experiments involved ^{35}S-sulphate as the radiolabelled precursor, some used UDP-[^{14}C]glucose. The latter was later shown to give a number of radioactive lipids which migrated at the same place as SQDG on thin-layer chromatography and could give rise to misleading results in vitro (Roy and Harwood, 1999).

Recently Sanda et al. (2001) have shown that the SQD1 protein from *Arabidopsis thaliana* (which is analogous to the bacterial SQDB protein) catalyses the formation of UDP-sulphoquinovose from UDP-glucose and sulphite. Bearing in mind that these proteins have sequence similarity to UDP-glucose epimerase (Benning and Somerville, 1992; Essigmann et al., 1998), the new results fit excellently to our scheme (Fig. 2). Pugh et al. (1995) gave two alternatives in their scheme. The first was to generate UDP-4-ketoglucose-5-ene directly from the UDP-4-ketoglucose intermediate on the epimerase. The second was that sulphite could either add to an enzyme-bound UDP-4-ketoglucose-5-ene or to a free UDP-glucose-5-ene as an intermediate. Since a single protein (SQD1) catalyses the complete sequence from UDP-glucose in vitro (Sanda et al., 2001), we would suggest that enzyme-bound intermediates are involved, with sulphite addition occurring to UDP-4-ketoglucose-5-ene (see Pugh et al., 1995).

It is notable that elucidation of the biosynthesis pathway for SQDG formation had its origins in pure chemistry; postulated and refined using biochemistry and later confirmed using molecular biology. Truly an inter-disciplinary venture!

4.2 Catabolism of SQDG

Although, as mentioned above, SQDG is a major component of the biological sulphur cycle (Harwood and Nicholls, 1979) rather little is known of its catabolism. Nevertheless, SQDG breakdown occurs at huge rates on an annual basis due to the turnover of photosynthetic membranes during senescence of photosynthetic tissues. The generally accepted view (Harwood, 1980b) is that SQDG is hardly metabolised by higher plants apart from initial hydrolysis by acyl hydrolases (Burns et al., 1979) and glycosidases. Although, Lee and Benson (1972) have detected some catabolites of sulpho-quinovose (SQV) in two plant tissues, it is considered widely that liberation of sulphate from SQV is due to microbial activity (Martelli and Benson, 1964; Strickland and Fitzgerald, 1983; Pugh et al., 1997).

Recently, Roy et al. (2000) isolated some soil bacterial strains which were capable of growing on SQV as their sole carbon source. Isolates of *Agrobacterium* sp and *Klebsiella* sp accumulated sulphate in the growth media but several isolates of *Pseudomonas* spp did not. In the latter cases an accumulating compound, identified as sulpholactate, was found. *Kledbsiella* accummulated sulphopropanediol as an intermediate which was subsequently broken down to sulphate (AB Roy, MJE Hewlins,

AJ Ellis, GF White and JL Harwood, unpublished data). The results suggest that bacteria may catabolise SQV via the equivalent of sulphoglycolytic pathway, where sulphonate derivatives substitute for phosphorylated ones. If that turns out to be true, it will be ironic in the light of the interest in a sulphoglycolytic pathway as a favoured suggestion for SQDG synthesis (see Davies et al., 1966) over thirty years ago!

5. SQDG FUNCTION

5.1 Results of In Vitro Experiments

The presence of the unique lipid, SQDG, in photosynthetic membranes suggests an important function there. Indeed, Benson (1964) and, later, Trosper and Sauer (1968) suggested that this sulpholipid might assist in the orientation of chlorophyll molecules (now know to be pigment-protein complexes) in the membranes. Examination of correlations between the appearance of chlorophyll and the presence and concentration of SQDG in maize photosynthetic tissues revealed that the sulpholipid increased during greening of etiolated tissue and during chloroplast maturation (Leech et al., 1973). Anderson (1975) considered the chlorophyll-protein complexes to be surrounded by boundary layers made up partly from the phytyl chains of chlorophyll molecules and partly from boundary anionic lipid (particularly SQDG) components of the membranes. Moreover, SQDG has been suggested to be connected with xanthophyll metabolism. Thus, the violaxanthin de-epoxidase independent conversion of violaxanthin into aurixanthin and some isomers which occurred in the presence of SQDG and PG could be explained by the acidic character of these lipids (Latowski et al., 2000).

SQDG has been found to be closely associated with certain membrane proteins. This association is probably partly stabilised by electrostatic interactions between its negatively charged sulphonate group and positive charges within the proteins. Thus, SQDG could bind the 35-kDa annexin in the chloroplast envelope in a Ca^{2+}-dependent manner. The binding of the spinach annexin to SQDG occurred at similar Ca^{2+} concentrations as those which promote the interaction of annexins to phospholipids in other membranes. Among chloroplast glycerolipids known to be accessible on the cytosolic face (outer leaflet) of the outer envelope membrane, SQDG and probably PG, would be the sole candidates for a putative Ca^{2+}-dependent interaction of annexin at the chloroplast surface (Seigneurin-Berny et al., 2000).

In some cases the (electostratic) interactions may be very strong, as suggested by the inability of saturated SQDG molecules associated with purified chloroplast CF_0-CF_1 ATPase to exchange with other acidic lipids (Pick et al., 1985). Because removal of SQDG from the protein complex caused inactivation, it was assumed that the lipid was an integral part of the complex. Barber and Gounaris (1986) have suggested that SQDG neither contributes significantly to the bulk lipid matrix nor is involved in "protein packing" like MGDG, but plays a more specific and intimate role in catalytic protein activities. Molecular species of SQDG involved in enzymatic activity are thought to contain a high amount of saturated acyl fatty acid chains and, perhaps, therefore act as "boundary" lipid. Use of antibodies suggested that SQDG and MGDG occur in high concentrations and are likely to be arranged in domains in the immediate vicinity of CF_1 (Haase et al., 1993). It was shown also that SQDG protects CF_1 against cold inactivation in the presence of some ATP. CF_1 is a latent enzyme that requires treatment by heat, trypsin or dithiothreitol for activation and normally contains negligible ATPase activity but exposure to heat (65°C), in the presence of ATP, caused high Ca^{++}-dependent ATPase activity to appear. Under these circumstances, the protein lost the capacity to stimulate photophosphorylation and to bind to the chloroplast membrane. In addition, CF_1 bound to membranes was found to be much more resistant to heat and cold than solubilised protein. Binding of CF_1 to membranes is promoted by lipids. SQDG fractions were considerably more effective than the galactolipid (GL) fraction. In addition, phospholipids (PL) were effective, but less than GL. It is interesting that mitochondrial coupling factor F_1 is also protected by PL and SQDG although, in that case, both were equally effective (Bennun and Racker, 1969; Livn and Racker, 1969).

It has been shown that SQDG added to heptane-extracted chloroplasts restored PS1 activity, thus suggesting its importance as a structural component of chloroplast membranes, for their photosynthetic activity. Both SQDG and GL also reactivated cyclic photophosphorylation (Krupa and Baszynski, 1977). SQDG was found in reaction centre (RC) preparations of PS II. These contained five polypeptides and only two lipids, MGDG and SQDG. In the PS II RC preparations only certain molecular species of SQDG were found, viz. palmitic, stearic and oleic residues made up most of the lipid present (Gounaris and Barber, 1985). Antibodies against DGDG and SQDG were found to react exclusively with the heterodimer D1/D2 of the PS II RC (Vijayann et al., 1998) and not with the other proteins of the complex (Siegenthaler, 1998). Therefore, it was concluded that bound SQDG and DGDG were localised at the surface of the native D1/D2 heterodimer (66 kDa). One suggestion was that bound SQDG might, therefore, hold the dimer together (de Kruijff et al., 1998; Vijayann et al., 1998).

Tight binding of certain lipids to defined peptides of the PS II core has a functional significance, since treatment with antibodies inhibits electron transport on the donor side of PS II (Radunz et al., 1984). However, data presented by Murata et al. (1990) showed that the MGDG was the only lipid found in the PS II RC. The MGDG was also highly saturated with myristate, palmitate and stearate as dominant acyl groups. The PS II core complex contained MGDG, DGDG and PG, but it lacked SQDG. These authors concluded that SQDG did not contribute to the activity of PS II.

Krupa et al. (1987) noted that SQDG was the major lipid in LHC II after treatment with galacto- and phospholipases. Since 20 to 25% of the complex was still present as an oligomer, it was argued that SQDG was involved in stabilising LHC II. Other data showed that LHC II-complexes of *Chlamydomonas reinhardtii* were enriched in SQDG with palmitate as the predominant acyl component (Sigrist et al., 1988). It was calculated that 40% of the thylakoid SQDG was tightly associated with LHC II which itself represented about half of the total thylakoid proteins. In particular, every LHC II-apoprotein was associated with two molecules of SQDG. It is interesting that LHC II-apoproteins isolated from the chlorophyll *b*-deficient mutant *pg-113* had the same lipid composition as those of wild type. SQDG was bound to the LHC II-apoproteins exclusively by hydrophobic interactions in both wild type and the chlorophyll *b*-deficient mutant. The high degree of saturation of bound SQDG (75% palmitate; 50% dipalmitoyl SQDG species) facilitated the interaction between this compound and protein. It was also shown by Remy et al. (1982) that the oligomeric form of the LHC II-complex is enriched in both SQDG and PG. Monomerisation of the complex resulted in loss of PG but not SQDG (Tremolieres et al., 1981). The lipid may play a role in establishing the correct tertiary structure thus allowing proper interaction between LHC II components. The trimeric organisation of LHC II polypeptides constitutes the basic ultrastructural and functional arrangement of the antennae around PS II (Tremolieres and Siegenthaler, 1998). In addition, intermolecular hydrophobic interactions were reported between the anionic SQDG and a 25 kDa polypeptide (Radunz et al., 1984) which was characterised by high levels of phosphorylation (Larsson and Andersson, 1985). The latter property suggested that SQDG might be involved in regulation of light energy distribution between the two photosystems via the phosphorylation of the mobile antennae.

Another suggestion is for a role of SQDG in PS II core complex dimerisation. The low molecular mass protein PsbH of the PS II core dimer contains a cluster of positively charged residues located at the stromal surface of its single transmembrane helix (Kruse et al., 2000). This region could be a site for binding PG and is also the region of reversible N-terminal phosphorylation (Michel et al., 1988). (However, there was no apparent effect on PS II assembly and function in a *Arabidopsis thaliana* mutant, lacking *trans*–hexadecenoic acid (McCourt et al., 1985)). Nevertheless, it is conceivable that SQDG could substitute for PG in the dimerization processes in the case of phosphate starvation. Concerning LHC II trimerization, information exists that PG and its *trans*-hexadecenoate component are not obligatory but appear to enhance the stabilization of oligomeric LHC II in rye thylakoids (Huner et al., 1987). In contrast, studies using *Synechocystis* sp. PCC6803 PG-defective mutants (disruption of *cdsA* or *pgsA* genes) revealed that PG was essential for maintenance of the chlorophyll-protein complex and normal conformation and activity of PS II. Experiments with *Chlamydomonas reinhardtii* grown under sulphur or phosphorus limitation showed that decreases in PG levels were not so obvious as the reduction in SQDG content under sulphur-limited conditions. In fact, PG remained at moderate levels. These data may reflect the fact that PG plays some essential role that can not be substituted by SQDG (Hagio et al., 2000; Sato et al., 2000). On the other hand, it is known that phosphorus deficiency can induce increased SQDG synthesis (see below and section 5.2). Thus, bearing in mind that anionic lipids in other biological situations are known to replace one another (Haines, 1983), together with results from mutant experiments, it seems conceivable that SQDG is able to partly substitute PG at least in the case of phosphorus deficiency.

A proposed structural role of SQDG is that the lipid might occupy regions in the leaflet opposite to MGDG inverted micelles or the region on the convex side of the bulge or cusp in corresponding models of lipid particles. Experiments showed that addition of SQDG to mixtures of MGDG and DGDG induced the appearance of paracrystalline arrays of 80-100 Å lipid particles which increased in number with higher SQDG ratios. The lipid would have a proportionally-larger polar head group, although SQDG is similar in molecular size to DGDG. However, due to the saturated nature of fatty acids at the *sn*-2 position, SQDG have a smaller hydrophobic cross section than DGDG and, thus, a slightly an inverted cone-shape. It was proposed that the orderly arranged lipid particles containing negatively charged SQDG might be the basis for the orderly arrangement of other thylakoid molecules functioning in photosynthesis (Sakai et al., 1983). Haines (1983) proposed that the lipid anionic headgroups of the energy transducing membranes could bind and conduct protons along the membrane surface. Anionic groups (such as the head-group of SQDG) were considered to form acid-anion dimers with two anions sharing a single proton via a hydrogen bond. Such putative acid-anion dimers may be protonated and deprotonated at a high rate, allowing protons to be conducted over long distances without exchanging with the bulk phase.

Several groups have suggested a role for SQDG in protein transport. It is known that transit peptides are necessary for targeting and translocation of chloroplast proteins into the chloroplast interior (Keegstra and Cline, 1999). They are known to be rich in basic amino acids. This property may account for their observed interactions with acidic lipids. Transit peptides are thought to interact with the lipid portion of the envelope membranes and, indeed, insert most efficiently into monolayers of PG, SQDG and MGDG suggesting that these lipid classes may be important for chloroplast targetting and transport (de Kruijff et al., 1998). The anionic lipids seem to be the strongest determinants for protein insertion. For example, the transit peptides of the small subunit of Rubisco (SSU) (van't Hof et al., 1991) and ferredoxin (Fd) (van't Hof et al., 1993; Pilon et al., 1995) interact with MGDG, SQDG and PG, whereas the transit peptide (tp) of outer chloroplast envelope membrane protein toc75 significantly inserted only into the negatively charged chloroplast lipids, SQDG and PG. In contrast the transit peptides of two stromal proteins, tpSSU (van't Hof et al., 1991; Pinnaduwage and Bruce, 1996) and tpFd (van't Hof et al., 1993), tpn75a (transit peptide of toc75) did not show a significant interaction with the chloroplast galactolipid MGDG because of the difference in affinity for MGDG between the transit peptides. The latter was caused by the distribution of hydroxylated amino acids in tpn75 as opposed to tpFd and tpSSU (Inoue et al., 2001). However, when experiments were performed in order to establish whether electrostatic or hydrophobic interactions were the more important in determining interactions with lipids, the data showed that insertion of proteins into monolayers of total lipid extracts from the chloroplast outer membrane was not reduced by high salt (masking the negative charge of the anionic lipids). These results suggested that hydrophobic interactions were dominant forces in determining interactions. A similar conclusion was reached for the interaction of peptides corresponding to different regions of the transit sequence of the small subunit of ribulose bis-phosphate carboxylase with outer membrane lipids (Van't Hof et al., 1991).

During interaction with membrane lipids in small unilamellar vesicles, peptide transit sequences can undergo a two-state random coil-helix transition (Horniak et al., 1993). Anionic lipids like PG or SQDG are strong α-helix promoters. They can convert up to 50% of the polypeptide into a helical conformation. The location of the induced α-helices within the overall peptide is not known. The only precise information concerns the transit peptide of ferredoxin from *Chlamydomonas reinhardtii* where Lancelin et al. (1994) argue that the α-helix is localised at the N-terminus. The proposed model assumes the newly synthesised precursor to interact initially via its transit sequence with cytosol-exposed lipid domains in outer membrane. The C-terminus of the transit sequence is suggested to interact with anionic lipids whereas the N-terminus interacts preferentially with chloroplast specific MGDG. These interactions are followed by insertion of both domains into the lipid layer.

SQDG in roots and non-photosynthesising tissues has been suggested to be involved with ATPase function. It has been reported that activity of the bean root ATPase, extracted with acetone, could be partially restored by addition of lipids. SQDG was the most efficient, whereas PC was only effective if it was enriched in linolenic acid (up to 76%). Small or insignificant effects were obtained with PE, PG and MGDG. DGDG and phosphatidylinositol were completely ineffective. The measured Q_{10} value of the activity of the enzyme reconstituted with PC was 1.3, whereas when reconstituted with SQDG it was 2.1 (compared with a Q_{10} value of 2.4 for the original ATPase) (Kuiper, 1972), suggesting that a lipid composition change induced by a shift of temperature could affect the ATPase temperature response in vivo. In addition, SQDG was proposed to be essential for an active microsomal Na^+/K^+-stimulated ATPases. A net negative surface charge may be a prerequisite to attract the highly mobile univalent cations Na^+ and K^+ and, thus, maintain the function of the lipoprotein complex (Kylin et al., 1972).

Other experiments also imply a regulatory role for SQDG in enzyme activity. Thus, the lipid-activated phospholipase A_2 (PLA_2) enzyme responded in a dose-dependent manner with more saturated SQDGs causing less activation. The activation itself took place at very low concentrations. Higher SQDG concentrations failed to increase activation further. In fact, they decreased activation. By using intrinsic fluorescence to reflect conformational changes in proteins due to substrate or ligand binding, it was found that the fluorescence intensity of pig pancreatic PLA_2 was drastically enhanced by SQDG. Such an increase was interpreted as representing the transition from the inactive to the hypothetical active state of the enzyme–bilayer complex (Vishwanath et al., 1996). The activity of another lipid-metabolising enzyme, the UDP-galactose:diacylglycerol galactosyltransferase, has also been shown to be enhanced with negatively charged lipids (PG most, PI and SQDG) (Coves et al., 1988; Li et al., 1997).

On the other hand, SQDG may lower or inactivate some enzyme activities. Thus, chlorophyllase activity assays show that anionic lipids (PG and SQDG) can associate with isolated enzyme preparations, thereby inactivating them in a co-operative fashion. The extent of this inactivation depends on the pH and ionic strength of the reaction medium and can be reversed by divalent cations (Lambers and Terpstra, 1985). In addition, inhibitory activity of aggregated SQDG was observed toward β-glucosidase. In this case, deacylated SQDG derivatives showed no inhibition and, therefore, it was assumed that inhibition by SQDG required the hydrophobic acyl groups perhaps to produce an aggregate and/or to interact with a hydrophobic site in the enzyme. DGDG suppressed the inhibition. The competition between SQDG and DGDG in mixed micelles were suggested to result in reduced access of SQDG to the β-glucosidase (Kurihara et al., 1996, 1997).

Extracts of SQDG and SQMG from lower plants have been shown to be potent inhibitors of DNA polymerases-α and -β in vitro and effective anti-neoplastic agents. These enzymes are essential for DNA replication and repair and, hence, it is not surprising that these enzyme inhibitors lead to cell death. SQDG and SQMG with longer acyl chains were more effective than those with shorter chains, especially SQMG containing 18C acids. The latter was 4–5-fold more effective than SQMG with a 14C acyl group. SQMG with unsaturated acyl moieties showed lower inhibition than saturated SQMG. The ^1H NMR spectrum of natural SQDG is comparable to that of synthetic (2S)-SQDG rather than that of the (2R) analogue. Both isomers were effective against DNA polymerase-α and -β. DNA polymerase-α was non-competitively inhibited. Both SQMG and SQDG were also effective as antagonists of both template-primer DNA-binding and nucleotide substrate-binding of DNA polymerase (Hanashima et al., 2000, 2001).

Data from other studies showed that SQDG isolated from a marine red alga, *Gigartina tenella*, inhibited eukaryotic DNA polymerases and HIV-reverse transcriptase type 1 in a dose-dependent manner (Ohta et al., 1998). Chemically synthesized SQMG and SQDG were also tested as mammalian DNA polymerases inhibitors in vitro using human cancer cell lines. The cell lines were arrested in early S-phase suggesting that SQMG/SQDG blocked the primary step of DNA replication by inhibiting DNA polymerase, possibly the α-type. A fluorescent SQMG analog was localized in the nucleus and on the nuclear surface, but the binding site seemed not to be the DNA/chromatin, suggesting that the SQMG or SQDG might interact with molecules located close to the DNA/chromatin and on the nuclear surface (Ohta et al., 2000). It was found that the mode of inhibition of DNA polymerase and human immunodeficiency virus reverse transcriptase (HIV RT) was competitive with the DNA template primer and was non-competitive with the substrate deoxythymidine triphosphate. In both cases the whole molecular structure is required for inhibition which was abrogated by loss of the sulphonate from position 6 of the hexose ring (Loya et al., 1998; Ogawa et al., 1998).

In addition to effects on proteins themselves, charged lipids can also regulate membrane properties in general. They suppress the formation of non-bilayer structures by imposing a bilayer arrangement on MGDG-enriched preparations (Quinn, 1998). The balance between the charged SQDG or PG and the uncharged galactolipids is probably controlled by the regulation of MGDG synthesis: the activity of UDP-galactose:diacylglycerol galactosyltransferase (as stated earlier) being altered by negatively-charged lipids (PG most, PI and SQDG)(Coves et al., 1988; Li et al., 1997).

In addition, when considering functions of SQDG in plants we should keep in mind the demonstration, from radiolabelling studies, that the saturated molecular species of SQDG have a much higher turnover than the more prevalent unsaturated ones (Heinz and Harwood, 1977). Thus it may be that fast turning-over species have a dynamic (enzyme-regulating) function while the slowly metabolised unsaturated species have a structural role (Harwood, 1988).

5.2 Experiments with SQDG–deficient Mutants

With the isolation of *sqd* genes required for SQDG biosynthesis in *R. sphaeroides* (Benning and Somerville, 1992; Rossak et al., 1995) and in *Synechococcus* sp. PCC7942 (Güler et al., 1996) it was possible to create SQDG-deficient null-mutants by gene disruption and thereby study SQDG function in vivo. A null-mutant of *Rhodobacter sphaeroides* completely lacking SQDG, obtained by disruption of its *sqdB* gene showed the same growth rates as wild–type algae under optimal growth conditions (150 μmol $m^{-2}s^{-1}$). Changing light and temperature parameters (15μmol $m^{-2}s^{-1}$ and 34^{0}C; 150μmol $m^{-2}s^{-1}$ and 40 ^{0}C and 150μmol $m^{-2}s^{-1}$ and 10 ^{0}C) showed that growth rates and yields for the wild-type and mutant cells were indistinguishable under these conditions.

In addition, the ratios of various components of the photosynthetic electron transport chain, as well as the electron transfer rates during cyclic electron transport were not altered (Benning et al., 1993).

Examination of the lipids isolated from *Chlamydomonas* mutant *hf2* and its parents (137c mt+) showed that SQDG was undetectable in *hf2* (Sato et al., 1995). A study with this SQDG-deficient mutant showed that it had a decreased growth rate (down 17%) probably because of an approximately 22% decrease in photosynthetic activity. Chl *a/b* ratios of the parent and *hf2* were indistinguishable. PS II activity was 42% lower in *hf2* than in its parent, whereas the rate of PS I electron transport was relatively unaffected. Light-saturated PS II activity was lower in *hf2* which indicated that the activity of PS II RC was decreased. Replotting the PS II activity as a relative value showed coincidence of the initial slopes. This indicated that *hf2* matched its parent in the antenna Chl size and PS II activity. This was interpreted to mean that either 40% of the PS II RC was inactivated or that all RCs were affected and their activity was reduced by 40%. The rate of light energy transfer to the PS II RC from LHC II was also lowered. The authors considered that the absence of SQDG reduced the activity of the PS II RC complex with little effect upon PS I and LHC II. There was little change in the content of the PS II core complex or in the PS I complex and LHC II. Since LHC II possesses almost all the Chl b, the constant Chl a/b is in agreement with this idea. Because the spectroscopic properties and stoichiometry of the pigment protein complexes were no different to the wild type, it was proposed that SQDG might play a modulating role rather than being absolutely necessary for the structure of PS II. Lipid analyses showed that SQDG was concentrated in the PS II core complex (13.7 mol% against <3.1) and two types of LHC II (17.8 mol% in LHC II-A and 23.6 mol% in LHC II-B), but not in the PS I complex. Thus, SQDG could contribute to the activity of the PS II RC by association with the PS II core complex while SQDG bound to LHC II may modulate the PS II RC function because the PS II core complex and LHC II are organised into the PS II complex in vivo (Sato et al., 1995). However, because the *Chlamydomonas* mutant was not genetically clean, some of the interpretations may have to be revised in the future.

However, in agreement with the results for *R. sphaeroides* discussed above, a SQDG-deficient *Synechococcus* sp. PCC7942 null mutant showed the same growth rates under optimal phototrophic growth conditions as the wild type, suggesting that SQDG was not essential for oxygenic photosynthesis. Normal growth of the mutant under optimal conditions, of course, does not exclude the possibility of a subtler role for SQDG in oxygenic photosynthesis under different conditions in vivo, *e.g.* high photon flux densities. Nevertheless, the rates of oxygen evolution as a function of proton flux density curves were nearly identical for the wild type and mutant. The similarity of emission spectra following chlorophyll *a* excitation at 440 nm indicated that neither the chlorophyll *a* binding to the core antenna proteins CP43 and CP47, nor the excitation energy transfer to the RC was affected. The loss of the anionic sulpholipid in the mutant was compensated by an increase in PG. Maintaining a certain level of total anionic lipids in the membranes seemed to be crucial for the organism, because reduction in PG under phosphate limitation in the wild type was compensated by an increased level of SQDG. The SQDG-deficient mutant however, would be unable to respond in the same way to phosphate limitation. It was concluded that SQDG did not play a specific role in oxygenic photosynthesis but it was suggested that SQDG could act as a substitute for anionic phospholipids under phosphate-limiting growth conditions (Guler et al., 1996).

It is interesting, that in SQDG-deficient mutants, the amount of negatively charged lipids was kept constant by an increase of PG (Sato et al., 1995; Benning, 1998; Selstam, 1998). Conversely, SQDG was important under conditions of phosphate starvation, when a SQDG-deficient mutant of *R. sphaeroides* ceased growth much earlier than the wild type. In the wild type, PG was replaced as a boundary lipid by the anionic glycolipid SQDG. Such an adaptation was not possible for the SQDG-deficient mutants (Benning, 1998). Similar results were obtained with the SQDG-deficient mutant SY-SQDB. In both cases it appeared that maintaining the net level of highly anionic lipids is crucial for maintaining maximum growth rates. The lack of SQDG became important under phosphate limitation when PG availability for thylakoid biogenesis or function was restricted. Therefore, it has been assumed that SQDG did not have a specific role in oxygenic photosynthesis (Vijayan et al., 1998).

In order to understand the molecular basis for the increase in the relative amounts of sulpholipid in the *pho1* (phosphate-transport) mutant and phosphate-deprived wild type *Arabidopsis thaliana*, the amounts of *SQD1* mRNA, and SQD1 protein were determined by Essigmann et al. (1998). The amount of *SQD1* mRNA in the *pho1* mutant and in the wild type grown under phosphate limitation was drastically increased. The increase in SQD1 protein appeared to be proportional to the increase in *SQD1* mRNA. Thus, a decrease in the internal phosphate concentration caused by the specific mutation in the *pho1* strain or by limiting the external phosphate supply induced an increase in the expression of *SQD1*, a gene encoding a protein essential for SQDG biosynthesis (sections 3 and 4.1). This, in turn, correlated with the increase in the content of SQDG. Thus, several photosynthetic organisms have been shown to exhibit inverse changes in the levels of PG and SQDG under phosphorus-limiting conditions or for mutations impaired in SQDG synthesis. Therefore, it has been assumed that photosynthetic organisms require a certain amount of total anionic lipids and that SQDG and PG can substitute for each other to some extent, if necessary. However, a mutant defective in the CDP-diacylglycerol synthase gene constructed from a cyanobacterium *Synechocystis* sp. PCC6803 and designated SNC1, required PG supplementation for growth. These results indicate that exogenous PG was essential, for the SNC1 mutant which was defective in PG synthesis. Moreover, the SQDG level in SNC1 remained almost constant, despite incorporation of various amounts of exogenously supplemented PG or after a decrease in the PG content after the shift to PG-free medium (Hagio et al., 2000). Experiments with the mutant *Synechocystis* (defective in *cdsA* or *pgsA* genes) were interpreted to show that PG was essential for cell growth and was needed for maintenance of the chlorophyll-protein complex levels and normal conformation and activity of the PS II complex (Sato et al., 2000). Thus, in the above case, it was not adequate to simply substitute PG with SQDG although this appeared to be the consequence of growing wild type *Chlamydomonas reinhardtii* under phosphorus deficiency (Sato et al., 2000).

A presumed function for the observed changes in lipid composition following growth on phosphorus-deficient media would be conservation of phosphorus, which is needed for many essential metabolites, by reducing the amount of phosphorus in membrane lipids (Essigmann et al., 1998). This reflects the fact that about one-third of total organic phosphorus in leaves of *A. thaliana* is contained within each of the phospholipid, nucleic acid and other phosphorus ester fractions (Poirier et al., 1991). If this hypothesis is correct a strong reduction in SQDG content caused by antisense expression of the *SQD1* cDNA would inhibit growth under phosphate-limiting conditions because the reduction in PG could not be compensated by an equivalent rise in SQDG. Indeed, SQD2 knockout mutants show similar characteristics to the bacterial knockout lives with regard to SQDG function (Yu et al., 2002).

More recently, SQDG was discovered in the non-photosynthetic, root nodule-forming bacterium, *Sinorhizobium meliloti*. This discovery raised the possibility of a function for SQDG in root nodule symbiosis. However, a SQDG-deficient mutant of *S. meliloti* disrupted in *sqdB* was capable of inducing functional nodules and did not show any obvious impairment of function under different conditions. Thus far, no specific role can be assigned to bacterial SQDG in either nodule-associated or free-living cells (Weissenmayer et al., 2000). The *pho1* mutant of *Arabidopsis* has been shown to respond to phosphate deficiency in its leaves by decreasing the amount of PG. While in the wild-type, the lipid pattern was almost insensitive to changes in light intensity, PG was reduced especially under high light in the mutant. The corresponding SQDG increase correlated with increased amounts of the SQD1 protein, which indicated an actively-mediated process. Inspection of the fatty acid compositions of PG and SQDG revealed a slight increase in the proportion of oleate in PG and a decrease for oleate in SQDG in the *pho1* mutant plants. Irrespective of the light conditions, the sum of SQDG and PG remained nearly unchanged in the *pho1* mutant as compared to the wild-type, suggesting that the total amount of negatively-charged lipids was important. Despite alterations in their ultrastructure, mutant thylakoids showed virtually no changes in photosynthetic electron transfer, O_2 evolution and excitation energy allocation to the RCs. It was assumed that PG deficiency can, at least partially, be compensated for by another anionic lipid (SQDG). This would allow a favourable thylakoid lipid composition for photosynthesis under restricted phosphate availability (Härtel et al., 1998).

6. ADAPTATION

Before discussing adaptive responses in plants, we must raise the possibility that many reactions to unfavourable environmental conditions may have a common basis. Thus, the expression of many genes is influenced by oxidative cellular processes. Furthermore, a common observation in stress metabolism is the marked formation of active oxygen species such as superoxide, hydrogen peroxide and hydroxide radicals. A major target of oxidative damage by such species are lipids, especially unsaturated species. Moreover, studies on lipid responses during stress have shown quantitative and qualitative changes in SQDG. In the following sections, we will consider a number of different stress responses and try to evaluate whether SQDG can play a role in adaptation.

6.1 Chilling Resistance

Many studies of lipid changes caused by temperature have been made with cyanobacteria because these prokaryotic organisms are homogeneous in culture and respond very quickly to temperature shifts. Growth following a temperature lowering was accompanied by an increase in MGDG and SQDG and a decrease in DGDG in *Anacystis nidulans*. SQDG was enriched in palmitate at the *sn*-1 and *sn*-2 positions and with palmitoleate at the *sn*-1 position. Almost all of the palmitate at the *sn*-1 position of SQDG was converted to palmitoleate at lower growth temperatures. On the other hand, for *Anabaena variabilis* growing at lower temperatures, the level of SQDG was stable with only a small conversion of palmitate to palmitoleate. However, the amounts of oleate and linoleate at the *sn*-1 position of all chloroplast lipids was decreased and replaced with α-linolenate (Sato et al., 1979; Wada and Murata, 1998). Low temperatures also cause changes in higher plant lipid compositions. For example, results of Kuiper (1972) showed that bean roots grown at 15°C contained a lower amount of SQDG (12% against 18%) than roots grown at 23°C. In the cyanobacterium *Spirulina platensis*, low temperatures caused a decrease in MGDG which was compensated by an increase in SQDG while DGDG and PG remained constant. The authors examined molecular species and found increases in the more unsaturated species (e.g. 18:3/16:1) at low temperatures while the more saturated species (e.g. 16:0/16:0) decreased. Desaturation of palmitate at the *sn*-2 position was stimulated by low temperatures, especially in SQDG (Quoc and Dubacq, 1997).

Data from our earlier investigations showed that, during autumn hardening, SQDG accumulation took place in one-year-old apple shoot bark and wood and was especially striking for a hardy variety (*Malus boccata* Borh) (Fig. 3) (Okanenko and Protsenko, 1977). In another study, Oquist (1982) found a 2-fold increase in the SQDG content in pine thylakoid preparations during the autumn. Its content remained high during the winter and then lowered in spring. At the same time, SQDG from pine in the winter was enriched in linoleate whereas palmitate was dominant in the summer. However, these changes may reflect a response to low growth temperatures rather than true adaptation.

Murata and Hoshi (1984) concluded that SQDG had no role in chilling resistance. In the plants they examined that only *Z. mays* and *O. sativa* (*Poaceae*, 18:3 plants)

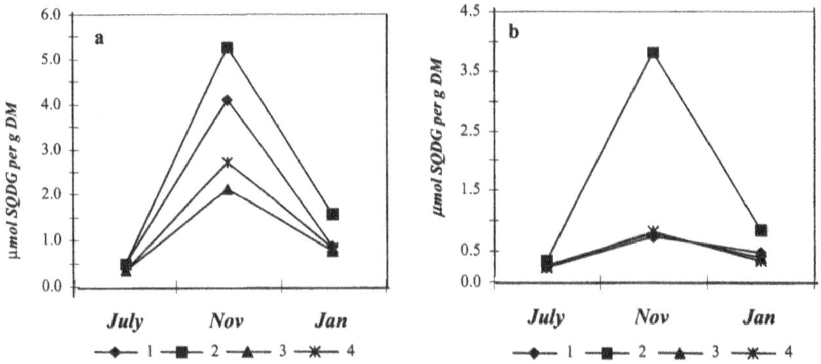

Figure 3. Seasonally–induced changes in SQDG content in one–year–old apple tree shoot bark (a), and wood (b). 1 - Antonivka (resistant variety); 2 - most resistant Sibirian apple tree (Malus boccata Borch); 3 - Renet Semirenko (moderate resistance); 4 - Kalvil snigovyj (moderate resistance)

contained significant amounts of saturated SQDG species which, therefore, might phase-separate and cause membrane damage at lower growth temperatures. They also suggested that the finding that saturated fatty acids (mainly 16:0) were more abundant at the *sn*-1 position of SQDG in chilling-sensitive plants but at the *sn*-2 position in chilling-resistant plants, could be a reflection of whether these plants were 16:3 and 18:3 species, i.e. the SQDG species would have an evolutionary origin. In another study Murata and Yamaya (1984) found that only in sponge cucumber did phase separation begin at 15°C for SQDG. Nevertheless, SQDG from chilling sensitive species contained up to three times more stearic acid (18:0) than resistant plants. Furthermore, Kenrick and Bishop (1986) concluded that, if the primary event in chilling sensitivity of higher plants is a phase transition in bulk chloroplast membrane lipids, then not only PG, but also SQDG might be involved. In particular, the main difference between *Carica papaya* (tropical origin) and *Carica pubescens* (adapted to temperate climates) was in their SQDG species (57% and 36% saturated, respectively), and not in PG. Their results showed that linolenic acid was abundant in the SQDG from chilling-tolerant species whereas palmitate was a major component in temperature-sensitive plants.

6.2 Drought Resistance

Drought stress often has two components - heat and water deficit. Concerning the first, results from several labs indicate that superoptimal temperatures cause an increase of SQDG content in most plants studied. For example, *Atriplex lentiformis* plants grown in coastal and desert (Death Valley) regions accumulated SQDG by 260% and 64%, respectively at high growth temperatures. Changes in the fatty acid (FA) patterns for the SQDG of Death Valley plants were seen without any significant changes in FA composition of SQDG in coastal plants (Pearcy, 1978). When *Nerium oleander* was grown at 45°C rather than 20°C, the percentage of palmitate in SQDG increased from about 45% to 55% with a concomitant decrease in the percentage of linoleate from about 31 to 19%. The main changes in molecular species were an increase in the proportion of dipalmitoyl-SQDG from 12 up to 20% and a decrease in that of linolenoyl-palmitoyl SQDG from 40 to 30% at the higher growth temperature. At the same time the T_c increased from 19±3°C to 26±3 °C. Taking into account the fact that dipalmitoyl- SQDG undergoes a transition at 42°C, it was suggested that this molecular species could be a major lipid component involved in phase transitions in the thylakoid polar lipids (Orr and Raison, 1987). Our data using wheat confirmed that growth at high temperature induced sulpholipid accumulation but only in the chloroplasts of drought resistant (DR) species. In contrast, in the only sensitive species (DS) (Myronivska 808) there was a decrease in SQDG content. For resistant wheat species, maximum levels of SQDG were found after growth at 45°C, above which the content in chloroplasts declined. In contrast (Fig. 4a) the drought-sensitive variety, Myronivska 808, exhibited a decline in SQDG at all higher temperatures. These results were confirmed by studies of leaf lipid composition (Okanenko and Taran, 1989).

Water deficit caused an increase in SQDG in drought-resistant wheat species but a reduction in drought-sensitive varieties (Quartacci et al., 1995). For tolerant plants, palmitate was enhanced while palmitate, oleate and linolenate decreased. The opposite was found in sensitive plants. Thus, the SQDG content increased in tolerant plants with a parallel rise in saturation while changes in sensitive plants were exactly the opposite. Another study showed that the exposure of rye to drought (Pancratova and Karimova, 1984) led to an increase in phospholipid, MGDG and SQDG contents in leaves of plants when compared to watered controls. When water deficit was combined with higher growth temperatures, the results for different species of wheat (Fig. 4b) were broadly similar to those obtained by heat alone (Fig. 4a). Again most drought-resistant varieties showed increases in SQDG content up to 45°C or 50°C, whereas the drought-sensitive Myronivska showed a decline in its SQDG at all higher temperatures. The SQDG fatty acid composition was characterised by a decrease in palmitoleate and an increase in stearate for both resistant and sensitive varieties whereas linoleate and linolenate decreased in sensitive but increased in resistant varieties at higher temperatures (Okanenko and Taran, 1989). It was interesting to compare these

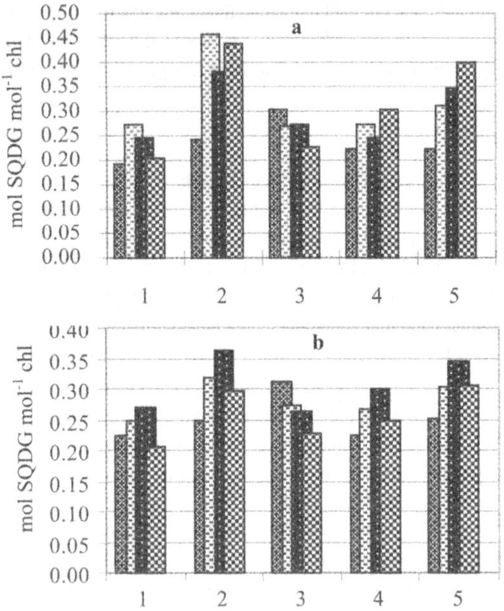

Figure 4. The SQDG content of chloroplasts of heat–stressed wheat plants. (a) 1st column, control; 2nd, 40°C; 3rd, 45 °C; 4th, 50 °C. The SQDG content of chloroplasts of heat and drought–stressed wheat plants; (b) 1st column, control; 2nd, heating 45°C; 3rd, WD; 4th, heating 45°C+WD: 1 - Odeska 1; 2 - Bezosta 1; 3 - Myronivska 808; 4 - Parus; 5 – Koral. WD stands for water deficit

laboratory results with ones obtained in field experiments using artificial irrigation during drought. An increase in SQDG was observed at two stages of development (when the first 'branches' are formed from stem buds and when the seed contents contain a white watery liquid) for wheat exposed to drought (by 73.7 and 51.1% respectively) (Taran *et al.*, 2000).

The importance of SQDG during drought exposure was also indicated from experiments on rehydration of air-dried cells of the desiccation-tolerant filamentous cyanobacterium *Nostoc commune*. The radiolabeled pool sizes of PG and SQDG reached steady-state within several minutes, whereas the two abundant membrane glycolipids, MGDG and DGDG, achieved uniform labeling only within 2 hours (Taranto et al., 1993). This rapid response may be connected with the ability of this cyanobacterium to tolerate desication.

6.3 Oxidative Stress

Bearing in mind that oxidative stress is a component part of the reaction of plants to many other stresses, any changes in lipid composition caused under such circumstances are of special significance. Unfortunately, there is little data available.

During the first phase (8h) of oxidative injury when plants are exposed to high concentration of ozone, there is a loss of pigments and lipids (mainly MGDG with some DGDG). The loss in chloroplast lipids is accompanied by a small increase of MDA (malondialdehyde) content and in TAG and DAG. The second phase of oxidative injury is characterised by a massive destruction of pigments and begins with a drastic fall in MGDG and a smaller decrease in DGDG and PC which is accompanied by significant increases in TAG, DAG and malondialdehyde (Sakaki, 1998). However, it is interesting that the anionic lipids, SQDG and PI, were unaffected during the period of ozone exposure (in spinach leaves, at least). Lipid changes similar to those in spinach were also observed in several plant species, and in broad bean leaves, a relative increase in SQDG took place. Because both galactolipids were significantly destroyed during ozone exposure, the SQDG content expressed as mol% of the total glycolipids increased up to 45mol% (depending upon species) (Sakaki et al., 1985, 1994).

The above experiments were performed at high ozone concentrations which are only approached under exceptionally bad environmental conditions. When moderate ozone levels were used in experiments with peas, a large decrease in all the chloroplast glycosylglycerides, including SQDG, was observed. It was also noticeable that the fatty acid pattern of SQDG was changed with a reciprocal decrease in α–linolenate and an increase in linoleate (Carlsson et al., 1994).

6.4 Other Types of Stress

There are a number of other types of stress that have been examined with regard to plants. These include both abiotic and biotic stress. Examples include elevated atmospheric CO_2, high salinity, heavy metals and microbial attack.

Some experiments on elevated CO_2 suggest that SQDG may be affected. For example, increased CO_2 partial pressure (700 ppm) caused a slight increase in SQDG content in *Nicotiana tabacum* (from 4.9 up to 6.5 % of total lipids). A simultaneous decrease in DGDG was also found, whereas the main thylakoid membrane lipid, MGDG, remained unaffected and the total amount of glycolipids was constant under these conditions. The saturation index of the total lipid extract increased due to a rise in palmitate and a decrease in the trienoic fatty acids (hexadecatrienoate and linolenate). The average chain length of fatty acids remained unchanged (Radunz et al., 2000). Another study with the alga *Spirulina platensis* showed that its total lipid content was doubled on a dry weight basis following long-term treatment but only increased 15-20% after short-term treatment with 10% CO_2. A small increase in the relative amount of MGDG and a decrease of PG was found, whereas DGDG and SQDG remained stable. Elevated CO_2 exposure induced a decrease of fatty acid unsaturation and increase in the 16C/18C ratio for total lipids. A noticeable decrease in γ-linolenate (which was significant in the MGDG and SQDG fractions) was found after short-term exposure to elevated CO_2. The highest concentration of palmitate was found in SQDG. CO_2 stress activated the synthesis of palmitic acid and suppressed desaturation and elongation of fatty acid chains (Mouradian et al., 1998). These changes in fatty acid metabolism caused by elevated CO_2 should be considered in relation to other alterations which have been studied in detail in wheat (e.g. Williams et al., 1998).

Several laboratories have reported that salinity can affect lipid composition but their results are sometimes contradictory. Early results showed that the SQDG content was stable (relative to chlorophyll) during adaptation of barley seedlings to high concentrations of sodium chloride while the MGDG content of thylakoid membranes decreased considerably. The latter decrease meant that the relative percentage of SQDG compared to total thylakoid lipids increased by 30% (Muller and Santarius, 1978). In wheat roots, increased salt levels (calcium sulphate) resulted in higher (two-fold increase) levels of SQDG, while the reverse occurred in lipids from oat roots (Kuiper et al., 1974). It may be interesting to note in this connection (and in view of the possible involvement of SQDG in ATPase activity: Section 5.1) that Ca^{2+} activation of the root ATPase is important in wheat (Kylin and Kahr, 1973) whereas in oat roots the ATPase activity is mainly increased by Mg^{2+}. However, treatment of sugar beet with sodium sulphate did not confirm the suggestion that SQDG was involved in salt-adaptation (Stuiver et al., 1984). On the other hand, SQDG ranged from 5 in spinach to 20 molecules in the halotolerant alga *Dunaliella salina* per CF_0-CF_1 ATPase complex and this lipid could not be readily exchanged (Pick et al., 1985) suggesting that it was intimately associated with the complex. Preliminary experiments (A. Okanenko et al., unpublished data) revealed that SQDG accumulation in *Salicornia europaea* L. was possibly connected with its ability to accumulate high concentrations of salts in its tissues. In contrast, SQDG levels were stable in *Artemisia arenaria* DC during changes in osmotic pressure caused by photoassimilate accumulation and decreased in *Atriplex pedunculata* during the elimination of salts from this organism. Thus, any involvement of SQDG in salt adaptation seems to be dependent upon the various mechanisms of salt tolerance used by different organisms.

The quantity of heavy metals in soil is often increased greatly in industrial areas and when they are taken up by plants, they can induce both toxic and adaptive responses. There have been few studies but the limited information available confirms that lipid composition can be affected (Harwood, 1998). Cd exposure induced a decrease in MGDG and SQDG contents in *Brassica napus* leaves which was accompanied by a sharp increase in extra-chloroplastic lipids (PC and PI) (Youssef et al., 1998).

Preliminary experiments (A. Okanenko et al, unpublished data) showed that lead, supplied at various concentrations to hydroponic cultures, caused a decrease in SQDG in wheat seedling leaves and roots. Thus, most heavy metals tested cause a decrease of SQDG content. This may be due to inhibitory effects on the biosynthetic pathway for SQDG formation or, indirectly, because heavy metals are toxic for photosynthesis and, therefore, interfere with thylakoid lipid metabolism. However, for organisms which are growing in habitually-polluted regions, the picture may be different. Thus, while acute exposure of the moss, *Rhytidiadelphus squarrosus*, to low levels (1-10 μM) of lead nitrate did not change the radiolabelling of SQDG significantly, populations gathered from lead-polluted soils showed more labelling of chloroplast lipids than moss from unpolluted areas (Guschina and Harwood, 2002). Because of the known inhibition by heavy metal pollution of chloroplast metabolism and function, including in *R. squarrosus* (Brown and Wells, 1990), the increased metabolism of chloroplast lipids in moss from lead-polluted regions may represent an adaptive response. Thus, membrane lipids, damaged as a result of heavy metal pollution, could be replaced at higher rates and any detrimental effect on photosynthesis minimised.

A few experiments suggest that virus infection can affect membrane lipids. In studying barley stripe mosaic virus strains and poa semilatent hordeivirus, a close relationship was found between the severity of symptoms and the extent of lipid changes. Losses of MGDG, PG, SQDG and chlorophyll as well as relative increases in phosphatidylcholine (PC), phosphatidylethanolamine and phosphatidylinositol contents were observed (Adam and Nagy, 1989). In our experiments, a relative accumulation of SQDG in the total glycolipids in wheat plants infected by *Puccinia graminis* (Taran et al., 2000) and kidney bean plants infected by potato *X* potexvirus (PXV) or tobacco mosaic (TMV) virus (Fig. 5) was observed. The latter phenomenon was confirmed by a significant increase of the SQDG/DGDG ratio (used because DGDG was the most stable component in the experiments) (Senchugova et al., 1999). These changes could be useful for the plant if the action of SQDG in inhibiting viral development (and against DNA-polymerase and reverse transcriptase activity (Section 5.1); Ohta et al., 1998, 1999), also apply to in vivo situations.

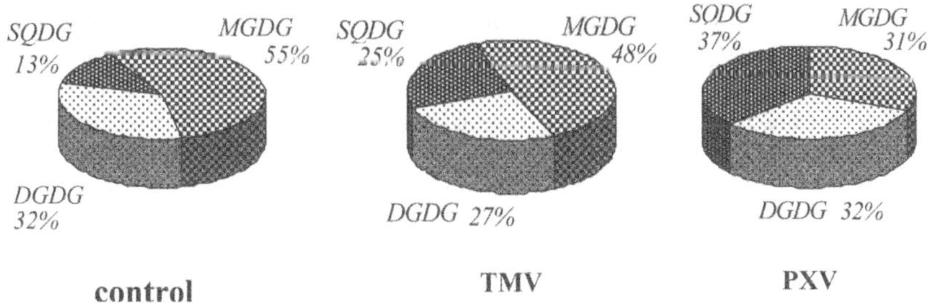

Figure 5. Glycolipid composition (as mol%) of kidney bean plants infected with tobacco mosaic virus (TMV) or potato x *virus (PXV).*

7. CONCLUSIONS

It will be noted from the previous sections of this chapter that the quality and/or quantity of SQDG changes in response to a number of stresses. Furthermore, this lipid has been proposed to be involved in the function of important plant processes such as stabilising D_1/D_2 dimers and LHC II or in the activity of ATPase (Pick et al., 1985). In addition, the constant ratio of thylakoid lipids (including the presence of SQDG) in all oxygen-evolving photosynthetic organisms points to a strong evolutionary advantage for the lipid. In particular, these latter observations suggest that SQDG is an important membrane component in photosynthetic organisms.

On the one hand, SQDG might have a specific function, such as its localisation on the surface of the native heterodimer D_1/D_2 (Vijayan et al., 1998) where it might stabilise the dimer structure (De Kruijtt et al., 1998). By contrast, it may play more general roles such as in facilitating electron transport (Siegenthaler and Tremolieres, 1998) or in ensuring a bilayer structure for a MGDG-enriched membrane (Coves et al., 1988; Li et al., 1997). The latter functions may be connected also with the anionic nature of SQDG and explain why PG and SQDG can replace each other under extreme phosphate (or sulphate) limiting conditions (e.g. Guler et al., 1996). The two types of functions – specific for enzyme activity and general for membrane properties – would explain why different molecular species have very different turnovers (e.g. Heinz and Harwood, 1977). It would also explain why saturated molecular species, such as are associated with the ATPase complex (Pick et al., 1985) have high turnovers whereas the bulk (unsaturated) molecular species are relatively stable (see Harwood, 1980b).

At first sight, the ability of mutants to grow when their SQDG is significantly depleted or even absent (section 5) would imply that SQDG does not play an essential role in nature. If that were the case, then why do cyanobacteria, various algae and both lower and higher plants have approximately the same thylakoid lipid composition? We would argue that, although it may be possible to grow organisms under 'ideal' conditions in the laboratory, in vivo such organisms are not only in competition with each other but also have to face stress conditions. Under these circumstances, even a 10% increase in growth can have a big advantage. This can explain why SQDG-mutants often show significant impairment of growth when stressed.

One recent interesting development has been the evaluation of SQDG and SQMG for medical purposes. Inhibition of HIV-1 infection (Gustafson et al., 1989) and anti-tumour activity (Ohta et al., 2001; Quasney et al., 2001) are notable in this context. It may be that new types of drugs, which are based on the plant sulpholipid will be useful in the future. In the meantime, we are left with many observations about SQDG which are intriguing, although some of these functional aspects are being elucidated gradually by the use of null mutants of pacteria plants. While the biosynthetic pathway, which was such an enigma for many years (Harwood, 1980b; Mudd and Kleppinger-Sparace, 1987) has now been solved (Pugh et al., 1995; Sanda et al., 2001), there are other aspects of metabolism that remains to be delineated. The future looks very interesting for research on this unique membrane lipid!

8. REFERENCES

Adam A and Nagy PD (1989) Variations of membrane polar lipids of barley leaves infected with three strains of barley stripe mosaic virus and with poa semilatent virus. *Plant Sci* **61:** 53—59

Anderson JM (1975) The molecular organization of chloroplast thylakoids. *Biochim Biophys Acta* **416:** 191—235

Barber GA (1963) The formation of uridine diphosphate L-rhamnose by enzymes of the tobacco leaf. *Arch Biochem Biophys* **103:** 276—282

Barber J and Gounaris K (1986) What role does sulpholipid play within the thylakoid membrane? *Photosynth Res* **9:** 239—249

Benning C (1998) Membrane lipids in anoxygenic photosynthetic bacteria. In: Siegenthaler P-A and Murata N (eds) Lipids in Photosynthesis: Structure, Function and Genetics. Advances in Photosynthesis, Vol 6, pp 83–101. Kluwer Academic Publisher, Dordrecht

Benning C and Somerville CR (1992) Identification of an operon involved in sulfolipid biosynthesis in *Rhodobacter sphaeroides. J Bacteriol* **174:** 6479—6487

Benning C, Beatty JT, Prince RC and Somerville CR (1993) The sulfolipid sulfoquinovosyldiacylglycerol is not required for photosynthetic electron transport in *Rhodobacter sphaeroides* but enhances growth under phosphate limitation. *Proc Natl Acad Sci USA* **90:** 1561—1565

Bennun A and Racker E (1969) Partial resolution of the enzymes catalyzing photophosphorylation. IV. Interaction of coupling factor I from chloroplasts with components of the chloroplast membrane. *J Biol Chem* **244:** 1325—1331

Benson AA (1963) The plant sulpholipid. *Adv Lip Res* **1:** 387—394

Benson AA (1964) Plant membrane lipids. *Annu Rev Plant Physiol* **15:** 1—16

Benson AA, Daniel H and Wiser R (1959) A sulpholipid in plants. *Proc Natl Acad Sci USA* **45:** 1582—1587

Brown DH and Wells JM (1990) Physiological effects of heavy metals on the moss *Rhytidiadelphus squarrosus. Ann Bot* **66:** 64—647

Browse J and Somerville CR (1991) Glycerolipid metabolism: biochemistry and regulation. *Annu Rev Plant Physiol Plant Mol Biol* **42:** 467—506

Burns DD, Galliard T and Harwood JL (1979) Purification of acyl hydrolase enzymes from the leaves of *Phaseolus multiflorus. Phytochemistry* **18:** 1793—1797

Bychek-Guschina IA (1997) Study of biochemical aspects of lichen symbiosis 1 Lipids and fatty acids in cultured lichen symbionts. *Biohimiya* **62:** 574—580 (in Russian)

Carlsson AS, Hellgren LI, Sellden G and Sandelius AS (1994) Effects of moderately enhanced levels of ozone on the acyl lipid composition of leaves of garden pea. *Physiol Plant* **91:** 754—762

Cedergreen RA and Hollingsworth RI (1994) Occurence of sulphoquinovosyl diacylglycerol in some members of the family *Rhizobiaceae. J Lipid Res* **35:** 1452—1461

Coves J, Joyard J and Douce R (1988) Lipid requirement and kinetic studies of solubilized UDP-galactose:diacylglycerol galactosyltransferase activity from spinach chloroplast envelope membranes. *Proc Natl Acad Sci USA* **85:** 4966—4970

Davies WH, Mercer EI and Goodwin TW (1966) Some observations on the biosynthesis of the plant sulpholipid by *Euglena gracilis. Biochem J* **98:** 369—373

De Kruijff B, Pilon R, van't Hof R and Demel R (1998) Lipid-protein interactions in chloroplast protein import. In: Siegenthaler P-A and Murata N (eds) Lipids in Photosynthesis: Structure, Function and Genetics. Advances in Photosynthesis, Vol 6, pp191—208. Kluwer Academic Publishers, Dordrecht

Douce R, Block MA, Dorne A-J and Joyard J (1984) The plastid envelope membranes: their structure, composition and role in chloroplast biogenesis. *Subcell Biochem* **10:** 1—84

Essigmann B, Guler S, Narang RA, Linke D and Benning C (1998) Phosphate availability affects the thylakoid lipid composition and the expression of *SQD1*, a gene required for sulpholipid biosynthesis in *Arabidopsis thaliana. Proc Natl Acad Sci USA* **95:** 1950—1955

Gabriel O (1973) Biological mechanisms involved in the formation of deoxy sugars: enzymatic hydrogen mediation. *Adv Chem Ser* **117:** 387—410

Gounaris K and Barber J (1985) Isolation and characterization of a photosystem II reaction centre lipoprotein complex. *FEBS Lett* **188:** 68—72

Guler S, Essigmann B and Benning C (2000) A cyanobacterial gene, *sqdx*, required for biosynthesis of the sulfolipid sulfoquinovosyldiacylglycerol. *J Bacteriol* **182:** 543—545

Guler S, Seeliger A, Härtel H, Renger G and Benning C (1996) A null mutant of *Synechococcus* sp PCC7942 deficient in the sulfolipid sulfoquinovosyl diacylglycerol. *J Biol Chem* **271:** 7501—7507

Guschina IA and Harwood JL (2002) Lipid metabolism in the moss *Rhytidiadelphus squarrosus* (Hedw) Warnst. from lead–contaminated and non–contaminated populations. *J Exp Bot* **53:** 455—463

Gustafson KR, Cardelina II JH, Fuller RW, Weislow OS, Kiser RF, Snader KM, Patterson GML and Boyd MR (1989) AIDS-Antiviral SL from cyanobacteria (blue-green algae). *J Nat CancerInst* **81:** 1254—1258

Haase R, Unthan M, Couturier P, Radunz A and Schmid GH (1993) Determination of glycolipid, sulpholipid and phospholipids in the thylakoid membrane. *Z Naturforsch* **48c:** 621—631

Hagio M, Gombos Z, Varkonyi Z, Masamoto K, Sato N, Tsuzuki M and Wada H (2000) Direct evidence for requirement of phosphatidylglycerol in photosystem II of photosynthesis. *Plant Physiol* **124:** 795—804

Haines TH (1971) The chemistry of the sulpholipids. *Prog Chem Fats Other Lipids* **11:** 297—345

Haines TH (1973) Sulpholipids and halosulpholipids In: Erwin JA (ed) Lipids and Biomembranes of Eukaryotic Organisms, pp197—232. Academic Press, New York

Haines TH (1983) Anionic lipid headgroups as a proton-conducting pathway along the surface of membranes: A hypothesis. *Proc Natl Acad Sci USA* **80:** 160—164

Hanashima S, Mizushina Y, Ohta K, Yamazaki T, Sugawara F and Sakaguchi K (2000) Structure-activity relationship of a novel group of mammalian DNA polymerase inhibitors, synthetic sulfoquinovosylacylglycerols. *Jpn J Cancer Res* **91:** 1073—1083

Hanashima S, Mizushina Y, Yamazaki T, Ohta K, Takahashi S, Sakaguchi K and Sugawara F (2001) Synthesis of sulfoquinovosylacylglycerols, inhibitors of eukaryotic DNA polymerase α and β. *Bioorg Med Chem* **9:** 367—376

Hartel H, Essigmann B, Lokstein H, Hoffmann-Benning S, Peters-Kottig M and Benning C (1998) The phospholipid-deficient *pho1* mutant of *Arabidopsis thaliana* is affected in the organization, but not in the light acclimation, of the thylakoid membrane. *Biochim Biophys Acta* **1415:** 205—218

Harwood JL (1975) Synthesis of sulphoquinovosyl diacylglycerol by higher plants. *Biochim Biophys Acta* **398:** 224—230

Harwood JL (1980a) Plant acyl lipids: structure, distribution and analysis. In: Stumpf PK and Conn EE (eds) The Biochemistry of Plants, Vol 4, pp 1—55. Academic Press, New York

Harwood JL (1980b) Sulpholipids. In: Stumpf PK and Conn EE (eds) The Biochemistry of Plants, Vol 4, pp 301—320. Academic Press, New York

Harwood JL (1988) Glycerolipid synthesis. In: Harwood JL and Walton RJ (eds). Plant Membranes: Structure, Assembly and Function, pp 113—127. The Biochemical Society, London

Harwood JL (1998) Membrane lipids in algae. In: Siegenthaler P-A and Murata N (eds) Lipids in Photosynthesis: Structure, Function and Genetics. Advances in Photosynthesis, Vol 6, pp 53—64. Kluwer Academic Publushers, Dordrecht

Harwood JL and Jones AL (1989) Lipid metabolism in algae. *Adv Bot Res* **16:** 1—53

Harwood JL, Morris CA, Kille P, Price LJ and Cole DJ (1998). Life and stress: a plant's standpoint. In: Sanchez J, Cerda–Olmedo E and Martinez-Force E (eds) Advances in Plant Lipids Research, pp 483—489. Universidad de Sevilla, Spain

Harwood JL and Nicholls RG (1979) The plant sulpholipid: a major component of the sulphur cycle. *Biochem Soc Trans* **7:** 440—447

Heinz E (1993) Recent investigation on the biosynthesis of the plant sulpholipid. De Kok LJ, Stulen I, Rennenberg H, Brunold C and Rauser WE (eds) Sulfur Nutrition and Sulfur Assimilation in Higher Plants, pp 165—173. SPB Academic Publishing, The Hague

Heinz E and Harwood JL (1977) Incorporation of carbon dioxide, acetate and sulphate into the glycerolipids of *Vicia faba* leaves. *Hoppe-Seyler's Z Physiol Chem* **358:** 897—908

Hofmann M and Eichenberger W (1997) Lipid and fatty acid composition of the marine brown alga *Dictyopteris membranacea. Plant Cell Physiol* **38:** 1046—1052

Hoppe W and Schwenn JD (1981) *In vitro* biosynthesis of the plant sulpholipid: on the origin of the sulphonate group. *Z Naturforsch* **36c:** 820—826

Horniak L, Pilon M, Van't Hof R and De Kruijff B (1993) The secondary structure of the ferredoxin transit sequence is modulated by its interaction with negatively charged lipids. *FEBS Lett* **334:** 241—246

Huner NPA, Krol M, Williams JP, Maissan E, Low PS, Roberts D and Thomson JE (1987) Low temperature development induces a specific decrease in trans-Δ3-hexadecenoic acid content which influences LHC II organization. *Plant Physiol* **84:** 12—18

Inoue K, Demel R, de Kruijff B and Keegstra K (2001) The N-terminal portion of the preToc75 transit peptide interacts with membrane lipids and inhibits binding and import of precursor proteins into isolated chloroplasts. *Eur J Biochem* **268:** 4036—4043

Janero DR and Barrnett R (1981) Cellular and thylakoid-membrane glycolipids of *Chlamydomonas reinhardtii* 137+. *J Lipid Res* **22:** 1119—1125

Joyard J, Mareshal E, Miege C, Block MA, Dome AJ and Douce R (1998) Structure, distribution and biosynthesis of glycerolipids from higher plant chloroplasts. In: Siegenthaler PA and Murata N (eds) Lipids in Photosynthesis: Structure, Function and Genetics. Advances in Photosynthesis 6, pp 21—52. Kluwer Academic Publishers, Dordrecht

Kaya K, Sano T, Watanabe MM, Shiraishi F and Ito H (1993) Thioic O-acid ester in sulpholipid isolated from freshwater picoplankton cyanobacterium, *Synechococcus* sp. *Biochim Biophys Acta* **1169:** 39—45

Keegstra K and Cline K (1999) Protein import and routing systems of chloroplasts. *Plant Cell* **11:** 557—570

Kenrick J and Bishop D (1986) The fatty acid composition of phosphatidylglycerol and sulphoquinovosyldiacylglycerol of higher plants in relation to chilling sensitivity. *Plant Physiol* **81:** 946—948

Kleppinger-Sparace KF, Mudd JB and Bishop DG (1985) Biosynthesis of sulphoquinovosyldiacylglycerol in higher plants: The incorporation of $^{35}SO_4$ by intact chloroplasts. *Arch Biochem Biophys* **240:** 859—865

Krupa Z and Baszynski T (1977) Participation of sulphoquinovosyldiacylglycerol in the reconstruction of photosystem I activity of heptane-extracted chloroplasts. *Bulletin de l'academie Polonaise des sciences. Serie des sciences biologiques* Cl II **XXV:** 409—413

Krupa Z, Huner NPA, Williams JP, Maissan E and James DR (1987) Development at cold-hardening temperatures. The structure and composition of purified rye light harvesting complex II. *Plant Physiol* **84:** 19—24

Kruse O, Hankamer B, Konczak C, Gerle C, Morris E, Radunz A, Schmid GH, and Barber J (2000) Phosphatidylglycerol is involved in the dimerization of Photosystem II. *J Biol Chem* **275:** 6509—6514

Kuiper PJC (1972) Temperature response of adenosine triphosphatase of bean roots as related to growth temperatures and to lipid requirement of the adenosine triphosphatase. *Physiol Plant* **26:** 200—205

Kuiper PJC, Kahr M, Stuiver CEE and Kylin A (1974) Lipid composition of whole roots and Ca^{2+}, Mg^{2+}-activated adenosine triphosphatases from wheat and oat as related to mineral nutrition. *Physiol Plant* **32:** 33—36

Kurihara H, Mitani T, Kawabata J and Hatano M (1997) Inhibitory effect on the β-glucosidase reaction by the aggregated state of sulfoquinovosyldiacylglycerol. *Biosci Biotechnol Biochem* **61:** 536—538

Kurihara H, Tada S, Takahashi K and Hatano M (1996) Digalactosyldiacylglycerol suppression of inhibition by sulfoquinovosyldiacylglycerol of β-glucosidase. *Biosci Biotechnol Biochem* **60:** 932—934

Kylin A and Kahr M (1973) The effect of magnesium and calcium ions on adenosine triphosphatases from wheat and oat roots at different pH. *Physiol Plant* **28:** 452—457

Kylin A, Kuiper PJC and Hansson G (1972) Lipids from sugar beet in relation to the preparation and properties of (sodium-potassium)-activated adenosine triphosphatases. *Physiol Plant* **26:** 271—278

Lambers JW and Terpstra W (1985) Inactivation of chlorophyllase by negatively charged plant membrane lipids. *Biochim Biophys Acta* **831:**225—35

Lancelin J-M, Bally A, Arland GJ, Blackedge M, Gans P, Stein M and Jacquot JP (1994) NMR structure of ferredoxin chloroplastic transit peptide from *Chlamydomonas reinhardtii* promoted by trifluoroethanol in aqueous solution. *FEBS Lett* **343:** 261—266

Larsson UK and Andersson B (1985) Different degrees of phosphorylation and lateral mobility of two polypeptides belonging to the light-harvesting complex of photosystem II. *Biochim Biophys Acta* **809:** 396—402

Latowski D, Kostecka A and Strzalka K (2000) Effect of monogalactosyldiacylglycerol and other thylakoid lipids on violaxanthin de-epoxidation in liposomes. *Biochem Soc Trans* **28:** 810—812

Lee RF and Benson AA (1972) The metabolism of glyceryl [^{35}S] sulphoquinovoside by the coral tree, *Erythrina crista–galli* and alfalfa, *Medicago sativa*. *Biochim Biophys Acta* **261:** 35—37

Leech RM, Rumsby MG and Thomson WW (1973) Plastid differentiation, acyl lipid, and fatty acid changes in developing green maize leaves. *Plant Physiol* **52:** 240—245

Lehmann J and Benson AA (1964) The plant sulpholipid. Sulfosugar synthesis from methyl hexoseenides. *J Am Chem Soc* **86:** 4469—4472

Li L, Karlsson OP and Wieslander A (1997) Activating amphiphiles cause a conformational change of the 1,2-diacylglycerol transferase from *Acholeplasma laidlavii* membranes according to protcolytic digestion. *J Biol Chem* **272:** 29602—29606

Livn A and Racker E (1969) Partial resolution of the enzymes catalyzing photophosphorylation. Interaction of coupling factor I from chloroplasts with ribonucleic acid and lipids. *J Biol Chem* **244:** 1332—1338

Loya S, Reshef V, Mizrachi E, Silberstein C, Rachamin Y, Carmeli S and Hizi A (1998) The inhibition of the reverse transcriptase of HIV-1 by the natural sulfoglycolipids from cyanobacteria: contribution of different moieties to their high potency. *J Natl Products* **61:** 891—895

McCourt P, Browse J, Watson J, Arntzen CJ and Somerville CR (1985) Analysis of photosynthetic antenna function in a mutant of *Arabidopsis thaliana* (L.) lacking trans-hexadecenoic acid. *Plant Physiol* **78:** 853—858

Makewicz A, Gribi C and Eichenberger W (1997) Lipids of *Ectocarpus fasciculatus* (Phaeophyceae). Incorporation of [1-^{14}C]oleate and the role of TAG and MGDG in lipid metabolism. *Plant Cell Physiol* **38:** 952—960

Martelli H and Benson AA (1964) Sulfocarbohydrate metabolism. 1. Bacterial production and utilisation of sulfoacetate. *Biochim Biophys Acta* **93:** 169—171

Michel H, Hunt DF, Shabanowitz J and Bennett J (1988) Tandem mass spectrometry reveals that three photosystem II proteins of spinach chloroplasts contain N-acetyl-O-phosphothreonine at their NH_2 termini. *J Biol Chem* **263:** 1123—1130

Mouradian E, Klyachko-Gurvich G and Pronina N (1998) Lipid metabolism of *Spirulina platensis* under CO_2 stress In: Sanchez J, Cerda-Olmedo E and Martinez-Force E (eds) Advances in Plant Lipids Research, pp 511—513. Universidad de Sevilla, Spain

Mudd JB, Dezacks R and Smith J (1980) Studies on the biosynthesis of sulphoquinovosyl diacylglycerol in higher plants. In: Mazliak P, Benveniste P, Costes C and Douce R (eds) Biogenesis and Function of Plant Lipids, pp 57—66. ElsevierScience Publisher, Amsterdam

Mudd JB and Kleppinger-Sparace KF (1987) Sulpholipids. In: Stumpf PK and Conn EE (eds) The Biochemistry of Plants, Vol. 9, pp 275—289. Academic Press, New York

Muller M and Santarius KA (1978) Changes in chloroplast membrane lipids during adaptation of barley to extreme salinity. *Plant Physiol* **62:** 326—329

Murata N and Hoshi H (1984) Sulfoquinovosyl diacylglycerols in chilling sensitive and chilling resistant plants. *Plant Cell Physiol* **25:** 1241—1245

Murata N and Yamaya J (1984) Temperature-dependent phase behavior of phosphatidyl glycerols from chilling-sensitive and chilling-resistant plants. *Plant Physiol* **74:** 1016—1024

Murata N, Higash S-I and Fugimura Y (1990) Glycerolipids in various preparations of photosystem II from spinach chloroplasts. *Biochim Biophys Acta* **1019:** 261—268

Murata N and Siegenthaler PA (1998) Lipids in photosynthesis: an overview. In: Siegenthaler PA and Murata N (eds) Lipids in Photosynthesis: Structure, Function and Genetics. Advances in Photosynthesis 6, pp 3—20. Kluwer Academic Publishers, Dordrecht

Ogawa A, Murate T, Izuta S, Takemura M, Furuta K, Kobayashi J, Kamikawa T, Nimura Y and Yoshida S (1998) Sulfated glycoglycerolipid from an archaebacterium inhibits eukaryotic DNA polymerase α, β and retroviral reverse transcriptase and affects methyl methanesulfonate cytotoxicity. *Int J Cancer* **76:** 512—518

Ohta K, Hanashima S, Mizushina Y, Yamazaki T, Saneyoshi M, Sugawara F and Sakaguchi K (2000) Studies on a novel DNA polymerase inhibitor group, synthetic sulfoquinovosylacylglycerols: inhibitory action on cell proliferation. *Mutat Res* **467:**139—152

Ohta K, Mizushina Y, Hirata N, Takemura M, Sugawara F, Matsukage A, Yoshida S and Sakaguchi K (1998) Sulfoquinovosyldiacylglycerol, KM043, a new potent inhibitor of eukaryotic DNA polymerases and HIV-reverse transcriptase type 1 from a marine red alga, *Gigartina tenella. Chem Pharm Bull* (Tokyo) **46:** 684—686

Ohta K, Mizushina Y, Yamazaki T, Hanashima S, Sugawara F and Sakaguchi K (2001) Specific interaction between an oligosaccharide on the tumor cell surface and the novel antitumor agents, sulfoquinovosylacylglycerols. *Biochem Biophys Res Commun* **288:** 893—900

Okanenko A and Protsenko D (1977) Apple-tree shoot bark lipid composition dynamics. *Physiologiya i biohimiya kulturnih rastenij* **9:** 80—85 (in Russian)

Okanenko A and Taran N (1989) High temperature and water deficit action upon winter wheat chloroplast lipid composition. In: Ostrovskaya LK (ed) Factori sredi i organizatsiya vtorichnogo processa photosynteza, pp 120—126. Naukova dumka, Kiev (in Russian)

Oquist G (1982) Seasonally induced changes in acyl lipids and fatty acids of chloroplast thylakoids of *Pinus silvestris. Plant Physiol* **69:** 869—875

Orr G and Raison J (1987) Compositional and thermal properties of thylakoid polar lipids of *Nerium oleander* L in relation to chilling sensitivity. *Plant Physiol* **84:** 88—92

Pancratova S and Karimova F (1984) Irrigation influence upon water exchange and winter rye leaf lipid composition. Kazan, pp 7. Deponin VINITY, N 5537-84 Dep (in Russian)

Pearcy R (1978) Effect of growth temperature on the fatty acid composition of the lipids in *Atriplex lentiformis* (Torr) Wats. *Plant Physiol* **61:** 484—486

Pick U, Gounaris K, Weiss M and Barber J (1985) Tightly bound sulpholipids in chloroplast CF_0-CF_1. *Biochim Biophys Acta* **808:** 415—420

Pilon M, Wienk H, Sips W, de Swaaf M, Talboom I, van 't Hof R, de Korte-Kool G, Demel R, Weisbeek P and de Kruijff B (1995) Functional domains of the ferredoxin transit sequence involved in chloroplast import. *J Biol Chem* **270:** 3882—3893

Pinnaduwage P and Bruce BD (1996) In vitro interaction between a chloroplast transit peptide and chloroplast outer envelope lipids is sequence-specific and lipid class-dependent. *J Biol Chem* **271:** 32907—32915

Poirier Y, Thoma S, Somerville C and Schiefelbein J (1991) A mutant of *Arabidopsis* deficient in xylem loading of phosphate. *Plant Physiol* **97:** 1087—1093

Pugh C, Roy AB, Hawkes T and Harwood JL (1995) A new pathway for the synthesis of the plant sulpholipid, sulphoquinovosyldiacylglycerol. *Biochem J* **309:** 513—519

Pugh C, Roy AB, White GF and Harwood JL (1997) In: Williams HP, Khan MU and Lem NW (eds) Physiology Biochemistry and Molecular Biology of Plants, pp 104—106. Kluwer Academic Publishers, Dordrecht

Quartacci MF, Pinzino C, Sgherri C and Navari-Izzo F (1995) Lipid composition and protein dynamics in thylakoids of two wheat cultivars differently sensitive to drought. *Plant Physiol* **108:** 191—197

Quasney ME, Carter LC, Oxford C, Watkins SM, Gershwin ME and German JB (2001) Inhibition of proliferation and induction of apoptosis in SNU-1 human gastric cancer cells by the plant sulfolipid, sulfoquinovosyldiacylglycerol. *J Nutr Biochem* **12:** 310—315

Quinn PJ (1998) The role of lipids in stability of plant membranes In: Sanchez J, Cerda-Olmedo E and Martinez-Force E (eds) Advances in Plant Lipid Research, pp 361—366. Universidad de Sevilla, Spain

Quoc KP and Dubacq JP (1997) Effect of growth temperature on the biosynthesis of eukaryotic lipid molecular species by the cyanobacterium *Spirulina platensis*. *Biochim Biophys Acta* **1346:** 237—46

Radunz A, Bader K and Schmid G (1984) Influence of antisera to sulphoquinovosyl diglyceride and to β-sitosterol on the photosynthetic electron transport in chloroplasts from higher plants In: Siegenthaler PA and Eichenberger W (eds) Structure, Function and Metabolism of Plant Lipids, pp 479—484. **Elsevier Science Publishers,** Amsterdam

Radunz A, Alfermann K and Schmid GH (2000) State of the lipid and fatty acid composition in chloroplasts of *Nicotiana tabacum* under the influence of an increased CO_2 partial pressure of 700 ppm. *Biochem Soc Trans* **28:** 885—887

Rawyler A and Siegenthaler PA (1985) Transverse localization of monogalactosyldiacyl glycerol and digalactosyldiacylglycerol in spinach thylakoid membranes. *Biochim Biophys Acta* **815:** 287—298

Rawyler A, Unitt MD, Giroud C, Davies H, Mayor J-P, Harwood JL and Siegenthaler A (1987) The transmembrane distribution of galactolipids in chloroplast thylakoids is universal in a wide variety of temperate climate plants. *Photosyn Res* **11:** 3—13

Remy R, Tremolieres A, Duval JC, Ambard-Bretteville F and Dubacq JP (1982) Study of the supramolecular organization of light-harvesting chlorophyll protein (LHCP). *FEBS Lett* **137:** 271—275

Rossak M, Schafer A, Xu N, Gage DA and Benning C (1997) Accumulation of sulphoquinovosyl-1-O-dihydroxyacetone in a sulpholipid-deficient mutant of *Rhodobacter sphaeroides* inactivated in *sqdC*. *Arch Biochem Biophys* **340:** 219—230

Rossak M, Tietje C, Heinz E and Benning C (1995) Accumulation of UDP-sulphoquinovose in a sulpholipid-deficient mutant of *Rhodobacter sphaeroides*. *J Biol Chem* **270:** 25792—25797

Roughan PG and Batt RD (1968) Quantitative analysis of sulfolipid (sulfoquinovosyl diglyceride) and galactolipids (monogalactosyl and digalactosyl diglycerides) in plant tissues. *Anal Biochem* **22:** 74—88

Roy AB and Harwood JL (1999) Re-evaluation of plant sulpholipid labelling from UDP-[^{14}C]glucose in pea chloroplasts. *Biochem J* **344:** 185—187

Roy AB, Ellis AJ, White GF and Harwood JL (2000) Microbial degradation of the plant sulpholipid. *Biochem Soc Trans* **28:** 781—783

Russel GB (1966) The routine measurement of lipid-bound sulfoquinovose in plants. *Anal Biochem* **14:** 205—214

Saidha T and Schiff JA (1989) The role of mitochondria in sulfolipid biosynthesis by *Euglena* chloroplasts. *Biochim Biophys Acta* **1001:** 268—273

Sakai WS, Yamamoto HY, Miyazaki T and Ross JW (1983) A model for chloroplast thylakoid membranes involving orderly arrangements of negatively charged lipidic particles containing sulphoquinovosyldiacylglycerol. *FEBS Lett* **158:** 203—207

Sakaki T (1998) Photochemical oxidants: toxicity. In: de Kok LJ and Stulen I (eds) Responses of Plant Metabolism to Air Pollution and Global Change, pp 117—129. Backhuys Publishers, Leiden

Sakaki T, Ohnishi J, Kondo N and Yamada M (1985) Polar and neutral lipid changes in spinach leaves with ozone fumigation. Triacylglycerol synthesis from polar lipids. *Plant Cell Physiol* **26:** 253—262

Sakaki T, Tanaka K and Yamada M (1994) General metabolic changes in leaf lipids in response to ozone. *Plant Cell Physiol* **35:** 53—62

Sanda S, Leustek T, Theisen MJ, Garavito RM and Benning C (2001) Recombinant *Arabidopsis* SQD1 converts UDP-glucose and sulfite to the sulfolipid head group precursor UDP-sulfoquinovose *in vitro. J Biol Chem* **276:** 3941—3946

Sassaki GL, Gorin PAJ, Tisher CA and Iacomini M (2001) Sulfonolipids from lichenized basidiomycete *Dictyonema glabratum*: isolation, NMR, and ESI-MS approaches. *Glycobiology* **11:** 345—351

Sato N, Murata N, Miura G and Veta N (1979) Effect of growth temperature on lipid and fatty acid composition in the blue-green algae *Anabaena variabilis* and *Anacystis nidulans*. *Biochim Biophys Acta* **572:** 19—28

Sato N, Hagio M, Wada H and Tsuzuki M (2000) Environmental effects on acidic lipids of thylakoid membranes. *Biochem Soc Trans* **28:** 912—914

Sato N, Somoke K, Tsuzsuki M and Kawaguchi A (1995) Impared PS II in a mutant of *Chlamydomonas reinhardtii* defective in sulphoquinovosyldiacylglycerol. *Eur J Biochem* **234:** 16—23

Seifert U and Heinz E (1992) Enzymatic characteristics of UDP–sulphoquinovose:diacylglycerol sulphoquinovosyltransferase from chloroplast envelopes. *Bot Acta* **105:** 197—205

Seigneurin-Berny D, Rolland N, Dorne AJ and Joyard J (2000) Sulfolipid is a potential candidate for annexin binding to the outer surface of chloroplast. *Biochem Biophys Res Commun* **272:** 519—524

Selstam E (1998) Development of thylakoid membranes with respect to lipids In: Siegenthaler PA and Murata N (eds) Lipids in Photosynthesis: Structure, Function and Genetics. Advances in Photosynthesis Vol 6, pp 209—224. Kluwer Academic Publishers, Dordrecht

Senchugova NA, Taran NYu and Okanenko AA (1999) Virus impact upon bean photo-synthesising tissue lipid composition. *Arch Phytopath Pflanz* **32:** 471—477

Shibuya I and Benson AA (1961) Hydrolysis of α-sulphoquinovosides by β-galactosidase *Nature* **192:** 1186—1187

Shipley GG, Green JP and Nichols BW (1973) The phase behavior of monogalactosyl, digalactosyl and sulphoquinovosyl diglycerides. *Biochim Biophys Acta* **311:** 531—544

Siegenthaler PA (1998) Molecular organization of acyl lipids in photosynthetic membranes of higher plants. In: Siegenthaler PA and Murata N (eds) Lipids in Photosynthesis: Structure, Function and Genetics. Advances in Photosynthesis, Vol 6, pp 120—144. Kluwer Academic Publishers, Dordrecht

Siegenthaler PA and Tremolieres A, (1998) Role of acyl lipids in function of photosynthetic membranes in higher plants. In: Siegenthaler PA and Murata N (eds) Lipids in Photosynthesis: Structure, Function and Genetics. Advances in Photosynthesis, Vol 6, pp145—173. Kluwer Academic Publishers, Dordrecht

Sigrist M, Zwellenberg C, Giroud CH, Eichenberger W and Boschetti A (1988) Sulfolipid associated with light-harvesting complex associated with photosystem II apoproteins of *Chlamydomonas reinhardtii*. *Plant Sci* **58:** 15—23

Simidjiev I, Stoylova S, Amenitsch H, Javorfi T, Mustardy L, Laggner P, Holzenburg A and Garab G (2000) Self-assembly of large, ordered lamellae from non-bilayer lipids and integral membrane proteins in vitro. *Proc Natl Acad Sci USA* **97:** 1473—1476

Snipes CE, Brillinger G-U, Sellers L, Mascaro L and Floss HG (1977) Stereochemistry of the dTDP–glucose oxidoreductase reaction. *J Biol Chem* **252:** 8113—8117

Strickland TC and Fitzgerald JW (1983) Mineralisation of sulfur in sulfoquinovose by forest soils. *Soil Biol Biochem* **15:** 347—349

Stuiver CEE, de Kok LJ, Santers JMO and Kuiper PJC (1984) The effect of Na_2SO_4 on the lipid composition of sugar beet plants. *Z Pflanzenphysiol* **114:** 187—191

Sundby C and Larsson C (1985) Transbilayer organization of the thylakoid galactolipids. *Biochim Biophys Acta* **813:** 61—67

Taran N, Okanenko A and Musienko M (2000) Sulpholipid reflects plant resistance to stress-factor action. *Biochem Soc Trans* **28:** 924—926

Taranto PA, Keenan TW and Potts M (1993) Rehydration induces rapid onset of lipid biosynthesis in desiccated *Nostoc commune* (Cyanobacteria). *Biochim Biophys Acta* **1168:** 228—237

Tremolieres A, Dubacq JP, Ambard-Bretteville F and Remy R (1981) Lipid composition of chlorophyll-protein complexes. *FEBS Lett* **130:** 27—31

Tremolieres A and Siegenthaler PA (1998) Reconstraction of photosynthetic structures and activities with lipids. In: Siegenthaler PA and Murata N (eds) Lipids in Photosynthesis: Structure, Function and Genetics. Advances in Photosynthesis, Vol 6, pp 175—189. Kluwer Academic Publishers, Dordrecht

Trosper T and Sauer K (1968) Chlorophyll *a* interactions with chloroplast lipids *in vitro*. *Biochim Biophys Acta* **162:** 97—105

van't Hof R, Demel RA, Keegstra K and de Kruijff B (1991) Lipid-peptide interactions between fragments of the transit peptide of ribulose-1,5-bisphosphate carboxylase/oxygenase and chloroplast membrane lipids. *FEBS Lett* **291:** 350—354

van't Hof R, van Klompenburg W, Pilon M, Kozubek A, de Korte-Kool G, Demel RA, Weisbeek PJ and de Kruijff B (1993) The transit sequence mediates the specific interaction of the precursor of ferredoxin with chloroplast envelope membrane lipids. *J Biol Chem* **268:** 4037—4042

Vijayan P, Routaboul J-M and Browse J (1998) A genetic approach to investigating membrane lipid structure and photosynthetic function. In: Siegenthaler PA and Murata N (eds) Lipids in Photosynthesis: Structure, Function and Genetics. Advances in Photosynthesis, Vol 6, pp 263—285. Kluwer Academic Publishers, Dordrecht

Vishwanath BS, Eichenberger W, Frey FJ and Frey BM (1996) Interaction of plant lipids with 14 kDa phospholipase A_2 enzymes. *Biochem J* **320:** 93—99

Voss R, Radunz A and Schmid GH (1992) Binding of lipids onto polypeptides of the thylakoid membrane. Galactolipids and sulpholipid as prosthetic groups of core peptides of the Photosystem II complex. *Z Naturforsch* **47c:** 406—415

Wada H and Murata N (1998) Membrane lipids in Cyanobacteria. In: Siegenthaler P-A, Murata N, (eds) Lipids in Photosynthesis: Structure, Function and Genetics. Advances in photosynthesis, Vol 6, pp 65—81. Kluwer Academic Publishers

Weissenmayer B, Geiger O and Benning C (2000) Disruption of a gene essential for sulfoquinovosyldiacylglycerol biosynthesis in *Sinorhizobium meliloti* has no detectable effect on root nodule symbiosis. *Mol Plant Microbe Interact* **13:** 666—672

Williams M, Robertson EJ, Leech RM and Harwood JL (1998) The effects of elevated atmospheric CO_2 on lipid metabolism in leaves from mature wheat plants. *Plant Cell Environ* **21:** 927—936

Youssef NB, Zarrouk M, Daoud D, Lemal F, Ouariti O, Ghorbal MH and Cherif A (1998) Membrane lipid changes in *Brassica napus* induced by cadmium chloride. In: Sanchez J, Cerda-Olmedo E and Martinez-Force E (eds) Advances in Plant Lipid Research, pp 534—537. Universidad de Sevilla, Spain

Yu B, Xu C and Benning C (2002) *Arabidopsis* disrupted in SQD2 encoding sulfolipid synthase is impaired in phosphate-limited growth. *Proc Natl Acad Sci USA* **99:** 5732—5737

9. AFFILIATIONS

School of Biosciences, Cardiff University, PO Box 911, Cardiff CF10 3US, U.K. (JH), National Taras Shevchenko University, Plant Physiology & Ecology Department, Kyiv, 01033, Ukraine (AAO)

For correspondence - JL Harwood: harwood@Cardiff.ac.uk

AA Okanenko: a_okanenko@hotmail.com

M. TAUSZ, G. GULLNER, T. KÖMIVES AND D. GRILL

THE ROLE OF THIOLS IN PLANT ADAPTATION TO ENVIRONMENTAL STRESS

Abstract. Thiols (-SH groups) are the main form of reduced sulphur in plants present as protein thiols (cys residues) or low molecular weight thiols, the most prominent is the tripeptide glutathione (GSH). Due to the redox properties of the thiol groups these compounds play important roles in stress responses of plants, since stress is nearly always associated with an enhanced formation of reactive oxygen species (ROS). Because of its antioxidative properties, GSH is directly involved in protection of cells from ROS damage and in regeneration of oxidized substrates. GSH is kept in the active reduced state by glutathione reductase (GR). In the following chapter, responses of GSH and GR system to natural abiotic stresses (drought, low and high temperatures, high light, UV, salinity, anoxia), biotic stresses (pathogen infection), and air pollution impact are discussed. The reviewed results show that plant resistance to stress is commonly correlated with increased GSH contents and GR activities. Damage to plant tissues is often marked by a breakdown of the GSH system. Besides its role as protectant, GSH is thought to play a role in cellular signalling and stress sensing. Apart from GSH, small protein thiols (thioredoxins) are essential for the adjustment of metabolism to environmental conditions. Most recent results suggest that thioredoxins can be directly induced by ROS stress. The role of thiols in regulation of plant stress responses, e.g. by inducing defense enzymes and compounds, is discussed.

1. THIOLS IN PLANTS

A good part of the sulphur incorporated into organic molecules in plants is located in thiol (-SH) groups in proteins (cys-residues) or non-protein thiols. Due to their particular redox properties, thiol groups can be oxidized forming disulphides (S-S groups). In proteins, thiol/disulphide status controls the three-dimensional molecule structure and in many enzymes thiol groups form the active centres (Packer, 1995). Thioredoxins are small proteins important for the redox regulation systems in chloroplasts, but also in other cell compartments (Schürmann and Jacquot, 2000). More recently, another group of small thiol proteins, the glutaredoxins which were well known as redox components in bacteria and mammal systems, have been characterized in plants (Rouhier et al., 2002). Non-protein thiols comprise e.g. coenzyme A, the ubiquitous acyl-group-acceptor or non-protein thiol tripeptides, such as glutathione (GSH, γ-glutamyl-cysteinyl-glycine) and its homologues. GSH is not only a central compound in plant sulphur metabolism, but also an important cellular redox buffer.

Redox reactions play a prominent role in cellular responses to environmental stress. Biochemically, the metabolism of cells under stress is generally characterized by an increased formation of reactive oxygen species (ROS) within the cell metabolism (Elstner and Osswald, 1994; Foyer and Noctor, 2000). ROS comprise potentially toxic compounds such as hydrogen peroxide, superoxide anion free radical, hydroxyl free radical or singlet oxygen, all of them able to oxidatively damage proteins, lipids, nucleic acid and pigments (oxidative stress). Stress-dependent ROS formation may also contribute to sensing and signalling of environmental stress impacts on plants. In

Yash P. Abrol and Altaf Ahmad (eds.), Sulphur in Plants, 221—244.

response to biotic stress (pathogen attacks), ROS bursts potentially kill the pathogen or in the hypersensitive defense, host cells.

Even under non-stress conditions, the presence of ROS in cells is an inescapable feature of life in an oxygen atmosphere (Elstner and Oßwald, 1994). Plants have evolved antioxidant defense systems to keep ROS under control. Low molecular weight antioxidant compounds, such as ascorbate, tocopherol and glutathione scavenge ROS or regenerate oxidatively damaged substrates in concerted actions with enzymic protection systems. The following chapter discusses the role of thiol compounds (with special emphasis on GSH redox system) in these defense systems and their involvement in plant responses to environmental stress.

2. MECHANISMS OF STRESS-INDUCED 'ROS' PRODUCTION IN PLANT CELLS

In plant cells, the most important locations of ROS production are the chloroplasts. An imbalance between the consumption of reductant (NADPH) in carbon fixation, and the need of the electron transport chain for the regenerated electron acceptor at the PS I site (NADP) leads to the transfer of electrons to alternative acceptors, predominantly molecular oxygen. Many environmental stress factors inhibit the function of the Calvin cycle (i.e., the CO_2 fixation and the NADPH and ATP consumption). For example, lack of CO_2 prevents carbon fixation when stomata close upon drought stress. At the same time, light driven electron transport remains fully active, which leads to an over-reduction of the electron transport chains. The univalent reduction of O_2 by leaking electrons yields the superoxide anion $O_2^{-\bullet}$ (Fig. 1). Superoxide is a free radical with potentially destructive effects on proteins, lipids, nucleic acids and pigments. For example, it may initiate free radical chain reactions in membrane lipids which produce lipid hydroperoxides with detrimental effects to the function and structure of biomembranes. Superoxide is detoxified enzymatically by superoxide dismutase (SOD, E.C. 1.15.1.1). This reaction forms H_2O_2 which is enzymically detoxified by the activities of peroxidases. Peroxidases in the plant chloroplasts use mainly ascorbate as reductant. Alternatively, H_2O_2 is also detoxified by catalase, which is located in peroxisomes and catalyses the disproportion of H_2O_2 to the products O_2 and H_2O. A direct transfer of excess excitation energy from the photosystems to molecular oxygen yields highly reactive singlet oxygen, an excited state of the molecule. Free radical chain reactions may also yield the hydroxyl free radical ($OH^\bullet$), the most reactive of all these molecules. Unlike the superoxide and hydroxyl radicals, hydrogen peroxide and singlet oxygen are not free radicals, but nevertheless highly reactive oxygen species (ROS) potentially dangerous to cell metabolism.

In green tissues the chloroplast is the most important, but not the only site of stress related ROS production. The electron transport chains in the mitochondria are also a potential source for side-reactions producing ROS. Although the total rates of the mitochondrial ROS production are considered to be much lower than in the chloroplast, values are rarely cited (Foyer and Noctor, 2000).

Photorespiration is an important adaptive pathway protecting the photosynthetic apparatus under excess light (Asada, 1999). In the course of this reaction sequence, H_2O_2 is produced in peroxisomes where it is scavenged by high activities of catalase

(Fig. 1). Stress situations or deficits in the catalase systems can lead to oxidative stress caused by photorespiratory H_2O_2. Photorespiratory activity also yields glycine which is, in turn, an important precursor in GSH biosynthesis.

When pathogens invade plant tissues, host cells respond by an oxidative burst, i.e. a sudden evolution of H_2O_2 in the apoplastic space. The aggressive agent kills pathogen cells and signals the infection to other host cells (Lamb and Dixon, 1997). The plant apoplast is well equipped to keep H_2O_2 under control, as lignification reactions require H_2O_2. Peroxidases catalyze lignification and, on the other hand, may confer protection against H_2O_2. Nevertheless, a further increase of ROS as in the oxidative burst means oxidative stress in the apoplast and a threat primarily to the plasmalemma structures.

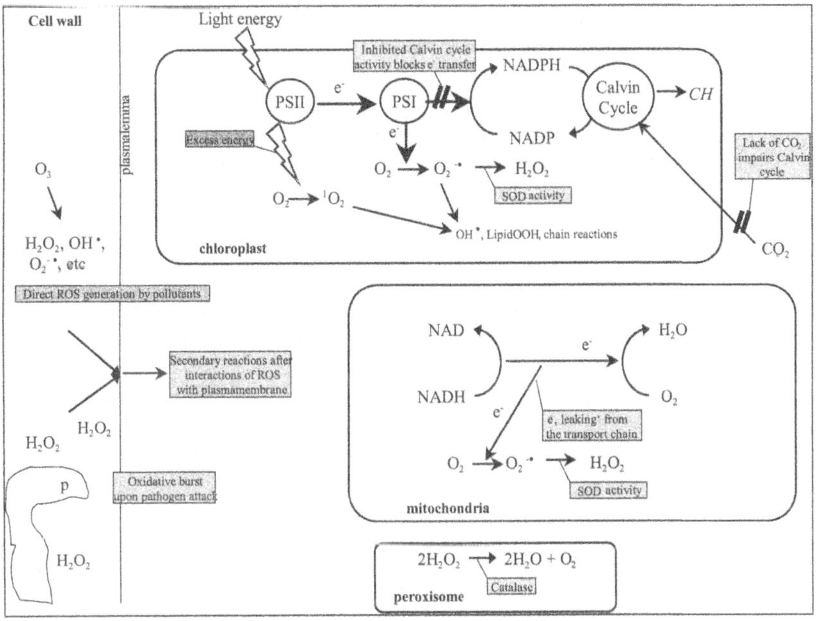

Figure 1. Some sites and ways of ROS production in plant cells under stress
CH = Carbohydrates, p = pathogen

A comparable situation occurs under the impact of oxidative air pollutions, such as ozone, which enter the apoplastic space in leaf tissues through stomata. Contact with apoplastic water causes the formation of a variety of ROS which attack the plasmalemma. These examples show that the compartmentation of ROS generation may be different in various stress situations.

With increasing stress, cell metabolism is severely disrupted. When membranes and cell compartments are destroyed, the chain reactions described above occur at membrane lipids and proteins, and can further destroy cellular organization. In the course of such destructive reaction the formation of free radicals is augmented in all cellular compartments (Elstner and Oßwald, 1994).

3. PARTICIPATION OF THIOLS IN PROTECTIVE PATHWAYS

Thiols can participate in different parts of the defence systems:

(1) Glutathione acts as chemical antioxidant. In this process, GSH reduces an oxidized substrate while it is itself oxidized to GSSG. The antioxidant properties in vivo are dependent on the rate constants of the direct reaction of the antioxidant with substrates as well as the concentration of the antioxidant. Taking into account these considerations, the importance of GSH as a chemical scavenger of ROS is supposed to be rather low (Rennenberg and Brunold, 1994).

(2) The regeneration of ascorbate, the major water-soluble antioxidant in plant cells, requires glutathione as an electron donor. In that reaction, GSH is oxidized to GSSG and regenerated by glutathione reductase (GR, E. C. 1.6.4.2) using NADPH as a reductant. This enzymatic pathway (ascorbate-glutathione-cycle) is supposed to be the primary way of removing hydrogen peroxide and simultaneously regenerating the electron acceptor NADP (Noctor and Foyer, 1998). The capacity of this cycle depends on enzyme activities and the concentration of antioxidants. Changes in turnover rates may manifest themselves in changes of the redox ratios of dehydroascorbate /ascorbate or GSH/GSSG.

(3) Due to its antioxidative properties GSH may play a role as a stabilizer of biomembranes. Membrane lipids are primary targets for ROS reactions; lipid peroxidation forms acyl peroxides, which can be removed by GSH (McKersie and Leshem, 1994). However, the most important compound playing this role in plant cells is the lipophilic antioxidant α-tocopherol, which is regenerated by ascorbate (Fryer, 1992).

(4) Thiol-disulphide exchange is of general importance in enzyme activations, protein protection, and cellular regulation processes (Kunert and Foyer, 1993). GSH may repair the damage caused by ROS because protein disulphides formation is a primary consequence of oxidative attack. Protein disulphide bonds can be reduced by GSH. The reduction state of protein thiol groups is a decisive factor in redox sensing in the chloroplast, which is crucial for the flexible and fast adaptation of the photosynthetic apparatus to rapidly changing environmental conditions (Vener et al., 1998).

(5) A non-enzymic reduction of dehydroascorbate (the oxidized form of ascorbate) by GSH is possible at a pH greater than 7 and GSH concentrations above 1 mmol L^{-1}. Such a milieu exists in the stroma of illuminated chloroplasts with a pH about 8 and sufficient GSH concentrations (McKersie and Leshem, 1994). The importance of this reaction in vivo is not clear.

(6) GSH may serve as a substrate for peroxidase reactions. In plant systems, increased glutathione peroxidase (GPX, E.C. 1.11.1.9) activities were measured upon stress impacts, but some of the reported activities may be due to the side reactions of glutathione-S-transferases (GST, E.C. 2.5.1.18, see below). In contrast to seleno-cysteine containing animal GPX, plant GPX as characterized in *Citrus sinensis* by Eshdat et al. (1997) has a normal Cys in the active centre. The best substrates for this enzyme are phospholipid hydroperoxides rather than H_2O_2, making the *Citrus* enzyme a phospholipid-hydroperoxide glutathione peroxidase (E.C. 1.11.1.12) rather than a GPX. The natural role of this protein may be the removal of toxic

lipid peroxidation products, which are results of oxidative stress in cells. However, specific activity of the plant enzyme is much lower than that of homologue animal GPXs (Eshdat et al., 1997). In pea and *Arabidopsis,* it was shown that genes for a chloroplast targeted GPX exist and respond to stress (Mullineaux et al., 1998).

(7) As a substrate of glutathione-S-transferases (GST, E.C. 2.5.1.18), GSH is responsible for the detoxification of xenobiotics or other adverse organic compounds from the cytoplasm (Schröder, 2001). The glutathione conjugates of these substances are sequestered to the vacuoles and, hence, removed from the cytoplasm. Some GSTs are also able to catalyze the reaction of GSH with products of lipid peroxidation chains and function as repair enzymes in oxidative stress situations. If peroxidation products are lipid hydroperoxides, GSTs mimic peroxidase (GPX) activity.

(8) In animals, GSH plays a decisive role in longer-range nitric oxide signalling. In blood vessels, the NO-signal is transported by conjugation with glutathione forming nitroso-glutathione (GSNO, Durner and Klessig, 1999). NO (a free radical), too, has been shown to participate in stress sensing and signalling in plants (Durner and Klessig, 1999). It is speculated that GSNO formation may also occur in plants and the product may serve in remote signalling processes.

(9) Thiols are essential in the heavy metal response of plants, since glutathione is the precursor of phyochelatin synthesis. Phytochelatins bind dangerous heavy metals in complexes and, hence, remove them from metabolism. For details see Verkliej et al. (this book) and Cobbett (this book).

4. SPECIFIC STRESS RESPONSES

Responses of the thiol system in plants were observed in response to virtually all stress situations, such as natural abiotic stresses, biotic stresses (pathogen), or pollutant impacts. Numerous and partly contradicting reports exist on the responses of the glutathione system of plants to environmental stresses. Recent reviews on the subject are written by Polle (1997), Noctor and Foyer (1998) and Grill et al. (2001).

4.1 Natural Stresses

4.1.1 Drought

Most plants minimize water loss by stomatal closure, which at the same time causes a lack of CO_2 in the chloroplasts. This leads to increased oxidative stress in illuminated green tissues (Smirnoff, 1993).

At very low relative water contents (below 40%), an oxidation of the GSH as well as ascorbate pool was measured in barley leaves (Smirnoff, 1993). Although the plants recovered when re-watered immediately, the relative water content is near the survival limit for these species. This situation may reflect biochemical damages to the tissues rather than protective reactions. Initial responses to drought were investigated in a short-term study on *Pinus canariensis*, a drought resistant pine species, which effectively prevents water loss by stomatal closure and maintaining high RWC. After complete stomatal closure, changes in photostress marker pigments (xanthophylls) occurred and the GSH/GSSG ratio dropped. These changes were accompanied by light dependent changes in total GSH concentrations (Tausz et al., 2001). In contrast, drought stressed wheat cultivars showed an increased GSH/GSSG ratio together with a loss of total GSH. The more sensitive cultivar exhibited a corresponding increase in GR activity (Loggini et al., 1999). A loss of GSH and an increase in GR activity was also reported in osmotically-induced drought stress ("physiological drought" due to hypertonic external solution) in rice hydrocultures (Boo and Jung, 1999).

Recently, Broin et al. (2000) showed that severe drought not only leads to an oxidation of the glutathione pool in potato plants, but simultaneously induced a special chloroplast thioredoxin. Since thioredoxins are important redox regulators of photosynthesis, it is presumed that this molecule has a role in ROS scavenging or protein repair in stressed chloroplasts.

Poikilohydric plants or tissues, which are able to endure large water losses to very low RWCs and quickly resume life functions after wetting, respond differently. Lichens and mosses, but also ferns and few higher plants ("resurrection plants") as well as resting stages such as many seeds are desiccation tolerant. Poikilohydry is related to the ability to maintain or quickly resume the active reduced state of antioxidants during desiccation and resurrection. In the moss *Tortula ruralis,* the GSH pool got oxidized and the activities of GR, GPX, and also GST increased if the moss was slowly dehydrated, but no changes were found upon rapid dehydration (Dhindsa, 1987, 1991). Slowly dehydrated plants quickly reduced their GSH pool upon wetting and the activities of the enzymes resumed control level. Quickly dehydrated plants showed a transient increase in GSSG and enzyme activities after wetting, and a subsequent return to control level within the next hours. A quick reduction within minutes of highly oxidized glutathione pools (70 to 90 % of total) upon wetting were also found for some lichen species (Kranner and Grill, 1997). The resurrection plant *Boea hygroscopica* (one of the few higher plants that show poikilohydry) showed an increase in GSH contents upon dehydration, which was interpreted as a possible protection of proteins from oxidation (Navari-Izzo et al., 1997). Desiccation tolerant seed tissues also reactivate their glutathione system when physiological activity is resumed upon wetting. Observations on spores of *Neurospora crassa* and dormant seeds of *Triticum*, *Hordeum*, *Pisum* and *Lycopersicum* reported the same response (see review Kranner and Grill, 1997). The resting seeds contained highly oxidized glutathione pools, which were quickly reduced upon imbibition. The capacity to endure high oxidation of the glutathione pool and quickly reduce it upon re-hydration seems to be decisive in desiccation tolerance.

4.1.2 Light regime

Light regime modulates thiol systems in plant tissues. Sun-exposed spruce needles contain higher glutathione concentrations than shade needles of the same tree (Grill et al., 1987). Corresponding results were found among others in sun and shade leaves of the evergreen oak, *Quercus ilex* (García-Plazaola et al. 1999a). Chloroplasts from pea leaves grown under high light conditions exhibited higher GR activities than those under low light (Gillham and Dodge, 1987). However, GSH levels remained unchanged in that experiment, a result also reported for sun and shade-grown beech leaves (Polle,

1997). In etiolated wheat seedlings, a rapid increase of GSH within the first hours was observed upon illumination (Mattagajasingh and Kar, 1989). In non-etiolated wheat seedlings, a transfer from low to high light did not affect GSH concentrations, but GR activities increased transiently (Mishra et al., 1995). Glutathione contents underwent light dependent diurnal changes with the highest concentrations around midday (Schupp and Rennenberg, 1988). This is consistent with the fact that glutathione synthesis itself is higher in illuminated leaves (Noctor et al., 1999). The observed diurnal variations are at least partly due to a shift in the synthesis-export balance in the needles, since higher export rates were observed in the night (Schupp et al., 1992). Needles of *Pinus ponderosa* did not show changes in the GSH content when detached branches (with limited possibilities for glutathione export from the needles) were allowed to recover overnight from stressful conditions (Tausz et al., 1999a). In whole plants as well, daily changes of the glutathione content in spruce needles were not found in all cases (Tegischer et al., 2002). It seems possible that the potential of higher glutathione synthesis in the light is only activated under additional stress conditions. An important regulating factor might be the supply of photorespiratory glycine, which may accumulate under stressful conditions in the light (Asada, 1999). Glycine is a substrate for GSH synthesis and may be able to promote or limit the GSH formation (Noctor et al., 1999).

Since the reduction of GSSG by GR is dependent on NADPH, which is mainly provided by the photosynthetic electron transport, a light-dependence of the GSH/GSSG ratio might be expected. Some studies reported a more reduced glutathione pool in the light, but in other cases the GSH pool became more reduced during a dark period (Tausz et al., 1999a). Because non-green tissues maintain their GSH redox pool in a reduced state with the necessary reducing power probably supplied by the pentose phosphate cycle, a direct dependence of the GSH/GSSG redox ratio on the light regime is unlikely.

4.1.3 Low temperature

Low temperature can induce different kinds of stress to plant tissues: (i) Photo-chilling is a situation where the Calvin cycle is slowed down or inhibited by low temperatures (depending on the plant species, often at 5 °C and below, higher for sensitive species such as maize), but absorbed light energy still drives the electron transport. This leads to the situation of an overproduction of ROS in the chloroplasts as described in Fig. 1. (ii) Frost is detrimental to tissues mainly due the formation of ice crystals, which removes water from the liquid phase and causes damages to proteins and the ultrastructure of cells. The level of damage is dependent on the plant species, the developmental and phenological stage, the organ and tissues, and the degree of stress. Frost resistant tissues allow the formation of ice crystals in the apoplast, but not in the symplast. This will cause removal of liquid water from the cells causing a situation of physiological drought. ROS are involved in effects of freezing stress, participating in lipid peroxidation, protein destruction, and collapses of the antioxidant defence systems (Polle, 1997).

Chilling has been shown to have a strong impact on the antioxidative systems, in particular in plants sensitive to chilling (often agricultural plants originating from warmer climates, Kocsy et al., 2001). Maize responds to chilling at temperatures as high as 11 °C with an increased production of GSH and other antioxidants whereas the GSH/GSSG ratio dropped upon short-term chilling in young plants. Chilling-sensitive cultivars further increased their GSH concentrations upon long-term chilling and their GSH/GSSG ratios were lower than those of more tolerant lines (Hodges et al., 1996). Field experiments showed that the chilling tolerance of maize cultivars may be correlated to the capacity of increasing the GSH concentrations and GR activities under stress conditions (Leipner et al., 1999).

The freezing tolerance of plant tissues is dependent on their physiological hardening stage. Under field conditions, a gradual adaptation to lower temperatures is accompanied by various physiological changes including increased antioxidant protection involving responses of the glutathione system which allow the plants to assume a new homoeostatic level (Polle, 1997). Many studies reported increased GSH levels in winter, including spruce needles (Esterbauer and Grill, 1978; Doulis et al., 1993), pine needles (Anderson et al., 1992), in winter wheat (Sagisaka et al., 1991) and in the living bark of apple trees (Kuroda and Sagisaka, 1998). The signal for the increase in GSH concentrations seems to be the lower temperature and not the decrease of the light period length (Herbinger et al., 1999). Winter hardening is also marked by increased activities of GR in spruce needles (Esterbauer and Grill, 1978; Doulis et al., 1993). Glutathione reductase activity in pea leaves was induced by cold treatment (Edwards et al., 1994). A direct causal connection between cold tolerance and GR activity was established by Foyer et al. (1995). In this study, transgenic poplar plants containing higher GR activity were significantly more cold tolerant than controls. In contrast, a direct causal connection of elevated GSH levels without increased GR activities to frost tolerance could not be established (Polle, 1997). Contrasting results were found on the evergreen Mediterranean oak species *Quercus ilex*. In leaves of this species, glutathione concentrations did not undergo seasonal changes, but only GR activities increased in sun leaves during winter adaptation (García-Plazaola et al., 1999b).

4.1.4 High temperature

High temperature effects on the glutathione system of plants have been scarcely studied. It has been shown that ROS are involved in high temperature damage to plants, and excised maize roots strongly intensified GSH synthesis upon a heat shock of 40 °C (Nieto-Sotelo and Tuan-Hua, 1986), but effects on the glutathione system of intact plants are largely unknown (Polle, 1997).

4.1.5 UV radiation

UV radiation transmits more energy per quantum than visible light and is potentially dangerous to organisms. The most important form of UV in the environment is UV-B (from 290 to 315 nm wavelength) because the UV-B doses are expected to increase due to the depletion of the protective stratospheric ozone layer. UV-A (315-400 nm) is only a minimal stress factor for plants (it is more important in the fuelling of atmospheric photo-smog reactions such as tropospheric ozone formation), whereas the highly destructive UV-C (220-290 nm) is quantitatively screened by the atmosphere. Plants respond with stress symptoms to elevated UV-B doses and ROS seem to be participate in these effects. UV-B absorption may lead to direct production of ROS at the cell surfaces (e. g. directly at the plasmalemma) and not only act by increasing the ROS production in chloroplasts like most other stress impacts. In most UV experiments, the activities of glutathione-related enzymes (glutathione peroxidase and glutathione reductase) remained largely unaffected (Willekens et al., 1994; Rao et al., 1996). GSH concentrations did not increase in spruce upon exposure to near-ambient UV-B doses as a single stress factor (Polle, 1997). More recently, results on maize suggested that UV-B does not change the pool size or redox state of GSH, but accelerates the turnover of this pool and modifies the circadiane rhythm of GSH concentrations (Masi et al., 2002). In this species, increased GR activities were observed (Masi et al., 1995). Increased concentrations of cysteinyl-glycine, a low-molecular weight thiol, which is a possible degradation product of glutathione, pointed to an increased glutathione turnover (Masi et al., 2002). Longer-term studies on Scots pine showed that elevated UV-B exposure affected older needles (decreased glutathione concentrations and higher oxidation state), but not young needles (Laakso et al., 2001). It was concluded that UV-B effects on conifers are accumulative.

4.1.6 Salinity

Plants experience salinity as a multiple stress: (1) Through inhibition of Calvin cycle enzymes, salinity induces oxidative stress in plant cells due to an imbalance in light driven electron transport and carbon fixation; and (2) Due to osmotic changes, high salinity can cause water loss in cells and induces physiological drought stress in the plant tissues. Responses of the glutathione systems to salt exposure have been reported: For example, in *Arabidopsis thaliana*, GSH and also cysteine levels increase markedly upon exposure to salt stress (Barroso et al., 1999) indicating a strongly induced GSH synthesis. In rice, salinity caused an induction of GR (Kaminaka et al., 1998). In the latter case, this induction was apparently connected to the hormonal stress response of plants, namely to the synthesis of abscisic acid, which is also related to drought and chilling stresses. The GPX response demonstrated in *Citrus* is also induced by salt stress, possibly directly as a result of the stress-induced formation of phospholipid-hydroperoxides (the best substrate for this enzyme) which occurred only in less tolerant cells (Avsiankretchmer et al., 1999).

4.1.7 Anoxia

Anoxia is the absence of oxygen in the environment. It is a common stress factor to plant organs submerged in water, such as rhizomes, roots and stems of wetland plants. Quite paradoxically, ROS and oxidative stress mechanisms are involved in anoxia response, as is shown by the production of H_2O_2 upon anoxic stress (Blokhina et al., 2001). Anoxic stress and the re-oxygenation response involve a degradation of the cellular glutathione concentrations in rhizomes, especially in anoxia-sensitive plants (Blokhina et al., 2000). The response sequence is still unclear, however, it is suggested that cellular sensing of hypoxia (low oxygen status) and antioxidant responses may have common signalling mechanisms (Semanza, 1999).

4.2 Biotic Stresses

Plants come into frequent contact with bacterial, fungal and viral pathogens. If the preformed and inducible plant defense mechanisms are inappropriate, the pathogen invasion is successful and disease ensues (compatible plant-microbe interactions). In many cases the microbial invasion fails and the damage caused by the pathogen remains restricted (incompatible interactions). Incompatibility can be due to preformed plant structural barriers or toxic compounds, or to the mobilization of a wide array of defense mechanisms, which are able to prevent the establishment and spread of the pathogen (Hammond-Kosack and Jones, 1996). One of the most important defense responses is the hypersensitive reaction (HR), which is usually characterized by the rapid appearance of necrotic spots containing dead cells at sites of attempted pathogen ingress. The localized host cell death probably contributes to pathogen limitation. HR is usually associated with an oxidative stress (Lamb and Dixon, 1997). Other important defense responses include the reinforcement of plant cell wall by deposition of lignin and hydroxyproline-rich glycoproteins, the induction of hydrolytic enzymes (such as chitinases and glucanases) and the biosynthesis of antimicrobial phytoalexins. In recent years, it has been found that GSH plays important regulatory roles in these defense responses (Fig. 2) (Foyer and Rennenberg, 2000; Gullner and Kömives, 2001).

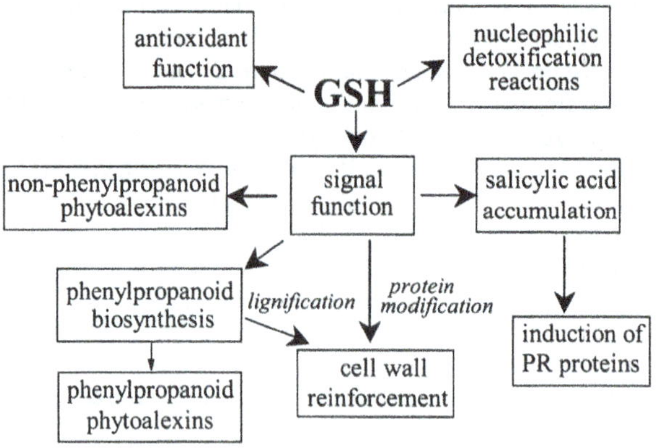

Figure 2. Functions of glutathione in plants infected with microbial pathogens

4.2.1 Changes of endogenous GSH level in infected plants

Pathogen infections can significantly alter the endogenous GSH levels in infected plants. Increase of GSH and homoglutathione levels in alfalfa and bean cells after treatment with a fungal elicitor has been demonstrated by Edwards et al. (1991). Increased non-protein thiol levels were found in mildewed leaves of a resistant and a moderately barley susceptible cultivar infected with the obligate biotrophic fungus *Blumeria graminis* f. sp. *hordei* (formerly known as *Erysiphe graminis* f. sp. *hordei*)

(El-Zahaby et al., 1995). Markedly increased GSH levels were found in the leaves and apoplasts of a resistant barley isoline after inoculation with avirulent races of *B. graminis* f. sp. *hordei*. In leaves and apoplasts of a susceptible barley line the GSH content decreased, but the redox state was unchanged in both cases (Vanacker et al, 1998a; Vanacker et al., 2000). Similar results were obtained in three oat lines, which expressed different degrees of race non-specific resistance to infection with *B. graminis* f. sp. *avenae* (Vanacker et al., 1998b). A massive increase of the total glutathione (GSH+GSSG) level was observed in leaves of tomato plants carrying the resistance genes *Cf-2* or *Cf-9* after the injection of an elicitor prepared from the fungal pathogen *Cladosporium fulvum*, mainly due to the accumulation of GSSG. In the compatible interaction only a slight increase of total glutathione levels was observed. Marked increases in the level of GSSG are strong evidence that the elicitor treatment results in severe oxidative stress in the incompatible interactions (May et al., 1996a). The supposedly protective role of cellular GSH was challenged by results obtained with the *Arabidopsis thaliana cad2-1* mutant whose total GSH content was only 30% of the wild-type level. This mutant still exhibited an almost normal resistance level against an avirulent strain of the fungus *Peronospora parasitica* (May et al., 1996b). Marked decreases of GSH levels were observed in oat leaves infected with virulent strains of the necrotrophic fungal pathogen *Drechslera* (*Helminthosporium*) (Gönner and Schlösser, 1993) and in tomato leaves infected with the necrotrophic fungal pathogen *Botrytis cinerea* (Kuzniak and Sklodowska, 2001). However, elevated GSH levels were found in areas surrounding rotted tissues of bean leaves infected with *B. cinerea* (Muckenschnabel et al., 2001). Interestingly, the accumulation of elemental S and transient increases in the concentrations of sulphate, GSH and cysteine were observed in tomato plants inoculated by the fungal vascular pathogen *Verticillium dahliae* (Williams et al., 2002). Tobacco mosaic virus (TMV) infection led to a substantial elevation of GSH levels in *Nicotiana tabacum* L *cv.* Xanthi-nc plants both locally (in the infected leaves) and systemically. The injection of salicylic acid into tobacco leaves also resulted in elevated foliar GSH content. It was supposed that the systemic induction of GSH-related antioxidative systems contributes to the development of systemic acquired resistance against a subsequent infection (Fodor et al., 1997). Summarizing the above reports it seems that the accumulation of GSH occurs mainly in incompatible plant-pathogen interactions, probably due to the increased demand of plants for antioxidative defense.

4.2.2 Protection against oxidative stress in infected plants by GSH

Pathogen infections cause rapid oxidative stress in plant tissues (Lamb and Dixon, 1997). Plant cells have to avoid excessive oxidative damage in non-invaded tissues around pathogen entry sites in order to exert effective defense responses. Probably this is the reason for the accumulation of GSH in plant cells as response to the oxidative stimulus (May et al., 1996a; Foyer and Rennenberg, 2000).

GST isoenzymes are known to participate in numerous detoxification reactions, and some isoenzymes exhibit also peroxidase activity (Schröder, 2001). Several studies suggested that the role of GSTs in infected plants is the decomposition of fatty acid hydroperoxides, which are major products of oxidative membrane damage (Bartling et al., 1993; Vollenweider et al., 2000). Wheat plants infected with the non-pathogen *Erysiphe graminis* f. sp. *hordei* (barley powdery mildew) showed local, induced resistance against a second infection with the pathogen *Erysiphe graminis* f. sp. *tritici*. One of the genes activated simultaneously with the onset of resistance was shown to encode a GST isoenzyme (Dudler et al., 1991; Mauch and Dudler, 1993). GST inductions were observed also in potato inoculated with the late blight fungus *Phytophthora infestans* (Hahn and Strittmatter, 1994), in barley cultivars after inoculation with *B. graminis* f. sp. *hordei* (El-Zahaby et al., 1995), and in *Arabidopsis thaliana* inoculated with the downy mildew pathogen *Peronospora parasitica* (Rairdan et al., 2001). Considerably induced GST activities were detected also in incompatible plant virus-interactions (Gullner et al., 1995; Fodor et al., 1997; Gullner et al., 1999). In bacterium infected plants, the induction of GST activity was reported in both compatible and incompatible interactions (Mittal and Davis, 1995; Vollenweider et al., 2000; Venisse et al., 2001).

4.3 Pollutants

Low molecular weight thiols are involved in many plant responses to environmental pollutants. Sulphur pollution involves thiols in their role as intermediates of sulphur metabolism, oxidant pollutants promote the formation of ROS challenging the redox properties of thiols, heavy metals are complexed by the formation of phytochelatins synthesized from glutathione and xenobiotics detoxification (more details thereon in Schröder, 2001) involves conjugation to glutathione. Plant metal chelators (phytochelatins) are discussed in detail by Rauser (2001), Verkliej et al. (this book) and Cobbett (this book).

4.3.1 Sulphurous air pollutants

Main sulphurous air pollutants are SO_2 and H_2S, both gases of phytotoxic capacities (see Agrawal, 2002 this book), but also possible sulphur source usable for plant growth requirements (De Kok and Tausz, 2001). Atmospheric sulphur taken up and metabolized may appear as thiols in the plant tissues (De Kok, 1990; De Kok et al., 1998). In general, SO_2 and H_2S exposure result in enhanced thiol levels in shoot and in roots, which can already be observed within a few hours after the start of the exposure. Since H_2S enters the sulphur assimilation pathway at the redox state of sulphide, it is incorporated directly into cysteine without costly reduction. Hence, thiol accumulation is generally higher upon H_2S than upon SO_2 exposure at equal atmospheric levels. Stress-dependent variation in thiol level is generally due to changes in GSH content (or its homologues), the predominant thiol present in plant tissue (>90%). However, upon exposure to SO_2 and H_2S also high levels of cysteine and γ-glutamyl-cysteine in darkness may accumulate in the shoot, though the impact of SO_2 and H_2S on the composition of the thiol pool strongly varies between species. In roots the thiol accumulation is generally exclusively due to enhanced GSH levels (De Kok et al., 1998). After cessation of the exposure, the thiol level decreases very rapidly and normal values are reached again within one or two days.

The physiological background of the altered composition of the thiol pool in the shoot upon exposure to SO_2 and H_2S is not yet solved. Apparently, if the sulphur is directly supplied to the shoot and the regulation of sulphate uptake by the roots is bypassed, then there is no strict regulation of composition of the thiol pool in the shoot (De Kok et al., 1998). The absorbed atmospheric sulphur may in part be metabolized in another subcellular compartment, outside the chloroplast where sulphur is assimilated under normal conditions (Hell, 1997). At this site there might be a shortage in substrate for the further metabolism of the excessively formed cysteine (γ-glutamyl-cysteine in darkness) into GSH. Furthermore, it has been observed that the accumulation of γ-glutamyl-cysteine in the dark can be prevented by adding glycine directly to the leaf tissue which results in GSH accumulation (Buwalda et al., 1990). A similar shortage of glycine for GSH synthesis was observed in plants wherein the level of γ-glutamyl-cysteine synthetase was overexpressed (Noctor and Foyer, 1998).

Thiol compounds such as GSH are assumed to fulfil signal functions in the regulation of uptake and reduction of sulphur in plants (Anderson and McMahon, 2001). Under normal conditions, the sulphate uptake is in tune with the metabolic need of sulphur for growth. The uptake may be regulated by changes in activity and/or the level of expression of sulphate transporter protein by negative feedback from sulphate itself or modulated by changes in levels by reduced sulphur compounds including GSH (Hell, 1997; Anderson and McMahon, 2001). SO_2 and H_2S impact studies are a promising tool to elucidate the signals involved in the regulatory aspects of the uptake, transport and reduction of sulphate in plants, and in the interactions between shoots and roots. Exposure of plants to sulphurous air pollutants may repress the uptake of sulphate by the roots and its further transport to the shoots (Herschbach et al., 1995; Westerman et al., 2000). It is still unclear to what extent an enhanced GSH level in the roots is the trigger of the repression of sulphate uptake. Without regard of a normal sulphur supply to the roots, plants are able to switch in part to atmospheric H_2S taken up by the shoots as sulphur source for plant growth. However, the pattern of increase in GSH levels in roots upon H_2S exposure (see above) appeared to be not in tune with the repression of sulphate uptake (Westerman et al., 2000). Some plants may already accumulate sulphate in the shoot upon exposure to relative low levels of SO_2 or H_2S (DeKok, 1990; De Kok et al., 1998). It still needs to be resolved whether the accumulated sulphate originates from either the oxidation of absorbed atmospheric sulphur gases in the plant foliage or it reflects a poor shoot/root interaction in regulatory control of the sulphate uptake by the roots and its transport to the shoot.

It has been postulated that an enhancement of the GSH, GR and SOD levels would have an adaptive value in the protection of plants against the toxic effects of SO_2. Part of the absorbed SO_2 would initiate a superoxide mediated free radical chain oxidation in the chloroplast, resulting in a burst of ROS which could be the basis for the injurious effects of SO_2. Indeed, in isolated chloroplasts in vitro addition of relatively high levels of sulphite or sulphide resulted in a superoxide-triggered oxidation of these reduced sulphur compounds upon illumination (Ghisi et al., 1990), but light-induced oxidation of sulphite was only substantial in broken chloroplasts and was diminished by addition of ROS scavengers including GSH (Ghishi et al., 1990). The sulphite/sulphide-induced superoxide mediated free radical chain reaction may have significance at acute high levels of sulphurous air pollutants. At normal ambient pollutant levels, it is unlikely that high levels of sulphite occur in the chloroplast, since it would be metabolized in the assimilatory sulphate reduction pathway with high affinity (De Kok and Tausz, 2001).

Chronically enhanced GSH levels have even been suggested as one of the causes of the phytotoxicity of SO_2 (Grill et al., 1979). In this view, it would result in a disruption of cellular metabolic regulation processes. The role of GSH in its interaction with proteins is well known (Kunert and Foyer, 1993). The GSH redox system may also have an important role in environmental sensing and stress signalling within cells, within or between tissues, or even between organs (May et al., 1998). Enhanced GSH levels beyond the normal metabolic control can have serious effects on the antioxidant system which was demonstrated in transgenic plants with constitutively increased GSH concentration in chloroplasts. This did not stimulate antioxidant defence but, quite in contrast, promoted oxidative stress in chloroplasts (Creissen et al., 1999).

4.3.2 Oxidative air pollution

Experimental fumigation with ozone, the major oxidative air pollutant, often increased GSH concentrations in plant foliage. Recent findings on these aspects reviewed by Polle (1998) and De Kok and Tausz (2001). Ozone fumigation is also able to induce the enzymes of the glutathione-ascorbate cycle (Sehmer et al., 1998) which supports a primary role of the antioxidant activity in response to ozone. Changes in the redox state of the GSH/GSSG pools can also be induced by ozone, in particular in sensitive cultivars, and under high ozone concentrations (more than 100 $\mu g\ m^{-3}$). In tobacco, the well-known and extensively studied ozone sensitive cultivar BelW3 showed a more pronounced oxidation of the GSH redox pool than more resistant varieties and the GR activity declined (Sandermann, 1996). Spinach leaves showed a nearly complete oxidation of the GSH pool upon acute ozone stress (Luwe et al., 1993). However, the reports of ozone effects on the glutathione systems are very divergent (Polle, 1998) and depend on many factors. As an example, spruce trees showed no response of the anti-oxidative systems to a long-term fumigation with about double-ambient ozone concentrations without regard of measurable effects on cell divisions (Wonisch et al., 1999).

In field studies, it is often difficult to isolate ozone impact as a single stress factor. In the San Bernardino Mountains (Southern California), the pollutant gradient is reverse to the gradient of natural stresses. *Pinus ponderosa* trees under higher natural stress contained more GSH, which was a confirmation of the altitudinal gradients found in spruce at altitude gradients in the Alps (Polle and Rennenberg, 1992). On the other hand, the reduction state of the glutathione pool was significantly lower at the high ozone plots (Tausz et al., 1999a), indicating the crucial role of GSH regeneration under ozone stress (Luwe, 1996; Tausz et al., 1999b). Exposure of transgenic poplar plants containing constitutively enhanced GSH contents (Foyer et al., 1998) to acute ozone levels revealed that increased pool sizes of GSH did not ameliorate ozone resistance. Ozone resistance did not correlate with GSH content or redox state in this system. Although manipulated elevated GR activity in the chloroplasts prevented the ascorbate pool from being oxidized, this did not lead to increased ozone resistance (Foyer et al., 1995).

On its way towards the plasmamembrane, ozone has to pass through the cell wall. The cell wall contains antioxidative compounds that are able to scavenge ROS, but in contrast to ascorbate, GSH seems to be unimportant in this apoplastic 'first line of defense' (Polle, 1998). Ascorbate is mainly regenerated through the action of the enzymic ascorbate-glutathione cycle, which involves cytoplastic GSH indirectly involved in the apoplastic detoxification capacity of ascorbate. Dehydroascorbate formed in the apoplastic cell wall space has to be reimported to the cytoplasm to be reduced therein (Horemans et al., 1998). There are some indications that the regeneration of apoplastic ascorbate is a time-consuming process and, therefore, an oxidation of the tissue GSH pool may occur with a certain lag phase after ozone episodes (Luwe, 1996) or the redox state of the glutathione pool may change upon long-term exposure (Tausz et al., 1999b). A recent field study under Mediterranean climate conditions showed that needle thiol content is a more accurate marker for long-term ozone exposure of pines than ascorbate, the latter being rather correlated with short-term exposure (Calzada et al., 2001).

The explanation of many divergent results may be found in the multiple roles of GSH in ozone stress response. Even in its direct role as antioxidant in the scavenging ROS, it is only one part of a complex system, and the specific conditions may require adjustments in the glutathione system or rather in other parts of the defence systems. The glutathione redox pool can play a role in environmental sensing and signalling of stress situations (Foyer and Noctor, 2001). In particular, transient changes of the GSH/GSSG redox state may be the signal for the induction of several defense reactions. GSH is also involved in the pathogen defence reactions that have similarities with ozone responses of plant cells (Sandermann, 1996).

5. REGULATION ASPECTS

Since responses of the thiol system to stress seem to be defensive, the question about control mechanisms arises. (1) Regulatory potential of thiols towards other cellular components and (2) control of the changes in the thiol system itself (glutathione sysnthesis and glutathione reduction state) are of interest.

5.1 Control of the Thiol System Over Other Cell Defence Reactions

5.1.1 Induction of other components of the antioxiodant defense systems

In an experiment with detached pine branches, high GSH/GSSG ratios decreased mRNA concentrations of CuZn-SOD, a major antioxidant enzyme in the cytosol. Expression of this enzyme seems to be de-repressed under oxidized conditions i.e. at lower GSH/GSSG ratios. SOD activity and isozyme composition itself remained unchanged indicating additional regulations at the transcription and/or protein level. GR activity was increased by low GSH/GSSG ratios, but the corresponding mRNA remained unaffected. GSH seems to have differential redox control on some enzymes of the antioxidant defense (Wingsle and Karpinski, 1996). Although the idea of a transient change in the GSH/GSSG ratio as a start signal for many cell defense reactions is intriguing (Foyer and Noctor, 2001), the situation is still unclear.

5.1.2 Reinforcement of plant cell wall

The composition of cell wall can be markedly altered by various environmental stimuli. The transcription of genes encoding cell wall hydroxyproline-rich glycoproteins was markedly stimulated by a fungal elicitor and also by exogenous GSH in bean (Wingate et al., 1988) and in maize (Garcia-Muniz et al., 1998) cell suspensions. It was also shown that the treatment of bean or soybean cells with fungal elicitor or GSH led to a rapid insolubilization of preexisting proline and hydroxyproline rich structural proteins in the cell wall. This very rapid insolubilization process should markedly increase the effectiveness of the cell wall as a protective barrier to microbial ingress (Bradley et al., 1992).

5.1.3 Induction of pathogenesis-related (PR) proteins

Pathogenesis-related (PR) proteins are plant proteins induced in pathological or related situations (Walter et al., 1996). Although some (such as β-1,3-glucanases and chitinases) have antimicrobial hydrolytic activities, it is unclear how most PR proteins contribute to disease resistance. In bean plants, the transcription of *Ypr10* genes encoding intracellular ribonuclease-like PR-10 proteins were induced by exogenous GSH (Walter et al., 1996). In addition, the expression of a gene encoding an acidic chitinase PR protein was activated by exogenous GSH in tomato callus culture (Bettini et al., 1998). Salicylic acid, which is an essential signal component in plant disease resistance, can induce the expression of several classes of PR-protein gene families (Hammond-Kosack and Jones, 1996). Exposure of Xanthi-nc tobacco leaf discs to L-2-oxothiazolidine-4-carboxylic acid (OTC), a synthetic cysteine precursor promoting GSH formation in the plant, led to a considerable increase in the level of free salicylic acid and a massive accumulation of its bound form, β-glucosylsalicylic acid (Gullner et al., 2001). These results confirm that GSH has important regulatory roles in the defense reactions of infected plants.

5.1.4 Phytoalexin formation

Phytoalexins are antimicrobial compounds that are synthesized by plants in response to microbial attacks and various abiotic stress effects. Several reports demonstrated the accumulation of phytoalexins in various GSH-treated plants (Table 1). Exogenous GSH was found to induce also the transcription of genes encoding enzymes participating in phytoalexin biosynthesis, such as chalcone synthase (CHS, E.C. 2.3.1.74.) and phenylalanine ammonia lyase (PAL, E.C. 4.3.1.5.) (Dron et al., 1988; Wingate et al., 1988). According to the above results it was supposed that GSH elicits signal transduction processes leading to the induction of defense reactions following infection. However, Edwards et al. (1991) found that treatments of bean or alfalfa cells with OTC did not elicit phytoalexin response in contrast to fungal elicitor treatments, which led to elevated GSH levels. In addition, they observed that changes in endogenous GSH content occurred too slowly to be consistent with a role in the initiation of the elicitation response. Interestingly, the dramatic reduction of GSH levels achieved in carrot cell cultures by L-buthionine-[S,R]-sulphoximine led to markedly increased levels of both H_2O_2 and the phytoalexin 6-methoxymellein (Guo et al., 1993). The above results showed that in spite of numerous studies, the role of GSH in the elicitation of phytoalexin production has remained elusive. It seems that the disturbance of the cellular redox balance may lead to phytoalexin accumulation.

5.1.5 Cell division

Sánchez-Fernández et al. (1997) found high GSH concentrations in the meristematic tissue of growing *Arabidopsis* root tips and regarded them beneficial for dividing cells. Higher resolution epifluorescence techniques revealed that cell nuclei contain high GSH concentrations (Müller et al., 2002). However, feeding of exogenous GSH to growing roots induced chromosomal damages in the dividing cells, possibly through disruption of thiol equilibrium (Zellnig et al., 2000). These results indicate a decisive role of thiols not only in gene expression but also in cell division control of plants.

Table 1. Accumulation of phytoalexins in plant tissues treated with exogenous glutathione

Phytoalexin	Plant tissue	GSH concentrations applied	References
coumestrol	Soybean cotyledon	0.01-3 mM	Graham and Graham, 1996
genistein	Soybean cotyledon	0.03-3 mM	Graham and Graham, 1996
maackiain	Chickpea root	2-10 mM	Armero et al., 2001
medicarpin	Chickpea root	2-10 mM	Armero et al., 2001
pisatin	pea epicotyl	50 mM	Yamada et al., 1989
sativan	*Lotus corniculatus*	2-10 mM	Robbins et al., 1991
vestitol	*Lotus corniculatus*	2-10 mM	Robbins et al., 1991

5.1.6 Regulation of ion channels

GSH has been shown to modulate certain ion channels in tonoplasts of higher plants, such as sugar beet and *Posidonia oceanica* (Carpaneto et al., 1999). This ability may be involved in concerted stress responses requiring turgor controlled mechanisms like stomatal responses.

5.2 Regulation of the Thiol System under Stress

GSH concentrations and GSSG regeneration are apparently under redox control, possibly by the GSH/GSSG ratio. Exogenous hydrogen peroxide is able to increase GSH production without increasing transcripts of the enzymes involved. Specifically, γ-glutamyl-cysteine (γ-EC) synthase protein translation is repressed by high GSH/GSSG ratios. Since γ-EC synthetase is the key enzyme for GSH synthesis, this means a direct control of GSH production (Xiang and Bertrand, 2000). Sulphur reduction as a pre-requisite for thiol production, can also be oxidatively induced. In maize roots, expression of APS-reductase and ATP-sulphurylase (but not cysteine-synthetase) increased upon decreased GSH/GSSG ratio. Corresponding enzyme activities also increased, whereas most other chloroplastic enzyme activities are reductively induced rather than oxidatively (Flückinger et al, 2000). At least some isozymes of GR are induced by low GSH/GSSG ratios (Mullineaux and Creissen, 1997). Such regulations guarantee high GSH concentrations upon stress and enable a high GSH/GSSG state.

6. MANIPULATION OF THIOLS AND STRESS RESISTANCE

Several lines of evidence have proved the correlation between artificially elevated endogenous GSH levels and increased disease resistance. Treatments of tomato and melon plants with dinitroaniline herbicides led to considerably increased endogenous GSH levels in the roots and concomitantly to induced resistance against the fungal pathogens *Fusarium oxysporum* f.sp. *lycopersici* and f.sp. *melonis* (Bolter et al., 1993). The herbicide safener flurazol, which elevated the cellular GSH levels in *Sorghum* plants, significantly decreased the incidence of downy mildew infections (Kömives and Dutka, 1989). The recently developed synthetic resistance inducer bendicar, which can provide long-lasting protection against powdery mildew and other pathogens, also considerably increased the GSH levels in soybean cells (Knörzer et al., 1999). The elevation of endogenous GSH levels in Xanthi-nc tobacco leaf discs by the synthetic cysteine precursor OTC resulted in increased resistance against TMV. The OTC treatment markedly suppressed the development of necrotic disease symptoms and decreased the foliar virus concentration (Gullner et al., 1999).

Chemical manipulation of the GSH levels in chilling sensitive maize corroborated the involvement of GSH in chilling tolerance. For example, when herbicide safeners were used to increase GSH levels, the shoots were more chilling resistant. Application of buthionine sulphoximine (BSO), a specific inhibitor of γ-EC synthetase (and, hence, GSH synthesis) resulted in more sensitive shoot (Kocsy et al., 2001).

Transgenic plants with manipulated GSH levels (through overexpression of γ-EC-synthetase, a key enzyme of GSH synthesis (Foyer et al., 1998; Creissen et al., 1999) or increased GR activities (Foyer et al., 1995) are now available. Experiments with these plants and natural stresses or air pollution (ozone) are somewhat contradictory until now. They may or may not be more resistant (Strohm et al., 1995; Foyer et al., 1998) or even more sensitive to oxidative stress (Creissen et al., 1999) when whole plant systems are observed.

7. REFERENCES

Anderson J, Chevone B and Hess J (1992) Seasonal variation in the antioxidant system of eastern white pine needles. *Plant Physiol* **98**: 501—508

Anderson JW and McMahon PJ (2001) The role of glutathione in the uptake and metabolism of sulfur and selenium. In: Grill D, Tausz M and DeKok LJ (eds) Significance of Glutathione in Plant Adaptation to the Environment, Kluwer Handbook Series of Plant Ecophysiology, pp 57—100. Kluwer Academic Publishers, Dordrecht

Armero J, Requejo R, Jorrin J, López-Valbuena R and Tena M (2001) Release of phytoalexins and related isoflavonoids from intact chickpea seedlings elicited with reduced glutathione at root level. *Plant Physiol Biochem* **39**: 785—795

Asada K (1999) The water-water cycle in chloroplasts: scavenging of active oxygens and dissipation of excess photons. *Ann Rev Plant Physiol Plant Mol Biol* **50**: 601—640

Avsiankretchmer O, Eshdat Y, Guetadahan Y and Benhayyim G (1999) Regulation of stress-induced phosholipid hydroperoxide glutathione-peroxidase expression in *Citrus*. *Planta* **209**: 469—477

Barroso C, Romero LC, Cejudo FJ, Vega JM and Gotor C (1999) Salt-specific regulation of the cytosolic O-acetylserine(thiol)lyase gene from *Arabidopsis thaliana* is dependent on abscisic acid. *Plant Mol Biol* **40**: 729—736

Bartling D, Radzio R, Steiner U and Weiler EW (1993) A glutathione S-transferase with glutathione-peroxidase activity from *Arabidopsis thaliana*. Molecular cloning and functional characterization. *Euro J Biochem* **216**: 579—586

Bettini P, Cosi E, Pellegrini MG, Turbanti L, Vendramin GG and Buiatti M (1998) Modification of competence for in vitro response to *Fusarium oxysporum* in tomato cells. III. PR-protein gene expression and ethylene evolution in tomato cell lines transgenic for phytohormone-related bacterial genes. *Theo Appl Genet* **97**: 515—583

Blokhina OB, Virolainen E, Fagerstedt KV, Hoikkala A, Wähälä K and Chirkova T (2000) Antioxidant status of anoxia-tolerant and -intolerant plant species under anoxia and reaeration. *Physiol Plant* **109**: 396—403

Blokhina OB, Chirkova TV and Fagerstedt KV (2001) Anoxic stress leads to hydrogen peroxide formation in plant cells. *J Exp Bot* **52**: 1179—1190

Bolter C, Brammall RA, Cohen R and Lazarovits G (1993) Glutathione alterations in melon and tomato roots following treatment with chemicals which induce disease resistance to *Fusarium* wilt. *Physiol Mol Plant Pathol* **42**: 321—336

Boo YC and Jung J (1999) Water deficit-induced oxidative stress and antioxidative defenses in rice plants. *J Plant Physiol* **155**: 255—261

Bradley DJ, Kjellbom P and Lamb CJ (1992) Elicitor- and wound-induced oxidative cross-linking of a proline-rich plant cell wall protein: a novel, rapid defense response. *Cell* **70**: 21—30

Broin M, Cuiné S, Peltier G and Rey P (2000) Involvement of CDSP 32, a drought-induced thioredoxin, in the response to oxidative stress in potato plants. *FEBS letters* **467**: 245—248

Buwalda F, Stulen I, De Kok LJ and Kuiper PJC (1990) Cysteine, γ-glutamyl-cysteine and glutathione contents of spinach leaves as affected by darkness and application of excess sulfur. II. Glutathione accumulation in detached leaves exposed to H_2S in the absence of light is stimulated by the supply of glycine to the petiole. *Physiol Plant* **80**: 196—204

Calzada I, Mesanza J, Casado H and Castillo FJ (2001) Biochemical changes in needles of *Pinus radiata* D.Don trees in relationship to long-term ozone exposure indices. *Environ Poll* **114**: 325—335

Carpaneto A, Cantué AM and Gambale F (1999) Redox agents regulate ion channel activity in vacuoles from higher plant cells. *FEBS Letters* **442**: 129—132

Creissen G, Firmin J, Fryer M, Kular B, Leyland N, Reynolds H, Pastori G, Wellburn F, Baker N, Wellburn A and Mullineaux P (1999) Elevated glutathione biosynthetic capacity in the chloroplasts of transgenic tobacco plants paradoxically causes increased oxidative stress. *Plant Cell* **11**: 1277—1291

De Kok LJ (1990) Sulfur metabolism in plants exposed to atmospheric sulfur. In: Rennenberg H, Brunold C, De Kok LJ and Stulen I (eds) Sulfur Nutrition and Sulfur Assimilation in Higher Plants: Fundamental, Environmental and Agricultural Aspects, pp 111—130. SPB Academic Publishing, The Hague

De Kok LJ, Stuiver CEE and Stulen I (1998) Impact of atmospheric H_2S on plants. In: De Kok LJ and Stulen I (eds) Responses of Plant Metabolism to Air Pollution and Global Change, pp 51—63. Backhuys Publishers, Leiden

DeKok LJ and Tausz M (2001) The role of glutathione in plant reaction and adaptation to air pollutants. In: Grill D, Tausz M and DeKok LJ (eds) Significance of glutathione in plant adaptation to the environment, Kluwer Handbook Series of Plant Ecophysiology, pp 185—208. Kluwer Academic Publishers, Dordrecht

Dhindsa RS (1987) Glutathione status and protein synthesis during drought and subsequent rehydration in *Tortula ruralis*. *Plant Physiol* **83**: 816—819

Dhindsa RS (1991) Drought stress, enzymes of glutathione metabolism, oxygen injury, and protein synthesis in *Tortula ruralis*. *Plant Physiol* **95**: 648—651

Doulis AG, Hausladen A, Mondy B, Alscher RG, Chevone BI, Hess JL and Weiser RL (1993) Antioxidant response and winter hardiness in red spruce (*Picea rubens* Sarg.). *New Phytol* **123**: 365—374

Dron M, Clouse SD, Dixon RA, Lawton MA and Lamb CJ (1988) Glutathione and fungal elicitor regulation of a plant defense gene promoter in electroporated protoplasts. *Proc Natl Acad Sci USA* **85**: 6738—6742

Dudler R, Hertig C, Rebmann G, Bull J and Mauch F (1991) A pathogen-induced wheat gene encodes a protein homologous to glutathione S-transferases. *Mol Plant-Microbe Interact* **4**: 14—18

Durner J and Klessig DF (1999) Nitric oxide as a signal in plants. *Curr Opi Plant Biol* **2**: 369—374

Edwards R, Blount JW and Dixon RA (1991) Glutathione and elicitation of the phytoalexin response in legume cell cultures. *Planta* **184:** 403—409

Edwards EA, Enard C, Creissen G and Mullineaux PM (1994) Synthesis and properties of glutathione reductase in stressed peas. *Planta* **192:** 137—143

Elstner EF and Oßwald W (1994) Mechanisms of oxygen activation during plant stress. *Proc Roy Soc Edinburgh* **102B:** 131—154

El-Zahaby HM, Gullner G and Király Z (1995) Effects of powdery mildew infection of barley on the ascorbate-glutathione cycle and other antioxidants in different host-pathogen interactions. *Phytopathol* **85:** 1225—1230

Eshdat Y, Holland D, Faltin Z and Ben-Hayyim G (1997) Plant glutathione peroxidases. *Physiol Plant* **100:** 234—240

Esterbauer H and Grill D (1978) Seasonal variation of glutathione and glutathione reductase in needles of *Picea abies*. *Plant Physiol* **61:** 119—121

Flückinger H, Jecklin S, Ballmoos PV, Suter M and Brunold C (2000) Effect of hydrogen peroxide induced oxidative stress in maize roots. In: Brunold C, Rennenberg H, De Kok LJ, Stulen I and Davidian JC (eds) Sulfur Nutrition and Sulfur Assimilation in Higher Plants, pp 401—403. Paul Haupt, Berne

Fodor J, Gullner G, Ádám AL, Barna B, Kömives T and Király Z (1997) Local and systemic responses of antioxidants to tobacco mosaic virus infection and to salicylic acid in tobacco. Role in systemic acquired resistance. *Plant Physiol* **114:** 1443—1451

Foyer C, Souriau N, Perret S, Lelandais M, Kunert KJ, Pruvost C and Jouanin L (1995) Over-expression of glutathione reductase but not glutathione synthetase leads to increases in antioxidant capacity and resistance to photoinhibition in poplar (*Populus tremula x P. alba*) trees. *Plant Physiol* **109:** 1047—1057

Foyer C, Kingston-Smith AH, Harbinson J, Arisi ACM, Jouanin L and Noctor G (1998) The use of transformed plants in the assessment of physiological stress responses. In: De Kok LJ and Stulen I (eds) Responses of Plant Metabolism to Air Pollution and Global Change, pp 251—261. Backhuys Publishers, Leiden

Foyer C and Noctor G (2000) Oxygen processing in photosynthesis: a molecular approach. *New Phytol* **146:** 359—388

Foyer C and Rennenberg H (2000) Regulation of glutathione synthesis and its role in abiotic and biotic stress defence. In: Brunold C, Rennenberg H, De Kok LJ, Stulen I and Davidian JC (eds) Sulfur Nutrition and Sulfur Assimilation in Higher Plants, pp 127—153. Paul Haupt, Bern

Foyer C and Noctor G (2001) The molecular biology and metabolism of glutathione. In: Grill D, Tausz M and De Kok LJ (eds) Significance of Glutathione in Plant Adaptation to the Environment, pp 27—56. Kluwer Academic Publishers, Dordrecht

Fryer MJ (1992) The antioxidant effects of thylakoid Vitamin E (α-Tocopherol). *Plant Cell Environ* **15:** 381—392

Garcia-Muniz N, Martinez-Izquierdo JA and Puigdomenech P (1998) Induction of mRNA accumulation corresponding to a gene encoding a cell wall hydroxyproline-rich glycoprotein by fungal elicitors. *Plant Mol Biol* **38:** 623—632

García-Plazaola JI, Artetxe U and Becerril JM (1999a) Diurnal changes in antioxidant and carotenoid composition in the mediterranean schlerophyll tree *Quercus ilex* during winter. *Plant Sci* **143:** 125—133

García-Plazaola JI, Artetxe U, Dunabeitia MK and Becerril JM (1999b) Role of photoprotec-tive systems of holm-oak (*Quercus ilex*) in the adaptation to winter conditions. *J Plant Physiol* **155:** 625—630

Ghisi R, Dittrich APM and Heber U (1990) Oxidation versus reductive detoxification of SO_2 by chloroplasts. *Plant Physiol* **92:** 846—849

Gillham D and Dodge A (1987) Chloroplast superoxide and hydrogen peroxide scavenging systems from pea leaves: seasonal variations. *Plant Sci* **50:** 105—109

Gönner MV and Schlösser E (1993) Oxidative stress in interactions between *Avena sativa* L. and *Drechslera* spp. *Physiol Mol Plant Pathol* **42:** 221—234

Graham TL and Graham MY (1996) Signaling in soybean phenylpropanoid responses. Dissection of primary, secondary, and conditioning effects of light, wounding and elicitor treatments. *Plant Physiol* **110:** 1123—1133

Grill D, Esterbauer H and Klösch U (1979) Effect of sulfur dioxide on glutathione in leaves of plants. *Environ Pollut* **17:** 187—194

Grill D, Pfeifhofer H and Esterbauer H (1987) Further investigations on the thiol content of Norway spruce needles. *Phyton Ann Rei Bot* **27:** 311—317

Grill D, Tausz M and De Kok LJ (eds) (2001) Significance of Glutathione in Plant Adaptation to the Environment. Kluwer Academic Publishers, Dordrecht

Gullner G, Kömives T and Gáborjányi R (1995) Differential alterations of glutathione S-transferase enzyme activities in three sorghum varieties following viral infection. *Z Naturforschung C* **50:** 459—460

Gullner G, Tóbiás I, Fodor J and Kömives T (1999) Elevation of glutathione level and activation of glutathione-related enzymes affect virus infection in tobacco. *Free Rad Res* **31:** S155—S161

Gullner G and Kömives T (2001) The role of glutathione and glutathione-related enzymes in plant-pathogen interactions. In: Grill D, Tausz M and De Kok LJ (eds) Significance of Glutathione in Plant Adaptation to the Environment, pp 207—239. Kluwer Academic Publishers, Dordrecht

Gullner G, McKenna A and Enyedi AJ (2001) Elevation of endogenous glutathione level leads to salicylic acid accumulation in tobacco leaf discs. 5th Conference on Oxygen, Free Radicals and Oxidative Stress in Plants, Nice, France. Abstract volume p L-33

Guo ZJ, Nakagawara S, Sumitani K and Ohta Y (1993) Effect of intracellular glutathione level on the production of 6-methoxymellein in cultured carrot (*Daucus carota*) cells. *Plant Physiol* **102:** 45—51

Hahn K and Strittmatter G (1994) Pathogen-defence gene *prp1-1* from potato encodes an auxin-responsive glutathione S-transferase. *Eur J Biochem* **226:** 619—626

Hammond-Kosack KE and Jones JDG (1996) Resistance gene-dependent plant defense responses. *Plant Cell* **8:** 1773—1791

Hell R (1997) Molecular physiology of plant sulphur metabolism. *Planta* **202:** 138—148

Herbinger K, Tausz M and Grill D (1999) Are winter adaptive changes in glutathione and chlorophyll contents in spruce needles triggered by low temperatures or short days? *Phyton Ann Rei Bot* **39:** 171—175

Herschbach C, De Kok LJ and Rennenberg H (1995) Net uptake of sulfate and its transport to the shoot in spinach plants fumigated with H_2S or SO_2: Does atmospheric sulfur affect the 'inter-organ' regulation of sulfur nutrition? *Bot Acta* **108:** 41—46

Hodges MD, Andrews CJ, Johnson DA and Hamilton RI (1996) Antioxidant compound responses to chilling stress in differentially sensitive inbred maize lines. *Physiol Plant* **98:** 685—692

Horemans N, Asard H and Caubergs RJ (1998) Carrier-mediated uptake of dehydroascorbate into higher plant plasma membrane vesicles shows trans-stimulation. *FEBS Lett* **421:** 41—44

Kaminaka H, Morita S, Nakajima M, Masumura T and Tanaka K (1998) Gene cloning and expression of cytosolic glutathione-reductase in rice (*Oryza sativa* L.). *Plant Cell Physiol* **39:** 1269—1280

Knörzer OC, Lederer B, Durner J and Böger P (1999) Antioxidative defense activation in soybean cells. *Physiol Plant* **107:** 294—302

Kocsy G, Galiba G and Brunold C (2001) Role of glutathione in adaptation and signalling during chilling and cold acclimation in plants. *Physiol Plant* **113:** 158—164

Kömives T and Dutka F (1989) Effects of herbicide safeners on levels and activities of cytochrome P-450 and other enzymes in corn. In: Hatzios KK and Hoagland RE (eds) Crop Safeners for Herbicides, pp 129—145. Academic Press, San Diego

Kranner I and Grill D (1997) Desiccation and the subsequent recovery of cryptogamics that are resistant to drought. *Phyton Ann Rei Bot* **37:** 139—150

Kunert KJ and Foyer C (1993) Thiol/disulfide exchange in plants. In: De Kok LJ, Stulen I, Rennenberg H, Brunold C and Rauser WE (eds) Sulfur Nutrition and Assimilation in Higher Plants; Regulatory, Agricultural and Environmental Aspects, pp 139—152. SPB Academic Publishing, The Hague

Kuroda H and Sagisaka S (1998) Metabolic and enzymatic responses associated with oxidative stress in plants acclimatized to cold environments. *Recent Res Dev Agric Biol Chem* **2:** 395—410

Kuzniak E and Sklodowska M (2001) Ascorbate, glutathione and related enzymes in chloroplasts of tomato leaves infected by *Botrytis cinerea*. *Plant Sci* **160:** 723—731

Laakso K, Kinnunen H and Huttunen S (2001) The glutathione status of mature Scots pines during the third season of UV-B radiation exposure. *Environ Poll* **111:** 349—354

Lamb C and Dixon RA (1997) The oxidative burst in plant disease resistance. *Annu Rev Plant Physiol Plant Mol Biol* **48:** 251—275

Leipner J, Fracheboud Y and Stamp P (1999) Effect of growing-season on the photosynthetic apparatus and leaf antioxidative defenses in 2 maize genotypes of different chilling tolerance. *Environ Exp Bot* **42:** 129—139

Loggini B, Scartazza A, Brugnoli E and Navari-Izzo F (1999) Antioxidative defense system, pigment composition, and photosynthetic efficiency in two wheat cultivars subjected to drought. *Plant Physiol* **119:** 1091—1099

Luwe M (1996) Antioxidants in the apoplast and symplast of beech (*Fagus sylvatica* L.) leaves: seasonal variations and responses to changing ozone concentrations in air. *Plant Cell Environ* **19:** 321—328

Luwe M, Takahama U and Heber U (1993) Role of ascorbate in detoxifying ozone in the apoplast of spinach (*Spinacia oleracea*) leaves. *Plant Physiol* **101:** 969—976

Masi A, Ferretti M, Passera C and Ghisi R (1995) Antioxidative responses in *Zea mays* seedlings grown under UV-B radiation. *Plant Physiol Biochem* (Special Issue): 264—265

Masi A, Ghisi R and Ferretti M (2002) Measuring low-molecular-weight thiols by detecting the fluorescence of their SBD-derivatives: application to studies of diurnal and UV-B induced changes in *Zea mays* L. J *Plant Physiol* **159:** 499—507

Mattagajasingh SN and Kar M (1989) Changes in the antioxidant system during greening of etiolated wheat leaves. *J Plant Physiol* **134:** 656—660

Mauch F and Dudler R (1993) Differential induction of distinct glutathione-S-transferases of wheat by xenobiotics and by pathogen attack. *Plant Physiol* **102:** 1193—1201

May MJ, Vernoux T, Leaver C, van Montagu M and Inzé D (1998) Glutathione homeostasis in plants: implications for environmental sensing and plant development. *J Exp Bot* **49:** 649—667

May MJ, Hammond-Kosack KE and Jones JDG (1996a) Involvement of reactive oxygen species, glutathione metabolism, and lipid peroxidation in the *Cf*-gene-dependent defense response of tomato cotyledons induced by race-specific elicitors of *Cladosporium fulvum*. *Plant Physiol* **110:** 1367—1379

May MJ, Parker JE, Daniels MJ, Leaver CJ and Cobbett CS (1996b) An *Arabidopsis* mutant depleted in glutathione shows unaltered responses to fungal and bacterial pathogens. *Mol Plant-Microbe Interact* **9:** 349—356

McKersie BD and Leshem YY (1994) Stress and stress coping in cultivated plants. Kluwer Academic Publishers, Dordrecht

Mishra NP, Fatma T and Singhal GS (1995) Development of antioxidantive defense system of wheat seedlings in response to high light. *Physiol Plant* **95:** 77—82

Mittal S and Davis KR (1995) Role of the phytotoxin coronatine in the infection of *Arabidopsis thaliana* by *Pseudomonas syringae* pv. *tomato*. *Mol Plant-Microbe Interact* **8:** 165—171

Muckenschnabel I, Williamson B, Goodman BA, Lyon GD, Stewart D and Deighton N (2001) Markers for oxidative stress associated with soft rots in French beans (*Phaseolus vulgaris*) infected by *Botrytis cinerea*. *Planta* **212:** 376—381

Müller M, Weidner W, De Kok LJ and Tausz M (2002) Differential effects of H_2S on cytoplasmic and nuclear thiol concentrations in different tissues of *Brassica* roots. *Plant Physiol Biochem* (in press)

Mullineaux PM and Creissen GP (1997) Glutathione reductase: regulation and role in oxidative stress. In: Scandalios JG (ed) Oxidative Stress and the Molecular Biology of Antioxidant Defenses, pp 667—713. Cold Spring Harbor Laboratory Press, New York

Mullineaux PM, Karpinski S, Jiménez A, Cleary SP, Robinson C and Creissen GP (1998) Identification of cDNAs encoding plastid-targeted glutathione peroxidase. *Plant J* **13:** 375—379

Navari-Izzo F, Meneguzzo S, Loggini B, Vazzana C and Sgherri CLM (1997) The role of the glutathione system during dehydration of *Boea hygroscopica*. *Physiol Plant* **99:** 23—30

Nieto-Sotelo J and Tuan-Hua DH (1986) Effect of heat shock on the metabolism of glutathione in maize roots. *Plant Physiol* **82:** 1031—1035

Noctor G and Foyer C (1998) Ascorbate and glutathione: keeping active oxygen under control. *Annu Rev Plant Physiol Plant Mol Biol* **49:** 249—279

Noctor G, Arisi ACM, Jouanin L and Foyer C (1999) Photorespiratory glycine enhances glutathione accumulation in both the chloroplastic and cytosolic compartments. *J Exp Bot* **50:** 1157—1167

Packer L (ed) (1995) Biothiols. Part A Monothiols and Dithiols, Protein Thiols, and Thiyl Radicals. Methods in Enzymology Vol 251. Academic Press, San Diego

Polle A and Rennenberg H (1992) Field studies on Norway spruce trees at high altitudes: II. Defence systems against oxidative stress in needles. *New Phytol* **121:** 635—642

Polle A (1997) Defense against photoxidative damage in plants. In: Scandalios JG (ed) Oxidative Stress and the Molecular Biology of Antioxidant Defenses, pp 623—666. Cold Spring Harbor Laboratory Press, NY

Polle A (1998) Photochemical oxidants: uptake and detoxification mechanisms. In: De Kok LJ and Stulen I (eds), Responses of Plant Metabolism to Air Pollution and Global Change, pp 95—116. Backhuys Publishers, Leiden

Rairdan GJ, Donofrio NM and Delaney TP (2001) Salicylic acid and NIM1/NPR1-independent gene induction by incompatible *Peronospora parasitica* in Arabidopsis. *Mol Plant-Microbe Interact* **14:** 1235—1246

Rao MV, Paliyath G and Ormrod DP (1996) Ultraviolet-B and ozone-induced biochemical changes in antioxidant enzymes of *Arabidopsis thaliana*. *Plant Physiol* **110:** 125—136

Rauser WE (2001) The role of glutathione in plant reaction and adaptation to excess metals. In: Grill D, Tausz M and DeKok LJ (eds) Significance of Glutathione in Plant Adaptation to the Environment, Kluwer Handbook Series of Plant Ecophysiology, pp 123—154. Kluwer Academic Publishers, Dordrecht

Rennenberg H and Brunold C (1994) Significance of glutathione metabolism in plants under stress. *Prog Bot* **55:** 142—156

Robbins MP, Hartnoll J and Morris P (1991) Phenylpropanoid defence responses in transgenic *Lotus corniculatus*. 1. Glutathione elicitation of isoflavan phytoalexins in transformed root cultures. *Plant Cell Rep* **10:** 59—62

Rouhier N, Gelhaye E and Jacquot J-P (2002) Exploring the active site of plant glutaredoxin by site-directed mutagenesis. *FEBS letters* **511:** 145—149

Sagisaka S, Matsuda Y, Okuda T and Ozeki S (1991) Comparative studies of the changes in enzymatic activities in hardy and less hardy cultivars of winter wheat in late fall and in winter under snow. *Soil Sci Plant Nutr* **37:** 543—550

Sánchez-Fernández R, Fricker M, Corben LB, White NS, Sheard N, Leaver CJ, Van Montagu M, Inzé D and May MJ (1997) Cell proliferation and hair tip growth in the *Arabidopsis* root are under mechanistically different forms of redox control. *Proc Natl Acad Sci USA* **94:** 2745—2750

Sandermann H (1996) Ozone and plant health. *Annu Rev Phytopath* **34:** 347—366

Schröder P (2001) The role of glutathione and glutathione-S-transferases in plant reaction and adaptation to xenobiotics. In: Grill D, Tausz M and DeKok LJ (eds) Significance of Glutathione in Plant Adaptation to the Environment, Kluwer Handbook Series of Plant Ecophysiology, pp 155—184. Kluwer Academic Publishers, Dordrech

Schupp R and Rennenberg H (1988) Diurnal changes in the glutathione content of spruce needles (*Picea abies* L.). *Plant Sci* **57:** 113—117

Schupp R, Schatten T, Willenbrink J and Rennenberg H (1992) Long distance transport of reduced sulfur in spruce (*Picea abies* L.). *J Exp Bot* **43:** 1243—1250

Schürmann P and Jacquot J-P (2000) Plant thioredoxin systems revisited. *Annu Rev Plant Physiol Plant Mol Biol* **51:** 371—400

Sehmer L, Fontaine V, Antoni F and Dizengremel P (1998) Effects of ozone and elevated atmospheric carbon dioxide on carbohydrate metabolism of spruce needles - catabolic and detoxification pathways. *Physiol Plant* **102:** 605—611

Semanza GL (1999) Perspectives on oxygen sensing. *Cell* **98:** 281—284

Smirnoff N (1993) The role of active oxygen in the response of plants to water deficit and desiccation. *New Phytol* **125:** 27—58

Strohm M, Jouanin L, Kunert K-J, Pruvost C, Polle A, Foyer C and Rennenberg H (1995) Regulation of glutathione synthesis in leaves of transgenic poplar (*Populus tremula x P. alba*) over-expressing glutathione synthetase. *Plant J* **7:** 141—145

Tausz M, Bytnerowicz A, Arbaugh MJ, Weidner W and Grill D (1999a) Antioxidants and protective pigments of *Pinus ponderosa* needles at gradients of natural stresses and ozone in the San Bernardino Mountains in California. *Free Rad Res* **31:** S113—120

Tausz M, Bytnerowicz A, Weidner W, Arbaugh MJ, Padgett P and Grill D (1999b) Changes in free radical scavengers describe the susceptibility of *Pinus ponderosa* to ozone in Southern Californian forests. *Water Air Soil Poll* **116:** 249—254

Tausz M, Wonisch A, Peters J, Jiménez MS, Morales D and Grill D (2001) Short-term changes in free-radical scavengers and chloroplast pigments in *Pinus canariensis* needles as affected by mild drought stress. *J Plant Physiol* **158:** 213—219

Tegischer K, Tausz M, Wieser G and Grill D (2002) Tree-age and needle-age dependent variations of antioxidants and photoprotective pigments in spruce needles at the alpine timberline. *Tree Physiol* **22:** 591—596

Vanacker H, Carver TLW and Foyer CH (1998a) Pathogen-induced changes in the antioxidant status of the apoplast in barley leaves. *Plant Physiol* **117:** 1103—1114

Vanacker H, Foyer CH and Carver TLW (1998b) Changes in apoplastic antioxidants induced by powdery mildew attack in oat genotypes with race non-specific resistance. *Planta* **208:** 444—452

Vanacker H, Carver TLW and Foyer CH (2000) Early H_2O_2 accumulation in mesophyll cells leads to induction of glutathione during the hypersensitive response in the barley-powdery mildew interaction. *Plant Physiol* **123:** 1289—1300

Vener AV, Ohad I and Andersson B (1998) Protein phosphorylation and redox sensing in chloroplast thylakoids. *Curr Opi Plant Biol* **1:** 217—223

Venisse JS, Gullner G and Brisset MN (2001) Evidence for the involvement of an oxidative stress in the initiation of infection of pear by *Erwinia amylovora*. *Plant Physiol* **125:** 2164—2172

Vollenweider S, Weber H, Stolz S, Chételat A and Farmer EE (2000) Fatty acid ketodienes and fatty acid ketotrienes: Michael addition acceptors that accumulate in wounded and diseased *Arabidopsis* leaves. *Plant J* **24:** 467—476

Walter MH, Liu JW, Wünn J and Hess D (1996) Bean ribonuclease-like pathogenesis-related protein genes (*Ypr10*) display complex patterns of developmental, dark-induced and exogenous-stimulus-dependent expression. *Eur J Biochem* **239:** 281—293

Westerman S, De Kok LJ and Stulen I (2000) Interaction between metabolism of atmospheric H_2S in the shoot and sulfate uptake by the roots of curly kale (*Brassica oleracea* L.). *Physiol Plant* **109:** 443—449

Willenkens HW, Van Camp W, Van Montagu M, Inzé D, Langebartels C and Sandermann H (1994) Ozone, sulfur dioxide, ultraviolet B have similar effects on mRNA accumulation of antioxidant genes in *Nicotiana plumbaginifolia* L. *Plant Physiol* **106:** 1007—1014

Williams JS, Hall SA, Hawkesford MJ, Beale MH and Cooper RM (2002) Elemental sulfur and thiol accumulation in tomato and defense against a fungal vascular pathogen. *Plant Physiol* **128:** 150—159

Wingate VPM, Lawton MA and Lamb CJ (1988) Glutathione causes a massive and selective induction of plant defense genes. *Plant Physiol* **87:** 206—210

Wingsle G and Karpinski S (1996) Differential redox regulation by glutathione of glutathione reductase and CuZn-superoxide dismutase gene expression in *Pinus sylvestris* L. needles. *Planta* **198:** 151—157

Wonisch A, Müller M, Tausz M, Soja G and Grill D (1999) Simultaneous analyses of chromosomes in root meristems and of the biochemical status of needle tissues of three different clones of Norway spruce trees challenged with moderate ozone levels. *Eur J For Path* **29:** 281—294

Xiang C and Bertrand D (2000) Glutathione synthesis in *Arabidopsis*: multilevel controls coordi-nate responses to stress. In: Brunold C, Rennenberg H, De Kok LJ, Stulen I and Davidian JC (eds.) Sulfur Nutrition and Sulfur Assimilation in Higher Plants, pp 409—412. Paul Haupt, Berne

Yamada T, Hashimoto H, Shiraishi T and Oku H (1989) Suppression of pisatin, phenylalanine ammonia-lyase mRNA and chalcone synthase mRNA accumulation by a putative pathogenicity factor from the fungus *Mycosphaerella pinodes*. *Mol Plant-Microbe Interact* **2:** 256—261

Zellnig G, Tausz M, Pešec B, Grill D and Müller M (2000) Effects of glutathione on thiol redox systems, chromosomal aberrations, and the ultrastructure of meristematic root cells of *Picea abies* (L.) Karst. *Protoplasma* **212:** 227—235

8. ACKNOWLEDGEMENTS

Work in the laboratory of M. Tausz and D. Grill was supported by the Austrian Science Foundation (FWF) and the COST Action 829. Work of G. Gullner and T. Kömives was supported by grants from the Hungarian Scientific Research Fund (OTKA T 032108 and T 029121). Our cooperation was initiated by an Austrian-Hungarian bilateral project (A23/2000).

9. AFFILIATIONS

Institute of Plant Physiology, University of Graz, Schubertstraße 51, 8010 Graz, Austria (MT, DG); Plant Protection Institute, Hungarian Academy of Sciences, 1525 Budapest, POB 102, Hungary (GG, TK)
For correspondence - *M Tausz*: michael.tausz@uni-graz.at

A. SIRKO, A. BŁASZCZYK, F. LISZEWSKA,
K. KAZIMIERCZUK AND D. GAGANIDZE

GENETIC ENGINEERING OF OXIDATIVE STRESS RESISTANCE IN PLANTS

Abstract. In this chapter, a brief description of oxidative stress inducing agents in plants and antioxidant systems of plants to defend themselves against environmental challenges is followed by an extensive review of genetic engineering strategies used to modify stress tolerance. Our main focus is on changes in the content and redox state of glutathione, a major low-molecular-weight thiol-containing compound and a scavenger of reactive oxygen species. A number of investigators have been successful in showing elevation of the glutathione level in transgenic plants and enhancement of resistance to various environmental stresses. In some cases, however, despite increased levels of glutathione, no alteration in resistance was observed. In fact, there was even high sensitivity of transgenic plants to the stresses. In addition, other approaches and targets, with no obvious link to glutathione and sulphur metabolism, are described in the attempt to demonstrate the complexity of stress adaptation and possibilities of multiple tactics to engineer plants with improved oxidative stress tolerance. Changing in abundance of enzymes directly involved in detoxification as well as modifications of signalling and regulatory cascades are among the reported approaches.

1. DEFINITION OF OXIDATIVE STRESS

Various environmental stresses limit the productivity of field-grown plants. Plant pathogens such as viruses, bacteria, fungi and insects are main cause of biotic stresses. Abiotic stresses include drought, salinity, heat, cold, excess light, flooding, xenobiotics, air pollutants, heavy metals, mineral deficiency and mineral toxicity (Bartosz, 1997; Dat, 2000). Since plants are immobile, their ability to adapt to the environment is crucial. They have developed the mechanisms to sense and respond to stresses. Reactive oxygen species (ROS), called also active oxygen species (AOS) or reactive oxygen intermediates (ROI), are a significant part of the machinery that triggers defense responses (Neill et al., 2002; Vranova et al., 2002).

ROS are normal products of cell metabolism and are produced in various compartments of unstressed cells. There are several known sources of ROS in plants. The most common include the leakage of electrons to O_2 from electron transport chains in the chloroplasts and mitochondria, photorespiration and xanthin oxidase in the peroxisomes, and oxidases and peroxidases in the cell wall. Main ROS formed in living organisms are: superoxide radical ion ($O_2^{\cdot-}$), hydrogen peroxide (H_2O_2) and hydroxyl radical ($^{\cdot}OH$). Reactions of these ROS with other cellular compounds can lead to the production of other reactive species, like alkylhydroperoxides (formed from unsaturated organic substrates) or peroxinitrate and nitrogen dioxide radical (formed from nitric oxide). In contrast to the normal conditions, stresses cause the sudden burst of ROS production in plant tissues that shifts the equilibrium between the rate of formation and removal of ROS. It leads to the secondary oxidative stress, which can be defined as a

Yash P. Abrol and Altaf Ahmad (eds.), Sulphur in Plants, 245—263.

disruption of cellular redox homeostasis. Sometimes, it is hard to distinguish whether oxidative stress is a direct consequence of action of the damaging agents or a secondary phenomenon resulting from a primary damage (Bartosz, 1997). High concentrations of ROS result in phytotoxicity and tissue damage. At the cellular level the harmful effects of ROS are associated with lipid peroxidation, inactivation of enzymes and other functional proteins, base modifications and strand brakes of DNA (Van Breusegem et al., 2001). For example, thiol-containing enzymes participating in CO_2 fixation in the chloroplasts are the primary targets of ROS.

2. STRESS RESPONSES AND ANTIOXIDANT SYSTEMS IN PLANTS

Recently excellent progress has been made in understanding plant stress responses. One main focus is on genetic engineering approaches towards studying and improving mechanisms of oxidative stress resistance in plants. A brief account of current knowledge about plant stress responses is presented. For more information on signaling cascades, see Nurnberger and Scheel (2001), Knight and Knight (2001) and Xiong and Zhu (2001).

Plant reactions to environmental stresses involve a few steps, as illustrated in Fig.1.

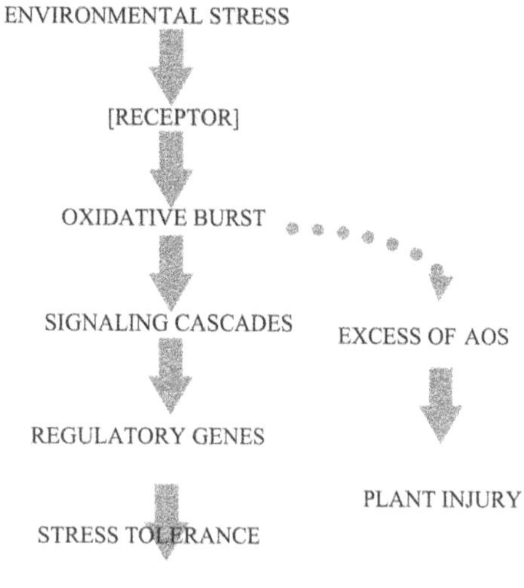

Figure 1. Simplified scheme of plant responses to environmental challenges

In general, the sequences of events appear to be common for all stresses. The initial step is sensing the stress by the respective receptors/sensors. So far, none of the receptors for abiotic stresses has been identified with certainty. However, it is commonly accepted that stress signals must be perceived by such proteins as receptor-like kinases, two-components histidine kinases or G-protein-coupled receptors (Xiong and Zhu, 2001). ROS production is accompanied (or even preceded) by the transient elevation of calcium in the cytoplasm (Reddy, 2001). Fast accumulation of ROS takes place shortly after perceiving the stress signal. The exact internal sources of ROS are uncertain in cases of most abiotic stresses. However, during biotic stress (pathogen attack) it is most probably NADPH-dependent oxidase (Nurnberger and Scheel, 2001), although other possible sources, like cell-wall peroxidases, exist (Bolwell et al., 2002). Not all signaling pathways resulting in ROS burst are identical. Different stresses may activate different sets of secondary messengers. They can initiate multiple cascades of responses resulting either in the increased stress tolerance or programmed cell death (Beers and McDowell, 2001). Such plant compounds as gibberellic acid and abscisic acid (Netting, 2000; Bethke and Jones, 2001), as well as nitric oxide (Wendehenne et al., 2001; Neill et al., 2002), salicylic acid, ethylene and jasmonate (Glazebrook, 2001; Howe and Schilmiller, 2002; Weber, 2002) participate in stress defense responses. Both, integration and interaction (cross-talk) can occur between different signaling cascades and molecules during transduction of the stress signal. In fact, signaling cascades form an interconnected network rather than truly specific stress-signaling responses (Knight and Knight, 2001). It is possible to achieve cross-tolerance against one stress upon exposure to other stress. For example, wounding can induce tolerance to ozone (Örvar et al., 1997) or treatment with hydrogen peroxide can increase tolerance against high-light stress (Gechev et al., 2002). Various phosphorylation/dephosphorylation pathways, including calcium dependent kinases (CDPKs) (Romeis et al., 2001), SNF1-like kinases (Mikolajczyk et al., 2000) and MAP kinases (Zhang and Klessig, 2001) are involved in stress signaling. In addition to the components that directly transduce stress signals, other molecules can regulate and provide stress-specific responses. These partners include modifiers (enzymes involved in such processes as protein lipidation, methylation, ubiquitination) and scaffold or adapter proteins.

Signal transduction cascades result in activation of transcription factors that regulate expression of stress-affected genes (Xiong and Zhu, 2001). There is no specific class of transcription factors participating in stress responses. Several kinds of them can regulate one group of stress-responsive genes or even several transcription factors of various classes can co-operatively activate the same gene. Some of the transcription factors can be activated on a transcriptional level by the stress treatment.

Stress-related transcription factors are responsible for increased expression of genes whose products, due to some distinctive enzymatic or structural functions, lead to modification of the cellular metabolism resulting in stress tolerance. Among the examples of genes that belong to this category are those involved in synthesis of antioxidants, osmoprotectants, heat shock proteins, chaperons, proteases preferentially degrading oxidized proteins, and enzymes needed for DNA repair and for reduction of oxidized thiol groups in proteins. Some of these products are stress-specific and some are common for most environmental stresses.

No efficient enzymatic defense can be used against $^{\cdot}OH$ due to the very reactive nature of this radical. Instead, plants have developed some strategies to prevent formation of $^{\cdot}OH$, such as the tight control of the availability of metal ions that may catalyze the conversion of H_2O_2 to $^{\cdot}OH$. Such role is played by several compounds, including ferritin (Briat et al., 1999), metallothioneins (Coyle et al., 2002) and phytochelatins (Cobbett, 2000a, b; Cobbett, 2002 this book).

Less aggressive ROS ($O_2^{\cdot-}$ and H_2O_2) can be detoxified by antioxidants located in different plant cell compartments. Antioxidant capacity is very much dependent on the severity of the stress as well as the plant species and its developmental stage. In the individual leaves it depends on the physiological state and position of the leaf. Plant antioxidants can be divided into three groups: (i) water soluble non-enzymatic compounds e.g. ascorbate, glutathione, flavonoids and polyphenols, (ii) lipid soluble non-enzymatic compounds e.g. tocopherol, carotenoids and xanthophylls and (iii) enzymes e.g. superoxide dismutases, catalases, glutathione transferases, various peroxidases, and peroxiredoxins.

Dismutation of $O_2^{\cdot-}$ to H_2O_2, which can later be removed by ascorbate-glutathione cycle, catalases and other enzymes, is conducted by a family of metalloenzymes called superoxide dismutases (SODs) (Alscher et al., 2002). SODs are classified according to their metal cofactor, as FeSOD, MnSOD, or Cu/ZnSOD. Different SOD isoenzymes are in different ways controlled developmentally and environmentally. They are located in various cell compartments. MnSOD occurs in mitochondria and peroxisomes, FeSOD in the chloroplasts, and Cu/ZnSOD in chloroplasts, cytosol and possibly in the extracellular space (Alscher et al., 2002).

The very important plant antioxidants necessary for reduction of H_2O_2 to H_2O are ascorbate (Smirnoff, 2000) and glutathione (Noctor et al., 2002). Both compounds can accumulate to millimolar concentrations in plant tissues. Oxidized ascorbate and glutathione are regenerated in a cyclic transfer of reducing equivalents called ascorbate-glutathione cycle (or Foyer-Halliwell-Asada cycle) that uses electrons derived from NADPH and involves four enzymes:ascorbate peroxidase, monodehydroascorbate reductase, dehydroascorbate reductase and glutathione reductase. The enzymes of the ascorbate-glutathione cycle have been found in different cell compartments, including chloroplasts, cytosol, mitochondria and peroxisomes (Del Rio et al., 2002; Noctor et al., 2002; Shigeoka et al., 2002). Ascorbate peroxidases (APXs) mediate the scavenging of H_2O_2 by ascorbate. They reduce H_2O_2 but cannot reduce lipid hydroperoxides and are very unstable in the absence of ascorbate. APX isoenzymes are located in stroma and thylakoid membranes of the chloroplasts, mitochondrial and other cellular membranes, peroxisomes and glyoxysomes, and in the cytosol.

Catalases (CATs) are heme proteins that catalyze the removal of free H_2O_2. They do not require reducing power and have high reaction rates but low affinities for H_2O_2. CATs remove the bulk of H_2O_2, while APX that have higher affinity for H_2O_2, allow for the scavenging of small amounts of H_2O_2 in more specific locations. There are three main isoforms of catalases differently localized and regulated (Van Breusegem et al., 2001).

Peroxiredoxins (PRX) are involved in the reduction of free H_2O_2, alkylhydroperoxides and reactive nitrogen peroxides. Plants contain several isoforms of these enzymes, which can be divided into four classes. Some of them have been recently identified as members of the antioxidant defense system (Lee et al., 2000; Dietz et al., 2002; Konig et al., 2002).

Antioxidant enzymes showing activities of phospholipid hydroperoxide glutathione peroxidase (PHGPX) and thioredoxin peroxidase (TPX) were recently shown to exist in various plants (Mullineaux et al., 1998; Avsian-Kretchmer et al., 1999; Li et al., 2000; Herbette et al., 2002). The enzymes react with alkyl, fatty acid and phospholipid hydroperoxides. Some of them can also be involved in detoxification of free H_2O_2.

Glutathione S-transferases (GSTs) play a well-defined role in plant detoxification reactions. Besides detoxification of xenobiotics through their binding to GSH, they have also glutathione peroxidase activity (GPX) and can be involved in reduction of lipid hydroperoxides (Gullner and Komives, 2001).

3. STUDYING STRESS RESISTANCE

3.1 Generation of Oxidative Stress in Experimental Conditions

Secondary oxidative stress is a consequence of various environmental stresses. In experimental conditions the plant exposure to the real abiotic stresses can be time-consuming and hard to perform. Therefore, several simple strategies have been developed for artificial generation of oxidative stress in plant tissues. Among them, floating of plant leaf discs on solutions of various oxidizing agents (Fig. 2) or spraying intact plants with such solutions belong to the easiest and most commonly applied.

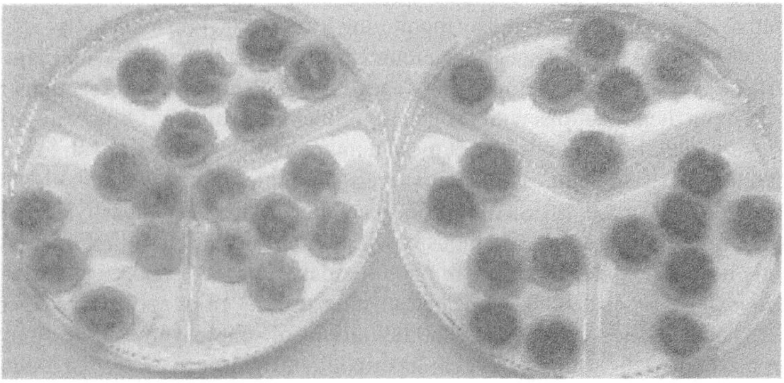

Figure 2. An example of the "floating experiment". The leaf discs have been floated on the solution of 0.5M H_2O_2 for 48 hours in the light. In this case on each plate six discs, 1.4 mm in diameter, cut out from three separate plants were tested. The differences in bleaching of chlorophyll are visible. Typical assays upon this type of stress include conductivity of the floating solution, chlorophyll and pheophytin contents as well as inhibition of photosynthesis in the leaf discs.

Methyl viologen (MV), an herbicide also known as paraquat, has been traditionally implicated in this type of experiments (Tepperman and Dunsmuir, 1990; Bowler et al., 1991). In light, MV accepts electrons from photosystem I and donates them to oxygen, thus, forming $O_2^{\cdot}$ within the chloroplasts. Therefore, the damage is observed first at chloroplastic membranes (Farrington et al., 1973) and next it is spread outward, causing damage to the cell membranes and ion leakage process (Slooten et al., 1995).

H_2O_2 is the most stable ROS and it can rapidly diffuse across cell membranes. Solution of H_2O_2 in different concentrations have been used in the floating experiments (Blaszczyk et al., 1999), in a spray form (Gechev et al., 2002), as an addition to the growth medium of plant cells cultures (Patra et al., 1997) or for seedlings immersion (Prasad et al., 1994). The subcellular place of the initiation of oxidative stress and nature of the stress and responses induced by H_2O_2 can differ from that induced by MV (Patra et al., 1997).

Environmental stresses can be mimicked in the laboratory conditions. Examples of stress conditions used by various researchers include such treatments of the seedlings as fumigation with ozone (Willekens et al., 1997) or toxic gaseous compounds (Youssefian et al., 1993, 2001), and exposure to xenobiotics (Gullner et al., 2001), cold (Roxas et al., 2000), freezing and heat (Kovtun et al., 2000). Germination of the seeds on solid media supplemented with heavy metals (Dominguez-Solis et al., 2001) or NaCl (Kovtun et al., 2000) also belong to the popular tests. Effects of biotic stresses can be studied either by using numerous elicitors or upon wounding (Vignutelli et al., 1998). In many cases, the investigators in a series of parallel tests verified resistance against several stresses.

3.2 Monitoring Oxidative Stress and Evaluation of the Level of Stress Resistance

Oxidative stress can be evaluated by monitoring ROS production in plant tissues. Direct analytical methods are not sensitive enough to give reliable results with plant tissue material. Several indirect methods that depend either on endogenous reactions caused by ROS or on their reactions with exogenous probes are available. Other ways of stress assessment include direct or indirect measurement of the expression of stress-induced genes and evaluation of the antioxidants levels. Examples of various methods are listed in Table 1. Most of the "stress indicators" can be measured in either invasive or non-invasive ways. Non-invasive methods based on high-resolution digital imaging have recently been used to localize sites of photo-oxidative stress responses in the leaves of *Arabidopsis* (Fryer et al., 2002). The investigators used a Peltier-cooled charge-coupled device camera and visualized photoinhibition of photosynthesis in individual chloroplasts, as well as singlet oxygen, superoxide and hydrogen peroxide production in leaf tissue. They also monitored activity of cytosolic ascorbate peroxidase promoter fused to a firefly luciferase reporter gene. Usefulness of non-invasive methods and their advantages are obvious, however, access to the equipment necessary to perform such experiments can be limiting for some investigators.

3.3 Using Transgenic Plants to Study Stress Responses

Genetic engineering approaches that have been useful in studying stress responses in plants include (i) construction of the fusions between stress-responsive promoters and the reporter genes in order to monitor changes in expression (Kovtun et al., 2000), (ii) introduction of the mutated alleles of genes resulting in dominant phenotype in order to study the effects of such modification (Harding et al., 1997) and (iii) reduction of some enzymes/compounds contents by antisense techniques, including glutathione reductase (Aono et al., 1995a), catalase (Chamnongpol et al., 1996; Willekens et al., 1997) and tocopherol (Grasses et al., 2001).

Aequorin, calcium-sensitive luminescent protein from the coelenterate *Aequorea victoria* can serve as an indicator of the level of cytosolic calcium. Some groups used aequorin-transformed tobacco plants or plant cells to study calcium-related stress signaling responses (Knight et al., 1991; Chandra et al., 1997; Cessna and Low, 2001).

Table 1. Examples of assays used to evaluate the level of oxidative stress and stress resistance

Endogenous indicators of AOS-generated damage
Evaluation of visible injury (Örvar et al., 1997)
Amount of chlorophyll/pheophytinized chlorophyll (Bowler et al., 1991; Gupta et al., 1993a; Van Camp et al., 1994)
Membrane leakage analysis by conductivity of the floating solution (Bowler et al., 1991; Gupta et al., 1993a; Willekens et al., 1997)
Lipid peroxidation level (O'Kane et al., 1996; Roxas et al., 2000)
Amount of protein carbonyls (Levine et al., 1990; Lee et al., 2000)
Rate of photosynthesis measured by CO_2 fixation and chlorophyll fluorescence (Tepperman and Dunsmuir, 1990; Gupta et al., 1993b; Creissen et al., 1999; Miyagawa et al., 2000;)
Exogenous indicators of AOS production
Determination of H_2O_2 level with horseradish peroxidase and fluorescent substrates (O'Kane et al., 1996; Creissen et al., 1999)
Determination of $OH^{\cdot}$ with fluorescent probes (Schopfer et al., 2001)
Determination of $O_2^{\cdot-}$ with fluorescent probes (Schopfer et al., 2001)
AOS detection with colorimetric probes (Ogawa et al., 1997)
Stress-related compounds/enzymes
GSH, GSSG and their ratio (Creissen et al., 1999)
Ascorbate, dehydroascorbate and their ratio (Roxas et al., 2000)
APX by Western blot (Miyagawa et al., 2000)
Pathogenesis-related protein PR-1 by Western blot (Chamnongpol et al., 1996)
Activity of SOD, GR, APX (Gupta et al., 1993b; Youssefian et al., 2001)
Activity of Rubisco, MDHAR, APX, DHAR, GR (Kingston-Smith and Foyer, 2000)
Activity of CAT, GR, APX (O'Kane et al., 1996)
Defense genes expression as stress indicators
Cu/Zn transcript (Youssefian et al., 2001)
APX transcript (Gupta et al., 1993b)

4. MAJOR TARGETS FOR ENGINEERED STRESS TOLERANCE

Current strategies of manipulating stress resistance in plants have been described in many recent reviews (Allen et al., 1997; Grover et al., 1999; Mullineaux and Creissen, 1999; Barthel, 2001; Stuiver and Custers, 2001). The potential to ameliorate stress tolerance by genetic engineering has so far been demonstrated on almost every step of the stress response. Most of the transgenes were introduced mainly to tobacco and *Arabidopsis*, however other plant species, including poplar, rice and maize, have also been genetically engineered and some of these modifications exerted a serious impact on stress resistance.

4.1 Scavengers of ROS and Associated Enzymes

Genes encoding enzymes involved in the ROS-scavenging systems have been the main targets for many investigators in their attempts to ameliorate stress tolerance. Enhancements of these systems were sufficient to protect plants from some stresses, but in few cases no effects or even detrimental effects have been reported (Tepperman and Dunsmuir, 1990; Creissen et al., 1995; Foyer et al., 1995). One possible explanation might be that H_2O_2 and $O_2^{\cdot-}$ are involved in mediating cell defense responses. Enhancing the antioxidant capacity might protect plants under specific conditions but it would also interfere with the signaling cascade involved in plant adaptation. Therefore, close control of ROS levels in various cellular compartments may offer better possibilities than just totally suppressing ROS accumulation in plants. Major targets of genetic engineering modifications which result in stress resistance either due to increasing ROS detoxification capacity or due to diminishing oxidative damage are schematically shown in Fig. 3.

4.1.1 Enzymes responsible for biosynthesis of cysteine and glutathione

Glutathione reacts chemically with many ROS and participates in detoxification of H_2O_2 in ascorbate-glutathione cycle through enzyme-catalyzed reactions. In addition, GSH is a substrate for glutathione sulphotransferases, a precursor of phytochelatins and a direct electron donor in reactions catalyzed by glutathione peroxidases (Noctor et al., 2002). Involvement of GSH in many systems engaged in scavenging of ROS makes it a good candidate for engineering of the oxidative stress tolerance.

Biosynthesis of cysteine is regarded as a key limiting step in the production of GSH, while availability of O-acetylserine (OAS) confines the amounts of cysteine made within plant cells (Noji and Saito, 2002 this book). OAS is produced from serine in a reaction catalyzed by serine acetyltransferase (SAT). Transgenic tobacco plants overproducing *Escherichia coli* SAT in either chloroplasts or cytosol, have been obtained in our laboratory (Blaszczyk et al., 1999). These plants not only had an increased level of cysteine and glutathione but they also were more resistant to the oxidative stress generated by H_2O_2. Biochemical analysis has shown that SAT-overproducing plants differ from the controls in some enzymatic activities and a chemical composition (Blaszczyk et al., 2002). Our conclusion was that in these plants some stress-related mechanisms had been induced even in the non-stress conditions. It is necessary to mention that despite elevated levels of cysteine and GSH no harmful effects, like necrosis or severe growth retardation, were observed in these transformants in the greenhouse conditions.

Overproduction of wheat O-acetylserine(thiol)lyase (OAS-TL) in transgenic tobacco conferred resistance not only to hydrogen sulphide (Youssefian et al., 1993) but also to MV (Youssefian et al., 2001). The increased resistance to MV and SO_3^{2-} toxicity has also been observed in transgenic tobacco producing spinach OAS-TL in either cytosol or chloroplasts (Noji et al., 2001). Moreover, the protecting effects were even higher in their cross-fertilized progeny containing enhanced OAS-TL in both compartments. An additional copy of a gene encoding cytosolic OAS-TL introduced to *Arabidopsis* plants significantly increased their ability to accumulate and tolerate higher concentrations of cadmium (Dominguez-Solis et al., 2001). None of the above papers has reported any phenotypic disadvantages due to the transgenes expression.

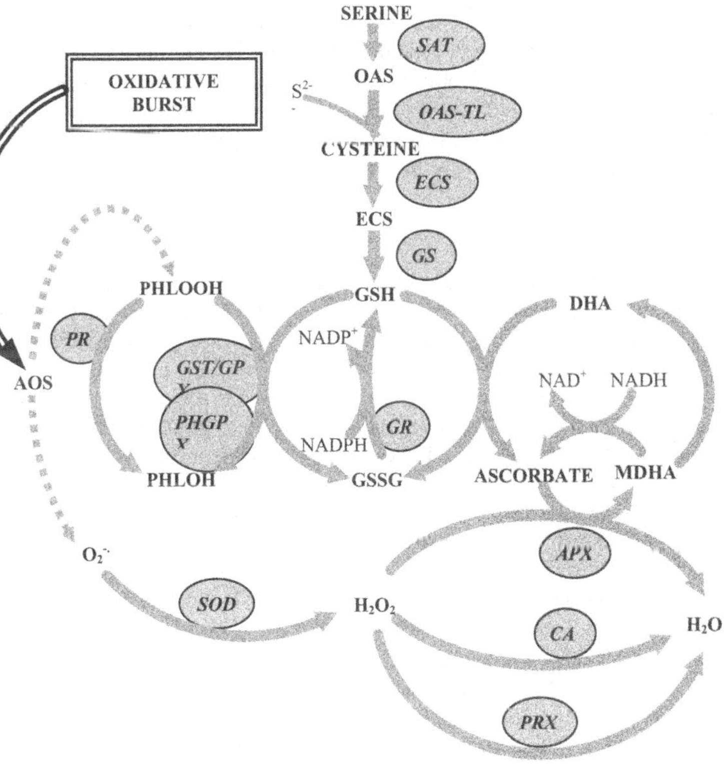

Figure 3. The reactions related to the ROS scavenging systems in plants. Only those enzymes that have been reported as targets of genetic engineering modifications are marked. Abbreviations are explained in the text and in the abbreviations list

GSH biosynthesis occurs in two steps catalyzed by γ-glutamyl-cysteine synthase (ECS) and glutathione synthase (GS), respectively. Bacterial genes encoding enzymes involved in GSH biosynthesis have been introduced into poplar (Noctor et al., 1998), mustard (Zhu et al., 1999) and tobacco (Creissen et al., 1999). In all these plants an additional ECS but not GS was sufficient to increase GSH synthesis. In poplar, no effects of either enzyme overproduction on the resistance against cadmium, MV, chilling and high-light stresses were observed (Foyer et al., 1995; Noctor et al., 1998). Instead, transgenic poplars overproducing ECS in either chloroplasts or cytosol had an elevated tolerance towards chloroacetanilide herbicides (Gullner et al., 2001). Contrary to the results obtained with transgenic poplar, overproduction of bacterial ECS in mustard (Zhu et al., 1999) and introduction of an additional copy of *AtGSH1* gene into *Arabidopsis* (Xiang et al., 2001) increased plants tolerance against cadmium. Creissen et al. (1999) reported that overproduction of bacterial ECS, but not GS, in tobacco chloroplasts resulted in the chlorotic/necrotic phenotypes when plants were grown under greenhouse conditions. The authors suggested that the observed damage might be the effect of the disruption of the redox-sensing process in the chloroplasts due to the rapid ROS scavenging by GSH. It is not clear why the detrimental effects of ECS overproduction were observed only in tobacco and not in other transformed plants. Several explanations of this phenomenon are possible, including species-specific differences in regulatory and stress signaling pathways, the various levels of achieved transgene expression and ECS activity, and dissimilar conditions of plants cultivation.

4.1.2 Enzymes related to glutathione metabolism and recycling

Maintenance of glutathione and ascorbate in the reduced state is critical for oxidative stress resistance (Noctor and Foyer, 1998). Capacity for glutathione recycling has been increased through heterologous expression of genes encoding glutathione reductase and amelioration of stress tolerance was reported in most cases. For example, decreased foliar damage after exposure to MV has been observed in tobacco plants producing *E. coli* glutathione reductase (GR) either in cytosol (Aono et al., 1991) or chloroplasts (Aono et al., 1993). In contrast, Foyer et al. (1991) reported lack of influence of cytosol-targeted bacterial GR on MV resistance in tobacco. Similarly, in transgenic poplar, bacterial GR targeted to cytosol did not confer MV-resistance. The same enzyme, however, targeted to the chloroplasts resulted in the increased tolerance to chilling stress in the high-light conditions (Foyer et al., 1995). Pea GR targeted either to cytosol or chloroplasts conferred plants resistance to MV and ozone (Broadbent et al., 1995). In summary, GR activity in some cellular compartments might be limiting for oxidative stress tolerance. Yet, the effects of overproduction vary between plant species and depend on the cellular compartment to which the protein had been targeted.

Plant GSTs belong to the heterogeneous family of enzymes that catalyze the conjugation of GSH to electrophilic sites of various compounds (Edwards et al., 2000; Pflugmacher et al., 2000). Their expression is highly responsive to environmental stresses (Chen and Singh, 1999). Over-expression of cDNA (Nt107) encoding enzyme with GST/GPX activity in transgenic tobacco resulted in the enhanced growth of the seedlings under both chilling and salt-stress conditions (Roxas et al., 1997, 2000).

Glyoxalase I from *Brassica juncea* has conferred salinity tolerance in transgenic tobacco (Veena et al., 1999). It has been proposed that the enzyme besides detoxification of methylglyoxal (which has been shown to decrease the rate of cell division) could also play a role in maintaining the glutathione homeostasis by recycling the glutathione "trapped" by methylglyoxal.

4.2 Other Enzymes Involved in Detoxification of AOS and Oxidized Products

The genes encoding peroxisomal APX from barley and *Arabidopsis* have been overexpressed in *Arabidopsis thaliana* (Shi et al., 2001) and tobacco (Wang et al., 1999), respectively. In both cases the protection against external stresses has been achieved. In addition, transgenic tobacco plants overproducing cytosolic APX have been more resistant to MV-generated damage (Pitcher et al., 1994; Webb and Allen, 1996), in contrast to the plants with chloroplast-targeted isoform (Pitcher et al., 1994). The last observation may be possibly explained by the fact that the activity of chloroplastic ascorbate peroxidase are extremely sensitive to oxidative stress generated by MV most probably due to instability of the enzyme in the absence of ascorbic acid (Mano et al., 2001).

Several research groups have studied the effects of overproduction of various SOD isoenzymes in different plant cell compartments. The results of the first experiments with chloroplastic Cu/ZnSOD from petunia were discouraging since no effects of overproduction of this enzyme in tobacco on tolerance against MV (Tepperman and Dunsmuir, 1990) or ozone (Pitcher et al., 1991) were observed. In contrast, overproduction of Cu/ZnSOD from pea resulted in the increased resistance of transgenic tobacco to MV (Gupta et al., 1993a) and higher resistance to the oxidative stress generated by exposing plants to low temperature at high light intensity (Gupta et al., 1993b). Besides, the transgenic plants had an increased APX activity. The amelioration of stress tolerance has also been associated with the chloroplastic overproduction of FeSOD isoenzymes in tobacco (Van Camp et al., 1996), maize (Van Breusegem et al., 1999) and alfalfa (McKersie et al., 2000). Overproduction of chloroplast-targeted MnSOD from *Nicotiana plumbaginifolia* in tobacco resulted in the reduction of ozone injury (Van Camp et al., 1994) and an increased tolerance to MV (Bowler et al., 1991; Slooten et al., 1995). The authors pointed out that the ratio between superoxide and hydrogen peroxide could be the factor determining whether SOD overproduction is beneficial or deleterious to the plants. Overproduction of SOD resulted in an increased tolerance to MV only if other components of the antioxidant systems were also present at higher levels. This conclusion has essentially been confirmed later (Kingston-Smith and Foyer, 2000) since the increased monodehydroascorbate reductase, dehydroascorbate reductase and GR activities were found in transgenic maize overproducing MnSOD from tobacco. Therefore, increased recycling of reduced ascorbate and glutathione was imposed by higher SOD activity, to compensate for the enhanced H_2O_2 production. Transgenic alfalfa plants containing MnSOD targeted to the chloroplasts had an improved winter survival in field conditions (McKersie et al., 1999). Alfalfa plants overproducing MnSOD in either chloroplasts or in cytosol had higher biomass production, however, their sexually crossed progeny overproducing MnSOD in two compartments, chloroplasts and mitochondria, were significantly lower in biomass production than any of their parents (Samis et al., 2002).

Finally, it is worth mentioning that the simultaneous overproduction of SOD and GR in the cytosol of transgenic tobacco resulting in the enhanced tolerance against MV has been reported (Aono et al., 1995b).

Catalase deficient tobacco plants that were generated through sense and antisense technology exhibited increased sensitivity to several abiotic stresses including MV, H_2O_2, salinity, ozone fumigation and high light (Willekens et al., 1997). In contrast, overproduction of catalase from *E. coli* in the chloroplasts of transgenic tobacco resulted in the increased resistance to drought, high light and MV stresses (Shikanai et al., 1998; Miyagawa et al., 2000).

Rice peroxiredoxin, overproduced in transgenic tobacco, enhanced antioxidant capacity of the plant leaf tissues against oxidative stress generated by H_2O_2 (Lee et al., 2000). The transgenic tobacco plants overproducing alfalfa aldose reductase were less sensitive to various stresses (MV, H_2O_2, heavy metal treatment and drought), most probably due to the elimination of the toxic lipid aldehydes by this enzyme (Oberschall et al., 2000).

4.3 Engineering Components of the Signaling Cascades

There are several reports on enhancing stress tolerance due to the overproduction of compounds of stress signaling pathway. For example, constitutively active yeast calcineurin improved salinity tolerance in tobacco (Pardo et al., 1998), constitutive overproduction of rice CDPK7 resulted in the increased tolerance to salt and cold stresses in transgenic rice plants (Saijo et al., 2000), and constitutively active tobacco MAPKKK, NPK1, led to the tolerance of transgenic tobacco to multiple stresses (oxidative, freezing, heat) due to the activation of stress activated MAPK cascade and protective genes expression (Kovtun et al., 2000). It is necessary to emphasize that phosphorylation pathways are complex and sometimes it is hard to obtain expected results.

4.4 Modification of Regulatory Factors

Plant transcription factors can be classified into several families according to the features of their conserved domains (Riechmann and Ratcliffe, 2000). Expression of stress-related genes can be regulated by factors from different classes. There are several reports about improving salinity, drought or freezing tolerance due to the overproduction of stress-specific transcription factors, which control the expression of a set of stress-adaptive proteins. Among the successful examples there are DREB1A from *Arabidopsis* (Liu et al., 1998; Kasuga et al., 1999), CBF1 from *Arabidopsis* (Jaglo-Ottosen et al., 1998), Alfin1 from alfalfa (Winicov and Bastola, 1999) and Tsi1 from tobacco (Park et al., 2001). The Tsi1 overexpression resulted in the improved tolerance not only to abiotic stress (salinity) but also to pathogens.

4.5 Other Stress-related Components with Specific Functions

In most cases, the investigators' efforts are directed towards ameliorating tolerance to particular stress conditions. Below, we shortly summarize some successful approaches that did not fit to the previous categories of transgenes improving stress tolerance.

Salinity is among the stresses that have attracted a lot of attention (Apse and Blumwald, 2002; Chen and Murata, 2002). Two main strategies have been used to specifically improve salinity tolerance: (i) increasing the level of osmoprotectants, called also compatible solutes, and (ii) changing ion homeostasis. The osmoprotectants include such small compounds as glycinebetaine, proline, mannitol, sorbitol, trehalose and fructan. These small metabolites stabilize proteins and cellular structure and they can increase the osmotic pressure of the cell. Engineered plants with increased level of osmoprotectants showed enhanced tolerance not only to the osmotic stress but also to other stresses, including drought, freezing and oxidative (Chen and Murata, 2002). The possible explanation of this phenomenon is that these metabolites can act as AOS scavengers. Many harmful effects of the osmotic stress can be caused by the influx of the sodium ions to the cytosol of the cells. Plants can protect themselves either by diminishing entry of sodium ions into the cells, by their extrusion out of the cells or by their vacuolar compartmentation (Serrano and Rodriguez-Navarro, 2001). Transgenic plants overproducing vacuolar sodium translocators (Na^+/H^+ antiporters) demonstrated usefulness of the latter strategy for amelioration of osmotic stress tolerance (Apse et al., 1999).

Improvement of aluminum stress resistance has been achieved in four transgenic lines of *Arabidopsis* due to the expression of four genes (Ezaki et al., 2000; Ezaki et al., 2001). These genes (previously shown to be induced by aluminium stress) encode the following proteins: (i) *Arabidopsis* blue copper-binding protein, (ii) tobacco GDP dissociation inhibitor, (iii) tobacco glutathione transferase and (iv) tobacco peroxidase. Mechanisms of action of the last two transgenes can be explained by participation of these enzymatic products in the diminishing of the oxidative damage caused by the stress (see Subchapter 4.1.2. and Fig. 3), while the expression of the first two transgenes resulted in the decrease of aluminium content in the root tips due to suppressed absorption or promoted release, respectively (Ezaki et al., 2001).

Overproduction of alfalfa ferritin in the chloroplasts of transgenic tobacco promoted not only iron and MV resistance but also resulted in the reduction of necrotic damage caused by viral and fungal infections (Deak et al., 1999). Similar results were obtained by overproduction of soybean ferritin in either chloroplasts or cytosol of tobacco (Van Wuytswinkel et al., 1999). Noticeably, the authors of the latter paper demonstrated that the transgenic plants behaved as iron deficient and activated iron transport systems.

Overproduction of the binding protein (BiP) from soybean resulted in increased tolerance of the transgenic tobacco to drought stress (Alvim et al., 2001). BiP is a member of molecular chaperons located in the endoplasmic reticulum. Most probably, enhanced accumulation of BiP raises the capacity of the cells to process protein folding, and these allow the cells to control the concentration of specific defense proteins more efficiently.

5. CONCLUSIONS

The majority of efforts to modify AOS scavenging systems have been focused on three enzymatic activities: GR, SOD and APX. Considerable changes in oxidative stress tolerance have been achieved due to the enhancement of these activities. However, plant stress responses, due to their genetic complexity, have not yet been completely understood. Antioxidant mechanisms involved during stress responses and their complex interactions are only now being elucidated. The success of many transgenic approaches has generally been limited by a lack of understanding of metabolic flux, compartmentation and function. It became clear that the stress signaling pathways are part of the complex networks and changes of one component have a huge impact on the others. In future, transformation with multiple, pyramided transgenes might provide desirable effects. For the best effects, carefully selected suitable transgenes must be targeted to the appropriate cellular compartments. The in depth knowledge of the stress-related mechanisms is essential to the success of these approaches. This can be achieved by combination of various non-standard approaches, including computer simulations of flux of the components (Polle, 2001) or applying high-throughput yeast two-hybrid screening system to study stress signal transduction pathways (Fang et al., 2002).

6. REFERENCES

Allen RD, Webb RP and Schake SA (1997) Use of transgenic plants to study antioxidant defences. *Free Rad Biol Med* **23:** 473—479

Alscher RG, Erturk N and Heath LS (2002) Role of superoxide dismutases (SODs) in controlling oxidative stress in plants. *J Exp Bot* **53:** 1331—1350

Alvim FC, Carolino SM, Cascardo JC, Nunes CC, Martinez CA, Otoni WC and Fontes EP (2001) Enhanced accumulation of BiP in transgenic plants confers tolerance to water stress. *Plant Physiol* **126:** 1042—1054

Aono M, Kubo A, Saji H, Natori T, Tanaka K and Kondo N (1991) Resistance to active oxygen toxicity of transgenic *Nicotiana tabacum* that expresses the gene for glutathione reductase from *Escherichia coli*. *Plant Cell Physiol* **32:** 691—697

Aono M, Kubo A, Saji H, Fujiyama K, Tanaka K and Kondo N (1993) Enhanced tolerance to photooxidative stress of transgenic *Nicotiana tabacum* with high chloroplastic glutathione reductase activity. *Plant Cell Physiol* **34:** 129—135

Aono M, Saji H, Fujiyama K, Sugita M, Kondo N and Tanaka K (1995a) Decrease in activity of glutathione reductase enhances paraquat sensitivity in transgenic *Nicotiana tabacum*. *Plant Physiol* **107:** 645—648

Aono M, Saji H, Sakamoto A, Tanaka K, Kondo N and Tanaka K (1995b) Paraquat tolerance of transgenic Nicotiana tabacum with enhanced activities of glutathione reductase and superoxide dismutase. *Plant Cell Physiol* **36:** 1687—1691

Apse MP and Blumwald E (2002) Engineering salt tolerance in plants. *Curr Opin Biotechnol* **13:** 146—150

Apse MP, Aharon GS, Snedden WA and Blumwald E (1999) Salt tolerance conferred by overexpression of a vacuolar Na^+/H^+ antiport in Arabidopsis. *Science* **285:** 1256—1258.

Avsian-Kretchmer O, Eshdat Y, Gueta-Dahan Y and Ben-Hayyim G (1999) Regulation of stress-induced phospholipid hydroperoxide glutathione peroxidase expression in citrus. *Planta* **209:** 469—477

Barthel D (2001) Targeting detoxification pathways: an efficient approach to obtain plants with multiple stress tolerance? *Trends Plant Sci* **6:** 284—286

Bartosz G (1997) Oxidative stress in plants. *Acta Physiol Plant* **19:** 47—64

Beers EP and McDowell JM (2001) Regulation and execution of programmed cell death in response to pathogens, stress and developmental cues. *Curr Opin Plant Biol* **4:** 561—567

Bethke PC and Jones RL (2001) Cell death of barley aleurone protoplasts is mediated by reactive oxygen species. *Plant J* **25:** 19—29

Blaszczyk A, Brodzik R and Sirko A (1999) Increased resistance to oxidative stress in transgenic tobacco plants overexpressing bacterial serine acetyltransferase. *Plant J* **20:** 237—243

Blaszczyk A, Sirko L, Hawkesford MJ and Sirko A (2002) Biochemical analysis of transgenic tobacco lines producing bacterial serine acetyltransferase. *Plant Sci* **162:** 589—597

Bolwell GP, Bindschedler LV, Blee KA, Butt VS, Davies DR, Gardner SL, Gerrish C and Minibayeva F (2002) The apoplastic oxidative burst in response to biotic stress in plants: a three-component system. *J Exp Bot* **53:** 1367—1376

Bowler C, Slooten L, Vandenbranden S, De Rycke R, Botterman J, Sybesma C, Van Montagu M and Inze D (1991) Manganese superoxide dismutase can reduce cellular damage mediated by oxygen radicals in transgenic plants. *EMBO J* **10:** 1723—1732

Briat JF, Lobreaux S, Grignon N and Vansuyt G (1999) Regulation of plant ferritin synthesis: how and why. *Cell Mol Life Sci* **56:** 155—166

Broadbent P, Creissen GP, Kular B, Wellburn AR and Mullineaux PM (1995) Oxidative stress responses in transgenic tobacco containing altered levels of glutathione reductase activity. *Plant J* **8:** 247—255

Cessna SG and Low PS (2001) Activation of the oxidative burst in aequorin-transformed Nicotiana tabacum cells is mediated by protein kinase- and anion channel-dependent release of Ca^{2+} from internal stores. *Planta* **214:** 126—134

Chamnongpol S, Willekens H, Langebartels C, Van Montagu M, Inze D and Van Camp W (1996) Transgenic tobacco with a reduced catalase activity develops necrotic lesions and induces pathogenesis-related expression under high light. *Plant J* **10:** 491—503

Chandra S, Stennis M and Low PS (1997) Measurement of Ca^{2+} fluxes during elicitation of the oxidative burst in aequorin-transformed tobacco cells. *J Biol Chem* **272:** 28274—28280

Chen TH and Murata N (2002) Enhancement of tolerance of abiotic stress by metabolic engineering of betaines and other compatible solutes. *Curr Opin Plant Biol* **5:** 250—257

Chen W and Singh KB (1999) The auxin, hydrogen peroxide and salicylic acid induced expression of the Arabidopsis *GST6* promoter is mediated in part by an ocs element. *Plant J* **19:** 667—677

Cobbett CS (2000a) Phytochelatin biosynthesis and function in heavy-metal detoxification. *Curr Opin Plant Biol* **3:** 211—216

Cobbett CS (2000b) Phytochelatins and their roles in heavy metal detoxification. *Plant Physiol* **123:** 825—832

Coyle P, Philcox JC, Carey LC and Rofe AM (2002) Metallothionein: the multipurpose protein. *Cell Mol Life Sci* **59:** 627—647

Creissen G, Reynolds H, Xue Y and Mullineaux P (1995) Simultaneous targeting of pea glutathione reductase and of a bacterial fusion protein to chloroplasts and mitochondria in transgenic tobacco. *Plant J* **8:** 167—175

Creissen G, Firmin J, Fryer M, Kular B, Leyland N, Reynolds H, Pastori G, Wellburn F, Baker N, Wellburn A and Mullineaux P (1999) Elevated glutathione biosynthetic capacity in the chloroplasts of transgenic tobacco plants paradoxically causes increased oxidative stress. *Plant Cell* **11:** 1277—1292

Dat J (2000) Dual action of the active oxygen species during plant stress responses. *Cell Mol Life Sci* **57:** 779—795

Deak M, Horvath GV, Davletova S, Torok K, Sass L, Vass I, Barna B, Kiraly Z and Dudits D (1999) Plants ectopically expressing the iron-binding protein, ferritin, are tolerant to oxidative damage and pathogens. *Nat Biotechnol* **17:** 192—196

Del Rio LA, Corpas FJ, Sandalio LM, Palma JM, Gomez M and Barroso JB (2002) Reactive oxygen species, antioxidant systems and nitric oxide in peroxisomes. *J Exp Bot* **53:** 1255—1272

Dietz KJ, Horling F, Konig J and Baier M (2002) The function of the chloroplast 2-cysteine peroxiredoxin in peroxide detoxification and its regulation. *J Exp Bot* **53:** 1321—1329

Dominguez-Solis JR, Gutierrez-Alcala G, Vega JM, Romero LC and Gotor C (2001) The cytosolic O-acetylserine(thiol)lyase gene is regulated by heavy metals and can function in cadmium tolerance. *J Biol Chem* **276:** 9297—9302

Edwards R, Dixon DP and Walbot V (2000) Plant glutathione S-transferases: enzymes with multiple functions in sickness and in health. *Trends Plant Sci* **5:** 193—198

Ezaki B, Gardner RC, Ezaki Y and Matsumoto H (2000) Expression of aluminum-induced genes in transgenic arabidopsis plants can ameliorate aluminum stress and/or oxidative stress. *Plant Physiol* **122:** 657—665

Ezaki B, Katsuhara M, Kawamura M and Matsumoto H (2001) Different mechanisms of four aluminum (Al)-resistant transgenes for Al toxicity in *Arabidopsis*. *Plant Physiol* **127:** 918—927

Fang Y, Macool DJ, Xue Z, Heppard EP, Hainey CF, Tingey SV and Miao GH (2002) Development of a high-throughput yeast two-hybrid screening system to study protein-protein interactions in plants. *Mol Genet Gen* **267:** 142—153

Farrington JA, Ebert M, Land EJ and Fletcher K (1973) Bipyridylium quaternary salts and related compounds. Pulse radiolysis studies of the reaction of paraquat radical with oxygen. Implications for the mode of action of bipyridyl herbicides. *Biochim Biophys Acta* **314:** 372—381

Foyer C, Lelandais M, Galap C and Kunert KJ (1991) Effects of elevated cytosolic glutathione reductase activity on the cellular glutathione pool and photosynthesis in leaves under normal and stress conditions. *Plant Physiol* **97:** 863—872

Foyer C, Souriau N, Perret S, Lelandais M, Kunert KJ, Pruvost C and Jouanin L (1995) Overexpression of glutathione reductase but not glutathione synthetase leads to increases in antioxidant capacity and resistance to photoinhibition in poplar trees. *Plant Physiol* **109:** 1047—1057

Fryer MJ, Oxborough K, Mullineaux PM and Baker NR (2002) Imaging of photo-oxidative stress responses in leaves. *J Exp Bot* **53:** 1249—1254

Gechev T, Gadjev I, Van Breusegem F, Inze D, Dukiandjiev S, Toneva V and Minkov I (2002) Hydrogen peroxide protects tobacco from oxidative stress by inducing a set of antioxidant enzymes. *Cell Mol Life Sci* **59:** 708—714

Glazebrook J (2001) Genes controlling expression of defense responses in Arabidopsis--2001 status. *Curr Opin Plant Biol* **4:** 301—308

Grasses T, Grimm B, Koroleva O and Jahns P (2001) Loss of alpha-tocopherol in tobacco plants with decreased geranyl reductase activity does not modify photosynthesis in optimal growth conditions but increases sensitivity to high-light stress. *Planta* **213:** 620—628

Grover A, Sahi C, Sanan N and Grover A (1999) Taming abiotic stresses in plants through genetic engineering: current strategies and perspective. *Plant Sci* **143:** 101—111

Gullner G and Komives T (2001) The role of glutathione and glutathione-related enzymes in plant-pathogen interactions. In: Grill D, Tausz M and De Kok L (eds) Significance of Glutathione in Plant Adaptation to the Environment, pp 207—239. Kluwer Academic Publishers, Dordrecht

Gullner G, Komives T and Rennenberg H (2001) Enhanced tolerance of transgenic poplar plants overexpressing gamma-glutamylcysteine synthetase towards chloroacetanilide herbicides. *J Exp Bot* **52:** 971—979

Gupta AS, Heinen JL, Holaday AS, Burke JJ and Allen RD (1993a) Increased resistance to oxidative stress in transgenic plants that overexpress chloroplastic Cu/Zn superoxide dismutase. *Proc Natl Acad Sci U S A* **90:** 1629—1633

Gupta AS, Webb RP, Holaday AS and Allen RD (1993b) Overexpression of superoxide dismutase protects plants from oxidative stress. *Plant Physiology* **103:** 1067—1073

Harding SA, Oh SH and Roberts DM (1997) Transgenic tobacco expressing a foreign calmodulin gene shows an enhanced production of active oxygen species. *EMBO J* **16:** 1137—1144

Herbette S, Lenne C, Leblanc N, Julien JL, Drevet JR and Roeckel-Drevet P (2002) Two GPX-like proteins from *Lycopersicon esculentum* and *Helianthus annuus* are antioxidant enzymes with phospholipid hydroperoxide glutathione peroxidase and thioredoxin peroxidase activities. *Eur J Biochem* **269:** 2414—2420

Howe GA and Schilmiller AL (2002) Oxylipin metabolism in response to stress. *Curr Opin Plant Biol* **5:** 230—236

Jaglo-Ottosen KR, Gilmour SJ, Zarka DG, Schabenberger O and Thomashow MF (1998) *Arabidopsis CBF1* overexpression induces *COR* genes and enhances freezing tolerance. *Science* **280:** 104—106

Kasuga M, Liu Q, Miura S, Yamaguchi-Shinozaki K and Shinozaki K (1999) Improving plant drought, salt, and freezing tolerance by gene transfer of a single stress-inducible transcription factor. *Nat Biotechnol* **17:** 287—291

Kingston-Smith AH and Foyer CH (2000) Overexpression of Mn-superoxide dismutase in maize leaves leads to increased monodehydroascorbate reductase, dehydroascorbate reductase and glutathione reductase activities. *J Exp Bot* **51:** 1867—1877

Knight H and Knight MR (2001) Abiotic stress signalling pathways: specificity and cross-talk. *Trends Plant Sci* **6:** 262—267

Knight MR, Campbell AK, Smith SM and Trewavas AJ (1991) Transgenic plant aequorin reports the effects of touch and cold-shock and elicitors on cytoplasmic calcium. *Nature* **352:** 524—526

Konig J, Baier M, Horling F, Kahmann U, Harris G, Schurmann P and Dietz KJ (2002) The plant-specific function of 2-Cys peroxiredoxin-mediated detoxification of peroxides in the redox-hierarchy of photosynthetic electron flux. *Proc Natl Acad Sci USA* **99:** 5738—5743

Kovtun Y, Chiu WL, Tena G and Sheen J (2000) Functional analysis of oxidative stress-activated mitogen-activated protein kinase cascade in plants. *Proc Natl Acad Sci USA* **97:** 2940—2945

Lee KO, Jang HH, Jung BG, Chi YH, Lee JY, Choi YO, Lee JR, Lim CO, Cho MJ and Lee SY (2000) Rice 1Cys-peroxiredoxin over-expressed in transgenic tobacco does not maintain dormancy but enhances antioxidant activity. *FEBS Lett* **486:** 103—106

Levine RL, Garland D, Oliver CN, Amici A, Climent I, Lenz AG, Ahn BW, Shaltiel S and Stadtman ER (1990) Determination of carbonyl content in oxidatively modified proteins. *Methods Enzymol* **186:** 464—478

Li WJ, Feng H, Fan JH, Zhang RQ, Zhao NM and Liu JY (2000) Molecular cloning and expression of a phospholipid hydroperoxide glutathione peroxidase homolog in *Oryza sativa. Biochim Biophys Acta* **1493:** 225—230

Liu Q, Kasuga M, Sakuma Y, Abe H, Miura S, Yamaguchi-Shinozaki K and Shinozaki K (1998) Two transcription factors, DREB1 and DREB2, with an EREBP/AP2 DNA binding domain separate two cellular signal transduction pathways in drought and low-temperature-responsive gene expression, respectively, in *Arabidopsis. Plant Cell* **10:** 1391—1406

Mano J, Ohno C, Domae Y and Asada K (2001) Chloroplastic ascorbate peroxidase is the primary target of methylviologen-induced photooxidative stress in spinach leaves: its relevance to monodehydroascorbate radical detected with in vivo ESR. *Biochim Biophys Acta* **1504:** 275--287

McKersie BD, Bowley SR and Jones KS (1999) Winter survival of transgenic alfalfa overexpressing superoxide dismutase. *Plant Physiol* **119:** 839—848

McKersie BD, Murnaghan J, Jones KS and Bowley SR (2000) Iron-superoxide dismutase expression in transgenic alfalfa increases winter survival without a detectable increase in photosynthetic oxidative stress tolerance. *Plant Physiol* **122:** 1427—1437

Mikolajczyk M, Awotunde OS, Muszynska G, Klessig DF and Dobrowolska G (2000) Osmotic stress induces rapid activation of a salicylic acid-induced protein kinase and a homolog of protein kinase ASK1 in tobacco cells. *Plant Cell* **12:** 165—178

Miyagawa Y, Tamoi M and Shigeoka S (2000) Evaluation of the defense system in chloroplasts to photooxidative stress caused by paraquat using transgenic tobacco plants expressing catalase from *Escherichia coli. Plant Cell Physiol* **41:** 311—320

Mullineaux P and Creissen G (1999) Manipulating oxidative stress responses using transgenic plants: successes and dangers. In: Altman A (ed) Plant Biotechnology and In Vitro Biology in the 21st century, pp 525—532. Kluwer Academic Publishers, Dordrecht

Mullineaux PM, Karpinski S, Jimenez A, Cleary SP, Robinson C and Creissen GP (1998) Identification of cDNA encoding plastid-targeted glutathione peroxidase. *Plant J* **13:** 375—379

Neill SJ, Desikan R, Clarke A, Hurst RD and Hancock JT (2002) Hydrogen peroxide and nitric oxide as signalling molecules in plants. *J Exp Bot* **53:** 1237—1236

Netting AG (2000) pH, abscisic acid and the integration of metabolism in plants under stressed and non-stressed conditions: cellular responses to stress and their implication for plant water relations. *J Exp Bot* **51:** 147—158

Noctor G, Arisi ACM, Jouanin L, Kunert KJ, Rennenberg H and Foyer C (1998) Glutathione: biosynthesis, metabolism and relationship to stress tolerance explored in transgenic plants. *J Exp Bot* **49:** 623—647

Noctor G, Gomez L, Vanacker H and Foyer C (2002) Interaction between biosynthesis, compartmentation and transport in the control of glutathione homeostasis and signalling. *J Exp Bot* **53**: 1283—1304

Noji M, Saito M, Nakamura M, Aono M, Saji H and Saito K (2001) Cysteine synthase overexpression in tobacco confers tolerance to sulfur-containing environmental pollutants. *Plant Physiol* **126:** 973—980

Nurnberger T and Scheel D (2001) Signal transmission in the plant immune response. *Trends Plant Sci* **6:** 372—379

Oberschall A, Deak M, Torok K, Sass L, Vass I, Kovacs I, Feher A, Dudits D and Horvath GV (2000) A novel aldose/aldehyde reductase protects transgenic plants against lipid peroxidation under chemical and drought stresses. *Plant J* **24:** 437—446

Ogawa K, Kanematsu S and Asada K (1997) Generation of superoxide anion and localization of CuZn-superoxide dismutase in the vascular tissue of spinach hypocotyls: their association with lignification. *Plant Cell Physiol* **38:** 1118—1126

O'Kane D, Gill V, Boyd P and Burdon R (1996) Chilling, oxidative stress and antioxidant responses in *Arabidopsis thaliana* callus. *Planta* **198:** 371—377

Örvar BL, McPherson J and Ellis BE (1997) Pre-activating wounding response in tobacco prior to high-level ozone exposure prevents necrotic injury. *Plant J* **11:** 203—212

Pardo JM, Reddy MP, Yang S, Maggio A, Huh GH, Matsumoto T, Coca MA, Paino-D'Urzo M, Koiwa H, Yun DJ, Watad AA, Bressan RA and Hasegawa PM (1998) Stress signaling through Ca^{2+}/calmodulin-dependent protein phosphatase calcineurin mediates salt adaptation in plants. *Proc Natl Acad Sci USA* **95:** 9681—9686

Park JM, Park CJ, Lee SB, Ham BK, Shin R and Paek KH (2001) Overexpression of the tobacco Tsi1 gene encoding an EREBP/AP2-type transcription factor enhances resistance against pathogen attack and osmotic stress in tobacco. *Plant Cell* **13:** 1035—1046

Patra J, Panda KK and Panda BB (1997) Differential induction of adaptive responses by paraquat and hydrogen peroxide against the genotoxicity of methyl mercuric chloride, maleic hydrazide and ethyl methane sulfonate in plant cells in vivo. *Mut Res* **393:** 215—222

Pflugmacher S, Schroder P and Sandermann H Jr (2000) Taxonomic distribution of plant glutathione S-transferases acting on xenobiotics. *Phytochemistry* **54:** 267—273

Pitcher LH, Brennan E, Hurley A, Dunsmuir P, Tepperman JM and Zilinskas BA (1991) Overproduction of petunia chloroplastic copper/zinc superoxide dismutase does not confer ozone tolerance in transgenic tobacco. *Plant Physiol* **97:** 452—455

Pitcher LH, Repetti P and Zilinskas BA (1994) Overproduction of ascorbate peroxidase protects transgenic tobacco plants against oxidative stress (abstract No. 623). *Plant Physiol* **105:** S-116

Polle A (2001) Dissecting the superoxide dismutase-ascorbate-glutathione-pathway in chloroplasts by metabolic modeling. Computer simulations as a step towards flux analysis. *Plant Physiol* **126:** 445--462

Prasad T, Anderson M, Martin B and Steward C (1994) Evidence for chilling-induced oxidative stress in maize seedlings and a regulatory role for hydrogen peroxide. *Plant Cell* **6:** 65--74

Reddy ASN (2001) Calcium: silver bullet in signaling. *Plant Sci* **160:** 381—404

Riechmann JL and Ratcliffe OJ (2000) A genomic perspective on plant transcription factors. *Curr Opin Plant Biol* **3:** 423—434

Romeis T, Ludwig AA, Martin R and Jones JD (2001) Calcium-dependent protein kinases play an essential role in a plant defence response. *EMBO J* **20:** 5556—5567

Roxas VP, Smith RK Jr, Allen ER and Allen RD (1997) Overexpression of glutathione S-transferase/glutathione peroxidase enhances the growth of transgenic tobacco seedlings during stress. *Nat Biotechnol* **15:** 988—991

Roxas VP, Lodhi SA, Garrett DK, Mahan JR and Allen RD (2000) Stress tolerance in transgenic tobacco seedlings that overexpress glutathione S-transferase/glutathione peroxidase. *Plant Cell Physiol* **41:** 1229—1234

Saijo Y, Hata S, Kyozuka J, Shimamoto K and Izui K (2000) Over-expression of a single Ca^{2+}-dependent protein kinase confers both cold and salt/drought tolerance on rice plants. *Plant J* **23:** 319—327

Samis K, Bowley S and McKersie B (2002) Pyramiding Mn-superoxide dismutase transgenes to improve persistence and biomass production in alfalfa. *J Exp Bot* **53:** 1343—1350

Schopfer P, Plachy C and Frahry G (2001) Release of reactive oxygen intermediates (superoxide radicals, hydrogen peroxide, and hydroxyl radicals) and peroxidase in germinating radish seeds controlled by light, gibberellin, and abscisic acid. *Plant Physiol* **125:** 1591—1602

Serrano R and Rodriguez-Navarro A (2001) Ion homeostasis during salt stress in plants. *Curr Opin Cell Biol* **13:** 399—404

Shi WM, Muramoto Y, Ueda A and Takabe T (2001) Cloning of peroxisomal ascorbate peroxidase gene from barley and enhanced thermotolerance by overexpressing in *Arabidopsis thaliana. Gene* **273:** 23—27

Shigeoka S, Ishikawa T, Tamoi M, Miyagawa Y, Takeda T, Yabuta Y and Yoshimura K (2002) Regulation and function of ascorbate peroxidase isoensymes. *J Exp Bot* **53:** 1305—1319

Shikanai T, Takeda T, Yamauchi H, Sano S, Tomizawa KI, Yokota A and Shigeoka S (1998) Inhibition of ascorbate peroxidase under oxidative stress in tobacco having bacterial catalase in chloroplasts. *FEBS Lett* **428:** 47—51

Slooten L, Capiau K, Van Camp W, Van Montagu M and Sybesma C (1995) Factors affecting the enhancement of oxidative stress tolerance in transgenic tobacco overexpressing manganese superoxide dismutase in the chloroplasts. *Plant Physiol* **107:** 737—750

Smirnoff N (2000) Ascorbic acid: metabolism and functions of a multi-facetted molecule. *Curr Opin Plant Biol* **3:** 229—235

Stuiver MH and Custers JHHV (2001) Engineering disease resistance in plants. *Nature* **411:** 865—868

Tepperman JM and Dunsmuir P (1990) Transformed plants with elevated levels of chloroplastic SOD are not more resistant to superoxide toxicity. *Plant Mol Biol* **14:** 501—511

Van Breusegem F, Slooten L, Stassart JM, Moens T, Botterman J, Van Montagu M and Inze D (1999) Overproduction of Arabidopsis thaliana FeSOD confers oxidative stress tolerance to transgenic maize. *Plant Cell Physiol* **40:** 515—523

Van Breusegem F, Vranova E, Dat JF and Inze D (2001) The role of active oxygen species in plant signal transduction. *Plant Sci* **161:** 405—414

Van Camp W, Willekens H, Bowler C, Van Montagu M, Inze D, Reupold-Popp P, Sanderman H Jr and Langebartels C (1994) Elevated levels of superoxide dismutase protect transgenic plants against ozone damage. *Biotechnology* **12:** 165—168

Van Camp W, Capiau K, Van Montagu M, Inze D and Slooten L (1996) Enhancement of oxidative stress tolerance in transgenic tobacco plants overproducing Fe-superoxide dismutase in chloroplasts. *Plant Physiol* **112:** 1703—1714

Van Wuytswinkel O, Vansuyt G, Grignon N, Fourcroy P and Briat JF (1999) Iron homeostasis alteration in transgenic tobacco overexpressing ferritin. *Plant J* **17:** 93—97

Veena, Reddy VS and Sopory SK (1999) Glyoxalase I from Brassica juncea: molecular cloning, regulation and its over-expression confer tolerance in transgenic tobacco under stress. *Plant J* **17:** 385--395

Vignutelli A, Wasternack C, Apel K and Bohlmann H (1998) Systemic and local induction of an Arabidopsis thionin gene by wounding and pathogens. *Plant J* **14:** 285—295

Vranova E, Inze D and Van Breusegem F (2002) Signal transduction duringoxidative stress. *J Exp Bot* **53:** 1227—1236

Wang J, Zhang H and Allen RD (1999) Overexpression of an Arabidopsis peroxisomal ascorbate peroxidase gene in tobacco increases protection against oxidative stress. *Plant Cell Physiol* **40:** 725—732

Webb RP and Allen RD (1996) Overexpression of pea cytosolic ascorbate peroxidase confers protection against oxidative stress in transgenic *Nicotiana tabacum*. *Plant Physiol* **111:** S-48

Weber H (2002) Fatty acid-derived signals in plants. *Trends Plant Sci* **7:** 217—224

Wendehenne D, Pugin A, Klessig DF and Durner J (2001) Nitric oxide: comparative synthesis and signaling in animal and plant cells. *Trends Plant Sci* **6:** 177—183

Willekens H, Chamnongpol S, Davey M, Schraudner M, Langebartels C, Van Montagu M, Inze D and Van Camp W (1997) Catalase is a sink for H_2O_2 and is indispensable for stress defence in C_3 plants. *EMBO J* **16:** 4806—4816

Winicov II and Bastola DR (1999) Transgenic overexpression of the transcription factor alfin1 enhances expression of the endogenous MsPRP2 gene in alfalfa and improves salinity tolerance of the plants. *Plant Physiol* **120:** 473—480

Xiang C, Werner BL, Christensen EM and Oliver DJ (2001) The biological functions of glutathione revisited in *Arabidopsis* transgenic plants with altered glutathione levels. *Plant Physiol* **126:** 564—574

Xiong L and Zhu J-K (2001) Abiotic stress signal transduction in plants: molecular and genetic perspectives. *Physiol Plant* **112:** 152—166

Youssefian S, Nakamura M and Sano H (1993) Tobacco plants transformed with the O-acetylserine (thiol) lyase gene of wheat are resistant to toxic levels of hydrogen sulphide gas. *Plant J* **4:** 759—769

Youssefian S, Nakamura M, Orudgev E and Kondo N (2001) Increased cysteine biosynthesis capacity of transgenic tobacco overexpressing an O-acetylserine(thiol) lyase modifies plant responses to oxidative stress. *Plant Physiol* **126:** 1001—1011

Zhang S and Klessig DF (2001) MAPK cascades in plant defense signaling. *Trends Plant Sci* **6:** 520—527

Zhu YL, Pilon-Smits EA, Tarun AS, Weber SU, Jouanin L and Terry N (1999) Cadmium tolerance and accumulation in Indian mustard is enhanced by overexpressing gamma-glutamylcysteine synthetase. *Plant Physiol* **121:** 1169—1178

7. ACKNOWLEDGEMENTS

The authors wish to thank to Prof. Grażyna Muszyńska, Dr. Grażyna Dobrowolska and Dr. Urszula Maciejewska for their critical comments. Work in the authors' laboratory has been supported by the State Committee for Scientific Research (KBN grant 6P04A07121)

8. AFFILIATIONS

Institute of Biochemistry and Biophysics, Polish Academy of Sciences, ul. Pawinskiego 5A, 02-106 Warsaw, Poland
For correspondence - *A. Sirko*: asirko@ibb.waw. pl

K.C. LAKKINENI, A. AHMAD AND Y.P. ABROL

HYDROGEN SULPHIDE: EMISSION AND UTILIZATION BY PLANTS

Abstract. Plants take up sulphur from both pedospheric and atmospheric sources. Both excessive and deficient conditions of sulphur level in plants causes imbalances in physiological processes, and thus, negatively affect the plant growth. To adapt to these conditions, plants possess physiological plasticity and developed different strategies, viz. avoidance, compartmentation and volatilization. Emission of volatile sulphur by plants appears to be responsible for maximum amount of sulphur that are thought to enter the atmosphere by natural process, and hydrogen sulphide is the major volatile sulphur-containing compound emitted by cells of higher plants. In case of sulphur deficient condition, which is prevalent in most agricultural soils world over, plants are also able to utilize atmospheric H_2S for growth. H_2S can, to a considerable extent, replace pedospheric sulphate as a source of sulphur. Thus, plants have potential for both synthesis and consumption of hydrogen sulphide. In this chapter, we will describe in detail the mechanism of H_2S emission and utilization by plants that will help to develop strategies for usefulness of the channeling of extra sulphur into specific sulphur pool in order to improve crop quality by using molecular biology tools.

1. INTRODUCTION

Hydrogen sulphide (H_2S) is the most reduced form of sulphur in nature. It attains importance in terms of its significant presence in the atmosphere mainly as a result of anthropogenic activity and biogenic emissions (Rennenberg, 1984; Bacci et al., 2000). Out of the total amount of sulphur that enters into the atmosphere through natural processes, majority constitutes volatile sulphur emitted by plants. H_2S is the major volatile S-containing compound emitted by cells of higher plants (Filner et al., 1984). Hydrogen sulphide in the atmosphere also attains importance in view of its availability to plants as a source of reduced sulphur. A number of studies have revealed that plant shoots are capable of utilizing the atmospheric H_2S to levels sufficient to meet the sulphur need of the whole plant (De Kok et al., 1997, 2000a; Stuiver and De Kok, 2001). Such findings hold practical implications especially in regions that are known to have widespread inadequacy of sulphur in the soils. The advent of cDNA clones for a variety of enzymes of sulphur metabolism further enhanced our understanding on the regulation of sulphate uptake and utilization as affected by atmospheric H_2S. Recent investigations revealed existence of good co-ordination between the metabolism of atmospheric H_2S taken by plant shoots and pedospheric sulphate (SO_4^{2-}) taken by plant roots (Westerman et al., 2000; De Kok et al., 2000b).

2. EMISSION OF HYDROGEN SULPHIDE BY PLANTS

Most studies conducted during 1980s and early 1990s on sulphur emissions are limited to the laboratory conditions where either detached leaves or leaf discs were used in the

Yash P. Abrol and Altaf Ahmad (eds.), Sulphur in Plants, 265—278.

experimentation. Sulphur, in the gaseous form, is emitted through the stomata of higher plants predominantly as H_2S. The content of H_2S emitted ranged from 0.06 to 2530 μmol kg^{-1} dw h^{-1} (Table 1, Ernst, 1990). A built up of excess sulphur in plant cells is attributed the primary cause of H_2S emission.

Table 1. Emission of hydrogen sulphide by plants under experimental conditions and their ecological relevance

Plant species	Emission rate (μmol kg^{-1} h^{-1})	Sulphur concentration in test condition	Sulphur speciation	Ref.
Cucumis sativus				
Young leaf	2530[a]	Maximum estimation	Sulphur	1
Mature leaf	235[a]	of 1000 μmol m^{-3} air	dioxide	
Lemna gibba	0.06–0.39	10–20 water	Sulphite	2
Lemna valdiviana	0.32–0.94			
Glycine max	2.11–8.45	0.04–1 μmol water	Na-sulphate	3
Picea abies 3yr seedling	0.07	250 mmol water	K-sulphate	4
Pinus sylvestris 3yr seedling	0.79	29.7 μmol m^{-3} air	Sulphur dioxide	5
Pinus sylvestris 15–25yr trees	5.28–21.1	1.6–6.2 μmol m^{-3} air	Sulphur dioxide	5
***Phaseolus* sp**	0.08–0.2	unknown	Probably sulphate	6

[a] assuming 90% water content in fresh material

Sources: [1]Sekiya et al. (1982); [2]Takemoto et al. (1986); [3]Winner et al. (1981); [4]Spaleny (1977); [5]Hallgren and Fredriksson (1982); [6]Filner et al. (1984)

Plants growing in sulphur-rich sites deal with a supply of sulphur, which far exceeds the requirement for optimal growth. Different species thriving in high sulphur-containing areas such as sulphataras, gypsum-bearing soils, and water-logged soils in pedosphere, all saline environments in hydrosphere, and vicinity of volcanoes, natural springs, burning lignite deposits and man-made all industrial areas, it is obvious that plants have adapted to excess sulphur through strategies viz., avoidance, compartmentalization and volatilization (Bartuli et al., 1997; Ernst, 1998). Avoidance is not the general mechanism by which plants cope with an external or internally generating excess of sulphur (Rennenberg, 1984). Once excess sulphur has entered a plant's cell, it may be stored directly as sulphate or after conversion into a storage form. When glycophytes and halophytes are exposed to sulphate in the rooting medium, sulphate concentration remains low in the vacuole (2-14 % of total S content) as long as the supply is low, but strongly increased at high external supply (van Diggelen, 1988). In the extreme sulphur-enriched environment of gypsum-bearing soils, more than 50 per cent of the plant S is stored in the form of sulphate, when the total plant S concentration exceeded 3 per cent (Boukhris and Lossaint, 1970, 1975). In *Moricandia arvensis* (a gypsophilous plant) about 92 per cent of total plant sulphur has been found as sulphate (Boukhris and Lossaint, 1975). Due to limitation of cellular storage capacity, the cell can either translocate sulphur to other cells and plant parts or release it into the environment after conversion into volatile compounds. Translocation takes place in both organic and inorganic forms. Experiments with tobacco, castor bean and cucurbit plants showed that glutathione is the predominant long-distance transport form of organic sulphur translocated in the phloem (Rennenberg et al., 1979; Bonas et al., 1982). In the petiole of leaves fumigated with [^{35}S] sulphur dioxide, substantial quantities of labeled organic sulphur were detected. Glutathione accounted for about 60 per cent of this fraction (Garsed and Read, 1977). Translocation of excess sulphur from one part of the plant to another, however, will only satisfy the plant's need to get rid of excess sulphur over a short period of time. A more suitable mechanism to prevent accumulation of toxic amount of sulphur appears to be its release into the environment. Because of a more rapid dilution, a release into the atmosphere, after volatilization, may be preferred to a release into the soil (Rennenberg, 1984).

Mechanisms leading to generation of hydrogen sulphide in the cell and its eventual release into the atmosphere are categorized in response to different sources of sulphur supply. Under laboratory conditions, H_2S emission was demonstrated in leaves supplied with excess sulphate (Sekiya et al., 1982), sulphite/sulphur dioxide (Hallgren and Fredriksson, 1982) or L-cysteine (Rennenberg, 1983). For each of these S sources a different path of H_2S synthesis was established (Rennenberg, 1984).

2.1 Hydrogen Sulphide Synthesis in Response to Sulphur Dioxide

Due to the widespread occurrence of sulphur dioxide in the atmosphere, its impact on the plant functioning has received most attention during the last two decades. Despite significant reduction of emission from anthropogenic source due to legislation, SO_2 is still the most important atmospheric sulphur compound taken up by plants. Sulphur dioxide is rapidly oxidized after entry into plants. Sulphate being the principal oxidation product which accumulated to high concentrations during chronic exposure of plants to SO_2 (Rennenberg, 1991). High internal sulphate concentration caused inhibition of growth of the plants. It may be considered as uncontrolled sulphur fertilization at the nutritional level and will, therefore, interact with processes that control the plant's sulphur nutrition. Several pieces of evidence suggest that emission of H_2S is a means by which plants get rid of excess sulphur and, thus, control sulphur nutrition. It was first observed by Materna (1966) that plants exposed to SO_2 in atmosphere re-emit part of the sulphur taken up. Sulphur accumulation in the leaves of *Geranium carolinianum* fumigated with sulphur dioxide was found to be 7 to 15 per cent lower than estimated from mass-balance calculations of sulphur dioxide fluxes (Taylor and Tingey, 1983). Numerous studies have shown that hydrogen sulphide is the predominant sulphur compound released by plants in response to SO_2 and that this release of H_2S is not restricted to a particular species, but is a general phenomenon in higher plants (Rennenberg, 1991). Published rates of hydrogen sulphide emission in response to SO_2 differ between 0.8 to 8.0 nmol sulphide h^{-1} cm^{-2} leaf area, but are usually between 1 and 2 nmol sulphide h^{-1} cm^{-2} leaf area (Hallgren and Fredriksson, 1982; Sekiya et al., 1982). These differences in the amount of H_2S emitted can at least partially be attributed to differences in the experimental conditions used. The amount of hydrogen sulphide emitted in response to SO_2 is positively correlated with the rate of influx of

SO_2 into the leaves, exhibiting a threshold of about 0.08 pmol SO_2 per cell over 15 min for the induction of measurable H_2S emission (Taylor and Tingey, 1983).

Hydrogen sulphide synthesis in plants in response to sulphur dioxide takes place in at least three possible following pathways.

Pathway I. H_2S emission may be the direct reduction of sulphide/sulphite. When cucurbit plants were exposed to unlabelled sulphate and fumigated with [^{35}S] sulphur, about 10 per cent of the absorbed radioactivity was incorporated into and released as H_2S. Analysis of the specific activities of SO_2 used for fumigation and the hydrogen sulphide emitted showed comparable values.

$$SO_2 \longrightarrow SO^-_3 \xrightarrow{h\nu} SH^- \overset{H^+}{\rightleftharpoons} H_2S$$

Upon exposure of cucurbit plants to [^{35}S] sodium sulphate and fumigated with unlabelled sulphur dioxide, only 0.2 per cent of the leaves' radioactivity (85 per cent of which was present in sulphate) was incorporated into H_2S. The specific activity of the hydrogen sulphide emitted was one order of magnitude smaller than the specific activity of the sulphate inside the sells (Sekiya et al., 1982).

Pathway II. Since a major part of the sulphur dioxide absorbed by the plant is oxidized to sulphate, H_2S synthesis in response to SO_2 exposure may include reduction of sulphate to sulphide. This reduction may be followed by a release of sulphide from carrier-bound sulphide, leading to emission of H_2S.

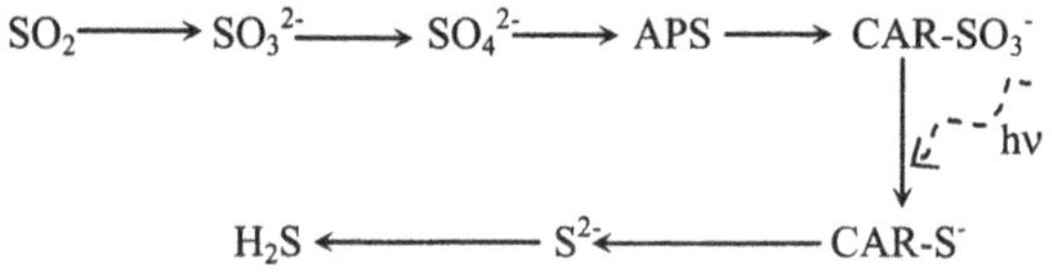

Pathway III. The third possible pathway of H_2S synthesis in response to SO_2 may be in the order of (i) reduction of sulphate to sulphide, (ii) incorporation of the sulphide moiety of carrier-bound sulphide into cysteine, and (iii) the subsequent degradation of cysteine to hydrogen sulphide.

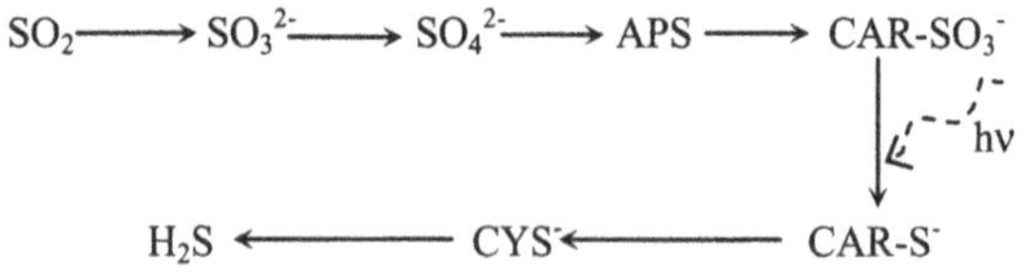

2.2 Hydrogen Sulphide Synthesis in Response to Sulphate

Plants growing in gypsum and salt marsh soils have to deal with a supply of sulphur in the root environment, which far exceed the requirement for optimal growth (Carlson and Forrest, 1982). When roots take up excess sulphate, the translocation of sulphate in the xylem may exceed the leaves' needs for sulphur. As storage of sulphur in the sink increases, the rate of transport will decrease. Therefore, over a longer period translocation of excess sulphur will be little more than a translocation of the plant's problem in getting rid of the excess sulphur taken up. Thus, the most suitable mechnism to overcome this problem is to release the excess sulphur into the atmosphere. In the experiments with leaf discs or detached leaves, Spaleny (1977) have reported that upon administration of sulphate near the site of reduction, hydrogen sulphide emission was usually much higher than if sulphate was fed to the root. The amount of H_2S emitted was found to be dependent on the sulphate concentration and the light fluence rate applied, but was of the same order of magnitude as emission in response to sulphur dioxide (Wilson et al., 1978). The rate of H_2S emission from sulphate can be inhibited by O-acetylserine, or by the precursor of this compound; acetyl Coenzyme and Coenzyme A. Filner et al. (1984) proposed that the extent to which the amount of reduced sulphate exceeds the amount of O-acetylserine available determines how much H_2S is emitted. This was supported by two experiments:

(i) 3-Fluropyruvate, an inhibitor of acetyl CoA (O-acetylserie precursor) synthesis by pyrivate dehydrogenase stimulated H_2S emission (Bisswanmder, 1981).

(ii) Emission of hydrogen sulphide in response to sulphate in continuous light goes through a series of transients. First it rises to a peak within about 1 hour, and then declines gradually. If the experiments were shifted from light to dark, then back to light, an overshoot peak followed by oscillating transients occurred (Sekiya et al., 1982). This was eliminated when O-acetylserine was supplied to leaf discs. Acetate does not work, nor does serine. These results suggest that H_2S is emitted when sulphite is produced in excess of the rate at which O-acetylserine is formed. It has also been suggested that immediately following a shift of leaf tissue from dark to light, acceleration of O-acetylserine production lags behind the acceleration of sulphide production, but eventually photosynthetic energy and photosynthetically generated carbon compounds increase the rate of O-acetylserine synthesis. This, therefore, results in a decreased rate of sulphide production and its subsequent release into the atmosphere as H_2S (Filner et al., 1984).

The three identified pathways of sulphate reduction leading to generation of H_2S are as follows:

(i) $SO^{-2}_{4} \longrightarrow APS \longrightarrow CAR\text{-}SO_3 \longrightarrow CAR\text{-}SH^- \longrightarrow SH^- \longrightarrow H_2S$

(ii) $SO^{-2}_{4} \longrightarrow APS \longrightarrow CAR\text{-}SO_3 \longrightarrow CAR\text{-}SH^- \longrightarrow CYS \longrightarrow SH^- \longrightarrow H_2S$

(iii) $SO^{-2}_{4} \longrightarrow APS \longrightarrow CAR\text{-}SO_3 \longrightarrow SO^-_3 \longrightarrow SH^- \longrightarrow H_2S$

2.3 Hydrogen Sulphide Synthesis in Response to Cysteine

Hydrogen sulphide emission in response to L-cysteine was reported in several studies (Harrington and Smith, 1980; Renennberg, 1983; Filner et al., 1984). Other amino acids like D-cysteine, L-cystine, O-Methyl-L-Cysteine, Cysteamine, L-Homocysteine, S-Methyl-L-cysteine, L-cysteic acid, DL-Cysteic acid, L-Serine and L-Methionine did not cause emission of hydrogen sulphide (Sekiya et al., 1982). As the hydrogen sulphide emitted is derived directly from the L-cysteine added, it appears to be generated (in leaf discs of cucurbitaceae) by the action of a pyrodoxal phosphate-dependent, L-cysteine-specific desulphydrase. Cysteine desulphydrase activity in cucurbits is enhanced by pre-incubation of leaf discs with L- or D-cysteine at millimolar concentrations (Rennenberg and Filner, 1983). The enzyme activity is also enhanced by pre-incubation with structural analogs of L-cysteine; cystine, however, it is not effective. Although stimulation of cysteine desulfydrase was half-maximal within 15 min of pre-incubation with L- or D-cysteine, the enhanced activity was not due to an activator formed upon pre-incubation or by inhibitor-presenting tissues not pre-incubated (Sekiya et al., 1982). Stimulation of cysteine desulphydrase activity in leaf discs is by itself light-independent; however, the pre-exposure of the entire plant to light or dark determines the extent to which stimulation by pre-incubation with L- or D-cysteine can take place. Darkness reduced the extent to which stimulation can occur, whereas exposure to light after a period of darkness restored the potential for stimulation of cysteine desulphydrase (Sekiya et al., 1982). These observations indicate that desulfhydration of L-cysteine is controlled by a compound that accomplishes such an inter-organ regulation of the potential for hydrogen sulphide emission in response to L-cysteine, which remains to be elucidated.

$$\text{Cys} \xrightarrow{\textit{Cyseteine desulphydrase}} SH^- \xrightarrow{H^+} H_2S$$
$$\searrow NH_4^+ + \text{Pyruvate}$$

Cyseteine desulphydrase was found to be responsible for the emission of H_2S by cucurbit plants in response to L-cysteine supply (Rennenberg and Filner, 1983). Pre-incubation of leaf discs in L-cysteine enhanced the activity of this enzyme, as judged by the assays of cell free extracts. The additional amount of cysteine desulfhydrase activity found after the pre-incubation with L-cysteine represented about 40-70 per cent of the leaves' total enzyme activity. Hydrogen sulphide can also arise from L-cysteine in cyanide dependent reaction catalyzed by the mitochondrial enzyme β-cyanoalanine synthase (Miller and Conn, 1980).

$$\text{L-cysteine} + \text{HCN} \xrightarrow{\beta\textit{-cyanoalanine synthase}} \beta\text{-cyanoalanine} + H_2S$$

Both L-cysteine desulfhydrase and β-cyanoalanine synthase enzymes known to have different catalytic functions in the metabolism, however, yield the common product H_2S. Apparently, the pathways leading to H_2S generation and its subsequent emission are part of an intracellular sulphur cycle (Rennenberg, 1984), which may operate to maintain cysteine concentration at a safer level in plant cells.

2.4 Regulation of H_2S Emission in Plants

Emission of hydrogen sulphide is regulated by various factors, viz. nitrogen nutrition, light, age of plants, and several metabolites.

2.4.1 Nitrogen nutrition

Studies on interaction between nitrogen and sulphur metabolism at molecular, cellular and whole plant level are widely published. However, limited information is available on co-regulation of N and S metabolism in relation to H_2S emission. O-acetylserine, a product of nitrogen metabolism was suggested to have regulatory control on the emission of H_2S in response to sulphate supply (Rennenberg, 1983). Availability of O-acetylserine to sulphate fed cucurbit cells declined the rate of H_2S emission by incorporating the sulphide released from the carrier into O-acetylserine to form cysteine. This indicated that the availability of O-acetylserine is the rate-limiting factor of cysteine synthesis in cells supplied with sulphate. In addition, alterations in light energy resulted in transients H_2S release, which coupled with changes in the rate of the biosynthesis of O-acetylserine (Rennenberg, 1983). Such adjustments seem to be suitable for the regulation of cysteine synthesis and may serve as a tool to co-ordinate sulphur and nitrogen metabolism in a constantly changing environment. N and S interaction in relation to H_2S emission was investigated in some crop species under laboratory conditions (Lakkineni et al., 1990). The enhancement in the emission of H_2S with an external supply of sulphate and cysteine was inhibited when nitrate was provided concomitantly with either sulphate or cysteine to mustard and groundnut leaf discs (Table 2). Similarly, supply of cysteine to *Brassica* leaves considerably induced the activity of the enzyme L-cysteine desulfhydrase (Lakkineni and Abrol, 1990). However, provision of NO_3^- in the incubation medium brought down the enzyme activity nearly by two fold. This phenomenon suggested a possibility that nitrogen interferes with the system responsible for H_2S emission at the level of substrate availability and diverts the sulphur that is being emitted to synthesize compounds containing N and S.

Table 2. Effect of sulphate, cysteine and nitrate on the emission of H_2S from the leaf discs of mustard, groundnut and wheat

Treatment	H_2S emission (nmol g^{-1} FW h^{-1})		
	Mustard	Wheat	Groundnut
Control (H_2O)	*0.801*	*1.265*	*0.245*
Sulphate (5 mM)	*1.151*	*1.852*	—
Cysteine (5 mM)	*1.108*	*2.187*	*0.804*
Sulphate + Nitrate (5 mM)	*0.814*	*1.288*	—
Cysteine + Nitrate (5 mM)	*0.722*	*2.634*	—

Source: Lakkineni et al. (1990)

2.4.2 Age of plants

Age of the plants also influenced the emission of H_2S. Young seedlings have a low potential for H_2S emission in response to sulphate, whereas 3-4 weeks old plants have high potential to emit H_2S in response to sulphate. The opposite was found true in the case of H_2S emission in response to L-cysteine. In addition, the potential for H_2S emission decreases with increasing age of individual leaf. In case of cucurbit plants, most of the potential for H_2S emission in respone to L-cysteine is built up during the early growth of plants, whereas high potential to emit H_2S in response to sulphate is developed at later growth stages. The extractable activity of cysteine desulfhydrase has also been shown to be under the developmental control. The developmental changes in cysteine desulfhydrase activity are strikingly similar to developmental changes in potential for H_2S emission in response to L-cysteine (Rennenberg and Filner, 1983).

2.4.3 Light

Light plays an important role in the synthesis of hydrogen sulphide from sulphate and sulphur dioxide. This was first reported by De Cormis (1968), and later confirmed by several workers (Wilson et al., 1978; Sekiya et al., 1982). Light is required to generate reductant, which may either be an intermediate in the photosynthetic electron chain, or the end product, NADPH. The instantaneous decline in emission rate at the onset of darkness coincided with generation of small and short-lived pool of the reactant, presumably reductant, leading to H_2S synthesis (Wilson et al., 1978).

3. UTILIZATION OF HYDROGEN SULPHIDE BY PLANTS

The occurrence of sulphurous gases in the atmosphere is largely attributed to volcanic activity, industry, vehicular traffic, agriculture and bio-industry. Besides sulphur dioxide, H_2S in the atmosphere forms a major component in the global sulphur cycle.

3.1 Sources of H_2S

Hydrogen sulphide originates predominantly from volcanoes, sulphur springs, anaerobic decomposition of S-containing organic matter and plant vegetation. Anthropogenic sources of H_2S include oil refineries, burning of coal, petroleum and natural gases, processes for producing pulp from wood, farming and animal production (De Kok et al., 2000b). The background levels of H_2S range from 0.3 ml L^{-1} (Kellogg et al., 1972; Beauchamp et al., 1983) in the less polluted areas to 0.1 µl L^{-1} in the polluted areas (Beauchamp et al., 1983).

3.2 Phytotoxicity of H_2S

Hydrogen sulphide reduces plant growth at relatively low levels (0.1 ml L^{-1}). However, a wide variation exists in susceptibility between plant species and cultivars (Maas et al., 1987; De Kok et al., 1998). The biochemical basis of H_2S toxicity has been ascribed to its reaction with a wide variety of NADH oxidizing enzymes, including heme-containing enzymes in a reaction similar to that of hydrogen cyanide (De Kok, 1989). The observed differences in susceptibility between the species and among cultivars are largely attributed to differences in H_2S deposition, plant morphology and cellular concentration of the absorbed H_2S (De Kok, 1989, De Kok et al., 1989; Stulen et al., 1990).

3.3 Kinetics of H_2S Deposition

Uptake of H_2S predominantly proceeds via the stomata. Both stomatal and internal (mesophyll) resistance determines the rate of H_2S uptake (Taylor et al., 1983; De Kok et al., 1989). The actual H_2S flux to shoots varied strongly between species and maximal H_2S fluxes were obtained in the order of 0.2 to 1.0 µl L^{-1} (Van der Kooij and De Kok, 1998). H_2S uptake by shoots showed saturation kinetics with respect to the atmospheric H_2S concentration, which can be described by the Michaelis-Menten equation. The rate of uptake of H_2S by shoots varies strongly between species and it coincides with the differences in sulphur need for growth (van der Kooij and De Kok, 1998). The internal resistance and hence the rate of uptake of H_2S is largely determined by the rate of metabolism of H_2S (Cope and Spedding, 1982), and is strongly affected by temperature (De Kok et al., 1991). For instance in spinach, the H_2S flux to shoots was 5-fold higher at 32 °C than at 1 °C (De Kok et al., 1991). The H_2S deposition velocity was found close only to that of diffusive conductance to H_2O vapour efflux at temperatures of 12 °C and higher and at H_2S concentrations >0.1 µl L^{-1} (De Kok et al., 1991). Evidently, the internal resistance in plants for H_2S gas flux appears to be determined by the rate of its metabolism in the plant (De Kok et al., 1998). The uptake kinetics may also be determined by the in situ activity of *O*-acetylserine (thiol)lyase (catalyzes the synthesis of cysteine), the degree of association of *O*-acetylserine (thiol)lyase with serine acetyltransferease (the latter catalyzes the synthesis of OAS) in the cysteine synthase complex, the affinity of *O*-acetylserine (thiol)lyase for sulphide and the in situ availability of *O*-acetylserine (OAS) (Stuiver and De Kok, 2001).

3.4 Metabolism of Atmospheric H_2S in Plants

Atmospheric H_2S taken up by the plants is directly metabolized into cysteine with high affinity and subsequently into other organic sulphur compounds (Fig. 1). Cysteine synthase is most likely directly involved in the fixation of atmospheric H_2S by the plants (De Kok et al., 1997). This is supported by the observation that the H_2S uptake kinetics could be influenced by the supply of its substrate at high atmospheric levels (Buwalda et al., 1992). O-acetylserine, the substrate for cysteine synthase, has been reported to play a limiting role in the assimilatory sulphate reduction (Brunold, 1993). There are circumstantial evidences that cysteine desulfhydrase might also be involved in the assimilation of H_2S (Schutz et al., 1991) as in vitro studies showed that this enzyme also may catalyze cysteine synthesis in the reverse direction, with sulphide and β-chloro-L-alanine as substrate (Poortinga et al., 1997). However, the significance of its in situ contribution to the metabolism of H_2S by plants needs further investigation.

Atmospheric H_2S deposition in plants generally results in a slight overload of the plant sulphur supply as is illustrated by an increased size and change in composition of the thiol pool in the shoot (Maas et al., 1987; De Kok et al., 1997). After some delay, H_2S exposure also may result in an enhanced thiol level in the roots, though to a lesser extent than that observed in shoots (Herschabach et al., 1995; Tausz et al., 1998). The increase in size of the thiol pool is dependent on the H_2S concentration, leaf age, the exposure temperature, the level of sulphur nutrition, and it, however, varies strongly between species (De Kok, 1989; Bosma et al., 1990; Buwalda et al., 1993; Poortinga and De Kok, 1997; Stuvier and De Kok, 1997). Glutathione is usually the most abundant thiol compound present in plant tissue (Rennenberg 1982, 1997, Herschbach et al., 2000). However, if the regulation of the uptake of sulphate by the roots is by-passed and sulphur is directly supplied to foliar tissue in either oxidized or reduced form viz., H_2S, then not only the size but also the composition of the thiol pool changes and in addition to glutathione, high levels of cysteine and other thiols may occur (Tausz et al., 1998). The physiological background underlying the changes in size and composition of the thiol pool upon H_2S exposure, however, needs to be resolved. O-acetylserine(thiol)lyase, the enzyme responsible for the metabolism of atmospheric H_2S into cysteine, is present in both chloroplasts and cytosol (Gotor et al., 1997; Heses et al., 1997; Hell, 1997). The increase in both size and altered composition of the thiol pool upon H_2S exposure could be due to an enhanced synthesis of cysteine in the cytosol, beyond the existing systems of feedback control of sulphate reduction and assimilation in the chloroplast. The use of transgenic plants overexpressing or depressing O-acetylserine (thiol)lyase in cytosol or chloroplast will be of immense help to study the changes in thiol composition upon H_2S exposure.

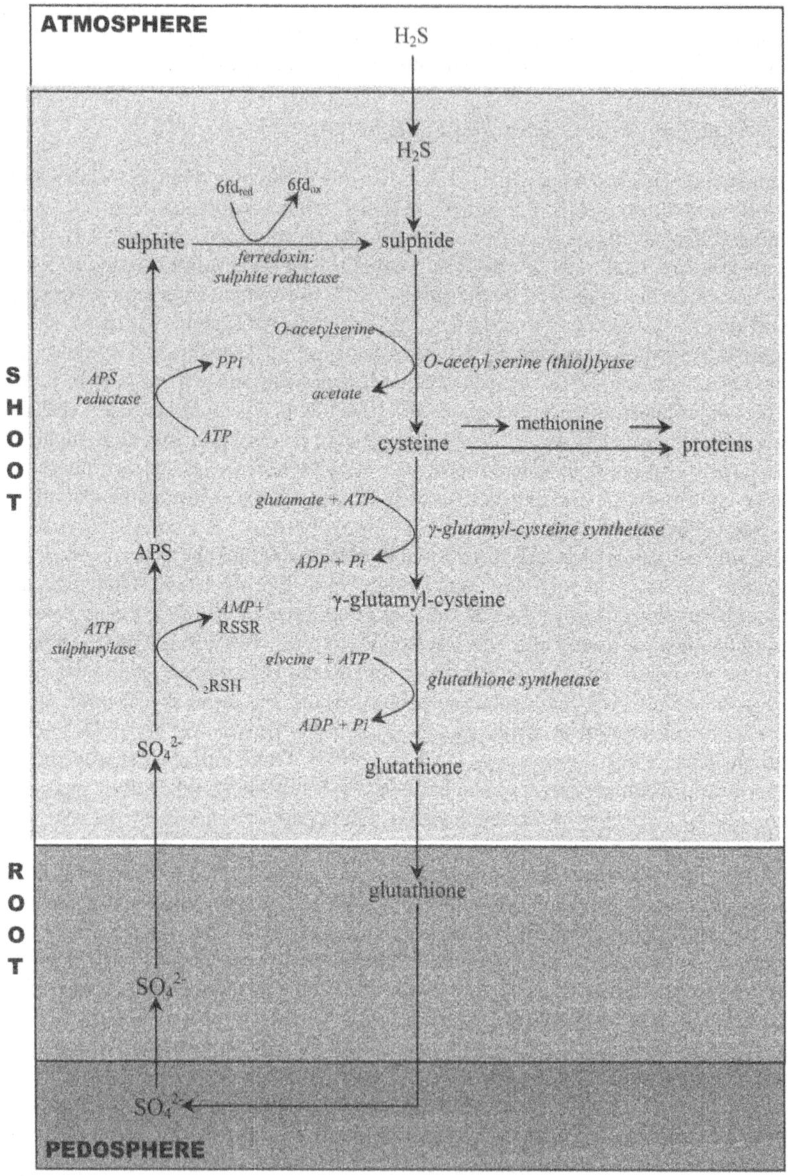

Figure 1. Metabolism of H_2S by plants

3.5 Uptake and Assimilation of Sulphate as Affected by H_2S Deposition

There is ample evidence to show that the atmospheric H_2S is utilized as a source of sulphur for plant growth (De Kok et al., 1997; Stuiver and De Kok, 1997). H_2S exposure resulted in a reduced uptake of sulphate by the roots and/or its transport to the shoots (Herschbach et al., 1995; De Kok et al., 1997). At an atmospheric H_2S level of 0.2 ml L^{-1} the sulphate uptake was reduced by about 40 per cent. The H_2S-induced reduction of sulphate uptake by the roots of *B. oleracea* was never higher than 60 per cent, not even at atmospheric H_2S levels higher than 0.2 m l L^{-1} and occurred within a few days after the onset of the exposure (Westerman et al., 2001a). In the absence of a sulphur source in the pedosphere, plants utilize H_2S as a sole sulphur source for growth (De Kok et al., 1991; 1997; 1998; Herschbach et al., 2000). At sufficient levels, H_2S abolishes sulphur deficiency symptoms, which occur upon sulphate-deprivation. From the H_2S uptake kinetics and the plant organic sulphur content, it was estimated that an atmospheric concentration of 0.2 μl L^{-1} H_2S was sufficient to meet the plant's sulphur demand to maintain growth under sulphate-deprived conditions (De Kok et al., 2000b). These findings clearly indicate that H_2S plays a significant role as source of atmospheric sulphur in the heavily polluted areas. However, it is still unclear which signals are directly involved in the down-regulation of both APS reductase in the shoots and the sulphate uptake by the roots. Feedback regulation of APS reductase may be triggered by glutathione. For the uptake by the roots and its transport to the shoot, both reduced sulphur compounds and the subcellular sulphate concentration may have significance in their regulation. From pilot studies, it became evident that the expression of the root sulphate transporters was affected by sulphate nutrition and H_2S exposure (Westerman et al., 2001b), though the level of expression could not directly be related to changes in the sulphate uptake by the roots. The discrepancy between the expression of the sulphate transporters and the sulphate uptake needs further evaluation. The significance of thiols as possible signal factors for the expression/ activity of the sulphate transporters in the partial down regulation of sulphate uptake upon H_2S exposure is currently under investigation.

4. REFERENCES

Bacci E, Gaggi C, Lanzillotti E, Ferrozzi S and Valli L (2000) Geothermal power plants at Mt. Amiata (Tuscany-Italy): Mercury and hydrogen sulphide deposition revealed by vegetation. *Chemosphere* **40:** 907—911

Bartuli C, Botre C, Botre F and Peccie G (1997) Atmospheric pllution originating from the interaction of different gaseous effluents. *J Environ Pathol Toxicol Oncol* **16:** 245—252

Beauchamp Jr RO, Bus JS, Popp JA, Boreiko CJ and Andjelkovich DA (1983) A critical review of the literature on hydrogen sulphide toxicity. *CRR Critical Rev Toxicol* **13:** 25—97

Bisswanmder H (1981) Substrate specificity of the pyruvate dehydrogenase complex from *Escherchia coli. J Biol Chem* **256:** 815—822

Bonas H, Schmitz K, Rennenberg H and Bergmann L (1982) Phloem transport of sulphur in *Ricinus. Planta* **155:** 82—88

Bosma W, Kamminga G and De Kok LJ (1990) H_2S-induced accumulation of sulfhydryl-compounds in the leaves of plants under field and laboratory exposure. In: Rennenberg H, Brunold CH, De Kok LJ and Stulen I (eds), Sulfur Nutrition and Sulfur Assimilation in Higher Plants: Fundamental, Environmental and Agricultural Aspects, pp 173—175. SPB Academic Publishing, The Hague

Boukhris M and Lossaint P (1970) Sur La teneur en soufre de quelques.plantes gypsophile de Tunisie. *Oecol Plant* **5:** 345—354

Boukhris M and Lossaint L (1975) Aspects écologiques de la nutrition minérale des plantes gypsicoles de Tunisie. *Rev Écol Biol Sol* **12:** 329—248

Brunold C (1993) Regulatory interactions between sulphate and nitrate assimilation. In: Renneberg H, Brunold C, De Kok LJ and Stulen I (eds) Sulfur Nutrition and Sulfur Assimilation in Higher Plants: Regulatory, Agricultural and Environmental Aspects, pp 61—75. SPB Academic Publishing, The Hague

Buwalda F, De Kok LJ and Stulen I (1992) The flux of atmospheric H_2S to spinach leaves can be affected by the supply of O-acetylserine. *Phyton* **32:** 15—18

Buwalda F, De Kok LJ and Stulen I (1993) Effects of atmospheric H_2S on thiol composition of crop plants. *J Plant Physiol* **142:** 281—285

Carlson Jr PR and Forrest J (1982) Uptake of dissolved sulfide by *Spartina alterniflora*: Evidence from natural sulfur isotopes abundance ratios. *Science* **216:** 633—635

Cope DM and Spedding DJ (1982) Hydrogen sulphide uptake by vegetation. *Atoms Environ* **16:** 349—353

De Cormis L (1968) Degagement d'hydrogene sulfure par des plantes soumises a une atmosphere contenant de l'anhydride sulfureux. *CR Acad Sci* **266D:** 683—685

De Kok LJ (1989) Responses of sulfur metabolism in plants to atmospheric hydrogen sulfide. *Phyton* **29:** 189—201

De Kok LJ, Stahl K and Rennenberg H (1989) Fluxes of atmospheric hydrogen sulfide to plant shoots. *New Phytol* **112:** 533—542

De Kok LJ, Rennenberg H and Kuiper PJC (1991) The internal resistance in spinach shoots to atmospheric H_2S deposition is determined by metabolism processes. *Plant Physiol Biochem* **29:** 463—470

De Kok LJ, Stuiver CE, Rubbing M, Westerman S and Grill D (1997) Impact of atmospheric sulfur deposition on sulfur metabolism in plants: H_2S as sulfur source for sulphur deprived *Brassica oleracea* L. *Bot Acta* **110:** 411—419

De Kok LJ, Stuiver CEE and Stulen I (1998) The impact of elevated levels of atmospheric H_2S on plants. In: De Kok LJ and Stulen I (eds), Responses of Plant Metabolism to Air Pollution and Global Change, pp 51—63. Backhuys Publishers, Leiden

De Kok LJ, Stuiver CEE, Westerman S and Stulen I (2000a) Atmospheric H_2S pollution- Deposition and impact on sulphur metabolism in plants. In: Yunus M, Singh N and De Kok LJ (eds), Environmental Stress: Indication, Mitigation, and Eco-conservation, pp 135—141. Kluwer Academic Publishers, Dordrecht

De Kok LJ, Westerman S, Stuiver CEE and Stulen I (2000b) Atmospheric H_2S as plant sulfur source: interaction with pedospheric sulfur nutrition - a case study with *Brassica oleracea* L. In: Brunold C, Rennenberg H, De Kok LJ, Stulen I and Davidian JC (eds), Sulfur Nutrition and Sulfur assimilation in Higher Plants: Molecular, Biochemical and Physiological Aspects, pp 41—55. Paul Haupt, Bern

Ernst WHO (1990) Ecological aspects of sulphur metabolism. In: Rennenberg H, Brunold CH, De Kok LJ and Stulen I (eds), Sulphur Nutrition and Sulphur Assimilation in Higher Plants: Fundamental, Environmental and Agricultural Aspects, pp 131—144. SPB Academic Publishing, The Hague

Ernst WHO (1998) Ecotypic variation and environmental adaptation to air pollution and global change. In: De Kok LJ and Stulen I (eds), Responses of Plant Metabolism to Air Pollution and Global Change, pp 217—232. Backhyus Publishers, Leiden

Filner P, Rennenberg H, Sekiya J, Bressan RA, Wilson LG, LeCureux L and Shimei T (1984) Biosynthesis and emission of hydrogen sulphide by higher plants. In: Koziol MJ and Whatley FR (eds), Gaseous Air Pollutants and Plant Metabolism, pp 291—312. Butterworth, London

Garsed SG and Read DJ (1977) Sulphur dioxide metabolism in soybean, Glycine max var. Biloxi. II. Biochemical distribution of $^{35}SO_2$ products. *New Phytol* **79:** 583—592

Gotor C, Cejudo FJ, Barroso C and Vega JM (1997) Tissue specific expression of ATCYS-3A, a gene encoding the cytosolic isoform of O-acetylseine(thiol)lyase in *Arabidopsis*. In: Cram WJ, De Kok LJ, Stulen I, Brunold and C and Rennenberg H (eds) Sulphur Metabolism in Higher Plants: Molecular, Ecophysiological and Nutritional Aspects, pp 285—288. Backhuys Publishers, Leiden

Hallgren JE and Fredriksson SA (1982) Emission of hydrogen sulfide from sulfur dioxide fumigated pine trees. *Plant Physiol* **70:** 456—459

Harrington HM and Smith IK (1980) Cysteine metabolism in cultured tobacco cells. *Plant Physiol* **65:** 151—155

Hell R (1997) Molecular physiology of plant sulfur metabolism. *Planta* **202:** 138—148

Herschbach C, De Kok LJ and Rennenberg H (1995) Net uptake of sulfate and its transport to the shoot in tobacco plants fumigated with H_2S or SO_2. *Plant Soil* **175:** 75—84

Herschbach C, van der Zalm E, Schneider A, Jouanin L, De Kok LJ and Rennenberg H (2000) Regulation of the uptake and transport of sulphate in wildtype and transgenic poplar trees overexpressing γ-glutamylcysteine synthetase as affected by atmospheric H_2S. *Plant Physiol* **124:** 461—473

Heses H, Lipke J, Altman T and Hofgen R (1997) Expression analysis and subcellular localization of cysteine synthase isoforms from *Arabidopsis thaliana.* In: Cram WJ, De Kok LJ, Stulen I, Brunold and C and Rennenberg H (eds) Sulphur Metabolism in Higher Plants: Molecular, Ecophysiological and Nutritional Aspects, pp 227—230. Backhuys Publishers, Leiden

Lakkineni KC, Nair TVR and Abrol YP (1990) S and N interaction in relation to H_2S emission in some crop species. In: Rennenberg H, Brunold C, De Kok LJ amd Stulen I (eds) Sulphur Nutrition and Sulphur Assimilation in Higher Plants: Fundamental, Environmental and Agricultural Aspects, pp 213—216. SPB Academic Publishing, The Hague

Maas FM, De Kok LJ, Peters JL and Kuiper PJC (1987) A comparative study on the effects of H_2S and SO_2 fumigation on the growth and accumulation of sulfate and sulfhydryl compounds in *Trifolium pratense* L., *Glycine max* Merr. and *Phaseolus vulgaris* L. *J Exp Bot* **38:** 1459—1469

Materna J (1966). Die Ausscheidung des durch die Fichtennadeln absorbiertes Schwefeldioxids. *Arch Forstwesen* **15:** 691—692

Miller JM and Conn EE (1980) Metabolism of hydrogen cyanide by higher plants. *Plant Physiol* **65:** 1199—1202

Poortinga AM and De Kok LJ (1997) Uptake of atmospheric H_2S by *Spinacia oleracea* L. and consequences for thiol content and composition in shoots and roots. In: Cram WJ, De Kok LJ, Stulen I, Brunold and C and Rennenberg H (eds) Sulphur Metabolism in Higher Plants: Molecular, Ecophysiological and Nutritional Aspects, pp 280—285. Backhuys Publishers, Leiden

Rennenberg H (1982) Glutathione metabolism and possible biological roles in higher plants. *Phytochemistry* **21**: 2771—2781

Rennenberg H (1983) The role of O-acetylserine in hydrogen sulphide emission from pumpkin leaves in response to sulfate. *Plant Physiol* **73:** 560—565

Rennenberg H (1984) The fate of excess sulfur in higher plants. *Annu Rev Plant Physiol* **35:** 121—153

Rennenberg H (1997) Molecular approaches to glutathione biosynthesis. In: Cram WJ, De Kok LJ, Stulen I, Brunold and C and Rennenberg H (eds) Sulphur Metabolism in Higher Plants; Molecular, Ecophysiological and Nutritional Aspects, pp 59—70. Backhuys Publishers, Leiden

Rennenberg H and Filner P (1983) Developmental changes in the potential for H_2S emission in cucurbit plant Plant *Physiol* **71:** 269—275.

Rennenberg H, Schmitz K and Bergmann L (1979) Long-distance transport of sulfur in *Nicotiana tabacum. Planta* **23:** 667—672

Rennenderg H (1991) The significance of higher plants in the emission of sulfur compounds from terrestrial ecosystems. In: Sharkey TD, Holland EA and Mooney HA (eds) Trace Gas Emission by Plants, pp 217-260, Academic Press, San Diego

Schütz B, De Kok LJ and Rennenberg H (1991) Thiol accumulation and cysteine desulfhydrase activity in H_2S-fumigated leaves and leaf homogenates of cucurbit plants. *Plant Cell Physiol* **32:** 733—736

Sekiya J, Wilson LG and Filner P (1982) Resistance to unjury by sufur dioxide. *Plant Physiol* **70:** 437—441

Spaleny J (1977) Sulphate transformation to hydrogen sulphide in spruce seedlings. *Plant Soil* **48:** 557—563

Stuiver CEE and De Kok LJ (1997) Atmospheric H_2S as sulphur source for sulphur deprived *Brassica oleracea* L. and *Hordeum vulgare* L. In: Cram WJ, De Kok LJ, Stulen I, Brunold and C and Rennenberg H (eds), Sulphur Metabolism in Higher Plants; Molecular, Ecophysiological and Nutritional Aspects, pp 293—294. Backuys Publishers, Leiden

Stuiver CEE and De Kok LJ (2001) Atmospheric H_2S as sulfur source for plant growth: Kinetics of H_2S uptake and activity of *O*-acetylserine(thiol)lyase as affected by sulfur nutrition. *Env Exp Bot* **46:** 29—36

Stulen I, Posthumus FS, Amancio S and De Kok LJ (1990) Why is H_2S not phytotoxic in monocots? *Physiol Plant* **79:** A123

Takemoto BK, Noble RD and Harrington HM (1986) Differential sensitivity of duckweeds (Lemnaceae) to sulphite. II. Thiol production and hydrogen sulphide emission as factors influencing sulphite phytotoxicity under low and high irradiance. *New Phytol* **103:** 541—548

Tausz M, van der Kooij TAW, Müller M, De Kok LJ and Grill D (1998) Uptake and metabolism of oxidized and reduced sulfur pollutants by spruce trees. In: De Kok LJ and Stulen I (eds), Responses of Plant Metabolism to Air Pollution and Global Change, pp 457—460. Backhuys Publishers, Leiden

Taylor GE Jr and Tingey DT (1983) Sulphur dioxide flux into leaves of *Geranium carolinianum* L.Evidence of non-stomal and residual resistance. *Plant Physiol* **72:** 237—244

Taylor GE, McLaughlin SB, Shriner DS and Selvidge WJ (1983) The flux of sulphur containing gases in to vegetation. *Atmos Environ* **17:** 789—796

van der Kooij TAW and De Kok LJ (1998) Kinetics of deposition of SO_2 and H_2S to shoots of *Arabidopsis thaliana* L. In: De Kok LJ and Stulen I (eds.), Responses of Plant Metabolism to Air Pollution and Global Change, pp 479—481. Backhuys Publishers, Leiden

van Diggelen J (1988) A comparative study on the ecophysiology of salt marsh halophytes. Ph.D. Thesis. Free University, Amsterdam

Westerman S, De Kok LJ and Stulen I (2000) Interaction between metabolism of atmospheric H_2S in the shoot and sulfate uptake by the roots of curly kale (*Brassica oleracea* L.). *Physiol Plant* **109:** 443—449

Westerman S, De Kok LJ, Stuiver CEE and Stulen I. (2001a) Sulfate uptake and utilization by two varieties of *Brassica oleracea* with different sulfur need as affected by atmospheric H_2S. *Phyton* **41:** 49—62

Westerman S, Stulen I, Stuter M, Brunold C and De Kok LJ (2001b) Atmospheric H_2S as sulfur source for *Brassica oleracea*: consequences for the activity of the enzymes of the assimilatory sulfate reduction pathway. *Plant Physiol Biochem* **39:** 425—432

Wilson LG, Bressan RH and Filner P (1978). Light dependent emission of hydrogen sulphide from plants. *Plant Physiol* **61:** 184—189

Winner WE, Smith CL, Koch GW, Mooney HA, Bewley JD and Krouse HR (1981) Rates of emission of H_2S from plants and patterns of stable sulphur isotope fractionation. *Nature* **289:** 672—673

5. AFFILIATIONS

201-2300 Grenet, St-Laurent, Montreal, QC, Canada H4L 4Y9 (KCL), Department of Environmental Botany, Hamdard University, New Delhi, India (AA, YPA)
For correspondence - *KC Lakkineni*: karunachand_lakkineni@lycos.com

M. AGRAWAL

PLANT RESPONSES TO ATMOSPHERIC SULPHUR

Abstract. Gaseous sulphur dioxide, particulate sulphates and sulphuric acid in wet deposition are the main anthropogenic sulphur species in the atmosphere. Atmospheric emissions of sulphur species result in an increase in acid deposition and tropospheric levels of sulphate aerosols. Anthropogenic emission of SO_2 showed a declining pattern in North America and Europe since 1980, however, emissions from South America, Africa and Asia are increasing dramatically. As sulphur is an essential element for plant growth, low concentrations of SO_2 have been shown to stimulate the growth and physiological responses, especially for plants growing in sulphur deficient soil, where SO_2 might be metabolised to fulfil the demand of sulphur as nutrient. However, the higher uptake of SO_2 turns toxic and has been reported to damage plants causing reduction in growth and productivity through interference with different physiological and metabolic processes. SO_2 gains entry into the leaves through stomata, and readily dissolves in the apoplastic water to produce mainly sulphite (SO_3^{2-}), bisulphite (HSO_3^-) and H^+ ions. The phytotoxicity due to SO_2 has been ascribed to the reactions of sulphite and bisulphite ions with various compounds. Most leaves have the capacity to detoxify sulphite and bisulphite by oxidising them to less toxic sulphate ions through a series of reactions, if the concentrations are not excessively high. The SO_2 absorbed on foliage may undergo into reductive conversion including assimilation into organic sulphur compounds and export from the leaves in the form of H_2S. The oxidative or reductive pathway of SO_2 in plants depends on the plant species, soil sulphur content, concentration and duration of SO_2 exposure and developmental stage of plants.

1. INTRODUCTION

Atmospheric sulphur has always been present throughout the course of Earth's history. Interest in this form has increased tremendously due to concern with environmental pollution, as sulphur pollution is one of the most recognized environmental problems throughout the world, which has generated general public concern also. Anthropogenic sources contribute a significant fraction of sulphur in the atmosphere since the beginning of the industrialization and urbanization. The main anthropogenic sulphur species released into the atmosphere are gaseous SO_2 and particulate sulphate. Locally high levels of H_2S may also be due to anthropogenic activity. The reduced sulphur compounds undergo oxidation in the atmosphere to form SO_2 which further undergo oxidation, primarily with hydroxyl radicals on the surface of particles, or in precipitation leading to formation of sulphate either in particles or in clouds and rain droplets. Gaseous SO_2 may also oxidized due to peroxylradical or ozone, or in the presence of metal catalysts. Rennenberg (1991) described that the contribution of atmospheric sulphur ranges from 38-89 Tg a^{-1} by marine environment and terrestrial ecosystems. The anthropogenic sulphur emissions are also in the same range, with a high degree of uncertainty. Fossil fuel combustion accounts for 80 to 85% of the total anthropogenic sulphur emission, with the remainder coming from additional sources such as certain metallurgical processes, petroleum refineries, coking of coal, manufacture of paper and pulp, other industrial processes and burning. The emission of sulphur from these sources is largely in the form of gaseous SO_2. H_2S is other

Yash P. Abrol and Altaf Ahmad (eds.), Sulphur in Plants, 279—294.

sulphurous form known to be phytotoxic (De Kok, 1990). H_2S emission and utilization has been dealt with by Lakkineni et al. (this book). The present review will be restricted to the consequences of SO_2 influx in plants.

2. PATTERN OF ATMOSPHERIC SULPHUR DEPOSITION

Before the industrial revolution, mean atmospheric S depositions were below 10 kg S ha^{-1} a^{-1}, but from 1890 to 1980, S deposition increased steadily by 0.47 kg S ha^{-1} a^{-1} (Daemmgen et al., 1998) due to increased utilization of fossil fuels along with industrial and urban developments. Sulphur emissions in North America and Europe have shown an upward trend from the early part of the last century into the decade of the 1970, but thereafter a declining pattern was observed due to control measures on sulphur emission in the region (Iversen et al., 1989). The sulphur emission, which was estimated as 17 Tg a^{-1} in 1970, went down to 12 Tg a^{-1} in 1985 in North America. Similarly in Western Europe the sulphur emission of 26 Tg a^{-1} was down to 21 Tg a^{-1} in 1988 (Iversen et al., 1989). Arends et al. (1997) reported approximately 3% a^{-1} decrease in summer sulphate concentration in Europe. However, emissions from Asia, South America and Africa are increasing dramatically. The future sulphur emissions estimated on the basis of per capita emission and population growth show increase of 6 times for South America, 7 times for Africa and 6.5 times for Asia (Galloway, 1989). On the basis of average emission density for sulphur (0.4 g m^{-2} a^{-1}), the ratio of anthropogenic to natural emissions is of order unity or slightly greater on a global basis but at highly industrialized areas, with high average emission density, the ratio is approximately 10.

Global emissions of SO_2 are 294 mt a^{-1}, of which 160 mt are of anthropogenic origin (UNEP/GEMS, 1991). About 90% of SO_2 emission arise from Europe, North America, and Asia in Northern Hemisphere. The approximate 1980 regional apportionment of fossil fuel sulphur emission is estimated as 16% for China and Japan, 28% for USA and Canada, 34% for Europe, 15% for the remainder of Northern Hemisphere and 7% for Southern Hemisphere (Hameed and Dignon, 1988). Global anthropogenic emissions of SO_2 are predicted to rise by 30% beyond year 2000 upon present day rates due to forecasts of increased coal combustion (Wellburn, 1994).

Global anthropogenic emissions of SO_2 will range from 86 to 106.5 Tg a^{-1} in 2010 and from 139 to 279.5 Tg a^{-1} in the year 2100 for the protocol and no protocol scenarios, respectively (Alcamo et al., 1995). Based on the RAINS model, SO_2 emissions in India show an increase of 321% i.e. from 4.48 mt to 18.87 mt under the base scenario during the period 1990-2020 (Panwar, 1998). The percentage of SO_2 emissions from coal and oil are 49 and 29%, respectively in 1990 and would increase to 50 and 40% in 2020 due to increase in share of oil consumption in energy production. Though SO_2 emission from developed countries has been reduced significantly, SO_2 is still the most important atmospheric sulphur form affecting the vegetation.

3. RESPONSES OF PLANT TO SULPHUR DIOXIDE

SO_2 is a unique form of anthropogenic S as its low doses may have beneficial effects on plant productivity but increased uptake causes depressing effects on plant growth.

3.1 Uptake

SO_2 uptake by vegetation may occur in three different ways: (i) dry deposition on cuticle, (ii) wet deposition on cuticle, and (iii) diffusion through stomata (Rennenberg and Polle, 1994). Dry deposition of SO_2 on the cuticle does not contribute significantly to the SO_2 influx into the plants due to extremely low velocity of SO_2 permeation (Lendzian and Kerstiens, 1991). Wet deposition of SO_2, though is higher than dry deposition, it is limited due to lowering of pH leading to reduced SO_2 solubilization (Fowler, 1985). SO_2 mainly gains entry into the leaves through stomata following the same diffusion pathway as CO_2. Black and Unsworth (1980) found that 90% of the SO_2 deposited in *Vicia faba* leaves entered through the stomata. SO_2 influx in plants was a linear function of atmospheric concentration (Tausz et al., 1998). SO_2 enters the stomatal cavity as a result of diffusion gradients maintained through the reaction of SO_2 with water to form bisulphite (HSO_3^-) and sulphite (SO_3^{2-}) ions. At the pH range of 4.5-6.8 of apoplastic space HSO_3^- formation predominates over SO_3^{2-}. Bisulphite accumulates in the apoplastic water of the cell well and converted to SO_4^{2-} as a result of sulphite oxidase requiring $O^{\cdot}_2$, H_2O_2 and monophenols that are catalyzed by peroxidases. These reactions also generate $HSO^{\cdot}_3$ radicals, which have damaging influence on cell membrane (Rennenberg and Pole, 1994). Accumulation of SO_4^{2-} was reported in the apoplastic space due to infiltration of leaves with HSO_3^- (Pfauz et al., 1990). The sulphate produced by apoplastic enzyme activity enters to the apoplastic pool of sulphate which is transported into the leaf via the transpiration stream leading to lower apoplastic sulphate concentration (Wolfenden et al., 1991). Sulphate can be taken up in symplasm by the cell via active carrier mediated transport (Cram, 1990). Sulphur dioxide can also enter the cytoplasm directly by diffusing through the cell membrane where it is converted to SO_4^{2-}. Both SO_4^{2-} and SO_2 can be incorporated into normal sulphur metabolic system of plants. Tausz et al. (1998) found a pronounced increase of sulphate content (50% of the control) in the needles of three years old spruce trees preconditioned for a week with 225 nl L^{-1} SO_2 in fumigation cabinets. Most of the studies showed sulphur accumulation in plants exposed to SO_2 in fumigation experiments as well as under field condition (Kaiser et al., 1993a; Agrawal and Singh, 2000).

3.2 Stomatal Conductance

The flux of the SO_2 to the leaf interior is a function of the concentration gradient between the atmosphere and the leaf interior, and the sum of the physical, chemical and biological resistances to diffusion from source to sink. The dose of SO_2 absorbed at leaf level is controlled by factors such as stomatal aperture and frequency, and, therefore, the plants exhibiting intrinsically higher stomatal conductance showed more susceptibility to SO_2 damage (Darrall, 1989). Winner et al. (1985) have reported that the absorption of SO_2 by the mesophyll tissue of leaf is proportional to the stomatal conductance. Variations in sensitivity to SO_2 have often been correlated with stomatal conductance. Increase in stomatal opening induced during SO_2 exposure is variously reported (Rao and Dubey, 1988; Singh et al., 1990; Deepak and Agrawal, 2001). In contrast, reduction in stomatal aperture was reported with increasing SO_2 dose leading to reduction in transpiration rate (Ashenden, 1979). More sensitive cultivars of soybean closed stomata quickly in response to SO_2 (Amundson and Weinstein, 1981). At natural ambient concentrations of SO_2 diurnal stomatal response is similar to that of movement of CO_2 into the leaves and H_2O outwards. Environmental conditions such as irradiant, vapor pressure deficit, atmospheric CO_2 concentration, soil moisture, etc strongly influence the rate of uptake through effects on stomata (Wolfenden and Mansfield, 1991).

Pollutant molecule once enters the stomatal cavity, the inward facing surface of the epidermis is suggested to be major site of uptake of SO_2 (Wolfenden and Mansfield, 1991). Neighbour et al. (1988) provided evidence of epidermal cell damage in birch trees exposed to 20-60 ppm SO_2 and NO_2. The surface of leaves from SO_2 exposed plants showed areas of wide open stomata surrounded by damaged epidermal cells. Pande (1985) applied 40, 80 and 120 ppb SO_2 over a three week period to five spring barley cultvars and found that the most tolerant cultivar, Midas, showed the lowest stomatal conductance and the most sensitive cultivar, Koru, showed the highest stomatal conductance. Increase in stomatal aperture is caused by changes in turgor of the guard cells due to the movement of K^+ accompanied by internal production of malate, a counter ion (Raschke, 1979). Sulphite mediated decrease in stomatal opening is also described (Rao and Anderson, 1983). Kimmerer and Kozlowski (1981) suggested two types of avoidance response to SO_2 at stomatal level shown by five clones of *Populus tremuloides* which differ in SO_2 susceptibility. Few clones showed passive avoidance, i.e. maintenance of a low gas exchange capability in the presence and absence of SO_2, and active avoidance, i.e. reduction of gas exchange upon SO_2 stress. In a field study, however, Bell and Mudd (1976) did not find consistent differences in stomatal conductance between resistant and susceptible cultivars of perennial ryegrass. These differences in stomatal response to SO_2 indicate that internal resistance are the result of solubilization of SO_2 in water surrounding mesophyll cells, the influx of SO_2 into mesophyll cells, its metabolism and the translocation of its metabolic products (Rennenberg, 1984). Taylor and Tingey (1983) have shown the existence of a diffusion barrier (the plasmalemma and or the chloroplast envelope) for SO_2 other than stomatal aperture in *Geranium carolinianum*. The mechanism and regulation of influx of SO_2 or its products into the mesophyll cells are not fully understood. However, sulphite influx at the rate of 1 μmol mg chlorophyll^{-1} h^{-1} concentration of 2 mmol sulphite was shown for isolated spinach chloroptast (Hampp and Ziegler, 1977).

3.3 Photosynthesis and Pigments

Primary site of SO_2 action is the chloroplast (Schiffgens-Gruber and Lutz, 1992). The effects of SO_2 exposure on photosynthesis (Ps) have been extensively studied both for short term and long term SO_2 exposures (Darrall, 1989; Saxe, 1991; Wolfenden and Mansfield, 1991; Okpodu et al., 1996; Verma et al., 2000). SO_2 adversely affects net photosynthesis in plants. Photosynthetic inhibition was normally found in short-term exposures when the concentration was above 400 ppmh and at the lower concentrations only after 2 hours or more of exposure (see review Saxe, 1991). In most of the cases high doses of SO_2 or prolonged exposures at even low concentrations are very inhibitory to Ps. Initial stimulation of Ps at very low concentration of SO_2 has also been reported in some plants (Atkinson et al., 1988), which was short lived and again turned inhibitory with the increase in duration of exposure. Darrall (1989) summarized that no change in Ps of different cultivars of barley and bean plants occurred below 200 and 300 ppb, respectively. The same was inhibited in wheat and oilseed only at 500 ppb or above concentration of SO_2. The inhibitory effect of SO_2 on photosynthesis may be irreversible or reversible depending upon the concentration of SO_2 applied and duration of exposure.

Different workers have proposed various mechanisms for the reduction in Ps due to SO_2 exposure. Anderson and Duggan (1977) reported that low level of SO_3^{2-} inhibited enzyme activated light modulation and consequently the light regulation of CO_2 fixation. Shimazaki et al. (1980) suggested that photosystem II activity might be inhibited in chloroplasts of spinach leaves treated with SO_2. The superoxide (O_{2-}) and other free radicals produced due to dissolution of SO_2 into the cell sap can attack the susceptible sites such as lipids or proteins in the plasma membrane and, thus, alter its permeability (Malhotra and Khan, 1984). Bisulphite compounds formed by SO_2 exert nonspecific effect on chloroplast membrane and disrupt fluxes of iron and cause leakage of photosynthetic products. Formation of glycollate bisulphite has been reported as one of the key mechanisms for Ps inhibition at enzymatic level (Tanaka et al., 1972).

Investigations on impacts of SO_2 at ultrastructural level showed swelling of thylakoid of chloroplast or reduction in grana lamellae (Malhotra and Khan, 1984). The injuries in thylakoids may be correlated with the decline in amount of chlorophyll and reduced CO_2 uptake. Reduction in chlorophyll content has been reported due to SO_2 pollution (Agrawal et al., 1982; Nandi et al., 1984; Verma and Agrawal, 1996). Such decrease has been attributed either to the inhibition of chlorophyll or its destruction (Malhotra and Khan, 1984). Tausz et al. (1996) found that 0.06 ppm SO_2 reduced the chlorophyll concentration in Norway spruce (*Picea abies* L.) plants. Carotenoids were found more sensitive to SO_2 as compared to the chlorophyll (Agrawal et al., 1982).

3.4 Respiration

Dark respiration either stimulated or remained unaltered due to acute SO_2 pollution (see review Saxe, 1991). The stimulated dark respiration was generally believed to reflect an increased repair rate. Respiration was enhanced for the supply of extra energy for quick oxidation of SO_3^{2-} to SO_4^{2-}. Hunt and Black (1988) have reported enhancement of dark respiration of *Vicia faba* in response to SO_2 under optimum environmental conditions, irrespective of SO_2 concentrations applied. Saxe (1991) summarized no change in dark respiration relative to viable leaf area, when plants were exposed to SO_2 for long duration. Katainen et al. (1987) observed stimulation of dark respiration even at low and medium SO_2 concentrations (11-159 ppb SO_2) for at least 1 month. The stimulation increased linearly from the beginning of exposure, and the effect increased with SO_2 exposure upto 120 ppb. Less stimulation was observed at 150 ppb than 120 ppb SO_2. Dark respiration was increased by 20, 19 and 23%, respectively in wheat cultivars M 234, HP 1209 and M 213 and by 29 and 17%, respectively in soybean cultivars PK 472 and Bragg upon exposure to 0.06 ppm SO_2 for 8 hrs daily for 70 days in open top chambers (OTCs) (Deepak, 1999).

3.5 Metabolic Pools

Alterations in metabolic pools of SO_2 exposed plants have also been studied extensively. Increase in free amino acid content of plants exposed to SO_2 has been reported prior to appearance of visible foliar injury (see review Malhotra and Khan, 1984). It has also been suggested that SO_2 interferes with enzymes regulating amino acid metabolism (Pierre and Queiroz, 1982) leading to qualitative and quantitative changes in amino acids. Such changes may reduce the protein content of SO_2 exposed plants (Deepak and Agrawal, 2001). Such a decrease could be attributed to hydrolysis of the existing proteins and also to reduced de novo synthesis (Khan and Malhotra, 1983). SO_2 also interferes with the fat, carbohydrate and nitrogen metabolism of plants. Beg and Farooq (1988) suggested that the process of metabolism is activated towards synthesis in sensitive plants after SO_2 exposure and, therefore, most of the biochemical constituents get accumulated in higher amounts. In contrast, Deepak (1999) observed reductions in starch and total non structural carbohydrate contents and enhancement in total soluble sugar and non-reducing sugar of wheat and soybean plant upon exposure to SO_2 (0.06ppm), whereas there was increase in total soluble sugars and non reducing sugars. Inhibition of photosynthetic CO_2 fixation by SO_2 is suggested to be the primary cause of carbohydrate reduction. Koziol and Jordan (1978) observed an initial increase in sugars and starch contents of bean leaves exposed to higher concentration of SO_2 followed by a decrease with the appearance of foliar injury symptoms. Increase in reducing sugar content and decrease in non-reducing sugar was observed during the development of injury symptoms in SO_2 exposed plants (Khan and Malhotra, 1983). Nandi et al. (1986) reported decrease in starch content and increase in reducing sugar content of *Oryza sativa* plants exposed to 670 to 1330 $\mu g\ m^{-3}$ SO_2 during tillering and flowering stages.

Lipid metabolism has also been found affected due to SO_2 exposure. A decrease in unsaturated fatty acids in pine needles and *Ulmus americana* leaves after SO_2 exposure has been summarized (Malhotra and Khan, 1984). Decrease in lipid content may be ascribed either due to degradation of lipids (Shimazaki et al., 1980) or inhibition of lipid synthesis (Malhotra and Khan, 1984). Increased lipase activity, peroxidation of fatty acid chains or a combination of the above was reported due to SO_2 exposure (Malhotra and Khan, 1984). However, these changes are dependent on concentration and duration of SO_2. Enhancement of phenol levels of SO_2 exposed plants has also been reported (Nandi et al., 1984, 1990), which may also stimulate the peroxidase activity (Nandi et al., 1990).

3.6 Antioxidants

Sulphur dioxide induces acidification of sub-cellular compartments and increases the oxidative stress (Hippeli and Elstner, 1996). The extent of damage due to acidification relates with the ability of cells to buffer the increasing acidity or to excrete the protons to the external media (Solvik, 1996). Cell buffering is carried out by wide variety of molecules both large (proteins) and small (amino acids) having free carboxyl, phosphate, amino or sulfhydryl groups. High doses of SO_2 lower down the buffering capacity, but a similar dose at low concentration did not (Nandi et al., 1990). It has been estimated that a reduction of chloroplast stroma pH by 0.5 unit may cause 50% decrease in net photosynthesis (Wellburn, 1994).

Oxidative stress is dependent on the efficiency of endogenous antioxidant system that scavange the active oxygen species formed during oxidation of SO_2. After exposure to SO_2 bisulphite is formed, which initiates intermediate free radical formation through a series of reaction before conversion to sulphates. The rate constants of individual reactions and competitions with other processes determined the concentrations of intermediates and their potential damage during oxidation of bisulphite. Strong evidences are there to link SO_2 tolerance with the activity of intracellular enzymes such as catalase (CAT), superoxide dismutase (SOD), and glutathione synthetase (GS), glutathione reductase (GR), glutathione peroxidase (GPX) and glutathione transferase (GST) (Tanaka et al., 1988; Ranieri et al., 1992), and also glutathione (GSH) and ascorbic acid (ASC) (Madamanchi and Alscher, 1991). Sandmann and Gonzales (1989) found that peroxidative processes increase in SO_2 exposed bean plants leading to suppression of fatty acids, chlorophylls, carotenoids and vitamin C. A higher vitamin C content completely protected the plants from peroxidation. Reduction in ASC content of wheat cultivar Mec and Chiarano by 36 and 13%, respectively was reported at 0.064 ppm SO_2 for four month (Ranieri et al., 1997). It has been found that higher constitutive level of ASC leads to resistance against SO_2 (Deepak, 1999). Ascorbic acid, being a scavanger of superoxide radical may inhibit SO_2 induced chlorophyll degradation (Nandi et al., 1984; Agrawal et al., 1986). Nandi et al. (!984) observed significant decrease of CAT activity and increase in peroxidase activity of rice plants. Peroxidase activity usually increase with the increase of sulphur content of SO_2 exposed plants (Pierre and Queiroz, 1982; Nandi et al., 1990). Jager et al., (1985) found that SO_2 tolerance was related to the activities of SOD and peroxidase. A similar result was also reported by Agrawal et al. (1986)

3.7 Growth and Yield

Localized interveinal symptoms in the form of bifacial necrotic areas are characteristic symptoms of acute injury to dicotyledonous plants by SO_2. The affected parts first appear greyish green and water soaked, which later become dry and change the color to reddish brown. In monocotyledonous plants acute injury caused by SO_2 develops as intervenal yellowish-white spots beginning at the leaf tip and then extends down the blade. Chronic injury to SO_2 develops as pale green or yellow green interveinal and bifacial areas.

The phenotypic response of plants to SO_2 is primarily a depression in normal growth and development (Last, 1982). Exposure of plants to 0.06 ppm SO_2 for 8 hrs daily for 70 days led to decrease of 13.4, 7.3 and 8.3% respectively in plant length wheat of cultivars HP 1209, M 234 and M 213 (Deepak, 1999). Davieson et al. (1990) found significant reductions in shoot length of wheat exposed to 0.05 ppm SO_2 for 8 hrs d^{-1} throughout the growing season. Murray and Wilson (1988) did not observe any significant negative effect on growth of wheat cultivar Halbeard exposed to 0.049 ppm SO_2. In contrast, significant reductions of plant length and number of tillers and ears of wheat cultivar Banks were observed upon SO_2 exposure of 0.042 ppm for 4 hrs d^{-1} for 79 days in OTCs, the reductions were 20,55 and 35% respectively in above parameters.

Inter varietal differences in growth response have also been reported for five varieties of barley (Pande, 1985). Low concentration of SO_2 was reported to enhance plant growth with respect to tiller production in winter wheat (Olszyk et al., 1986). Pande and Mansfield (1985) reported about 15% increase in plant height of winter barley upon exposure to SO_2. Leaf area, numbers of leaves, nodules and pods, and root and shoot lengths decreased significantly in soybean plants exposed to 0.15 ppm SO_2 for 4 hrs day^{-1} for 6 weeks (Rajput and Agrawal, 1994).

Many workers have reported the adverse effect of SO_2 on biomass accumulation and yield of many crop plants (Sprugel et al., 1980, Kress et al., 1986; Agrawal and Verma, 1997; Deepak and Agrawal, 1999, 2001). Murray and Wilson (1988) reported no change in biomass of wheat cultivar Halbeard treated with 0.04 ppm SO_2, but the shoot biomass of the same cultivar reduced at continuos exposure to 0.1 ppm SO_2. A similar result was obtained by Davieson et al. (1990) for wheat cultivar Eradu. Deepak (1999) observed significant reductions in leaf biomass of wheat cultivars, and in leaf and shoot biomass of soybean cultivars upon exposure to 0.06 ppm SO_2 for 8 hrs daily for 70 days. Reduction in shoot biomass was also reported in soybean, clover and bean plants exposed to SO_2 (Maas et al., 1987). Qifu and Murray (1993) observed stimulated shoot biomass but no change in root biomass in soybean plants exposed to 110 ppb SO_2 for 16 days whereas at 250 ppb root growth decreased, but had no effect on shoot growth.

Biomass partitioning was also modified due to SO_2 exposure. Reductions in relative growth rate and net assimilation rate with no change in leaf area ratio have been reported under SO_2 treatment (Bell, 1982; Rajput et al., 1995). This suggests that SO_2 has a greater adverse effect on assimilation capacity than the leaf area. Root shoot ratio was lowered down upon SO_2 exposure of wheat plants (Rajput and Agrawal, 1994). As the growth of underground plant parts is dependent on photosynthates translocated from the above ground parts, direct effects of SO_2 on the photosynthetic leaves are reflected in poor development of root system including nodule growth. When dry matter accumulation responds positively to small doses of pollutants, even than there is a reduction in the proportional allocation to roots (Pande and Mansfield, 1985; Wolfenden and Mansfield, 1991). To explore the action of SO_2 pollution on phloem translocation, Noyes (1980) exposed *Phaseolus vulgaris* plants to 100 ppb SO_2 and measured the $^{14}CO_2$ fixation and export. SO_2 had no effect on net photosynthesis but rate of export of ^{14}C label was reduced by 39% suggesting a direct effect of SO_2 on phloem loading. Minchin and Gould (1986) were able to demonstrate that phloem loading was immediately reduced after exposing wheat leaves to a high concentration of SO_2, but there was no effect on maize where bundle sheath cells protect the phloem from direct contact with SO_2.

Reduction in yield of crop plants due to SO_2 exposure has a frequent observed influence (Sprugel et al., 1980; Heagle et al., 1983; Olszyk et al., 1986; Agrawal and Agrawal, 2000). Rajput et al. (1995) have reported loss of wheat yield due to SO_2 exposure as a result of reduced number of grains and reduction in individual grain weight. Wilson and Murray (1990) have reported reduction in number of ears $plant^{-1}$ by 36.6% and individual ear weight by 26.7% in wheat cultivar Banks exposed to 0.042 ppm SO_2 for 79 day in OTCs. Bytnerowicz et al. (1987) also reported yield reduction in wheat cv Yecora Rajo exposed to 0.03 ppm SO_2 continuously for 118 days. Rice plants also showed significant reductions in grain yield and net primary productivity upon 1.5

hrs daily exposure to 0.25 and 0.5 ppm SO_2 for 20 days during tillering and 10 days during flowering (Nandi et al., 1985). Harvest index increased in wheat cultivars HP 1209 and M 213 and soybean cultivar PK 472 upon SO_2 exposure suggesting more allocation for photosynthates towards reproductive parts than vegetative (Deepak, 1999).

Forest trees also showed negative influence of SO_2 on growth and development. Garsed et al. (1979) reported different responses of seedlings of *Pinus sylvestris* (Scots pine), *Betula* (birch), *Quercus* (oak) and *Acer* (maple) to continuous fumigation with low SO_2 concentrations. Growth of *Pinus* declined significantly. The dry weights of roots and shoots of deciduous trees remained unaltered except for *Acer* where leaf weight and area were reduced. Stimulation of height and premature aging of leaves were observed for *Quercus* and *Acer*.

3.8 Fate of Sulphur Accumulation

Once atmospheric SO_2 has entered a cell, it may be stored after conversion as SO_4^{2-} or else it may be translocated out of the cell directly after conversion to other parts and / or released into the environment. Enhanced levels of sulphur in form of sulphate have been reported at exposure to low SO_2 concentrations (Deepak and Agrawal, 1999, 2001). Filner et al. (1984) have suggested that cytoplasmic sulphate oxidase activity may be mainly responsible for rapid conversion of SO_2 / sulphite to sulphate with increasing distances from S emitting sources, decreasing sulphate concentrations in foliar tissues. A significant positive correlation between the SO_2 concentration and foliar accumulation of total and inorganic S has been reported in various field studies around S emitting sources (Agrawal and Singh, 2000). Foliar SO_4^{2-} - S / organic S ratio increased at locations close to SO_2 emitting sources than distantly situated sites. This ratio is found a useful indicator of the state of forest trees under S pollution stress (Agrawal and Agrawal, 2000). Hüve et al. (1995) suggested that SO_3^{2-} or HSO_3^- ions produced after solubilization of SO_2 are oxidized to less toxic sulphate in course of detoxification of SO_2 in most of the plants. The sulphate may be slowly utilized in metabolic process or remain accumulated (Mass et al., 1987). SO_4^{2-} S represents the accumulation of S in excess of S assimilation. In a field study, three cultivars of wheat and two cultivars of soybean were exposed to 0.06 ppm SO_2 for 8 hrs daily from germination to maturity (Deepak, 1999). Among wheat cultivars M 234 showed maximum increment in foliar SO_4^{2-} concentration by 108% followed by 74% in HP 1209 and 68% in M 213. Similarly Bragg cultivar of soybean showed greater increment (90%) than PK 472 (73%) (Deepak and Agrawal, 2001). Among plant parts, leaves showed higher accumulation of SO_4^{2-} than rest of the plant parts. The accumulation of SO_4^{2-} increased with age.

The SO_4^{2-} accumulation depends on the plant species, soil sulphur content, SO_2 concentration and duration of exposure and on various environmental conditions (Rennenberg and Polle, 1994). Reductive conversion of SO_2 derived sulphite in form of H_2S also occurs in chloroplast with the help of sulphite reductase enzyme. Rennenberg (1989) showed that when leaf tissue is exposed to SO_2, light dependent rate of H_2S emission ranged from 0.8 to 8nmol h^{-1} cm^{-2} leaf area. Taylor and Tingey (1983) calculated that H_2S emission represents 7 to 15% of the SO_2 influx based on mass balance calculations. This pathway also prevents the SO_2 derived acidification of the cell. Filner et al. (1984) have shown that young leaves of cucurbit plants that absorbed SO_2 at higher rate were more resistant due to 100 fold higher emission rate of H_2S than mature leaves showing low SO_2 absorption. Kindermann et al. (1995) have also shown that under field condition, SO_2 tolerant conifers (*Picea pungens*) emitted higher H_2S than sensitive cultivar (*P. abies*).

4. FACTORS INFLUENCING THE PLANT RESPONSES TO SO_2

The variations in plant sensitivity to SO_2 were largely attributed to the high variability in genetic and ontogenic components and also to variable environmental conditions. Leaves and plants, which have just reached physiological maturity, are more sensitive to SO_2 as they have high gaseous exchange rate. Light intensity, relative humidity, temperature, wind speed and soil moisture are major environmental factors affecting the plant sensitivity to SO_2. Low light intensity and low temperature increased plant sensitivity to SO_2 (Sheppard, 1994). The sensitivity of plants is also reduced under salinity stress condition (Qifu and Murray, 1993). High nutrient content (NPK levels) of soil reduced the magnitude of injurious effects of SO_2 to soybean and wheat plants (Rajput and Agrawal, 1994, Verma and Agrawal, 1996; Agrawal and Verma, 1997). Plants grown under ample supply of water are more susceptible to SO_2 injury than those growing under a slight water stressed condition (Neighbour et al., 1988;Wolfenden and Mansfield , 1991). When the plants were slightly wilted the stomata closed and plants were able to tolerate increased concentration of SO_2 without being injured. Acid rain with high concentration of SO_4^{2-} made the plants more susceptible to frost (Sheppard, 1994). Synergism between SO_4^{2-} and H^+ ions in precipitation leads to more phytotoxic effects, whereas nitrate ions in precipitation ameliorate the effects through enhanced assimilation and growth dilution. Studies have suggested that responses of plants to SO_2 may be influenced by other pollutants (Agrawal et al., 1982; Pande and Mansfield, 1985; Davieson et al., 1990; Barnes and Wellburn, 1998). The combined exposure to O_3 + SO_2 results in greater than additive effects on plant growth and productivity (Agrawal et al., 1982; Barnes and Wellburn, 1998). Synergistic reduction has been reported for many growth parameters upon exposure to SO_2 and NO_2 combinations (Pande and Mansfield, 1985). Carbon dioxide plays protective role in reducing the damage caused by SO_2 (Deepak and Agrawal, 1999, 2001). Elevated level of CO_2 has ameliorated the toxic effects of SO_2 by increasing the photoassimilate availability and by reducing the SO_2 influx (Deepak and Agrawal, 2001). The greater photoassimilate availability also enabled the plants to better utilize the SO_2 as nutrient source. CO_2 enrichment may improve the ability of plants to tolerate the SO_2 stress. The absence of negative effects of SO_2 at high CO_2 concentration in wheat was ascribed to increase in the levels of ascorbate, glutathione and the redox state (Rao and De Kok, 1994).

5. CONCLUSIONS

Regional concentrations of SO_2 have decreased due to emission reduction in developed countries, but industrializing countries are increasing the SO_2 emission dramatically. Effects of SO_2 on plants vary from negative influence on carbon assimilation to carbon accumulation and translocation to different parts leading to significant reduction in growth and development. Plants have different mechanisms to compensate the negative effect of SO_2 including conversion to sulphate, synthesis of organic sulphur compounds, emission of H_2S, and internal detoxification of oxidizing intermediates produced during oxidation of SO_2. Recent interest in the effects of air pollution has mainly focussed on O_3, but SO_2 still remains an important pollutant due to its involvement in acid rain. Future work is required to establish long term effects of dry deposition of SO_2 directly on vegetation and its indirect effect through acid precipitation under field conditions on natural and seminatural plant communities. The studies on consequences of simultaneous exposure to combination of several atmospheric pollutants and global climate change variables will be particularly important in view of increasing concerns for the long range transport of pollutants. Dose response relationships should be explored for a wide range of plant species from different regions at realistic levels of sulphur dioxide. Magnitude of tolerance/compensation for excess sulphur in different plants may provide selection of tolerant breed for sulphur dioxide through genetic manipulation.

6. REFERENCES

Agrawal M and Agrawal SB (2000) Researches on air pollution impacts on Indian forests. In: Innes JL and Haron AH (eds) Air Pollution and the Forest of Developing and Rapidly Industrializing Regions, pp 165—187, CABI Publishing, Oxon

Agrawal M and Singh (2000) Impact of thermal power plant emission on foliar elemental concentrations in plants in a low rainfall tropical region. *Environ Monit Assess* **60:** 261—282

Agrawal M and Verma M (1997) Amelioration of sulphur dioxide phytotoxicity in wheat cultivars by modifying NPK nutrients. *J Env Manage* **49:** 231-244

Agrawal M, Nandi PK and Rao DN (1982) Effect of ozone and sulphur dioxide pollutants separately and in mixture on chlorophyll and carotenoid pigments of *Oryza sativa*. *Water Air* and *Soil Pollut* **18:** 459—454

Agrawal M, Nandi PK, Agrawal SB and Rao DN (1986) Superoxide dismutase: An indicator of SO_2 tolerance in plants. *Curr Sci* **56:** 35—36

Alcamo J, Krol M and Posch M (1995) An integrated analysis of sulphur emissions acid deposition and climate change. *Water Air* and *Soil Pollu* **85:** 1539—1550

Amundson RG and Weinstein LH (1981) Joint action of short term sulfur dioxide and nitrogen dioxide on foliar injury and stomatal behaviour in soybean. *J Env Qual* **10:** 204—206

Anderson LE and Duggan JX (1977) Inhibition of light modulation of chloroplast enzyme activity by sulfite: one of the lethal effects of SO_2. *Oecologia* **28:** 147—157

Arends BG, Baard JH and Ten Brink HM (1997) Trends in summer sulphate in Europe. *Atmos Environ* **31:** 4063—4072

Ashenden TW (1979) Effects of SO_2 and NO_2 pollution on transpiration on *Phaseolus vulgaris* L. *Environ Pollut* **18:** 45—49

Atkinson CJ, Robe SV and Winner WE (1988) The relationship between changes in photosynthesis and growth for radish plants fumigated with SO_2 and O_3. *New Phytol* **110:** 173—184

Barnes JD and Wellburn A (1998) Air pollutant combinations. In: De Kok LJ and Stulen I (eds) Responses of Plant Metabolism to Air Pollution and Global Change, pp 147—164, Backhuys Publishers, Leiden.

Beg MU and Farooq M (1988) Sulphur dioxide resistance of Indian trees II Experimental evaluation of metabolic profile. *Water Air* and *Soil Poll* **40:** 317—326

Bell JNB (1982) Sulfur dioxide and the growth of grasses. In: Unsworth MH and Ormrod DP (eds) Effects of Gaseous Air Pollution in Agriculture and Horticulture, pp 225—246, Butterworth Scientific, London

Bell JNB and Mudd CH (1976) Sulfur dioxide resistance in plants: a case study of Lolium perenne. In: Mansfield TA (ed) Effects of Air Pollutants on Plants, pp 87—103, Cambridge University Press, Cambridge

Black VJ and Unsworth MH (1980) Stomatal response to sulphur dioxide and vapour pressure deficit. *J Exp Bot* **31:** 667—677

Bytnerowiez A, Olszyk DM, Kats G, Dawson PJ, Wolff J and Thompson CR (1987) Effect of SO_2 on physiology, elemental content and injury development of winter wheat *Agric Ecosys Environ* **20:** 37—47

Cram WJ (1990) Uptake and transport of sulfate. In: Rennenberg H, Brunold C, De Kok LJ and Stulen I (eds) Sulfur Nutrition and Sulfur Assimilation in Higher Plants, pp 3—10, SPB Academic Publishers, The Hague

Daemmgen U, Walker K, Groenhage L and Jaeger HJ (1998) The atmospheric sulphur cycle. In: Schnug E (ed) Sulphur in Agro Ecosystems, Vol. 2, pp 75—114. Kluwer Academic Publishing, Dordrecht

Darrall NM (1989) The effect of air pollutants on physiological processes in plants. *Plant Cell Environ* **12:** 1—30

Davieson G, Murray F and Wilson S (1990) Effects of sulphur dioxide and hydrogen fluoride, singly and in combination, on growth and yield of wheat in open top chambers. *Agric Ecosys Environ* **30:** 317—325

De Kok LJ (1990) Sulfur metabolism in plants exposed to atmospheric sulfur. In: Rennenberg H, Brunold C, De Kok LJ and Stulen I (eds) Sulfur Nutrition and Sulfur Assimilation in Higher Plants, pp 110—113. SPB Academic Publishing, The Hague

Deepak SS (1999) Interactive effects of elevated levels of carbon dioxide and sulphur dioxide on selected plants. Ph. D thesis, Banaras Hindu University, Varanasi

Deepak SS and Agrawal M (1999) Growth and yield responses of wheat plants to elevated levels of CO_2 and SO_2, singly and in combination. *Environ Pollut* **104:** 411—419

Deepak SS and Agrawal M (2001) Influence of elevated CO_2 on the sensitivity of two soybean cultivars to sulphur dioxide. *Environ Exp Bot* **46:** 81—91

Filner P, Rennenberg H, Sekiya J, Bressan RA, Wilson LG, Lecureux L and Shimei C(1984) Biosynthesis and emission of hydrogen sulfide by higher plants. In: Kozlol MJ and Whatley FR (eds) Gaseous Air Pollutants and Plant Metabolism, pp 291—312, Butterworth, London

Fowler D (1985) Deposition of SO_2 onto plant canopies. In: Winner WE, Mooney HA and Goldstein RA (eds) Sulfur Dioxide and Vegetation, pp 389—402, Stanford University Press, Stanford

Galloway JN (1989) Atmospheric acidification: projections the future. *Ambio* **18:** 161—166

Garsed SG and Read DJ (1977) Sulphur dioxide metabolism in soybean, *Glycine max* var Biloxi I The effects of light and dark on uptake and translocation of $^{35}SO_2$. *New Phytol* **78:** 111—119

Garsed SG, Farrar JF and Rutler J (1979) The effects of low concentrations of sulphur dioxide on the growth of four broad leaved tree species. *J Appl Ecol* **16:** 217—226

Hameed S and Dignon J (1988) Changes in the geographical distributions of global emissions of NOx and SOx from fossil fuel combustion between 1966 and 1980. *Atmos Environ* **22:** 441—449

Hampp R and Ziegler I (1977) Sulfate and sulfite translocation via the phosphate translocator of the inner invelop membrane of chloroplasts. *Planta* **137:** 309—312

Heagle AS, Heck WW, Rawlings JO and Philbeck RP (1983) Effects of chronic doses of ozone and sulfur dioxide on injury and yield of soybean in open top field chambers. *Crop Sci* **23:** 1184—1191

Hippeli S and Elstner EE (1996) Mechanisms of oxygen activation during plant stress: Biochemical effects of air pollutants. *J Plant physiol* **148:** 249—257

Hüve K, Dittrich A, Kindermann G, Slovik S and Heber U (1995) Detoxification of SO_2 in conifers differing in SO_2 tolerance: A comparison of *Picea abies*, *Picea pungens* and *Pinus sylvestris*. *Planta* **195:** 578—580

Hunt GA and Black VJ (1988) Environmental stress: amplifier of physiological responses to SO_2 in plants. *Environ Pollut* **53:** 433—435

Iverson T, Saltbones J, Sandnes H, Eliassen A and Hov O (1989) Airborne transbondary transport of sulphur and nitrogen over Europe-model descriptions and calculations. EMEP MSCW Report 2/89, Norwegian Meteorological Institute, Oslo

Kaiser W, Dettrich A and Heber U (1993) Sulfate concentrations in Norway spruce needles in relation to atmospheric SO_2: A comparison of trees from various forests in Germany with trees fumigated with SO_2 in growth chambers. *Tree Physiol* **12:** 1—13

Katainen HS, Mäkinen E, Jokinen J, Karjalainena R and Kellomäki S (1987) Effects of SO_2 on the photosynthetic and respiration rates in Scots pine seedlings. *Environ Pollut* **46:** 241—251

Khan AA and Malhotra SS (1983) Protein biosynthesis in jackpine and its inhibition by sulphur dioxide. *Phytochem* **22:** 1325—1328

Kimmerer TW and Kozlowski TT (1981) Stomatal conductance and sulfur uptake of five clones of *Populus tremuloides* exposed to sulfur dioxide. *Plant Physiol* **67:** 990—995

Kindermann G, Hüve K, Slovik S, Lux H and Rennenberg H (1995) Emission of hydrogen sulphide by twigs of conifers- a comparison of Norway spruce (*Picea abies* (L) Karst.) scotch pine (*Pinus sylvestris* L.) and blue spruce (*Picea pungens* Engelm.). *Plant* and *Soil* **168/169:** 421—423

Koziol MJ and Jordan CF (1978) Change in carbohydrate levels in red kidney bean (*Phaseolus vulgaris* L) exposed to sulphur dioxide. *J Exp Bot* **29:** 1037—1043

Kress LW, Miller JE and Smith HJ (1986) Impact of O_3 and SO_2 on soybean yield. *Environ Pollut* **41:** 105—123

Last FT (1982) Effects of atmospheric sulphur compounds on natural and man made terrestrial and aquatic ecosystems. *Agric Environ* **7:** 299—381

Lendzian K and Kerstiens G (1991) Sorption and transport of gases and vapours in plant cuticles. *Rev Environ Contam Toxicol* **121:** 65—128

Maas FM, De Kok LJ, Strik Timmer W and Kuiper PJC (1987) Plant response to H_2S and SO_2 fumigation II Differences in metabolism of H_2S and SO_2 in spinach. *Physiol Plant* **70:** 722—728

Madamanchi N and Alscher R (1991) Metabolic basis for differences in sensitivity of two pea cultivars to sulfur dioxide. *Plant Physiol* **97:** 88—93

Malhotra SS and Khan AA(1984) Biochemical and physiological impact of major pollutants. In: Treshow M (ed) Air Pollution and Plant Life, pp 113—157. John Wiley & Sons, New York

Minchin PEH and Gould R (1986) Effects of SO_2 on phloem loading. *Plant Sci* **43:** 179—183

Murray F and Wilson S (1988) The joint action of sulphur dioxide and hydrogen fluoride on yield and quality of wheat and barley. *Environ Pollut* **55:** 239-249

Nandi PK Agrawal M and Rao DN (1984) SO_2 induced enzymatic changes and ascorbic acid oxidation in *Oryza sativa. Water Air* and *Soil Pollut* **21:** 25—32

Nandi PK, Agrawal M and Rao DN (1985) Effects of sulphur dioxide on nutrient status and growth of rice plants. *Ind J Air Pollut Cont* **6:** 5—14

Nandi PK, Agrawal M, Agrawal SB and Rao DN (1990) Physiological responses of *Vicia faba* plants to sulphur dioxide. *Ecosys Environ Safety* **19:** 64—71

Neighbour EA, Cottam DA and Mansfield TA (1988) Effect of sulphur dioxide and nitrogen dioxide on the control of water loss by birch (*Betula* spp). *New Phytol* **108:** 149—157

Noyes RD (1980) The comparative effects of sulphur dioxide on photosynthesis and translocation in bean. *Physiol Plant Pathol* **16:** 73—79

Okpodu CM, Alscher RG, Grabau EH and Cramer CL (1996) Physiological, biochemical and molecular effects of sulfur dioxide. *J Pl Physiol* **148:** 309—316

Olszyk DM, Bytherowicz A, Kals G, Dawson PJ, Wolf J and Thompson CR (1986) Effects of sulfur dioxide and ambient ozone on winter wheat and lettuce *J Environ Qual* **15:** 363—369

Pande PC (1985) An examination of the sensitivity of five barley cultivars to SO_2 pollution. *Environ Pollut* **37:** 27—41

Pande PC and Mansfield TA (1985) Responses of spring barley to SO_2 and NO_2 pollution. *Environ Pollut* **38:** 87—97

Panwar TS (1998) Acid rain-case study for India (Including RAINS-Asia model). Project report Acid rain and Emissions Reduction in Asia, Phase II. Asian Development Bank.

Pfauz H, Dietz KJ, Weinerth I and Oppmann B (1990) Detoxification of sulfur dioxide by apoplastic peroxidases. In: Rennenberg H, Bronold C, De Kok LJ and Stulen I (eds) Sulfur Nutrition and Sulfur Assimilation in Higher Plants, pp 229—233, SPB Academic Publishing, The Hague

Pierre M and Queiroz O (1982) Enzymic and metabolic changes in bean leaves during continuos pollution by subnecrotic levels of SO_2. *Environ Pollut* **25:** 41—51

Qifu MA and Murray F (1993) Effects of SO_2 and salinity on nitrogenase activity, nitrogen concentration and growth of young soybean plants. *Environ Exp Bot* **33:** 529—537

Rajput M and Agrawal M (1994) Responses of soybean plants to sulphur dioxide at varying soil fertility regimes. *Biotronics* **23:** 81—92

Rajput M, Agrawal M and Singh J (1995) Interaction of soil fertility levels and sulphur dioxide on growth, assimilate partitioning and yield of wheat *J Environ Biol* **16:** 265—275

Ranieri A, Castagna A, Lorenzini G and Soldatini GF (1997) Changes in thylakoid protein palterns and antioxidant levels in two wheat cultivars with different sensitivity to sulphur dioxide. *Environ Exp Bot* **37:** 125—135

Ranieri A, Durante, M, Volterrani, A Lorenzini G and Soldatini GF (1992) Effects of low SO_2 levels on superoxide dismutase and peroxidase isoenzymes in two different wheat cultivars. *Biochem Physiol Pflangen* **188:** 67—71

Rao M and Anderson LE (1983) Light and stomatal metabolism II effects of sulfite and arsenite on stomatal opening and light modulation of enzymes in epidermis. *Plant Physiol* **71:** 456—459

Rao MV and Dekok LJ (1994) Interactive effects of high CO_2 and SO_2 on growth and antioxidant levels in wheat. *Phyton* **36:** 279—290

Rao MV and Dubey PS (1988) Plant response against SO_2 in field conditions. *Asian Environ* **3:** 1—9

Raschke K (1979) Movement of stomata. In: Encyclopedia of Plant Physiology, pp 323—441, Springer Verlag, Berlin

Rennenberg H (1984) The fate of excess sulfur in higher plants. *Annu Rev Plant Physiol* **35:** 121—153

Rennenberg H (1989) Synthesis and emission of hydrogen sulfide by higher plants. In: Saltzman ES and Cooper WJ (eds) Biogenic Sulfur in the Environment, pp 44—57, Am Chem Soc, Washington DC.

Rennenberg H and Herschbash C (1996) Responses of plants to atmospheric sulphur. In: Yunus M and Iqbal M (eds) Plant Responses to Air Pollution, pp 285—293, John Wiley & Sons, Chichester

Rennenberg H and Polle A (1994) Metabolic consequences of atmospheric sulphur influx into plants. In: Alscher RG and Wellburn AR (eds) Plant Responses to the Gaseous Environment, pp 165—180, Chapman and Hall, London

Rennenberg H (1991) The significance of higher plants in the emission of sulfur compounds from terrestrial ecosystems, In: Sharkey TD, Holland EA and Mooney HA (eds) Trace Gas Emission by Plants, pp 217—260, Academic Press, San Diego

Saxe H (1991) Photosynthesis and stomatal response to polluted air and the use of physiological and biochemical responses for easy detection and diagnostic tools. *Adv Bot Res* **18:** 1—128

Schiffgens- Gruber A and Lutz C (1992) Ultrastructure of mesophyll cell chloroplasts of spruce needles exposed to O_3, SO_2 and NO_2 alone and in combination. *Environ Exp Bot* **32:** 243—254

Sheppard LJ (1994) Causal mechanisms by which sulphate, nitrate and acidity influence frost hardiness in red spruce: review and hypothesis. *New Phytol* **127:** 69—82

Sheppard LJ, Cape JN and Leith ID (1993) Influence of acidic mist on frost hardiness and nutrient concentrations in red spruce seedlings. II Frequency and dose. *New Phytol* **124:** 607—615

Shimazaki K, Sakaki T, Kondo N and Sugahara N (1980) Active oxygen participation in chlorophyll destruction and lipid peroxidation in SO_2 fumigated leaves of spinach. *Plant Cell Physiol* **21:** 1193—1204

Singh N, Singh SN, Srivastava K, Yunus M, Ahmad KJ, Sharma SC and Sharga AN (1990) Relative sensitivity and tolerance of some *Gladiolus* cultivars to sulphur dioxide. *Ann Bot* **65:** 41—44

Slovik S (1996) Chronic SO_2 and NOx pollution interferes with the K^+ and Mg^{2+} budget of Norway spruce trees. *J Plant Physiol* **148:** 276—286

Sprugel DG, Miller JE, Muller RN, Smith HJ and Xerikos PB (1980) Sulfur dioxide effects on yield and seed quality in field grown soybeans. *Phytopathol* **70:** 1129—1133

Sandmann G and Gonzales HG (1989) Peroxidative processes induced in bean leaves by fumigation with sulphur dioxide. *Environ Pollut* **56:** 145—154

Tanaka H, Takahashi T and Yatazawa M (1972) Experimental studies on sulfur dioxide injury in higher plants I: Formation of glyoxylate bisulfite in plant leaves exposed to sulfur dioxide. *Water Air Soil Pollut* **1:** 205—211

Tanaka K, Furusawa I, Kondo N and Tanaka K (1988) SO_2 tolerance of tobacco plants regenerated from paraquat tolerant cells. *Plant Cell Physiol* **29:** 743—746

Tausz M, De Kok LJ, Stulen I and Grill D (1996) Physiological responses of Norway spruce trees to elevated CO_2 and SO_2. *J Plant Physiol* **148:** 362—367

Tausz M, VanderKooij TAW, Müller M, De Kok LJ and Grill D (1998) Uptake and metabolism of oxidized and reduced sulfur pollutants by spruce trees. In: De Kok LJ and Stulen I (eds) Responses of Plant Metabolism to Air Pollution and Global Change, pp 457—460, Backhuys Publishers, Leiden

Taylor, Jr GE and Tingey DT (1983) Sulfur dioxide flux into leaves of *Geranium carolinianum*: Evidence for non stomal or residual resistance. *Plant Physiol* **72:** 237—244

UNEP/GEMS (1991) Urban Air Pollution.United Nations Environmental Programme, Nairobi

Verma M and Agrawal M (1996) Alleviation of injurious effects of sulphur dioxide on soybean by modifying NPK nutrients. *Agric Ecosys Environ* **57:** 49—55

Verma M, Agrawal M and Deepak SS (2000) Interactive effects of sulphur dioxide and mineral nutrient supply on photosynthetic characteristics and yield in four wheat cultivars. *Photosynth* **38:** 91—96

Wellburn A (1994) Air Pollution and Climate Change: The Biological Impact.2nd edition, Longman Scientific and Technical, Essex

Wilson SA and Murray F (1990) SO_2 induced growth reduction and sulphur accumulation in wheat. *Environ Pollut* **66:** 179—191

Winner WE, Mooney HA and Goldstein RA (eds) (1985) Sulphur Dioxide and Vegetation: Physiology, Ecology and Policy Issues, Standford University Press, Stanford

Wolfenden J, Pearson M and Francis I (1991) Effects of over winter fumigation with sulphur and nitrogen oxides on biochemical parameters and spring growth in red spruce (*Picea rubens* Sarg.). *Plant Cell Environ* **14:** 35—45

Wolfenden J and Mansfield (1991) Physiological disturbances in plants caused by air pollutants. *Proceed Royal Soc* **97B:** 117—138

7. ACKNOWLEDGEMENT

The author thankfully acknowledges the financial assistance from Department for International Development, U.K.

8. AFFILIATIONS

Department of Botany, Banaras Hindu University, Varanasi 221 005, India
For correspondence: *M. Agrawal*: madhoo@bhu.ac.in

J. IMSANDE

SULPHUR NUTRITION AND LEGUME SEED QUALITY

Abstract. Plants are the major food source for humans and other animals, providing carbohydrates, protein, lipids and vitamins. This chapter describes the very important role plants play in providing the protein required in animal diets. Protein accounts for approximately 10 to 15% of the dry weight of cereal grains, 20 to 25% of the dry weight of many legume seeds, and 40% of the dry weight of a typical soybean seed. Even though cereal grains and legume seeds are excellent sources of protein, many are deficient in at least some of the ten essential amino acids. Maize, for example, is deficient in both lysine and tryptophan, whereas legume seeds are frequently deficient in the sulphur amino acids, methionine and cysteine. Clearly, a plant must coordinate its effort to obtain reduced nitrogen and reduced sulphur if it is to synthesize large quantities of seed protein. On the other hand, the seed must contain sufficient nitrogen to nourish the developing seedling. Nitrogen-fixing legumes have evolved so that they can produce a large quantity of two different classes of seed storage proteins, those that are rich in both nitrogen and sulphur and others that are sulphur-deficient. Glycinins, which account for approximately 40% of soybean seed protein and are rich in both nitrogen and sulphur, represent the former class. On the other hand, the β-chain of soybean β-conglycinin lacks methionine and cysteine and normally accounts for approximately 10% of soybean seed protein. When a plant is grown on a sulphur-deficient regime, however, the β-chain of β-conglycinin can account for 30% of total seed protein. Mechanisms by which the plant regulates the relative abundance of these two protein classes are proposed. Also, efforts to enhance the relative abundance of the sulphur amino acids in seed storage proteins are described.

1. INTRODUCTION

Legume seeds have served as a human food source for many centuries. Indeed, *Vicia faba* L. (faba bean) is reported to have been used in Egypt already in the 18th century B. C. (Muntz et al., 1999). Today, legume seeds are used widely both as a protein source for human foods and for domestic animal feeds. Cereal grains also are used for these purposes; however the percentage of protein in cereal grains is usually 10 to 15 % of seed dry matter whereas the protein content of most legume seeds ranges from 20 to 25% (Derbyshire et al., 1976). Soybean seeds, on the other hand, are comparatively very rich in protein. The percentage of protein in soybean seeds frequently ranges from 37 to 45% (Serretti et al., 1994). Thus, on seed dry weight basis, a typical legume provides approximately 2-fold more protein than most cereal grains whereas soybean provides approximately 3-fold more protein (Table 1 and Fukushima, 1991b). Consequently soybean seeds and meal prepared from soybean seeds, have become prized food sources in many parts of the world. Because of the wide reliance on legumes as a protein source, this review of seed storage protein quality will focus largely on legumes, with a primary emphasis on soybean. Protein, of course, is the primary source of the dietary essential amino acids, which include the sulphur-containing amino acids methionine and cysteine (Fukushima, 1991a).

Yash P. Abrol and Altaf Ahmad (eds.), Sulphur in Plants, 295—304.

Table 1. Abundance and characterization of seed storage proteins

	Fraction of total seed protein in class (mean values)				
Seed source	**Protein content of seed (%)**	**Albumin (water soluble)**	**Globulin (dilute salt soluble)**	**Prolamin (dilute alcohol soluble)**	**Glutelin (acid or alkali soluble)**
Soybean	40	10	90	--	0
Lupin	30	10	90	--	0
Pea	23	21	66	--	12
Broad bean	23	20	50	--	15
Kidney bean	22	--	50	--	--
Peanut	26	15	70	--	10
Rice	9	Trace	6	4	88
Oat	11	1	80	13	5
Maize	10	Trace	6	54	39
Wheat	13	4	8	48	38
Barley	13	4	16	40	40
Rye	12	8	8	42	42

Soucre: Modified from Fukushima (1991b)

2. SEED STORAGE PROTEINS

2.1 Physical Properties and Abundance

All legume seeds and cereal grains contain groups of proteins that are classified as structural proteins, functional enzymes, and seed storage proteins. Structural proteins assist in determining seed size, shape and compartmentalization. Functional enzymes are required for germination and the initiation of seedling growth. Seed storage proteins are the most abundant class and generally account for approximately two-thirds of total seed protein. Seed storage proteins have been studied extensively (Osborne, 1924; Shewry and Casey, 1999). In a germinating seed, these proteins are a nutrient source for the developing seedling. The globular seed storage proteins from legumes, which are soluble in a saline solution, are divided into two general groups, legumins and vicilins. The molecular mass of a typical legumin, frequently referred to as 11S or 12S protein, is approximately 350 kDa, whereas the mass of a typical vicilin, sometimes referred to as a 7S protein, is approximately 200 kDa (Table 2; Derbyshire et al., 1976; Casey, 1999). The "S values" are sedimentation constants and are indicative of the molecular mass of these proteins; however, the S values will vary slightly depending upon experimen-tal conditions. Hence, the same protein may be referred to as 11S by one author and 12S by a second author. In soybean, the 11-12S legumins are designated "glycinins" (Kitamura et al., 1976; Nielsen et al., 1989) whereas the 7S globulins, the vicilins, are designated "β-conglycinins" (Coates et al., 1985). As stated, protein accounts for approximately 40% of the dry weight of a typical soybean seed and the seed storage proteins, the glycinins plus the β-conglycinins, account for approximately 70% of total soybean seed protein (Meinke et al., 1981). In all legumes both the legumins and vicilins (Table 1) are globular proteins composed of several smaller protein subunits, sometimes joined by disulphide bridges.

Table 2. Classification and molecular weight of seed storage protein

Species	**7S proteins**	**M x 10^{-3}**
Pisum sativum (pea)	Vicilin	150-190
Vicia faba (broad bean)	Vicilin	150
Vigna unguiculata (cowpea)	Vicilin	170
Pisum sativum (pea)	Convicilin	220-290
Vicia faba (broad bean)	Convicilin	150-210
Canavalia spp. (jack bean)	Canavalin	170
Phaseolus vulgaris (French bean)	Phaseolin	140-163
Glycine max (soybean)	β-Conglycinin	160-210
Lupinus spp. (lupins)	Conglutin β	185
	11-12S Proteins	
Pisum sativum (pea)	Legumin	330-450
Vicia faba (broad bean)	Legumin	320-400
Vigna unguiculata (cowpea)	Legumin	300-400
Arachis hypogea (peanut)	Arachin	300-350
Lupinus spp. (lupin)	Conglutin α	315
Helianthus annuus (sunflower)	Helianthinin	300-350
Cucurbita spp. (melons, etc)	Cucurbitin	325
Glycine max (soybean)	Glycinin	320-375
Brassica napus (oilseed rape)	Cruciferin	300

Source: Casey (1999)

2.2 Glycinins

The 11-12S glycinins (Table 2) are synthesized essentially throughout seed development whereas the 7S β-conglycinins are formed predominantly during late seed development (Meinke et al., 1981). As will be discussed later, this simple fact may help to explain why the glycinins contain a higher frequency of the sulphur amino acids than do the β-conglycinins (Table 3). As isolated from seed extracts, the glycinins are an oligomer of six similar subunits (Badley et al., 1975). Genes encoding five of these subunits have been cloned and sequenced (Nielsen et al., 1989; Xue et al., 1992). These five genes, termed Gy1, Gy2, Gy3, Gy4, and Gy5 encode proteins with molecular weights of 55.7 kDa, 54.4 kDa, 54.3 kDa, 63.7 kDa and 58.0 kDa, respectively. Collectively, the glycinins account for approximately 40% of total soybean seed protein. The deduced amino acid sequences of these proteins show a normal proportion of the 20 amino acids (Table 3; Nielsen et al., 1989; Fukushima, 1991b). Also, by chemical analysis the glycinins have been shown to contain a normal complement of the 20 amino acids and hence possess a good nutritive value (Fukushima, 1991b). The glycinins encoded by genes Gy1, Gy2, Gy3, Gy4, and Gy5 contain approximately 493, 481, 480, 564, and 513 amino acids residues per protein molecule, respectively. Because the number of amino acid residues per protein molecule varies from one species of protein to another, a direct comparison of nutritive value of several proteins can be confusing. Therefore, for ease of comparison, the frequency of each of the 20 amino acids is presented on the basis of one thousand amino acid residues of that protein (Table 3). Egg protein, the nutritive value standard, contains 28 methionine residues per thousand amino acid residues whereas soybean glycinin contains 18 methionine residues. The methionine contents of other legume seeds and cereal grains are as follows: pea legumin, 8; soybean β-conglycinin, 3; pea vicilin, 1.3; corn zein, 8; and wheat proteins, 1 and 6 methionine residues each per 1000 amino acid residues (Table 3). Likewise, egg protein contains 28 cysteine residues per thousand amino acid residues whereas soybean glycinin contains 14 cysteine residues. Again, in comparison, the cysteine content of other legumes seeds and cereal grains is as follows: pea legumin, 7; soybean β-conglycinin, 3; pea vicilin, 0.5; corn zein, 10; and wheat proteins 4 and 23 cysteine residues each per one thousand amino acid residues (Table 3). On the other hand, the legume proteins contain almost two-fold more nitrogen-rich amino acid residues, i.e., arginine, aspartate plus asparagine, and glutamate plus glutamine residues, than does the egg protein standard (Table 3). The amino acid compositions of a broad range of 11-12S seed storage proteins from legumes and cereals can be found in Shewry and Casey (1999).

2.3 β-conglycinins

Like the 11-12S glycinins, the 7S β-conglycinins (Tables 2 and 3) are encoded by a family of genes which have been cloned and sequenced (Coates et al., 1985; Harada et al., 1989). The β-conglycinins are trimeric, composed primarily of three protein subunits, namely the α-chain, the α'-chain and the β-chain. The β-conglycinins are composed of various combinations of these three subunits and there are at least seven different isomeric forms (Coates et al., 1985; Sebastiani et al., 1991). Growth conditions can alter the relative abundance of these seven isomers. Indeed, it is possible for the seed to construct a trimer containing three β-chains. The three subunits of the 7S β-conglycinins, which account for approximately 30% of total seed protein, differ in amino acid composition (Table 4). The α-subunit, which is composed of 605 amino acid residues (72 kDa), contains three methionine residues and five cysteine residues (Sebastiani et al., 1990; Sebastiani et al., 1991; Yoshino et al., 2001). The α' subunit, which is composed of 639 amino acid residues (76 kDa), contains six methionine residues and five cysteine residues. Thus, each of the sulphur amino acids accounts for less than one percent of the amino acid residues in these two chains (Sebastiani et al., 1990; Lelievre et al., 1992). The amino acid sequences of the α and α' chains are nearly identical, except for three insertions, consisting of 4, 12, and 19 amino acid residues, into the α' chain (Lelievre et al., 1992). The β-chain of β-conglycinin, which can account for 10% of total seed protein and is composed of 416 amino acid residues (53 kDa), lacks both methionine and cysteine as well as tryptophan (Table 4, Coates et al.,

Table 3. Amino acid compositions of seed storage proteins (amino acid residues per 1000 residues)

Amino acids[1]	Egg protein (standard)	Legumes				Corn zein	Whcat	
		11S globulin		7S globulin				
		Soybean glycinin	Pea legumin	Soybean β-conglycinin	Pea vicilin		ω-Gliadin	β-Gliadin
Glycine (G)	56	40	78	29	52	18	16	55
Alanine (A)	83	36	59	37	45	132	6	27
Valine (V)	79	51	58	51	57	41	5	48
Leucine (L)	88	72	83	103	103	189	34	72
Isoleucine (I)	55	47	45	64	55	39	38	42
Serine (S)	55	42	63	68	78	64	34	54
Threonine (T)	54	34	37	28	30	30	12	14
Tyrosine (Y)	34	41	22	36	23	24	11	33
Phenylalanine (F)	48	57	38	74	50	52	91	35
Tryptophan (W)	8	16	7	3	0.5	0	--	4
Proline (P)	48	54	51	43	44	101	229	169
Methionine (M)	28	18	8	3	1.3	8	1	6
Cysteine (C)	28	14	7	3	0.5	10	4	23
Lysine (K)	65	49	55	70	71	1	4	5.5
Histidine (H)	20	22	21	17	16	9	10	13
Arginine (R)	46	78	77	88	57	10	7	16
Aspartate (D+N[2])	101	131	117	141	134	49	5	24
Glutamate (E+Q[2])	108	180	176	205	182	213	501	389

Source: Modified from Fukushima (1991b)
[1]Single-letter abbreviations given in parentheses
[2]N is asparagine and Q is glutamine

Sulphur in Plants
Edited by Yash P. Abrol & Altaf Ahmad

Table 4. Amino acid compositions of β-conglycinin subunits

	Number of residues per molecule			Amino acid frequency mole %		
Amino acid	α	α'	β	α	α'	β
Glycine (G)	27	31	19	4.46	4.85	4.57
Alanine (A)	28	28	22	4.63	4.38	5.29
Valine (V)	27	32	23	4.46	5.01	5.53
Leucine (L)	54	49	44	8.93	7.67	10.58
Isoleucine (I)	31	29	26	5.12	4.54	6.25
Serine (S)	45	47	31	7.44	7.36	7.45
Threonine (T)	11	14	10	1.82	2.19	2.40
Tyrosine (Y)	15	15	12	2.48	2.35	2.88
Phenylalanine (F)	30	32	26	4.96	5.01	6.25
Tryptophan (W)	2	3	0	0.33	0.47	0.00
Proline (P)	40	35	21	6.61	5.48	5.05
Methionine (M)	3	6	0	0.50	0.94	0.00
Cysteine (C)	5	5	0	0.83	0.78	0.00
Lysine (K)	35	42	21	5.79	6.57	5.05
Histidine (H)	8	23	8	1.32	3.60	1.92
Arginine (R)	48	43	29	7.93	6.73	6.97
Asx (D+N)	70	70	54	11.57	10.95	12.98
Glx (E+Q)	126	135	70	20.82	21.13	16.83
Total	605	639	416	100.0	100.0	100.0

Sources: Sebastiani et al. (1990); Lelievre et al. (1992); Maruyama et al. (2001) and Yoshino et al. (2001).

1985; Maruyama et al., 2001). DNA sequencing data confirm that the β-chain of β-conglycinin lacks both methionine and cysteine (Maruyama et al., 2001). Interestingly, most of the methionine and cysteine residues in the α and α' chains reside very near the N-terminal ends of these two proteins. This region seems to be deleted in the smaller β-chain because good homology exists among the three chains over the remainder of the molecule (Lelievre et al., 1992).

Seeds from many plants contain low concentrations of small 2S proteins that are relatively rich in cysteine and methionine (Shewry and Pandya, 1999). Soybean also produces low molecular weight polypeptides that contain a disproportionately high methionine content (George and de Lumen, 1991; Paek et al., 2000), the relative abundance of these polypeptides is very low. Consequently, the available data strongly suggest that the increase or decrease in the methionine content of soybean seed protein is primarily determined by the relative abundance of the β-chain of β-conglycinin (Naito et al., 1995; Paek et al., 1997; Imsande and Schmidt, 1998; Paek et al., 2000; Imsande, 2001). The amino acid compositions of a broad range of 7S seed storage proteins from legumes and cereals can be found in Shewry and Casey (1999).

3. GROWTH CONDITIONS AND β-CHAIN ABUNDANCE

As stated, the glycinins are generally synthesized throughout seed development whereas the β-conglycinin chains are generally synthesized only during late seed development (Meinke et al., 1981). What regulates this sequential expression? Suppose that during the latter stages of seed development, and because of the high protein content of legume seeds, the availability of reduced sulphur, i.e., cysteine and methionine, becomes rate-limiting in protein synthesis. Thus, synthesis of those proteins containing cysteine and methionine residues would be curtailed and the seed, out of necessity, would preferentially synthesize those proteins that lack cysteine and methionine (i.e., the β-chain of β-conglycinin). This model is supported by the fact that nodulated legumes produce ammonia from dinitrogen and thus, are able to synthesize the eighteen amino acids lacking sulphur, and proteins lacking methionine and cysteine, when the plant is confronted with a deficiency of the sulphur amino acids. Indeed, the ability to produce ammonia from dinitrogen may help to explain why protein concentrations of legume seeds are generally two-fold greater than that of the typical non-nodulating cereal plant. On the other hand, there is no evidence that legumes have developed an increased capacity for the production of reduced sulphur. Experimental results described below tend to support this model.

Cereals and leguminous plants are known to degrade foliar proteins during seed formation. Subsequently, arginine, asparagine and glutamine are readily transported in the phloem to support seed development and seed storage protein synthesis (Pate, 1980). Sulphur amino acids, however, are transported primarily as glutathione, or homoglutathione, and S-methylmethionine (Sunarpi and Anderson, 1997; Bourgis et al., 1999; Anderson and Fitzpatrick, 2001). Due to its very reactive sulfhydryl group, free cysteine is seldom transported in either plants or animals. In mid-sized seeds, the concentration of homoglutathione, a tripeptide that contains a shielded cysteine residue, is approximately 1000-fold greater than that of methionine or cysteine (Sunarpi and Anderson, 1997). Although soybean pods and seeds contain an active ATP-sulphurylase, their modest concentration of free sulphate is soon depleted, (Sunarpi and Anderson, 1997; Sexton and Shibles, 1999; Anderson and Fitzgerald, 2001). Thus, during seed development, it appears that many plants, especially legumes, are less proficient at procuring and transporting the sulphur-containing precursors of protein synthesis than they are at procuring and transporting the nitrogenous precursors.

Sulphate deprivation during pod filling is reported to enhance the relative abundance of the β-chain of β-conglycinin (Gayler and Sykes, 1985). Likewise, providing excess reduced nitrogen (e.g., urea) to a soybean plant during pod filling has been shown to increase both seed yield and the abundance of the sulphur-deficient β-chain of β-conglycinin (Paek et al., 1997; Imsande, 1998; Imsande and Schmidt, 1998; Paek et al., 2000). Conversely, non-nodulated soybean plants, which are generally nitrogen deficient due to restricted nitrate uptake during pod filling (Imsande and Edwards, 1989; Imsande, 1998), produce a relatively low seed yield and those seeds contain only trace amounts of the β-chain of β-conglycinin (Ohtake et al., 1994; Krishnan et al., 2000). Thus, the abundance of sulphur-containing amino acid residues present in soybean seed storage proteins can be increased simply by decreasing soybean seed yield. This, however, is not a reasonable alternative for the soybean producer. Thus, many lines of research strongly suggest that well-nodulated soybean plant is more proficient in the procurement of reduced nitrogen than in the procurement of reduced sulphur during the latter stages of pod filling. It was recently reported, however, that sulphur assimilation by developing lupin cotyledons may be an important source of sulphur amino acids during the synthesis of storage proteins in maturing seeds (Tabe and Droux, 2001). Interestingly, the protein content of some species of lupin seeds is approximately 70% that of a typical soybean seed (Lilley, 1999). Other investigations are exploring carbon-flow models for understanding the integration of glycolysis and nitrogen assimilation (Smith et al., 2000; Imsande et al., 2001) as well as the ratio of seed storage proteins to seed lipids (Golombek et al., 2001).

In summary, the bulk of the available data strongly supports the view that a deficiency of free sulphur amino acids restricts the synthesis of the glycinins and promotes the preferential synthesis of the sulphur-deficient β-chain of β-conglycinin , during the later stages of seed development (Coates et al., 1985; Naito et al., 1994; Paek et al., 1997; Imsande and Schmidt, 1998; Paek et al., 2000). Indeed, seeds of many cereals contain "vicilin-like" 7S globulins that are sulphur-deficient (Kriz, 1999). In a less quantitative manner, the role of these 7S cereal proteins could be similar to that proposed above for the β-chain of β-conglycinin. Also, a dominant Scg-1 gene has been found in soybean that suppresses β-chain synthesis (Teraishi et al., 2001). One would predict, however, that suppression of β-chain synthesis by this dominant gene would lower both seed yield and seed protein concentration because of a sulphur deficiency. Hence, to improve the nutritional quality of soybean seed storage proteins, the plant must increase the availability of methionine and cysteine during the latter stages of seed storage development. Increased nutritive value of lupin seeds has been achieved by genetically transferring a sunflower seed albumin into *Lupinus angustifolis* L. (Molvig et al., 1997). The methionine content of the seeds produced by these transgenic plants was increased approximately 90%. Also, the 15 kDa maize zein gene, behind the β-phaseolin promoter for seed expression, has been inserted into soybean lines (Dinkins et al. 2001). Amino acid analysis of two of these lines indicated that the seed methionine content was increased approximately 15% and the cysteine content approximately 20% (Dinkins et al., 2001). Although the β-phaseolin promoter is seed-specific, it is possible that the zein gene was expressed early during seed development, and hence, the zein protein was synthesized before methionine became rate limiting. Also, an effort has been made to produce potato plants with an enhanced methionine content (Langille et al., 1998). This approach may be useful for nutritional improvement of cereal seeds as well.

4. METHIONINE ABUNDANCE AND REGULATION OF SEED COMPOSITION

Both in vitro and in planta, high levels of methionine inhibit synthesis of the β-chain of β-conglycinin (Holowach et al., 1984; Gayler and Sykes, 1985; Naito et al., 1994, 1995; Paek et al., 2000). Therefore, if a soybean plant were to over-produce methionine during late pod filling, perhaps the elevated concentration of methionine would inhibit synthesis of the β-chain without restricting protein synthesis or lowering seed yield.

Ethionine, a chemical analogue of methionine, is highly toxic to most organisms (Alix, 1982). Because of their similarity in chemical structure, an increased concentration of methionine lessens the toxic effect of a low concentration of ethionine. Thus, low concentrations of ethionine have been used successfully to select and identify methionine over-producing variants from mutagenized tissue culture lines (Gonzales et al., 1984; Langille et al., 1988; Madison and Thompson, 1988) and *Arabidopsis* seeds (Inaba et al., 1994). Recently, Imsande (2001) used these techniques to select soybean lines, the phenotypes of which are consistent with modest methionine over-production. Upon analysis, these lines produced a normal seed yield and a normal protein content. The abundance of the sulphur amino acids in these seeds, however, was increased by at least ten percent, as was the total sulphur content of the seeds. Thus, this technique should be successful in obtaining an increased sulphur amino acid content in other agronomically important crops that produce sulphur-deficient 7S globulins (Kriz, 1999).

The cloned soybean gene encoding the β-chain of β-conglycinin has been transferred to both a wild-type and to a methionine over-producing *Arabidopsis thaliana* plant (Naito et al., 1994, 1995). The wild-type transgenic plant readily synthesized the β-chain of β-conglycinin whereas the β-chain of β-conglycinin was formed only marginally by the methionine over-producing transgenic mutant (Naito et al., 1995). Regulation of methionine biosynthesis is discussed by Noji and Saito (this book). In short, the report by Chiba et al., (1999) provides a realistic mechanism by which the intracellular concentration of methionine can control the synthesis of cystathionine γ-synthase, the first enzyme specific for methionine biosynthesis in plants.

Mutant soybean tissue culture cell lines have been conditioned to accumulate approximately a 10-fold increase in intracellular methionine (Madison and Thompson, 1988). Moreover, these cell lines also accumulated approximately a 10-fold increase in S-methylmethionine (Madison and Thompson, 1988). In plants, S-methylmethionine reacts enzymatically with homocysteine to form two molecules of methionine (Ranocha et al., 2000). Approximately 90% of the methionine produced by a plant normally is consumed in transmethylation reactions that generate homocysteine. Thus, accumulation of S-methylmethionine is a very efficient way to store and generate methionine. Conceivably, enhanced storage of S-methylmethionine could become a component of the mechanism whereby plants acquire needed extra methionine during the latter stages of seed development.

In summary, plants can produce and maintain an elevated concentration of non-protein methionine. Also, plants can preferentially form proteins that are enriched in the sulphur-containing amino acids. Thus, it is possible for future investigators to produce plants that are more nutritious for man and other animals.

5. REFERENCES

Alix JH (1982) Molecular aspects of the in vivo and in vitro effects of ethionine, an analog of methionine. *Microbiol Rev* **46:** 281—295

Anderson JW and Fitzgerald MA (2001) Physiological and metabolic origin of sulphur for the synthesis of seed storage proteins. *J Plant Physiol* **158:** 447—456

Badley RA, Atkinson D, Hauser H, Odani D, Green JP and Stubbs JM (1975) The structure, physical and chemical properties of the soybean protein glycinin. *Biochim Biophys Acta* **412:** 214—228

Bourgis F, Roje S, Nuccio ML, Fisher DB, Tarczynski MC, Li C, Henderson C, Rennenberg H, Pimenta MJ, Shen T-L, Gage DA and Hansen AD (1999) S-methylmethionine plays a major role in phloem sulfur transport and is synthesized by a novel type of methyltransferase. *Plant Cell* **11:** 1485—1497

Casey R (1999) Distribution and properties of seed globulins. In: Shewry PR and Casey R (eds) Seed Proteins, pp 159—169. Kluwer Academic Publishers, Dordrecht

Chiba Y, Ishikawa M, Kijima F, Tyson RH, Kim J, Yamamoto A, Nambara E, Leustek T, Wallsgrove RM, and Naito S (1999) Evidence for autoregulation of cystathionine γ-synthase mRNA stability in Arabidopsis. *Science* **286:**1371—1374

Coates JB, Medeiros JS, Thanh VH and Nielsen NC (1985) Characterization of the subunits of β-conglycinin. *Arch Biochem Biophys* **243:** 184—194

Derbyshire E, Wright DJ and Boulter D (1976) Legumin and vicilin, storage proteins of legume seeds. *Phytochem* **15:** 3—24

Dinkins RD, Reddy MSS, Meurer CA, Yan B, Trick H, Thibaud-Nissen F, Finer JJ, Parrott WA and Collins GB (2001) Increased sulfur amino acids in soybean plants overexpressing the maize 15 kDa zein protein. *In Vitro Cell Develop Biol-Plant* **37:** 742—747

Fukushima D (1991a) Recent progress of soybean protein foods: Chemistry, technology, and nutrition. *Food Rev Int* **7:** 323—351

Fukushima D (1991b) Structure of plant storage proteins and their functions. *Food Rev Int* **7:** 353—381

Gayler KR and Sykes GS (1985) Effects of nutritional stress on the storage proteins of soybean. *Plant Physiol* **78:** 582—585

George AA and de Lumen BO (1991) A novel methionine-rich protein in soybean seed: Identification, amino acid composition, and N-terminal sequence. *J Agric Food Chem* **39:** 224—227

Golombek S, Rolletschek H, Wobus U and Weber H (2001) Control of storage protein accumulation during legume seed development. *J Plant Physiol* **158:** 457—464

Gonzales RA, Das PK and Widholm JM (1984) Characterization of cultured tobacco cell lines resistant to ethionine, a methionine analog. *Plant Physiol* **74:** 640—644

Harada JJ, Barker SJ and Goldberg RB (1989) Soybean β-conglycinin genes are clustered in several DNA regions and are regulated by transcriptional and posttranscriptional processes. *Plant Cell* **1:** 415—425

Holowach LP, Thompson JF and Madison JT (1984) Effects of exogenous methionine on storage protein composition of soybean cotyledons cultured in vitro. *Plant Physiol* **74:** 576—583

Imsande J (1998) Nitrogen deficit during soybean pod fill and increased plant biomass by vigorous dinitrogen fixation. *Euro J Agron* **8:** 1—11

Imsande J (2001) Selection of soybean mutants with increased concentrations of seed methionine and cysteine. *Crop Sci* **41:** 510—515

Imsande J and Edwards DG (1988) Decreased rates of nitrate uptake during pod fill by cowpea, green gram, and soybean. *Agron J* **80:** 789—793

Imsande J and Schmidt JM (1998) Effect of N source during soybean pod filling on nitrogen and sulfur assimilation and remobilization. *Plant Soil* **202:** 41—47

Imsande J, Berkemyer M, Scheibe R, Schumann U, Gietl C and Palmer RG (2001) A soybean plastid-targeted NADH-malate dehydrogenase: cloning and expression analyses. *Am J Botany* **88:** 2136—2142

Inaba KT, Fijiwara T, Hayashi H, Chino M, Komeda Y and Naito S (1994) Isolation of an Arabidopsis thaliana mutant, mto1, that over-accumulates soluble methionine. *Plant Physiol* **104:** 881—887

Kitamura K, Takagi T and Shibaski K (1976) Subunit structure of soybean 11S globulin. *Agr Biol Chem* **40:** 1837—1844

Krishnan HR, Jiang G, Krishnan AH and Wiebold WJ (2000) Seed storage protein composition of non-nodulating soybean (*Glycine max* (L.) Merr.) and its influence on protein quality. *Plant Sci* **157:** 191—199

Kriz AL (1999) 7S globulins. In: Shewry PR and Casey R (eds) Seed Proteins, pp 477—498. Kluwer Academic Publishers, Dordrecht

Langille AR, Lan Y and Gustine DL (1998) Seeking improved nutritional properties for the potato: Ethionine-resistant protoclones. *Am J Potato Res* **75:** 201—205

Lelievre J-M, Dickinson CD, Dickinson LA and Nielsen NC (1992) Synthesis and assembly of soybean β-conglycinin in vitro. *Plant Mol Biol* **18:** 259—274

Lilley GG (1999) Lupin Globulins. In: Shewry PR and Casey R (eds) Seed Proteins, pp 315—354. Kluwer Academic Publishers, Dordrecht

Madison JT and Thompson JF (1988) Characterization of soybean tissue culture cell lines resistant to methionine analogs. *Plant Cell Rep* **7:** 474—476

Maruyama N, Adachi M, Takahashi K, Yagasaki K, Kohno M, Takenaka Y, Okuda E, Nakagawa S, Mikami B, and Utsumi S (2001) Crystal structures of recombinant and native soybean β-conglycinin β homotrimers. *Eur J Biochem* **268:** 3595—3604

Meinke DW, Chen J and Beachy RN (1981) Expression of storage-protein genes during soybean seed development. *Planta* **153:** 130—139

Molvig L, Tabe LM, Eggum BO, Moore AE, Craig S, Spencer D and Higgins TJV (1997) Enhanced methionine levels and increased nutritive value of seeds of transgenic lupins (*Lupinus angustifolius* L.) expressing a sunflower seed albumin gene. *Proc Natl Acad sci USA* **94:** 8393—8398

Muntz K, Horstmann C and Schlesier B (1999) Vicia globulins. In: Shewry PB and Casey R (eds) Seed Proteins, pp 259—284. Kluwer Academic Publishers, Dordrecht

Naito S, Inaba-Higano K, Kumagai T, Kanno T, Nambara E, Fujiwara T, Chino M and Komeda Y (1994) Maternal effects of mto1 mutation, that causes over-accumulation of soluble methionine, on the expression of a soybean β-conglycinin gene promoter-GUS fusion in transgenic *Arabidopsis thatliana. Plant Cell Physiol* **35:** 1057—1063

Naito S, Yokota M, Inaba-Higano K, Nambara E, Fujiwara T, Hayashi H, Komeda Y and Chino M (1995) Expression of soybean seed storage protein genes in transgenic plants and their response to sulfur nutritional conditions. *J Plant Physiol* **145:** 614—619

Nielsen NC, Dickinson CD, Cho T-J, Thanh VH, Scallon BJ, Fisher RL, Sims TL, Drews GN and Goldberg RB (1989) Characterization of the glycinin gene family in soybean. *Plant Cell* **1:** 313—328

Ohtake N, Nishiwaki T, Mizukoshi K, Chinushi T, Takahashi Y and Ohyama T (1994) Lack of β-subunit of β-conglycinin in non-nodulating isolines of soybean. *Soil Sci Plant Nutr* **40:** 345—349

Osborne TB (ed) (1924) The Vegetable Proteins. 2nd edition, Longman, Green and Co., London

Paek NC, Imsande J, Shoemaker RC and Shibles R (1997) Nutritional control of soybean seed storage protein. *Crop Sci* **37:** 498—503

Paek NC, Sexton PJ, Naeve SL and Shibles R (2000) Differential accumulation of soybean seed storage protein subunits in response to sulfur and nitrogen nutritional sources. *Plant Prod Sci* **3:** 268—274

Pate JS (1980) Transport and partitioning of nitrogenous solutes. *Annu Rev Plant Physiol* **31:** 313—340

Ranocha P, Bourgis F, Ziemak MJ, Rhodes D, Gage DA and Hanson AD (2000) Characterization and functional expression of cDNAs encoding methionine-sensitive and -insensitive homocysteine S-methyltransferease from *Arabidopsis. J Biol Chem* **275:** 15962—15968

Sebastiani FL, Farrell LB, Schuler MA and Beachy RN (1990) Complete sequence of a cDNA of α subunit of β-conglycinin. *Plant Mol Biol* **15:** 197—201

Sebastiana L, Farrell LB, Vasquez M and Beachy RN (1991) Conserved amino acid sequences among plant proteins sorted to protein bodies and plant vacuoles. *Eur J Biochem* **199:** 441—450

Serretti C, Schapaugh Jr WT and Leffel RC (1994) Amino acid profile of high seed protein soybean. *Crop Sci* **34:** 207—209

Sexton PJ and Shibles RM (1999) Activity of ATP sulfurylase in reproductive soybean. *Crop Sci* **39:** 131—135

Shewry PR and Casey R (eds) (1999) Seed Proteins. Kluwer Academic Publishers, Dordrecht

Shewry PR and Pandya MJ (1999) The 2S albumin storage proteins. In: Shewry PR and Casey R (eds) Seed Proteins, pp 563—586, Kluwer Academic Publishers, Dordrecht

Smith CR, Knowles VL and Plaxton WC (2000) Purification and characterization of cytosolic pyruvate kinase from *Brassica napus* (rape seed) suspension cell cultures: implication for the integration of glycolysis with nitrogen assimilation. *Euro J Biochem* **267:** 4477—4485

Sunarpi and Anderson JW (1997) Allocation of S in generative growth of soybean. *Plant Physiol* **114:**687—693

Tabe LM and Droux M (2001) Sulfur assimilation in developing lupin cotyledons could contribute significantly to the accumulation of organic sulfur reserves in the seed. *Plant Physiol* **126:** 176—187

Teraishi M, Takahashi M, Hajika M, Matsunaga R, Uematsu Y and Ishimoto M (2001) Suppression of soybean β-conglycinin genes by a dominant gene Scg-1. *Theor Appl Genet* **103:** 1266—1272

Yoshino M, Kanazawa A, Tsutsumi K-I, Nakamura I and Shimamoto Y (2001) Structure and characterization of the gene encoding α subunit of soybean β-conglycinin. *Genes Genet Syst* **76:** 99—105

Xue Z-T, Xu M-L, Shen W, Zhuang N-L, Hu W-M and Shen SC (1992) Characterization of a Gy4 glycinin gene from soybean *Glycine max* cv. Forrest. *Plant Mol Biol* **18:** 897—908

6. AFFILIATIONS

Plant Genetics Group, Department of Agronomy, Iowa State University, Ames, Iowa 50011 USA

For correspondence: *jimsande@iastate.edu*

H.A. NAEEM AND F. MACRITCHIE

EFFECT OF SULPHUR NUTRITION ON AGRONOMIC AND QUALITY ATTRIBUTES OF WHEAT

Abstract. Global sulphur deficiency in arable lands is mainly due to phasing out of S-containing fertilizers and decrease in atmospheric deposition due to reduced emissions of SO_2 since the 1970's. Wheat has a low S requirement and addition of 10-20 kg/ha in most soil types is sufficient for maximum yield and optimum quality. Field experiments are the most direct and reliable methods of determining nutritional status of the soil. In addition to field trials, m any chemical indices have been investigated for diagnosis of S deficiency. Among these indices malate:sulphate ratio is a simple one step procedure for determination of soil S status. A malate:sulphate ratio of ≤ 1 indicates S sufficiency and ratio of >1 suggests S deficiency.

The effect of S deficiency on wheat proteins has been investigated by controlling different factors like environment, nutrient supply and genotypes. Actual occurrences of S deficiency in commercial crops like wheat have also been determined. The endosperm storage proteins or prolamins mainly determine end use properties of wheat flour. Wheat prolamins are classified on the basis of solubility and structural relationships. In structural classification prolamins are divided into three main classes – S-poor, S-rich and HMW prolamins. S availability does not affect total protein, individual protein components or their properties like electrophoretic mobility. It however does alter the proportions of different proteins, increases the gel protein weight and decreases its elastic modulus. S deficiency during plant growth results in an increase in the relative amounts of HMW-GS and omega-gliadins. This changed proportion increases the HMW/LMW-GS ratio and causes a shift in molecular weight distribution of glutenin to higher molecular weight. These changes in the amounts of different protein classes change the functional properties; dough becomes stronger with increased mixing requirements, less extensibility and reduced loaf volume. Grain S concentrations of 1.2 mg/g and N/S ratio of 17:1 appear to be critical for optimum quality. Maintaining a N/S fertilizer ratio of 3:1 also appears to be helpful in minimizing detrimental effects on quality. S applications would be needed in most parts of world to avoid yield losses, maintain optimum functional properties and nutritional value of this most important cereal.

1. INTRODUCTION

Sulphur (S) is an essential nutrient, both for plants and animals. It constitutes one of the macronutrients, nitrogen (N), phosphorus (P) and potassium (K), considered vital for proper plant growth and development (Syers et al., 1987). S uptake and assimilation in higher plants is one of the crucial factors determining crop yield, quality and even resistance to pests and environmental stresses. In a cascade of enzymatic steps, inorganic S is converted to nutritionally and functionally important S containing compounds like Cys, Met, several co-enzymes, thioredoxins, sulpholipids and vitamins i.e. biotin, thiamine and ferredoxin (Hell, 1997; Hell and Rennenberg, 1998; Saito, 2000). S is important in the formation of sulfhydryl (S-H) and disulphide bonds (S-S). These bonds are important for stabilization of protein structures and are responsible in part, for the unique viscoelastic properties of wheat gluten.

The importance of S as a plant nutrient has been recognized for a long time but active research started in the second half of the 20th century when widespread S

Yash P. Abrol and Altaf Ahmad (eds.), Sulphur in Plants, 305—322.

deficiencies were observed (Duke and Reisenauer, 1986). One of the main reasons for limited research on this important nutrient was the difficulty in measuring S in soils (Tillman, 1988). It is now widely recognized that S deficiency results in yield losses and quality deterioration. This chapter is concerned with an overview of the distribution of S deficiency, its amelioration, effect of other plant nutrients on S uptake and specifically its impact on processing quality of wheat.

2. CAUSES AND DIAGNOSIS OF SULPHUR DEFICIENCY

S requirement of crops varies with soil type and the crop grown. Many reviews on the occurrence of S deficiency have been published in the last decade. These reviews are based on specific region or country. A detailed global picture of S has been presented by the 'The Sulphur Institute of Washington' in the book entitled "Sulphur Outlook 2001". This book reviews comprehensively regional, political, technical and environmental issues and world events affecting the global S supply and demand, and projection till 2010. Some of the recent reviews include: on agro-ecosystems in Canada (Janzen and Ellert, 1998), S fertilizers for pastures (Boswell and Gregg, 1998) and S status of British arable land (Zhao et al., 1995).

S deficiency has been seen in many regions of the world in the final quarter of the last century. It depends on soil type, location, climatic conditions, and does not occur uniformly. The shift in the S balance of agricultural soils is due to:

1. Use of fertilizers high in N and P containing either low or no S at all.
2. Total consumption of N doubled from 1974 to 1990. However, the total consumption of S remained almost static at about 10 million tons during the same period (Ceccotti and Messick, 1994).
3. The industrialized countries have gradually cut down upon the atmospheric emission of SO_2 resulting in a massive decrease of atmospheric deposition of S in areas within the vicinity of industry. The decreasing trend in SO_2 emissions, as required by the international agreements is likely to continue at least to 2010. It is predicted that world fertilizer deficit will increase from 7.5 to over 11 million tons by 2010 (Ceccotti, 1996).
4. Introduction of high yielding crop varieties and increased cropping intensity resulted in increased removal of soil nutrients. This is especially true for S as the use of S containing fertilizers declined and high analysis nitrogen fertilizers like urea increased.
5. Other factors that may affect S availability are the soil pH (Freney and Williams, 1983), regular burning of vegetation that causes volatilization of S and farming system. S deficiencies may also occur when sulphate sorption capacity greatly exceeds that of adsorbed sulphate (Fox et al., 1983).

S requirements of wheat may roughly be defined as the minimum uptake of S associated with maximum grain yield and optimum quality. S deficiency often first manifests as a yellowing of younger leaves. Subsequently, these symptoms spread to the whole plant. Field experiments are the most direct and reliable methods of determining nutritional status of the soil. Field trials integrate many factors influencing supply of S and plant response. These tests are site-specific, time-consuming but provide the yardstick for all other diagnostic methods.

Several chemical indices are proposed for the determination of S deficiency in the soil and crops. Each of these tests has limitations with respect to ease of analysis and efficiency. Soil analysis provides useful information on available S in the soil but correlations between soil extractable S and yield are not reliable under field conditions of wet temperate regions (Jones, 1986; Zhao and McGrath 1994). Other diagnostic procedures put forth to resolve the issue, include determination of the N/S ratio (Rasmussen et al., 1977), the percentage of total S as sulphate (Spencer and Freney, 1980), the sulphate (Scaife and Burns, 1986), glutathione (Zhao et al., 1996, 1997), total S (Pinkerton, 1998), and malate:sulphate (Blake-Kalff et al., 2000). Limitations of these procedures lie with the plant growth stages for sampling, plant tissues analyzed, analytical methods used and growth environment i.e. whether the plants were grown in the field, greenhouse or growth chamber. Yield predictions based on analyses of plant tissues combined with crop and soil history are more reliable and provide efficient means for correcting nutrient deficiency. The use of sulphate:total S ratio involves twice as much work as that of sulphate or total-S alone. Moreover, the sulphate-S (numerator) is the major variable in the total S (denominator), thus reducing the sensitivity of the index. Malate:sulphate ratio proposed by Blake-Kalff et al. (2000) overcomes the limitations of growth stage, growth environment and plant tissues analyzed. It provides a simple one step testing procedure. A malate:sulphate ratio ≤ 1 indicates S sufficiency whereas a ratio of >1 suggests S deficiency at the time of sampling. However, its usefulness as a sound testing procedure for prediction of S deficiency needs further testing.

3. EFFECT OF SULPHUR DEFICIENCY ON WHEAT YIELD AND ITS AMELIORATION

Agronomic responses of crop growth and yield to the addition of S fertilizer are documented from most parts of the world, but are more extensive in regions of Western USA, Western Canada, Australia, New Zealand and Western Europe. Yield responses to S application depend upon available S levels in the field under trial and have been reported from 0-42% in wheat (Islam et al., 1999; De Ruiter and Martin, 2001). No response is indicative of S sufficiency. Yield increases of over one ton/ha on shallow calcareous soil in southern England are reported (Withers et al., 1995, 1997). Maximum response to S application in wheat is usually obtained between 10-20 kg/ha, which also corresponds to the optimum requirements of S for wheat (McGrath et al., 1996). Requirements as high as 49-60 kg/ha have also been reported for highly calcareous soils in India (Sakal et al., 1999a, 2000). This variation in S requirement springs from soil types, climate and the cropping system.

Several greenhouse and field experiments with wheat have indicated that effect of S is primarily on the number of grains per spike indicating that S deficiency either increases the mortality of florets or reduces the initiation of florets (Archer, 1974; Islam et al., 1999). Other yield components such as number of tillers and 1000-grain weight are less affected by S availability unless the deficiency is severe.

S deficiencies are generally alleviated by additions of S fertilizers in agricultural systems during early vegetative growth. The most widely used forms are sulphates and elemental S. The effectiveness of different forms of S including gypsum, potassium sulphate, elemental S and ammonium sulphate has been investigated in a number of studies but significant differences were not found (Chaplot et al., 1992; Nuttall et al., 1993; Sakal et al., 1999b). Similarly, non-significant differences are observed for fresh, residual and cumulative application of S to wheat. Residual and cumulative effects are significant only if initial S application was above 20 kg/ha. Although the nutrient status of soil is also important, the genotypes have a prominent role in the utilization of available resources (Sawyer and Ebelhar, 1995). Elemental S has the advantage over sulphates in reducing losses through leaching. Appropriate application methodology, long-term effects and interaction with soil pH are some of the aspects that need further investigation in the utilization of elemental S.

4. ASSIMILATION AND REDISTRIBUTION OF SULPHUR IN RESPONSE TO NITROGEN AVAILABILITY

Interrelationship of S and N supply on plant biochemistry has been extensively studied in rapeseed and soybean (Suarpi and Anderson, 1996, 1997a, 1997b; Blak-Kalff et al., 1998, 2000) and in cereals (Adiputra and Anderson, 1992, 1995; Fritzgerald et al., 1999a, 1999b, 2001; Kazuhara et al., 2000). Most studies are, however, limited to the vegetative growth stage. Plants take up S mainly by roots as sulphates from the soil solution. It is transported as such and in reduced form as Cys to the chloroplasts where it is incorporated into proteins. Transport from the sites of uptake in the roots to the sites of reduction and assimilation in the chloroplasts and to storage tissues as free amino acids or proteins involves numerous steps controlled by enzymes and long distance internal transport (Clarkson et al., 1993). This long distance transport of S and resource allocation is of particular importance in agriculture in relation to the incorporation of reduced S in the harvested part of the plant e.g. grains in cereals. Plants appear to possess the mechanisms to coordinate assimilatory sulphate and nitrate reduction so that the appropriate proportions of both S containing and other amino acids are available for protein synthesis (Bell et al., 1995b). When the amount of sulphate delivered to leaves exceeds that required for protein synthesis, sulphate accumulates in the leaves (Brunold, 1993). S deficiency induces a greatly expanded pool of free amino acids with a composition different from that of S sufficient plants and results in the accumulation of arginine, asparagine and *O*-acetyl-L-serine with a depressed level of S amino acids (Macinol, 1983; Macinol and Randall, 1987). These regulatory interactions between assimilatory sulphate and nitrate reduction are at transcriptional level (Lappartient et al., 1997; Prosser et al., 2001).

Nutrient requirement of growing plant changes with the growth stage especially at the onset of flowering. For example, barley plants grown in a medium at subnormal sulphate concentration (25 μM) did not show S deficiency symptoms prior to flowering (Adiputra and Anderson, 1992), indicating increased requirements after anthesis. Protein synthesis in the younger leaves or in the developing grains is severely affected by an unbalanced supply of either S or N. Insufficient supply of S results in the accumulation of excess soluble nitrogenous compounds like nitrates and asparagines (Dietz, 1989; Brunold, 1993). Nitrate levels could reach to the threshold where the crop is unsuitable for human use (Forman et al., 1985). In contrast, when the S supply relative to N exceeds the metabolic requirements, glutathione accumulates in chloroplasts (Rennenberg, 1984) and sulphate in the vacuoles of roots and leaves (Bell et al., 1995a). This vacuolar sulphate is redistributed only when concentration of sulphate in the cytoplasm falls to a very low level and is often too slow to compensate for the S deficiency (Clarkson et al., 1993; Bell et al., 1995a, b).

Sulphate is the major form of soluble S in the endosperm cavity and endosperm of wheat in plants grown under S sufficient conditions. But glutathione is the major form of soluble S in plants grown under S deficiency during rapid grain development conditions (Fritzgerald et al., 2001). Loss of S from vegetative tissues during the generative stage is well accounted for by an increase in S content of grains. In different conditions of S stress, different pools of S are mobilized (Fritzgerald et al., 1999a). The endosperm does not acquire and assimilate sulphates indefinitely. Eventually delivery and assimilation of sulphates cease. A detailed account of S metabolism may be found in Leustek et al. (2000) and chapters 4-7 in this book.

5. SULPHUR NUTRITION AND GENETIC ENGINEERING

Present data strongly suggests that regulatory mechanisms for S uptake and its redistribution are at transcriptional level (Lappartient et al., 1997; Prosser et al., 2001). The next step is identification of the genes controlling S metabolism, their number and the chromosome mapping. It is a long complicated process and the hexaploid nature of bread wheat further complicates the situation. Hawkesford (2000) suggested a two-level model for the improvement of S utilization (Fig. 1). Plants can absorb sulphate from the soil even when present in low micro molar range (Smith et al., 1997) indicating that extremely sensitive mechanism exists for S-uptake in plants. The modification of root system as the one that exists in xerophytes for increased water reserves in the root vacuoles or root system with increased proliferation would be one approach. Enhanced remobilization of vacuolar S reserves may be fruitful in manipulation of metabolic

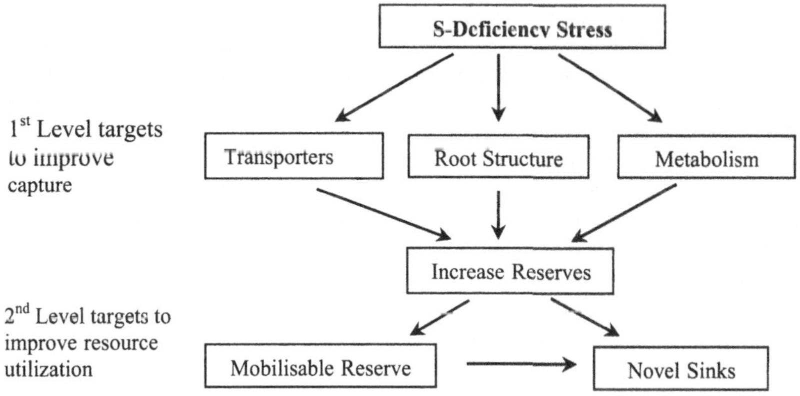

Figure 1. Strategies for the manipulation of the efficiency of sulphate uptake and utilization in crop plants. Possible targets to enhance the ability of a plant to respond to S-deficiency are split into two levels. Arrows indicate the inter-dependencies

processes. However, as pointed by Hawkesford (2000), little is known about the transporters catalyzing sulphate fluxes into and out of the vacuole.

6. EFFECTS OF SULPHUR SUPPLY ON WHEAT QUALITY ATTRIBUTES

The reduction of S application, mainly by the phasing out of S-containing fertilizers and the decrease in atmospheric deposition resulting from environmental controls has led to speculation that soil S deficiency could appear. The large glutenin polymers responsible for dough strength and elasticity are formed from glutenin subunits. These subunits polymerize via disulphide bonds as a post-translational modification, which occurs relatively slowly during grain development (Gupta et al., 1996). The possibility of S deficiency effects has spurred a number of studies, which have approached the problem from two perspectives. The first approach has sought to study the changes in protein composition and relate them to grain/flour properties when wheat is grown under controlled treatments of S and N fertilization. This aims to elucidate the effects of varying S availability on protein composition and end-use quality in order to be able to predict the consequences of S deficiency. The second approach has investigated whether S deficiency is actually occurring in commercial wheat crops and, if so, what are the ramifications. The two types of approaches are not completely separate and their objectives are similar. However, in order to simplify the picture, they will be considered separately here.

6.1 Classification of Wheat Proteins

Although starch is by far the most abundant component of wheat, protein dominates many of the properties that are important for end uses. This is because, in dispersed food systems, the continuous phase controls the properties and protein forms the continuous phase in products such as dough. Wheat contains a complex mixture of proteins. The two main classes are the gluten proteins (gliadins and glutenins) and the non-gluten proteins (albumins, globulins). The gluten proteins comprise two classes – gliadins, which are single chain, and glutenins, which are multiple chain proteins. In terms of molecular size, there are three main groups, the albumin/globulins, gliadins and glutenins with molecular weights in the ranges 20-30 K, 30-70 K and 100-many million K, respectively. Glutenins are polymers made up of two main sets of subunits. The high molecular weight glutenin subunits (HMW-GS) have molecular weights generally in the range 80-100 K and low molecular weight glutenin subunits (LMW-GS) have molecular weights in the range 40-55 K. The gliadins are classified in four groups based on their electrophoretic mobility in Acid-polyacrylamide gel electrophoresis (A-PAGE). The largest are the omega-gliadins (ω) with molecular weights of roughly 60-80 K. Then, in decreasing molecular size with some overlap are the gamma (γ), beta (β)- and alpha(α)- gliadins. The α- and β- gliadins are now believed to essentially belong to the same group.

Another classification of wheat proteins, based on primary structure relationships, considers the gluten proteins to be prolamins (Shewry et al., 1986). Prolamins are then divided into three classes – the S-poor, S-rich and HMW prolamins. The S-poor prolamins correspond to ω-gliadins. These proteins do not contain Cys residues. The

HMW prolamins correspond to HMW-GS and the S-rich prolamins include LMW-GS and α-, β- and γ-gliadins. Whereas the HMW- and LMW-GS polymerize through disulphide bonds, the S-rich gliadins usually have an even number of Cys residues and form only intramolecular disulphide bonds. This classification on the basis of different classes of prolamins is highly relevant to the present discussion on the effects of S availability.

6.2 Effects of S Availability on Protein Composition

Some preliminary analyses of how protein composition responded to variation of S content of the grain were reported by Moss et al. (1981) following earlier indications noted by Wrigley et al. (1980). These studies were subsequently extended to more detailed analyses of the samples from a factorial field experiment with the variety Olympic (Wrigley et al., 1984). It was shown that variation in S content of flour did not alter the individual proteins present or their properties such as electrophoretic mobility but only their relative amounts. The quantitative changes are illustrated in Figure 2 where densitometric scans of sodium dodecyl sulphate polyacrylamide gel (SDS-PAGE) are divided into five regions (A1-A5) corresponding to different protein classes.

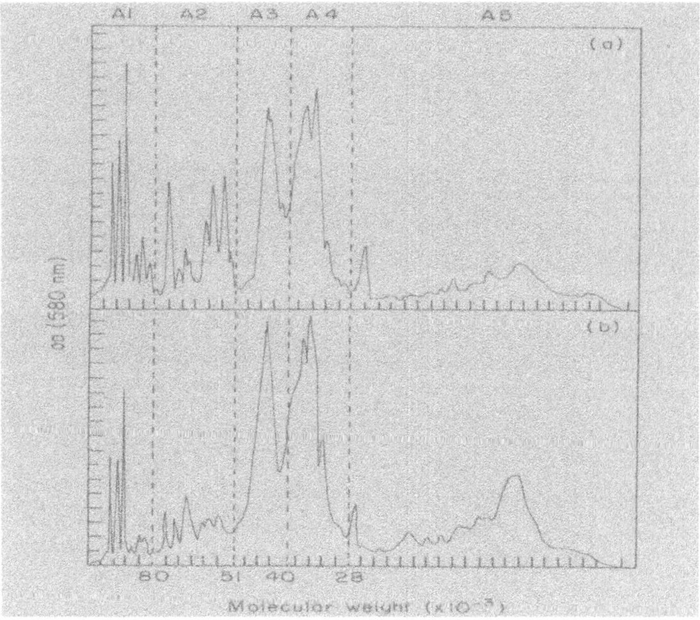

Figure 2. Densitometer scans of SDS gel electrophoregrams of polypeptides extracted with mercaptoethanol-SDS from sulphur-deficient flour (a, 0.083% S, 1.84% N) and from normal flour (b, 0.161% S, 1.96 % N). Regions of the densitometer scans were divided into five groups (A1-A5) as shown and the total area beneath the peaks in each group was shown. (Source: Wrigley et al., 1984)

S deficiency increases the relative amounts of proteins in regions A1 (mainly HMW-GS) and A2 (mainly ω-gliadins) and decreases the relative amounts of proteins in A3, A4 and A5. A3 contains LMW-GS, A4 has mainly α-, β- and γ-gliadins and A5 contains the albumins and globulins. These effects are directly related to the S contents of the different protein groups. HMW-GS and ω-gliadins are relatively S-poor whereas LMW-GS, α-, β- and γ-gliadins and albumins/globulins are relatively S-rich. Thus, when S availability declines, synthesis of the S-rich proteins is diminished relative to the S-poor proteins.

Protein characterization by electrophoresis of the flour samples from the variety Olympic has been extended to measurements by high performance liquid chromatography (HPLC) (MacRitchie and Gupta, 1993). Two modes of HPLC were used. Size-exclusion (SE-HPLC) separates wheat proteins according to molecular size into glutenins, gliadins and albumin/globulins. Reversed-phase (RP-HPLC) is useful for separation of gliadins or glutenin subunits. As S content of the flour decreased, the proportion of glutenin in the flour also decreased. LMW-GS is present in greater proportions (by a factor of 2-3) in normal flour than HMW-GS so that they contribute more to the glutenin content. An example of the effect of flour S content on the relative amounts of HMW-GS and LMW-GS is shown in Figure 3. As the flour S content was reduced from 0.146% to 0.075%, a dramatic shift in the HMW/LMW-GS ratio is seen to occur.

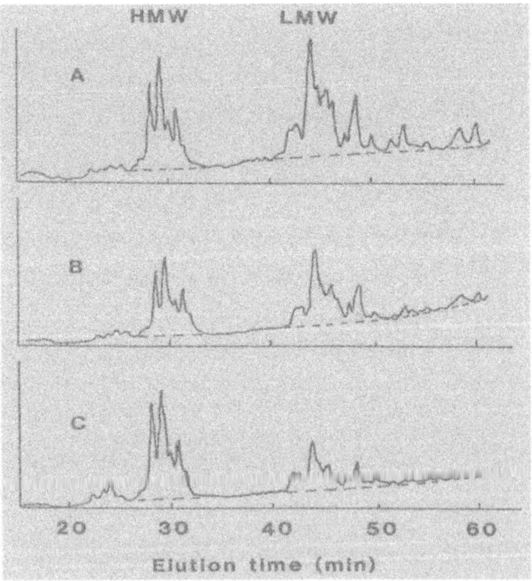

Figure 3. RP-HPLC profiles of glutenin reduced with mercaptoethanol from three flour samples of the variety Olympic varying in sulphur (S) and protein (P) contents. A, S = 0.146%, P = 10.4%; B, S = 0.100%, P = 7.8 %; C, S = 0.075 %, P = 9.7 % (From MacRitchie and Gupta, 1993).

6.3 Effects of S Availability on Functional Properties

One of the first extensive studies of the effect of varying levels of S fertilizer on grain and flour functionality was made by Moss et al. (1981). A wide range of functional properties was measured on a set of wheat samples from a factorial field experiment in which five S treatments and three nitrogen treatments were included (Randall et al., 1981). In this experiment, grain yield increased with increasing both S and N (Spencer and Freney, 1980). The relationship between grain S and nitrogen concentrations for the fourfold-replicated treatments is shown in figure 4. It is notable that high S concentration in the grain occurred only when both S and nitrogen were in good supply.

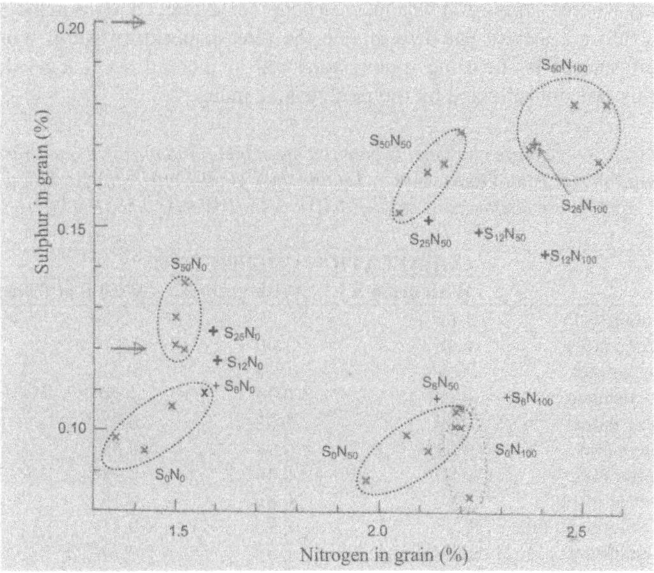

Figure 4. The sulphur and nitrogen concentrations of grain from fourfold-replicated treatments with combinations of sulphur (0, 6, 12.5, 25 and 50 kg sulphur/ha) and nitrogen (0, 50 and 100 kg/ha) fertilizers. The values of each of the four replicates are shown (x) for each combination of S_0 and S_{50} with N_0, N_{50} and N_{100}. For the remaining fertilizer combinations, only values for a composite sample (+) of each treatment are shown. The arrows at left indicate the limits of the range of grain sulphur contents reported by Schultz and French (1976). (Source: Moss et al., 1981)

The different quality measurements that were made are listed in Table 1 together with their correlation coefficients from linear regression with grain S, grain nitrogen and S at constant nitrogen. Correlation coefficients for nitrogen variation are generally in agreement with expectations. In the case of S, perhaps the most striking effect when grain S is decreased below a certain level is that dough properties change, the dough becoming stronger and less extensible. This is reflected in the high values of the correlation coefficients with extensograph parameters, negative with resistance and positive with extensibility. Some of the correlations shown in Table 1 appear to conflict. For example, there is a high positive correlation with pup-loaf volume but a negative one with loaf volume in a micro baking test. The micro baking test was optimized for variables such as dough mixing time. In contrast, the pup-loaf baking test used a constant dough development time, which would penalize the tougher doughs but may be more relevant to commercial bakery practice where doughs have to conform to strict tolerances, including mixing times. The correlation coefficients for dough development time are of opposite sign for the two recording dough mixers, being positive for the farinograph and negative for the mixograph. This must be related to the different mixing actions of the instruments. The farinograph forms a coherent dough immediately whereas, mixograph can have a delay for stronger doughs in incorporating the dough into a coherent mass to enable the pins to perform useful work. Grain hardness (measured by pearling index) appeared to increase as S level decreased, although this was not reflected by the particle size index.

Table 1. Linear correlation coefficients between the sulphur and nitrogen concentrations in the grain samples from Figure 4 (n = 33) and their quality attributes. Significance of correlation coefficients: 0.35 (P 0.05); 0.45 (P, 0.01); 0.55 (P 0.001)

	CORRELATION COEFFICIENT		
	With grain S	**With grain N**	**With S at Constant N**
Grain sulphur (%)	1.00		
Grain nitrogen (%)	0.48	1.00	
100-kernel weight	-0.66	-0.43	-0.57
Pearling resistance	-0.57	0.24	-0.80
Particle size index	0.20	-0.23	0.37
Flour protein (%)	0.54	0.98	0.35
Flour sulphur (%)	0.98	0.44	0.98
Flour water absorption	0.49	0.87	0.15
Dough development time	0.78	0.70	0.71
Dough breakdown	0.09	-0.72	0.73
Resistance to extension	-0.73	0.02	-0.85
(at 5 cm)			-0.64
Maximum resistance	-0.32	0.40	
Dough extensibility	0.95	0.52	0.83
Pup-loaf volume	0.70	0.85	0.83
Pup-loaf score	0.82	0.76	0.80
Mixograph peak	-0.82	-0.26	-0.82
development time			-0.17
Mixograph height	-0.06	0.17	
Mixograph work	-0.49	-0.08	-0.52
Micro-loaf volume	-0.41	0.14	-0.55

Source: Moss et al. (1981)

The changes in extensograph properties induced by S deficiency are relevant to flour quality for production of biscuits (cookies). Soft-grained wheat flours are preferred for biscuits with weaker dough properties than those required for bread making. Relatively low resistance to stretching and high dough extensibility is sought because biscuit doughs need to spread and not to shrink as happens with relatively elastic doughs. The American Association of Cereal Chemists has a standard cookie spread test to measure soft wheat flour quality. In regard to this, Moss et al. (1983) reported a significant correlation between cookie spread and flour S content in an Australian soft wheat variety Egret.

6.4 Utilization of Sulphur Variability to Explore Protein Composition-Functionality Relationships

One of the main ways to elucidate protein composition-functionality relationships for wheat flours has been to use different genotypes to impart variability in the protein composition, while maintaining constant environmental conditions. An alternative approach is to vary some environmental factor while maintaining the genotype the same. As protein composition is sensitive to S availability, this is one environmental factor that can be used. It thus provides a way of manipulating protein composition and has the great advantage that, by studying one variety, effects of genetic variation can be avoided. Some correlation coefficients between protein composition variables and flour quality parameters obtained from the study on Olympic flour, discussed above, are recorded in Table 2.

An unusual result was the high negative correlation of the percentage of glutenin in the protein with extensograph maximum resistance (R_{max}), an accepted measure of dough strength. In studies using different genotypes, a strong positive correlation is usually obtained albeit sometimes low (Singh et al., 1990; Gupta et al., 1992). This unusual result arises because S deficiency reduces the proportion of glutenin in the wheat protein due to the overriding contribution of the S-rich LMW-GS. However, simultaneously, the ratio of HMW/LMW-GS increases which is expected to cause the molecular weight distribution of glutenin to be shifted to higher molecular weights. This increase in molecular weight of glutenins shows up as a decrease in solubility. A relative measure of the molecular weight distribution can be obtained from the proportion of glutenin that is not extractable, termed the unextractable polymeric protein (UPP) (Gupta et al., 1993). In figure 5, a plot of UPP against the HMW/LMW-GS ratio for the set of Olympic flours grown at the different S and nitrogen fertilizer levels shows the relationship between these two parameters. The correlations shown in Table 2 confirm the close dependence of dough strength, measured by extensigraph R_{max}, on the molecular weight distribution, measured by UPP. It has been suggested that R_{max} is determined by the proportion of glutenin above a critical molecular weight while extensibility depends on the total proportion of glutenin with the provision that molecular weights should not be too high (Bangur et al., 1997; Larroque et al., 1999). The increased dough strength and decreased extensibility as S deficiency becomes more acute, is consistent with these considerations.

Grain hardness has been related to the nature of the adhesion between starch granules and the matrix (Greenwell and Schofield, 1986; Oda et al., 1992; Morris et al., 1994). Low molecular weight proteins (puroindolines) are supposed to be present at the surface of starch granules in soft wheat kernels causing poor adhesion. It may be that large protein molecules impart good adhesion so that glutenins would be concentrated

Table 2. Correlation coefficients for linear regression between some quality parameters and different measures of flour protein composition for 24 flour samples of the wheat variety Olympic grown under different sulphur/nitrogen fertilize rlevels. Rmax = extensograph maximum resistance; MDDT = mixograph dough development time; LV (rapid) = loaf volume in an optimized short time baking procedure; Ext = extensograph extensibility; FDDT = farinograph peak development time; LV (long ferm.) = loaf volume in a long fermentation baking procedure; FBD = farinograph breakdown; F.P. = flour protein content; PPP = percentage polymeric protein in the protein; UPP = percentage unextractable polymeric protein; FPP = percentage polymeric protein in the flour; HMW/LMW = ratio of HMW/LMW-GS. Significance of correlation coefficients: 0.40 (P 0.05); 0.52 (0.01); 0.63 (P 0.001).

Quality parameter	F.P.	PPP	UPP	FPP	HMW /LMW
Rmax	0.155	-0.721	0.824	-0.107	0.795
MDDT	-0.432	-0.364	0.785	-0.643	0.795
LV (rapid)	-0.168	-0.532	0.660	-0.413	0.650
Ext	0.570	0.336	-0.540	0.798	-0.591
FDDT	0.773	-0.068	-0.381	0.923	-0.344
LV (long ferm.)	0.865-	-0.123	-0.205	0.938	-0.122
FBD	0.647	0.875	0.559	-0.394	-0.790
Range	6.7-12.1	42.2-52.5	44.4-56.4	3.5-5.8	0.30-89

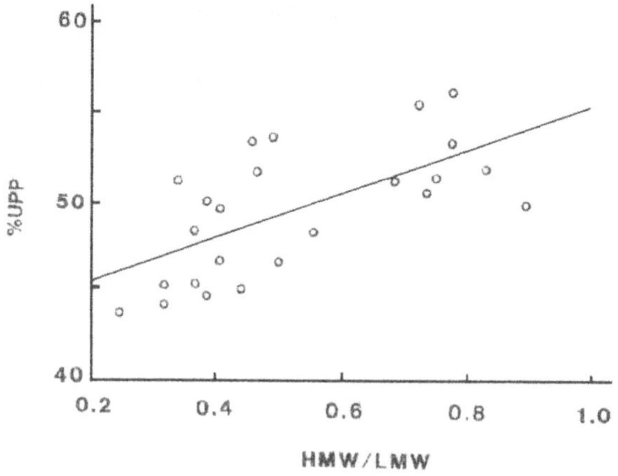

Figure 5. Plot of the percentage of unextractable polymeric protein (UPP) against the ratio of high to low molecular weight glutenin subunits (HMW/LMW-GS) for 24 flour samples of the wheat variety Olympic varying in sulphur and protein contents. Correlation coefficient for linear regression was 0.699 (P<0.001).

at the starch/matrix interface in hard wheats. Hard wheats are usually associated with strong dough properties and soft wheats with weaker dough properties. Although this may, to a large extent, be the result of targeting combined traits of grain hardness and dough strength in breeding, there may be some inherent relation between the two. If so, the larger glutenins giving rise to higher dough strength in S deficient grain may be contributing to the increased grain hardness.

6.5 Occurrence of S Deficiency: Effects on Quality

Studies of the negative effects of S deficiency on end-use quality of wheat such as that of Moss et al. (1981) alerted the scientific community to the possibility that the lower S supplies may show up as detrimental effects on quality in the wheat crop. A survey in the early 1980s showed no evidence of S deficiency in wheat (Byers et al., 1987). From that time, however, there has been mounting evidence for S deficiency in many areas around the world. The first changes in functionality of flours from field-grown wheat as a result of limiting S content in the UK were reported by Kettlewell et al. (1998). Three field experiments in 1986, 1987 and 1988 used late-season foliar-applied S and nitrogen fertilizers. Only grain from the 1988 season had values of S content and N/S ratio that could be considered to be deficient by the accepted thresholds. Multiple regression on combined data for the three seasons showed no effect of applied S on the extensograph resistance or the baking performance measured by bake test loaf volume. However, it was shown that dough extensibility increased with both increase of flour S and nitrogen concentration.

In wheat grain collected from the main cereal growing areas of Britain, the mean S content in the grain was found to be 1.72 mg/g for a large number of samples from the 1981 and 1982 crops (Byers et al., 1987) whereas the mean S content of the grain from samples collected in 1992 and 1993 was 1.35 mg/g (Zhao et al., 1995). Data from these studies are summarized in Table 3. The decrease in grain S status over a decade is evident. Although the varieties from the 1981-1982 study were superceded by different varieties in the 1992-1993 study, it was considered that the change was not due to this but to a genuine decline in S availability. Of course, S deficiency can be exacerbated by high nitrogen supply. In terms of criteria, it has been suggested that the critical value for S deficiency in wheat is 1.2 mg/g and a critical value for the N/S ratio is 17:1 (Randall et al., 1981). Thresholds for S deficiency in relation to baking quality have not been so clearly defined but they may be similar (Randall and Wrigley, 1986). When the S content of grain is below this value or the N/S ratio is above the critical value, detrimental effects on end-use quality can be expected. According to these criteria, no samples from the 1981 and 1982 samples of the study by Byers et al. (1987) were deficient in S but a relatively low but significant proportion of the 1992 and 1993 samples (Zhao et al., 1995) fell in the S deficient category.

Seven field experiments carried out on a premium hard winter wheat (variety Hereward) at four sites in the 1994/95 cropping season, investigated the effects of S application on bread making quality (Zhao et al., 1999a). Two levels of nitrogen and three levels of S were applied. Significant increases in loaf volume were observed for S applications, more so than for nitrogen. S did not increase grain protein content but did increase gel protein weight and decrease its elastic modulus. This reflects the role of S in increasing glutenin but decreasing the strength of the elastic network by shifting the HMW/LMW-GS ratio to lower values. Further field experiments carried out on three winter wheat varieties at three sites in the 1996-97 seasons confirmed the benefits of S applications to bread-making quality (Zhao et al., 1999b). This work showed minor effects of S on grain yield. Grain S levels were all above the threshold value of 1.2 mg/g although N/S ratios were quite high. The results tend to indicate that the figure of 1.2 mg/g of grain S is probably a reasonably reliable threshold in terms of yield but effects on dough properties and bread-making performance can be obtained at higher grain S levels.

Table 3. Ranges and means of grain S concentrations and N:S ratio for wheat grown in Great Britain in different years

	1993[a]	1992[a]	1982[b]	1981[b]
Number of samples	393	400	238	170
Sulphur (mg g^{-1})				
Range	0.80–1.67	0.54-2.09	1.33-2.14	1.30-2.03
Mean	1.26	1.43	1.72	1.71
Nitrogen/sulphur ratio				
Range	112.5-25.9	13.3-29.6	10.7-15.0	8.8-14.5
Mean	15.6	15.8	13.1	11.5

Sources: [a]Zhao et al. (1995); [b]Byers et al. (1987)

One of the effects of S deficiency is to increase dough mixing requirements and this can create problems in bakery production lines. In two field trials in New Zealand using four cultivars, Wooding et al. (1999) examined the nexus between dough strength (measured by extensograph R_{max}) and dough mixing requirements (measured by mechanical dough development work input, MDD WI, and mixograph dough development time). Effects of S fertilization in reducing mixing requirements were observed and some suggestions for reducing excessive mixing requirements without penalizing extensograph resistance were given. Another New Zealand study (Wooding et al., 2000) reported that maintaining the N/S fertilizer ratio at 3:1, minimized MDD WI for one cultivar while maintaining other dough properties. Values of grain S content and N/S ratio in these studies were slightly greater than 1.2 mg/g and less than 17:1, respectively even in the absence of S fertilization. It appears that S fertilizer can exert effects on dough properties when grain S levels are outside but close to the accepted threshold values for grain yield. Another field study in New Zealand confirmed that application of S fertilizer significantly reduced MDDT WI under conditions of high nitrogen management (De Ruiter and Martin, 2001).

7. CONCLUSIONS

S deficiency in wheat has increased in the last two decades, being particularly well documented in England but also observed in other countries. S concentrations in the grain of 1.2 mg/g and N/S ratios of 16:1 or 17:1 appear to be threshold values for S deficiency in regard to yield. However, changes in flour properties in response to S fertilization have been observed at higher values of grain S, particularly if the N/S ratio is high. S deficiency is exhibited initially by a shift in balance of dough properties towards higher strength and lower extensibility. The cause of this shift is a redistribution of proteins, S-poor proteins increasing at the expense of S-rich proteins. Between these two classes, the HMW-GS (S-poor) and LMW-GS (S-rich) are particularly relevant to functionality. A change to a higher ratio of HMW/LMW-GS in response to S shortage shifts the molecular weight distribution to higher molecular weights, leading to the higher strength and lower extensibility. These changes translate to extended dough mixing requirements and loss of baking performance, measured by bake-test loaf volume. S deficiency is likely to continue to increase. Dough properties and grain S levels need to be monitored and S applications need to be considered to avoid potential problems in the industry.

8. REFERENCES

Adiputra IGK and Anderson JW (1992) Distribution and redistribution of sulphur taken up from nutrient solution during vegetative growth. *Physiol Plant* **85:** 453—460

Adiputra IGK and Anderson JW (1995) Effect of nutrition on redistribution of sulphur in vegetative barley. *Physiol Plant* **95:** 643—650

Archer MJ (1974) A sand culture experiment to compare the effects of sulphur on five wheat cultivars (*Triticum aestivum* L.). *Aust J Agric Res* **25:** 369—380

Bangur R, Batey IL, McKenzie E and MacRitchie F (1997) Dependence of extensograph parameters on wheat protein composition measured by SE-HPLC. *J Cereal Sci* **25**: 237—241

Bell CI, Clarkson DT and Cram WJ (1995a) Partitioning and redistribution of sulphur during S-stress in *Macroptillium atropurpureum* cv. Siratro. *J Exp Bot* **46:** 73—81

Bell CI, Clarkson DT and Cram WJ (1995b) Sulphate supply and its regulation of transport in roots of a tropical legume *Macroptilium atropupureum* cv. Siratro. *J Exp Bot* **46:** 65—71

Blake-Kalff MMA, Harrison KR, Hawkesford MJ, Zhao FJ and McGrath SP (1998) Distribution of sulfur within oilseed rape leaves in response sulfur deficiency during vegetative growth. *Plant Physiol* **118:** 1337—1344

Blake-Kalff MMA, Harison KR, Hawkesford MJ, Zhao FJ and McGrath SP (2000) Diagnosing sulfur deficiency in field-grown oilseed rape (*Brassica napus* L.) and wheat (*Triticum aestivum* L). *Plant Soil* **225:** 95—107

Boswell CC and Gregg PEH (1998) Sulfur fertilizers for grazed pasture systems. In: Maynard DG (ed) Sulfur in the Environment, pp—134. Marcel Dekker, New York

Brunold CH (1993) Regulatory interactions between sulfate and nitrate assimilation. In: de Kok LJ, Stulen I, Rennenberg H and Brunold CH (eds) Sulfur Nutrition and Assimilation in Higher Plants: Regulatory, Agricultural and Environmental Aspects, pp 61—75. SPB Academic Publishing, The Hague

Byers M, McGrath SP and Webster R (1987) A survey of the sulphur content of wheat grown in Britain. *J Sci Food Agric* **38:**151—160

Ceccotti SP (1996) A global review of nutrient sulphur balance, fertilizers and the environment. *Agro-Food-Industry Hi-Tech* **7:** 18—22

Ceccotti SP and Messick DL (1994) Increasing plant nutrient sulphur use in European fertilizer industry. *Agro-Food-Industry Hi-Tech* **5:** 9—14

Chaplot PC, Jain GL and Bansal KN (1992) Effect of sulfur and phosphorus on sesame (*Sesamum indicum*) and its residual effect on wheat (*Triticum aestivum*). *Indian J Agron* **37:** 340—341

Clarkson DT, Hawkesford MJ and Davidian J (1993) Membrane and long-distance transport of sulphate. In: de Kok LJ, Stulen I, Rennenberg H and Brunold CH (eds) Sulfur Nutrition and Assimilation in Higher Plants: Regulatory, Agricultural and Environmental Aspects, pp 3—20. SPB Academic Publishing, The Hague

De Ruiter JM and Martin RJ (2001) Management of nitrogen and sulphur fertiliser for improved bread wheat (*Triticum aestivum*) quality. *New Zealand J Crop and Hort Sci* **29:** 287—299

Dietz KJ (1989) Leaf and chloroplast development in relation to nutrient availability. *J Plant Physiol* **134:** 544—550

Duke SH and Reisenauer HM (1986) Roles and requirements of sulfur in plant nutrition. In: Tabatabai, MA (ed) Sulfur in Agriculture, pp 123—168. American Society of Agronomy, Crop Science Society of America, Soil Science Society of America, Madison, Wisconsin

Forman D, Al-Dabbagh S and Doll R (1985) Nitrates, nitrites and gastric cancer in Great Britian. *Nature* **313:** 620—625

Fox RL, Asghar RM and Cable M (1983) Sulfate accretions in soils of the tropics. In: Blair GJ and Till AR (eds) Sulfur in S.E. Asian and S Pacific Agriculture, pp 39—53. Univ New England Press, Armidale

Freney JR and Williams CH (1983) The sulphur cycle in soil. In: Ivanov MV and Freney JR (eds) The Global Biogeochemical Sulphur, pp 129—201. John Wiley & Sons, Chichester

Fritzgerald MA, Uglade TD and Anderson JW (1999a) Sulphur nutrition changes the sources of S in vegetative tissues of wheat during generative growth. *J Exp Bot* **50:** 499—508

Fritzgerald MA, Uglade TD and Anderson JW (1999b) S nutrition affects the pools of S available to developing grains of wheat. *J Exp Bot* **50:** 1587—1592

Fritzgerald MA, Uglade TD and Anderson JW (2001) Sulphur nutrition affects delivery and metabolism of S in developing endosperms of wheat. *J Exp Bot* **52:**1519—1526

Greenwell P and Schofield JD (1986) A starch granule protein associated with endosperm softness in wheat. *Cereal Chem* **63:** 379—380

Gupta RB, Batey IA and MacRitchie F (1992) Relationship between protein composition and functional properties of wheat flours. *Cereal Chem* **69:** 125—131

Gupta RB, Khan K and MacRitchie F (1993) Biochemical basis of flour properties in bread wheats. I. Effects of variation in quality and size distribution of polymeric proteins. *J Cereal Sci* **18:** 23—44

Gupta RB, Masci S, Lafiandra D, Bariana HS and MacRitchie F (1996) Accumulation of protein subunits and their polymers in developing grains of hexaploid wheats. *J Exp Bot* **47:**1377—1385

Hawkesford MJ (2000) Plant responses to sulphur deficiency and the genetic manipulation of sulphate transporters to improve S-utilization. *J Exp Bot* **51:** 131—138

Hell R (1997) Molecular biology of plant sulfur metabolism. *Planta* **202:** 138—148

Hell R and Rennenberg H (1998) The plant sulfur cycle. In: Schnug E (ed) Sulphur in Agroecosystems, pp 135—137. Kluwer Academic Publishers, Dordrecht

Islam MR, Islam MS, Jahiruddin M and Hoque MS (1999) Effects of sulfur, zinc and boron on yield, yield components and nutrient uptake of wheat. *Pak J Sci Ind Res* **42:** 137—140

Janzen HH and Ellert BH (1998) Sulfur dynamics in cultivated temperate agroecosystems. In: Maynard DG (ed) Sulfur in the Environment, pp 11—94. Marcel Dekker Inc, New York

Jones MB (1986) Sulfur availability indexes. In: Tabatabai MA (ed) Sulfur in Agriculture, pp 549—566. American Society of Agronomy, Crop Science Society of America, Soil Science Society of America, Madison, Wisconsin

Kazuhara Y, Isobe A, Awazuhara M, Fujiwara T and Hayashi H (2000) Glutathione levels in phloem sap of rice plants under sulfur-deficient conditions. *Soil Sci Plant Nutr* **46:** 265—270

Kettlewell PS, Griffiths MW, Hocking TJ and Wallington DJ (1998) Dependence of wheat dough extensibility on flour sulphur and nitrogen concentrations and the influence of foliar-applied sulphur and nitrogen fertilisers. *J Cereal Sci* **28:** 15—23

Lappartient AG, Leustek T and Tourane B (1997) Are ATP sulphurylase mRNA and protein accumulated in roots of Arabidopsis following stress? In: Cram WJ, de Kok LJ, Stulen I, Brunold CH and Rennenberg H (eds) Sulfur Metbolism in Higher Plants: Molecular, Ecophysiological and Nutritional Aspects, pp 207—209. Backhuys Publishers, Leiden

Larroque O, Gianibelli MC and MacRitchie F (1999) Protein composition for pairs of wheat lines with contrasting dough extensibility. *J Cereal Sci* **29:** 27—31

Leustek T, Martin MN, Bick JN and Davies JP (2000) Pathways and regulation of sulfur metabolism revealed molecular and genetic studies. *Annu Rev Plant Physiol Plant Mol Biol* **51:** 141—165

Macinol PK (1983) Developmental changes in free amino acid pools and total protein amino acids of pea cotyledons (*Pisum sativum* L.). *Plant Physiol* **72:** 492—497

Macinol PK and Randall PJ (1987) Changes in the levels of major sulfur metabolites and free amino acids in pea cotyledons recovering from sulfur deficiency. *Plant Physiol* **83:** 354—359

MacRitchie F and Gupta RB (1993) Functionality-composition relationships of wheat flour as a result of variation in sulfur availability. *Aust J Agric Res* **44:** 1767—1774

McGrath SP, Zhao FJ and Withers PJA (1996) Development of sulfur deficiency in crops and its treatment. Proc Fertilizer Society No.379. The Fertilizer Society, Peterborough

Morris CF, Greenblat GA, Bettge AD and Malkawi HI (1994) Isolation and characterization of multiple forms of friabilin. *J Cereal Sci* **20:** 167—174

Moss HJ, Randall PJ and Wrigley CW (1983) Alteration to grain, flour and dough quality in three wheat types with variation in soil sulfur supply. *J Cereal Sci* **1:** 255—264

Moss HJ, Wrigley CW, MacRitchie F and Randall PJ (1981) Sulfur and nitrogen fertilizer effects on wheat. II. Influence on grain quality. *Aust J Agric Res* **32:** 213—26

Nuttall WF, Boswell CC, Sinclair AG, Moulin AP, Townley SLJ and Galloway GL (1993) The effect of time of application and placement of sulfur fertilizer sources on yield of wheat, canola, and barley. *Commu Soil Sci Plant Anal* **24:** 2193—202

Oda S, Komae K and Yasui T (1992) Relation between starch granule protein and endosperm softness in Japanese wheat (*Triticum aestivum*) cultivars. *Japanese J Breed* **42:** 161—165

Pinkerton A (1998) Critical sulfur concentrations in oilseed rape (*Brassica napus*) in relation to nitrogen supply and to plant age. *Aust J Exp Agric* **38:** 511—522

Prosser MI, Purves JV, Saker LR and Clarkson DT (2001) Rapid disruption of nitrogen metabolism and nitrate transport in spinach plants deprived of sulphate. *J Exp Bot* **52:** 113—121

Rasmussen PE, Raming RE, Ekin LG and Rhode CR (1977) Tissue analyses guidelines for diagnosing sulfur deficiency in white wheat. *Plant Soil* **46:** 153—163

Randall PJ, Spencer K and Freney JR (1981) Sulfur nitrogen fertilizer effects on wheat. I. Concentration of sulfur and nitrogen and the nitrogen to sulfur ratio in the grain in relation to yield response. *Aust J Agric Res* **32:** 203—212

Randal PJ and Wrigley CW (1986) Effects of sulfur supply on yield, composition and quality of grain from cereals, oilseeds and legumes. In: Pomeranz Y (ed) Advances in Cereal Science and Technology, Vol VIII, pp 171—206. American Association of Cereal Chemists, St Paul, Minnesota

Rennenberg H (1984) The fate of excess sulfur in higher plants. *Annu Rev Plant Physiol* **35:** 121—153

Saito K (2000) Regulation of sulfate transport and synthesis of sulfur-containing amino acids. *Curr Opin Plant Biol* **3:** 188—195

Sakal R, Singh AP, Sinha RB, Bhogal NS and Ismail MD (1999a) Impact of sulphur fertilisation in sustaining the productivity of rice-wheat cropping system. *Fert News* **44:** 49—52

Sakal R, Sinha RB and Singh AP (1999b) Effect of SSP and DAP on phosphorus and sulfur nutrition of crops in rice based cropping systems. *Fert News* **44:** 69—75

Sakal R, Sinha RB, Singh AP, Bhogal NS and Ismail MD (2000) Influence of sulphur on yield and mineral nutrition of crops in maize-wheat sequence. *J Indian Soc Soil Sci* **48:** 325—329

Sawyer JE and Ebelhar SA (1995) Sulfur nutrition of winter wheat varieties with till and no-till planting on highly weathered Alfisol soils. *Comm Soil Sci Plant Anal* **26:** 257—271

Scaife A and Burns IG (1986) The sulfate-S/total S ratio in plants as an index of their sulphur status. *Plant Soil* **91:** 61—71

Shewry PR, Tatham AS, Forde J, Kreis M and Miflin BJ (1986) The classification and nomenclature of wheat gluten proteins: a reassessment. *J Cereal Sci* **4:** 97—106

Singh NK, Donan GR and MacRitchie F (1990) Use of sonication and size-exclusion HPLC in the study of wheat four proteins. II. Relative quality of gluten as a measure of bread-making quality. *Cereal Chem* **67:** 161—170

Smith FW, Hawkesford MJ, Ealing PM, Clarkson DT, Vanden-Berg PJ, Belcher AR and Warrilow AGS (1997) Regulation of expression of a cDNA from barley roots encoding a high affinity sulphate transporter. *Plant J* **12:** 875—884

Spencer K and Freney JR (1980) Assessing the sulfur status of field grown wheat by plant analysis. *Agron J* **72:** 469—472

Sunarpi and Anderson JW (1996) Distribution and redistribution of sulfur supplied as [^{35}S] sulfate to roots during vegetative growth of soybean. *Plant Physiol* **110:** 1151—1157

Sunarpi and Anderson JW (1997a) Effect of nitrogen nutrition on the export of sulphur from leaves in soybean. *Plant Soil* **188:** 177—187

Sunarpi and Anderson JW (1997b) Inhibition of sulphur redistribution into new leaves of vegetative soybean by excision of the maturing leaf. *Physiol Plant* **99:** 538—545

Syers JK, Skinner RJ and Curtin D (1987) Soil and fertilizer sulphur in UK agriculture. Proc Fertilizer Society No. 379. The Fertilizer Society, Peterborough

Tillman RW (1988) A review of sulphur chemistry in soils. In: White, RE and Currie LD (eds) Proc Workshop Towards the More Efficient Use of Soil and Fertilizer Sulphur, pp 4—8. Occasional report No. 2, Fertilizer and Lime Research Centre, Massey University, Palmerston North

Withers PJA, Tytherleigh ARJ and O'Donnell FM (1995) Effect of sulphur fertilizers on the grain yield and sulphur content of cereals. *J Agric Sci* **125:** 317—324

Withers PJA, Zhao FJ, McGrath SP, Evans EJ and Sinclair AH (1997) Sulphur inputs for optimum yields of cereals. In: Gooding MJ and Shewry PR (eds) Aspects of Applied Biology 50, Optimizing Cereal Inputs: its Scientific Basis, pp191—198. The Association of Applied Biologists, Wellesbourne

Wooding AR, Kavale S, MacRitchie F and Stoddard FL (1999) Link between mixing requirements and dough strength. *Cereal Chem* **76:** 800—806

Wooding AR, Kavale S, MacRitchie F, Stoddard FL and Wallace A (2000) Effects of nitrogen and sulfur fertilizer on protein composition, mixing requirements and dough strength of four wheat cultivars. *Cereal Chem* **77:** 798—807

Wrigley CW, du Cros DL, Downie PG, Archer MJ and Roxburg CM (1980) The sulfur content of wheat-endosperm proteins and its relevance to grain quality. *Aust J Plant Physiol* **7:** 755—766

Wrigley CW, du Cros DL, Fullington JG and Kasarda DD (1984) Changes in polypeptide composition and grain quality due to sulfur deficiency in wheat. *J Cereal Sci* **2:** 15—24

Zhao, FJ and McGrath SP (1994) Soil extractable sulfate and organic sulphur and their availability to plants. *Plant and Soil* **164:** 243—250

Zhao, FJ, McGrath SP and Crosland AR (1995) Changes in the sulphur status of British wheat grain in the last decade, and its geographical distribution. *J Sci Food Agric* **68:** 507—514

Zhao FJ, Hawkesford MJ, Warrilow AGS, McGrath SP and Clarkson DT (1996) Responses of two wheat varieties to sulphur addition and diagnosis of sulphur deficiency. *Plant Soil* **181:** 317—327

Zhao FJ, Hawkesford MJ, Warrilow AGS, McGrath SP and Clarkson DT (1997) Diagnosis of sulphur deficiency in wheat. In: Cram WJ, de Kok LJ, Stulen I, Brunold CH and Rennenberg H (eds) Sulfur Metbolism in Higher Plants: Molecular, Ecophysiological and Nutritional Aspects, pp 349—351. Backhuys Publishers, Leiden

Zhao FJ, Salmon SE, Withers PJA, Evans EJ, Monaghan JM, Evans EJ, Shewry PR and McGrath SP (1999a) Variation in breadmaking quality and rheological properties of wheat in relation to sulfur nutrition under field conditions. *J Cereal Sci* **30:** 19—31

Zhao FJ, Salmon SE, Withers PJA, Evans EJ, Monaghan JM, Shewry PR and McGrath SP (1999b) Responses of breadmakig quality to sulphur in three wheat varieties. *J Sci Food Agric* **79:** 1865—187

9. AFFILIATION

Department of Grain Science and Industry, Kansas State University, Manhattan, KS 66506, USA

For correspondence – *F. MacRitchie: fim@wheat.ksu.edu*

K.C. WALKER AND E.J. BOOTH

SULPHUR NUTRITION AND OILSEED QUALITY

Abstract. Oilseed rape has a high requirement for sulphur and is particularly sensitive to any shortfall in sulphur supply. Yield responses of oilseed rape to sulphur supply have been reported in many countries. Soil sulphur content has been found to influence the level of yield response from a site and results indicate that correct sulphur nutrition is critical for double low varieties. An interaction between nitrogen and sulphur has been noted, indicating that there is a requirement to balance application of these nutrients for optimum response. The response of oil content to sulphur application is less consistent than that of yield, but increases in oil content appears more likely where a low sulphur status occurs. Small changes in fatty acid composition may occur. Differences in protein content according to sulphur application are usually small and associated with sulphur deficiency. Content of sulphur bearing amino acids may be reduced at low sulphur conditions.

Sulphur availability has a larger effect on the content of glucosinolates, sulphur containing secondary metabolites, in both the vegetation and the seed of oilseed rape plant. Site and variety type have a large effect on glucosinolate response to sulphur application, with low sulphur or deficient sites and high glucosinolate varieties being associated with the largest increases in glucosinolate content following sulphur application. Glucosinolates have been shown to be associated with biocidal effects, with glucosinolate levels raised by sulphur application being effective in reducing and delaying disease levels. However, glucosinolate levels raised within the plant foliage by sulphur supplementation could not fully substitute for fungicide application in terms of disease control and subsequent effects on yield. External application of glucosinolates allowed greater accuracy in timing and targeting of glucosinolates. Effectiveness against a range of organisms has been demonstrated, although development for use in the commercial situation may be difficult due to health and safetly implications of handling glucosinolate products in this form. Biofumigation, where soil borne pathogens and pests are suppressed by biocidal compounds released during the hydrolysis of glucosinolates during the breakdown of *Brassica* tissues offers an alternative means of delivery of glucosinolates to the target.

1. INTRODUCTION

Sulphur is an important nutrient for oilseed rape due to its association with yield and also a range of quality factors. Along with nitrogen it is required for the manufacture of protein which, in oilseed rape, contains relatively large quantities of sulphur containing amino acids compared to other crop plants. Sulphur is also required by *Brassicas* for the synthesis of the sulphur bearing compounds glucosinolates. The sulphur requirement of rapeseed is high compared to other crops with 20-30 kg S ha^{-1} removed in the seed whereas cereals remove only 5-15 kg ha^{-1} (Klessa and Sinclair, 1989). Consequently oilseed rape is particularly sensitive to any shortfall in sulphur supply. Sulphur is involved in the production of chlorophyll within the plant and deficiency symptoms appear initially as a slight paling of the leaves which then develop into interveinal chlorosis. Growth is stunted and the flower petals are smaller and paler. In many European countries the incidence of sulphur deficiency has increased dramatically. This is due to a number of factors including, principally, the declining levels of sulphur from industrial pollution which fall onto the land, the use of high nitrogen fertilisers which contain virtually no sulphur and the general

Y.P. Abrol and A. Ahmad (eds.), Sulphur in Plants, 323—339.

increase in yields which has resulted in greater sulphur off-takes. This chapter examines the effect of sulphur on yield and quality of oilseed rape and also describes the relationship between sulphur nutrient status within the plant and disease development.

2. YIELD

In 1988, Withers reported the first large increase in yield of oilseed rape (+ 0.5 t ha^{-1}) due to sulphur application in England (cited by Withers, 1989). In Scotland however, atmospheric sulphur depositions are lower and sulphur deficiency and yield responses have been observed since 1984, when the crop was first grown (Brokenshire et al., 1984). A trial carried out on a low sulphur soil in Aberdeenshire in 1984 (Anon, 1984), showed that yield was increased from 5.15 to 5.36 t ha^{-1} when 8 kg ha^{-1} sulphur was applied. A similar trial on a moderate sulphur status soil at Inverness showed increases in yield from 3.58 to 3.89 t ha^{-1} with an 8 kg ha^{-1} application of sulphur (Anon, 1984). In a subsequent trial on a low sulphur status soil at Laurencekirk, application of 8 kg S ha^{-1} was associated with a yield increase from 3.04 to 3.25 t ha^{-1} (Anon, 1985). Treatments of 0, 7, 14 and 28 kg S ha^{-1} in a further trial undertaken again on a low sulphur status soil at Laurencekirk resulted in yields of 1.75, 2.92, 3.25 and 3.36 t ha^{-1}, respectively and demonstrated that the increase in yield with sulphur application declined as the amount of sulphur applied was increased (AH Sinclair personal communications, 1988).

A programme of work conducted by McGrath and Zhao (1996) across the UK has shown that in 15 of the 39 oilseed rape trials a response to sulphur application was observed. The yield increase at responsive sites ranged from 10 to 327%.

Evidence from experiments conducted elsewhere confirmed that rises in yield can result from sulphur application. Trials by Wetter et al. (1970), showed that the mean yield for *B. napus, B. campestris, B. juncea* and *B. hirta* on a light grey soil in Canada was raised from 1.14 t ha^{-1} when no sulphur was applied to 1.59 and 1.82 t ha^{-1} when 20 and 80 kg S ha^{-1} were applied. Later work reported by Wetter et al. (1970) also showed that oilseed rape grown on black and dark brown soils did not respond to sulphur application. Nuttall et al. (1987) also found significant yield responses to sulphur application in trials undertaken in Canada. In a prelimi-nary trial, the mean yield of a range of rapeseed cultivars increased from 0.13 to 0.72 t ha^{-1} when 25 kg ha^{-1} sulphur were applied to a low sulphur status soil. A further trial showed that the yield meaned for four cultivars over three years rose highly significantly from 0.79 t ha^{-1} when no sulphur was applied to 1.17 and 1.22 t ha^{-1} when rates of sulphur treatments were 25 and 50 kg ha^{-1} respectively. Sulphur was again associated with significant yield increases in a trial where sulphur fertiliser applied at the rates of 0, 10, 25, 40 and 50 kg ha^{-1} resulted in mean yields of 0.71, 1.39, 1.69, 1.50 and 1.77 t ha^{-1} respectively.

Subsequent work by Nuttall et al. (1993) indicated that response to sulphur application varied from one site to another on Saskatchewan soils with sulphate S contents ranging from 17.9 – 36.5 kg S ha^{-1}. Recent work in Canada on sites which all showed severe sulphur deficiency symptoms indicated that application of nitrogen plus sulphur, and also the application of sulphur alone, increased yield (SS Malhi and KS Gill, personal communication, 2002). Increasing the rate of sulphur application from 15 to 30 kg ha^{-1} generally increased yield in this work.

A review by Tandon (1991) on sulphur and crop responses in India indicated that in 16 of the 25 rapeseed and mustard trials reported a 20% or greater rise in crop yields when sulphur was applied at rates of 20 to 250 kg S ha^{-1}. Yield increases were in the order of 144 – 566 kg ha^{-1}, with a mean of 30.0% and the magnitude of response to sulphur increased as available soil sulphur decreased.

Responsiveness of oilseed rape to sulphur varies according to site, season, form of sulphur and varietal type. Comparative trials by Booth et al. (1991b) conducted in Scotland indicate the importance of site in terms of predisposition to sulphur deficiency. These trials were all conducted within an area of north-east Scotland where sulphur deposition was less than 10 kg ha^{-1} $year^{-1}$ and both responsive and non-responsive sites were selected based on levels of extractable SO_4^{2-} in the respective soils coupled with observations of sulphur deficiency in previous crops. Responsive and non-responsive sites were selected in close proximity to each other to minimise agro-climatic effects and possible differences in atmospheric sulphur deposition. Trials were structured to compare 5 levels of sulphur (0, 8, 16, 32 and 64 kg ha^{-1}), 2 levels of nitrogen (150 and 250 kg ha^{-1}) all applied to 3 differing varietal types – Rafal, a high glucosinolate variety, Cobra, a low glucosinolate variety and Tapidor, a very low glucosinolate variety.

Results indicated that overall yield on the high sulphur site was 4.65 t ha^{-1} compared with only 2.39 t ha^{-1} on the adjacent low sulphur site, 100 metres away. At the low sulphur site, the application of foliar elemental sulphur had a significant effect on yield with responses up to the maximum sulphur level overall (Table 1). However, response to sulphur varied according to varietal type with the high glucosinolate variety Rafal being much less responsive (59% response) than the very low glucosinolate variety Tapidor (288% response) with Cobra coming somewhere in between (66% response).

Table 1. The effect of sulphur and nitrogen on yield of oilseed rape on a low sulphur

	Yield (t ha^{-1})					
Variety →	Rafal	Cobra	Tapidor	Rafal	Cobra	Tapidor
Nitrogen (kg ha^{-1}) →	150	150	150	250	250	250
Sulphur (kg ha^{-1}) ↓						
0	2.36	2.27	1.18	1.05	1.12	0.82
8	3.41	2.76	2.23	1.90	2.41	1.69
16	3.53	3.13	2.83	2.56	2.56	2.04
32	3.95	3.79	3.50	3.17	3.19	3.13
64	3.77	3.78	3.40	3.40	3.93	3.50
SE (internal means)	0.370	**SE (S mean)**	0.320	**SE (mean)**	0.261	

This indicates that correct sulphur nutrition is more critical for the double low varieties and adds weight to the hypothesis that varieties low in glucosinolates provide less capacity for sulphur storage (Schnug, 1989). In contrast, the high sulphur site showed a response to sulphur up to only 8 kg ha^{-1} (Table 2).

Table 2. The effect of sulphur and nitrogen on yield of oilseed rape on a high sulphur site 1989/90

	Yield (t ha^{-1})					
Variety →	Rafal	Cobra	Tapidor	Rafal	Cobra	Tapidor
Nitrogen (kg ha^{-1}) →	150	150	150	250	250	250
Sulphur (kg ha^{-1}) ↓						
0	3.91	4.66	4.80	5.00	5.01	4.52
8	4.75	4.64	4.60	5.47	5.03	4.75
16	4.39	4.83	4.84	4.76	4.86	5.09
32	4.68	4.85	4.61	5.50	4.70	5.02
64	3.99	4.64	4.79	5.11	4.32	5.08
SE (internal mean)	0.258	**SE (S mean)**	0.109	**SE (N mean)**	0.182	

Where no sulphur was applied, the yield from all treatments which had received 250 kg N ha^{-1} was considerably lower than the yield of treatments receiving 150 kg N ha^{-1}. The response to sulphur was larger for the higher nitrogen level with optimum sulphur application 32 kg ha^{-1} for the 150 kg N ha^{-1} level but 64 kg ha^{-1} for the higher nitrogen level. This interaction between nitrogen and sulphur indicates that maximum yield is only attained when the two nutrients are provided in a balanced way and has also been shown by Aulakh et al. (1980), Janzen and Bettany (1984) and Malhi and Gill (personal communication, 2002). The results also clearly signal that correct diagnosis of nutrient deficiency is vital. If sulphur deficiency is misdiagnosed as nitrogen deficiency and additional nitrogen is applied as a consequence, then the crop growth would be further adversely affected and a greater yield penalty would result.

In contrast, the high sulphur site showed a conventional response to nitrogen with higher yields obtained from the higher nitrogen treatment for all paired comparisons bar one.

3. RAPESEED OIL CONTENT AND COMPOSITION

Sulphur is known to be involved in the production of oil within the plant, the principal product of oilseed rape. The response of oil content to sulphur application is less consistent than that of yield. Trials indicate that sulphur application was associated with an increase in oil content where a particularly low sulphur status occurred (Appelqvist, 1968). In a review by Holmes (1980), it was found that where sulphur had relatively small effects on seed yield or none at all, sulphur application tended to depress oil content (Table 3). Recent work in Canada on sulphur deficient sites indicated an increase in oil content where sulphur was applied, although unlike yield, increasing the sulphur application from 15 to 30 kg ha^{-1} did not show a consistent response in oil content (Malhi and Gill, personal communication, 2002). Fismes et al. (1998) reported a nearly constant value of oil content in seed with an application of 0, 30 and 75 kg S ha^{-1}, despite an increase in yield compared to the control.

Table 3. Effect of sulphur application on oil content

Country	Conditions	Oil content (% of DM)[a]					References
Sweden	Pots (sand culture)	*0*	*0.5*	*1*	*2*	*8*	Appelqvist (1968)
	B napus, spring	30.2	36.3	34.4	31.8	33.3	
France	Rendzina	*0*	*50*	*50+*	*50*		Studer (1969)
	B napus, winter	48.0	47.1	47.6			Holmes (1980)
Canada	Grey wooded soil	*0*	*22*		*88*		Wetter et al. (1970)
	1962 *B napus*	49.8	48.4		46.2		
	B campestris	41.9	39.8		39.4		
		0	*22*		*88*	*132*	
	1963 *B napus*	47.2	44.8		46.0	43.7	
	B campestris	43.0	41.9		40.4	40.6	
	1964 Mean of four species[b] and two sites	39.3	38.5		38.9	38.0	
	1965 Mean of four species[b] and four sites	37.4	37.0		37.0	36.8	
UK	14 experiments	*0*		*37*	*74*		Holmes and Ainsley (1977)
	B. napus, spring	39.4		39.6			
UK	27 experiments	*0*		*55*			Holmes and Ainsley (1978)
	B. napus, winter	41.1		41.1			
India	*B. juncea*	*0*	*25*	*50*	*75*		Singh and Moolani 1970)
		37.1	39.2	39.6	35.3		Cited by Holmes (1980)
Poland	*B. napus*, winter	*0*	*7*	*11*	*15*		Babuchowski (1971),
		45.2	43.8	43.2	43.4		Cited by Holmes (1980)

[a]Italic numerals indicate Sulphur application rate (g/box for Swedish data; kg ha^{-1} for other data); [b]*B. napus*, *B. campestris*, *B. juncea* and *Sinapis alba*
Source: Holmes (1980)

Work conducted in India found that sulphur application resulted in an increase of oil content of oil crops tested. For rapeseed the mean oil increase, from 10 trials was 9.6% in rapeseed (Tandon, 1991). Other work from India with groundnuts also shows a significant response correlated to sulphur uptake. With rapeseed mustard, nitrogen and sulphur were both important for oil content, but nitrogen alone had little impact on oil content (Pasricha and Aulakh, 1991). Other studies reported by Tandon (1991) with rapeseed mustard showed an increase of oil in seeds by 17% and seed yield by 64% to give a combined oil yield increase of 91%. Work by Aulakh et al. (1980) showed an increase in oil content from additional sulphur application, however a study by Lakkineni and Abrol (1992) with pot trials of rapeseed showed no increase in oil content despite an increase in yield.

In the absence of sulphur, the fatty acid composition of rapeseed oil changed in pot trials undertaken by Appelqvist (1968). The oleic and ecosenoic acids increased whilst the erucic and to a lesser degree linolenic acid declined. Appelqvist hypothesised that this may be due to the inhibition of elongation of the fatty acid chain from 18 to 22 atoms when sulphur deficiency occurs. Ahmad and Abdin (2000) however, found that sulphur application increased the content of oleic acid and decreased the content of erucic acid.

4. PROTEIN AND AMINO ACID CONTENT

Rapeseed meal is a co-product in rapeseed production, and provides a useful and valuable source of protein for animal feed. Sulphur plays an important role in protein synthesis in the rapeseed plant. Extreme sulphur deficiency of rapeseed plants grown in soil-less culture was noted to depress protein content (Josefsson, 1970). In a review Holmes (1980) concluded that differences in protein content according to sulphur application in the field were usually small. Work in Canada by Wetter et al. (1970) showed an increase in protein content of less than 2% in one experiment in only one of four seasons work. This occurred on soils where low sulphur content would be expected.

In a range of trials undertaken in India, sulphur application resulted in a greater increase in grain protein of oilseeds compared to pulses (Tandon, 1991) although sulphur application did not always result in an increase. On soils which were deficient in nitrogen and sulphur, Aulakh et al. (1980) found that protein levels were reduced unless both nitrogen and sulphur were applied. A further review of work with rapeseed in India indicated that protein content increased by 2.8% with sulphur application (Pasricha and Aulakh, 1991).

Sulphur is a constituent of the amino acids methionine, cystine and cysteine. It has been found that sulphur deficiency is likely to reduce the content of these amino acids, thus, reducing the protein quality. Josefsson (1970) reported that the levels of sulphur amino acids methionine and particularly cystine were reduced in conditions of low sulphur. Arginine and aspartic acid were found in higher concentration at low sulphur levels. Work by Pasricha and Aulakh (1991) confirmed that the sulphur bearing amino acids methionine and cysteine were increased by sulphur application, whereas a study by Lakkineni and Abrol (1992) showed an increase in cysteine but not methionine with sulphur application.

5. GLUCOSINOLATES

Glucosinolates are sulphur containing secondary metabolites present throughout the plant in every *Brassica* species. They are products in which the plant stores both sulphur and energy and they also have anti-nutritive and biocidal properties. Consequently, they provide the plant with some defence mechanism against pests and diseases. The anti-nutritive effect of glucosinolates is well documented and requires reduced inclusion rates of rapeseed meal in animal feed rations in order to minimise any effects on animal metabolism

5.1 Glucosinolate Content of Vegetation

The effect of varying availability of several nutrients on glucosinolates contained within the vegetative parts of the plant was investigated as early as the 1940s. Pryor (1940) examined the effect of nitrogen, potassium and sulphur availability on *Brassica* species grown in pots and supplied with a liquid nutrient solution. Sulphur had the largest effect and resulted in a reduction in pungency, indicating a reduction in isothiocyanate concentration (derived from glucosinolates) for leaves of plants grown in sulphur deficient conditions. Experiments by Eaton (1942) showed that the content of the glucosinolate sinigrin in leaves of hydroponically grown mustard in conditions of no sulphur was considerably less than that of plants grown in the presence of sulphur. Freeman and Mossadeghi (1970) showed that the flavour of cruciferous vegetable, which is dependant on isothiocyanates derived from glucosinolates, was correlated with the sulphate concentration of the nutrient solution used to grow them. In experiments examining the effect of sulphur on glucosinolates in oilseed rape from the vegetative stage to maturity (Booth et al, 1991a), glucosinolates were higher in the vegetation of the high glucosinolate variety compared to the low glucosinolate variety (Table 4).

Table 4. Total glucosinolates in vegetation ($\mu M\, g^{-1}$ dry matter)

Sampling dates	***RAFAL*** 0 kg S ha $^{-1}$	32 kg S ha $^{-1}$	***COBRA*** 0 kg S ha $^{-1}$	32 kg S ha $^{-1}$
2 March	16.7	12.3	14.1	9.2
16 March	16.3	14.5	9.5	9.0
30 March	13.4	14.2	10.2	9.9
14 April	13.2	16.6	11.4	12.8
28 April	11.3	13.0	7.8	9.6
12 May	7.3	9.7	9.4	5.9
25 May	7.1	10.0	2.8	5.4
8 June	8.9	8.9	4.0	4.7
22 June	7.6	9.8	2.3	2.1
6 July	3.0	3.3	1.2	1.0
20 July	0.8	0.9	0.1	0.1
3 August	0.1	0.2	0.1	0.1

Glucosinolate concentration in the vegetation declined over time - particularly when the flowers and seeds were produced and glucosinolate concentration of the pods fell as that of seeds rose (Table 5).

It would appear that there is some redistribution of glucosinolates or glucosinolate precursors within the plant as maturity approaches. As proportions of the individual glucosinolates making up these totals also changes there may be an impact on plant/pathogen relationships. Sulphur application increases the glucosinolates level in the vegetative tissue and also increases the concentration in the flowers.

Table 5. Total glucosinolates in vegetation ($\mu M\ g^{-1}$ dry matter)

Sampling Dates										
Plant Part	***Floral***			***Pods***				***Seeds***		
	12 May	*25 May*	*8 June*	*22 June*	*6 July*	*20 July*	*3 August*	*6 July*	*20 July*	*3 August*
Rafal										
0 kg S ha^{-1}	15.2	17.8	26.6	17.1	4.1	3.1	0.3	30.7	48.2	55.1
32 kg S ha^{-1}	22.5	28.6	29.6	20.1	5.9	6.5	0.4	44.6	59.3	63.4
Cobra										
0 kg S ha^{-1}	11.3	9.0	8.9	2.9	1.7	0.3	0.0	4.7	12.7	17.1
32 kg S ha^{-1}	20.6	16.6	12.6	2.2	1.0	0.7	0.8	7.1	12.2	14.7

LSD ($P < 0.05$): 8.51 (Floral); 5.64 (Pods); 23.97 (Seeds)

5.2 Seed Glucosinolates

Work on spring rape grown in boxes of perlite demonstrated that seed harvested from plants given no or very low levels of sulphate fertiliser contained very small quantities of glucosinolates (Josefsson and Appelqvist, 1968). The glucosinolate content of seed from plants which had been grown with larger quantities of sulphate was progressively greater until reaching a relatively constant level. Josefsson (1970) later conducted a long term field trial on medium clay soil and found that slightly higher glucosinolate levels were obtained from plots that had received the highest sulphate fertiliser applications.

A series of field trials carried out by Wetter et al. (1970) in Canada on high glucosinolate varieties of oilseed rape showed that the application of sulphur fertilisers was associated with an increase in the content of glucosinolates. The average glucosinolate contents for three *Brassica* species (*B. napus, B. campestris* and *B. juncea*) rose from 4.25 to 7.80, 9.05 and 9.43 mg m^{-1} meal for sulphur treatments of 0, 22, 90 and 134 kg S ha^{-1}, respectively showing a diminishing response as greater amounts of sulphur were applied. A trial undertaken in another year at two different stations confirmed that sulphur application was associated with ever reducing increases in glucosinolate content. The mean values for four *Brassica* species tested (*B. napus, B. campestris, B. juncea* and *B. hirta*) were 9.68, 12.82,

13.18 and 13.73 mg g^{-1} meal for treatments of 0, 22, 90 and 134 kg S ha^{-1}, respectively.

Subsequent work undertaken in Canada determined that sulphur applied at 25 kg ha^{-1} as ammonium sulphate on a range of soil types in a series of trials had a significant effect on glucosinolates content of rapeseed in certain cases (Nuttall et al., 1987). In a series of further trials reported in the same work, sulphur application at 50 kg ha^{-1} was associated with a positive effect on glucosinolate content in all situations and was increased between 2.5 and 5.6 μM g^{-1} meal.

Samples of seed were taken from the yielded trials of Booth et al. (1991b) as described in Section 2.0 (Tables 1 and 2) and analysed for glucosinolates by HPLC. Individual glucosinolates were summed and the totals are presented in Table 6 and Table 7.

Site had a huge influence on the background glucosinolate level with means of 60, 15 and 8 μM g^{-1} seed for Rafal, Cobra and Tapidor respectively on the high sulphur site (Table 6), compared with 23, 6 and 5 μM g^{-1} for the same varieties on the low site (Table 7). Application of sulphur raised glucosinolate levels particularly for Rafal the high glucosinolate variety. Glucosinolate levels for Rafal at the low sulphur site were lower at the high nitrogen level indicating that the nitrogen sulphur imbalance affected glucosinolates as well as yield. A similar trend was observed for Cobra but the levels for Tapidor were too low for this to be confirmed.

Increases in the glucosinolate levels of the low glucosinolate varieties as described by Booth et al. (1991b) were much lower than responses shown elsewhere in the UK. The mean glucosinolate content of the varieties Cobra and Tapidor at the low sulphur site was 5.4 and 3.9 μM g^{-1} seed respectively when no sulphur was applied compared to 8.8 and 6.2 μM g^{-1} seed when 64 kg sulphur ha^{-1} was applied. This compared with work by Zhao et al. (1991) in which glucosinolate content of Cobra was 7 μM g^{-1} seed at 0 kg sulphur ha^{-1} and 18 and 23 μM g^{-1} seed at 50 and 100 kg S ha^{-1}, respectively. Work by Evans et al. (1991) showed an average rise of glucosinolates in Cobra of 7.2 μM g^{-1} seed in response to an application of 50 kg sulphur ha^{-1} where a response in yield was noted.

Table 6. Effect of sulphur and nitrogen on glucosinolate content (μM g^{-1} seed) of oilseed rape on a low sulphur site (1989/90)

Nitrogen (kg ha^{-1}) →	150			250		
Variety →	Rafal	Cobra	Tapidor	Rafal	Cobra	Tapidor
Sulphur (kg ha^{-1}) ↓						
0	25.4	6.6	4.0	7.3	4.1	3.9
8	27.0	6.2	3.9	5.5	3.4	4.2
16	27.9	6.8	4.1	7.1	4.5	3.5
32	38.0	7.9	4.9	18.2	7.6	5.2
64	46.0	9.6	4.9	29.8	7.9	7.6
SE (Internal mean)	**7.22**	**SE (S mean)**	**1.32**	**SE (N mean)**	**5.10**	

Table 7. Effect of sulphur and nitrogen on glucosinolate content (μM g^{-1} seed) of oilseed rape on a high sulphur site 1989/90

Nitrogen (kg ha^{-1})→ **Variety →**		150 Rafal	Cobra	Tapidor	250 Rafal	Cobra	Tapidor
Sulphur (kg ha^{-1})↓							
	0	66.2	14.2	5.2	69.0	14.3	6.9
	8	69.3	14.8	5.5	62.1	19.6	6.9
	16	71.6	11.0	9.4	70.8	16.3	7.0
	32	68.4	14.8	8.0	69.6	16.6	7.6
	64	75.0	13.4	8.7	69.5	18.6	10.5
SE (internal mean)		**4.10**	**SE (S mean)**	**2.99**	**SE (N mean)**		**2.90**

Glucosinolate content of seed was found to be higher from sulphur sufficient as compared to sulphur deficient sites from almost 30 field experiments conducted across the UK and reported by Zhao et al. (1997). The application of sulphur tended to increase the concentration of glucosinolates much more in sulphur deficient than sulphur sufficient conditions (Withers and O'Donnell, 1994).

Sulphur supply has an impact on the relative proportions of individual glucosinolates according to work by Zhao et al. (1994), with additional sulphur tending to be associated with an increase in the alkenyl group of glucosinolates, the predominant types in oilseed rape. These glucosinolates are derived from methionine, which has itself been found to be influenced by sulphur availability. Other work (Merrien et al., 1987; Booth, 1994) has, however, shown that the proportions of individual glucosinolates remained relatively constant over a wide range of sulphur applications. Only Tapidor, a variety with a very low level of glucosinolates tended to show differences in glucsoinolate profile following sulphur application at a sulphur deficient site (Booth, 1994). The total glucosinolate content of this variety was greatly depleted by sulphur deficiency and very small differences in individual glusoinolate concentration have an apparently large effect on the proportion of glucosinolates. It is considered that this in not indicative of a notable response of the individual glucosinolates in Tapidor to differences in sulphur availability.

6. GLUCOSINOLATES AND DISEASE RESISTANCE

6.1 Laboratory and Glasshouse Trials

In vitro, certain glucosinolates have an anti-fungal effect (Mithen et al., 1986, Peterka and Schlosser, 1989). Work by Luong et al. (1993) examined the impact of different sulphur environments on disease development on oilseed rape. Target diseases were inoculated *Alternaria brassicicola* and a natural infection of *Erysyphe cruciferarum*. Leaf samples were removed one month before inoculation, 10 days and 17 days after disease inoculation and analysed by HPLC for individual glucosinolates. The effect of sulphur on total summed glucosinolate levels is presented in Table 8. High sulphur status was associated with significantly higher

Table 8. The effect of sulphur on total glucosinolate content (μM g^{-1} DM)

Date sampled	*19 June*	*24 July*	*31 July*	*Mean*
High S	5.2	5.9	4.6	5.2
Low S	4.3	4.3	3.0	3.9
SE	0.72	0.72	0.72	0.41

levels of total glucosinolate. *Alternaria* symptoms were observed at the same time - 5 days after inoculation on both the high and low sulphur status plants (Table 9).

However, when assessed on 24 July, 10 days after inoculation, the low sulphur status plants had more than double the disease level of the high sulphur status plants. Within a week, these differences could no longer be detected. A natural infection of *Erysiphe cruciferarum* illustrated the disease prevention benefits of enhanced sulphur nutrition more clearly when ten times the level of disease was observed on 13 September on the low sulphur plots compared with the high sulphur treatments.

Table 9. Effect of sulphur on A. brassicicola infection (24, 31 July) and Friysphe cruciferarum infection (13 September) expressed as a percentage of leaf area infected

Date of Assessment	*24 July*	*31 July*	*13 September*
High S	1.5	2.4	2.5
Low S	3.2	2.6	25.4
SE	0.55	0.55	6.33

6.2 Field Trial Work on Glucosinolate and Disease Resistance

A series of field trials on oilseed rape was carried out in Scotland, where high yield responses to fungicide application are observed due to routine natural infections by light leaf spot (*Pyrenopeziza brassicae*). Sulphur was applied in the form of potassium sulphate, so that results would not be confounded by the direct fungicidal effect that the alternative foliar application of elemental sulphur may have. Potassium application was balanced for plots receiving low sulphur treatments. Levels of up to 160 kg sulphur ha^{-1} were applied.

At the moderate sulphur site, sulphur application had no effect on yield in the three seasons that the trial was carried out, whereas fungicide application was associated with significant increases in yield for two seasons (Booth and Walker, 1997). Data for the 1994/95 season is shown, and significant increases in yield were associated with fungicide application at the 80 and 160 kg S ha^{-1} treatments (Fig. 1). It was noted that fungicide had a major effect in reducing disease levels but sulphur had only a minor effect and this was only detected at the early disease assessments. At the low sulphur site, in the absence of applied sulphur, fungicide application had a very small positive effect in two of the four seasons of trials. In the other two seasons, fungicide treatment tended to depress yield when applied to plants which had received no sulphur as shown for the 1993/94 season (Fig. 2). Again it was noted that disease levels were lower on fungicide treated plants, but no differences in disease levels were noted following to sulphur application.

The trials demonstrated that raising glucosinolate levels within the plant foliage could not fully substitute for fungicide application in terms of disease control and subsequent effects on yield. Further work would be needed to fully elucidate the changes in metabolism associated with the reduction in yield in two of the four seasons when fungicide was applied to oilseed rape on sulphur deficient sites. The studies implied that adequate sulphur nutrition of the oilseed rape plant is essential for efficient utilisation of the fungicide input. Although glucosinolates have been shown to be fungitoxic in laboratory work, it may be possible that the affect of sulphur application in reducing and delaying disease may be due to some other additional affect rather than just through foliar glucosinolate enhancement. Sulphur is known to improve fitness of plant and it is not possible at the present time to categorically prove cause and affect.

These trials and other work in both glasshouse and field experiments, indicates that there does seem to be potential for the use as glucosinolates as biocides. However this is limited according to the glucosinolates presently found within oilseed rape, and it appears that manipulation of the content within the plant can only be partially successful.

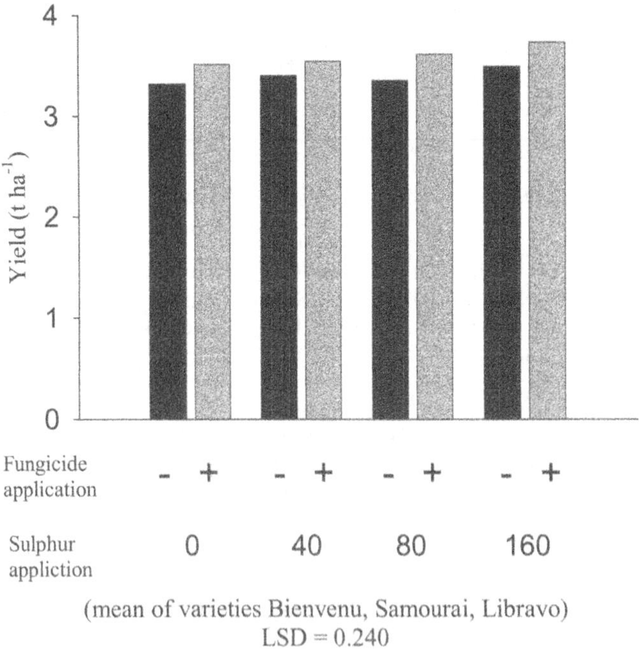

Figure 1. The effect of fungicide application and sulphur application on yield at a moderate sulphur site

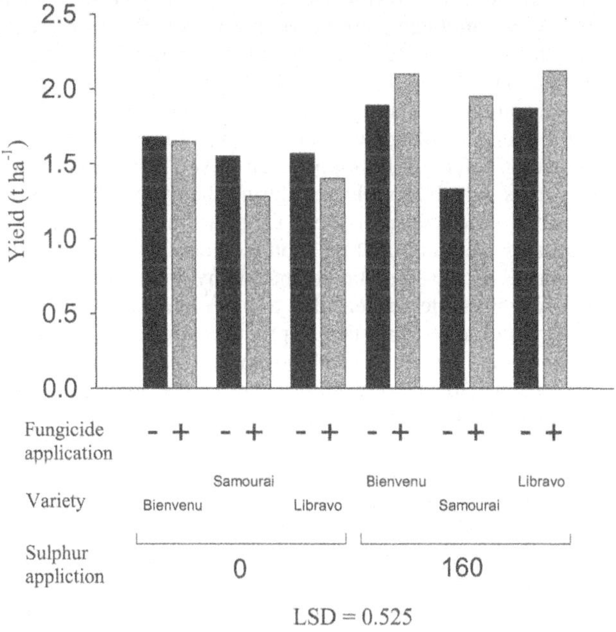

Figure 2. Effect of fungicide application and sulphur application on yield at a low sulphur site

7. USE OF SULPHUR AND *BRASSICA* CROPS AS NOVEL FORMS OF CROP PROTECTION

The development of the biorefinery process allowed the extraction of glucosinolates from *Brassica* seeds using enzyme treatments. This enabled the external application of these products direct to plants and therefore greater accuracy in timing and targeting of glucosinolates. Work has indicated that glucosinolate products derived from the biorefinery process can be effective in controlling clubroot (*Plasmodiophora brassicae*), a soil borne fungal disease, *Pythium*, a seedling disease responsible for damping off and also pests such as slugs (*Derocerus reticulatum*) and aphids (*Aulacorthum solani*) (Booth et al., 2000). The different glucosinolate profiles contained by various *Brassica* crops have been demonstrated to range in activity, with turnip oilseed rape (*B. campestris*), which contains a higher proportion of gluconapin (but-3-enyl glucosinolate) giving greater activity than the glucosinolates from *B. napus* oilseed rape, whose glucosinolate profile is dominated by progoitrin (2-hydroxy-but-3-enyl).

A major challenge arising from the use of glucosinolate products from the biorefinery process has been in development of a reliable method of delivery of the active glucosinolate products to the target. The glucosinolate products as derived from the biorefinery have proved unstable and variability from batch to batch, depending on the conditions of enzymic extraction has been noted. In addition, the level of care required in handling glucosinolate products in the form provided by the biorefinery means that there will be difficulty in their further development for use in the commercial situation.

The concept of biofumigation allows an alternative means of delivery of glucosinolates to the target with greater potential for utilisation by the 'end-user'. The term biofumigation refers to the suppression of soil borne pathogens and pests by biocidal compounds, which are released during the hydrolysis of glucosinolates during the breakdown of *Brassica* tissues. There are a number of commercially available soil sterilants including methyl bromide and Telone, the latter of which releases isothiocyanates as its active ingredient. Methyl bromide is being phased out and Telone causes considerable concern due to environmental hazards and risks to human health. The major active substances produced from glucosinolates are also isothiocyanates. However, the use of isothiocyantes in this form may give greater environmental acceptance. In addition the greater safety and ease of handling of glucosinolates within plant material will give far greater potential for further development.

There is great interest in biofumigation as a means of environment friendly pathogen and pest control in several countries around the world. In Australia several programmes of research are well developed. Work has indicated that 2-phenyl-ethyl glucosinolate found in the roots of Brassicas is the most toxic of glucosinolates against soil borne fungal pathogens and has been effective in reducing take-all infection (*Gaeumannomyces graminis* var. *tritici*) in wheat crops (Angus et al., 1994). The level of suppression was closely related to the levels of glucosinolates in the roots, which varied considerably according to the Australian canola (double low rapeseed) variety tested (Kirkegaard and Sarwar, 1999). Little information is currently available on the level and profiles of glucosinolates from the roots of oilseed rape varieties cultivated in Scottish conditions. Based on the Australian work it has been estimated that biofumigants can provide considerable financial benefits to the end-user over 2.5 million ha of land in SE Australia (Kirkegaard, 1999). Glucosinolate from the leaves and seed meal of *B. juncea* (Indian mustard) have also been shown to have particular activity as biofumigants (De Boer, 1999).

Work has indicated the success of biofumigants using Brassica cover crops with pineapple in Hawaii (Sipes and DeFrank, 1999). Australian work is also underway considering the use of Brassicas for biofumigation with cotton (Nehl, 2000), and in vineyards (Steel and McLeod, 1998). A review of the literature reveals that whilst selection of varieties for useful glucosinolate profiles is underway much work is needed to develop systems of utilisation of biofumigant crops to maximise glucosinolate activity, and this has been identified by several workers, for example Kirkegaard et al. (1999).

Interest from seed development companies is such that work is also ongoing to develop varieties of *Brassicas*, including the fast growing mustards (*B. juncea*), specifically for use as biofumigant break crops and 2 new biofumigant mustard cultivars have been released in Australia (Burton and Marcroft, 1999). A number of European seed companies are in the process of developing cultivars of oilseed rape with very different glucosinolate profiles with the intention of improving pathogen and pest resistance and biofumigation. Work is currently aiming to develop a system for utilisation of this plant material in biofumigation.

8. REFERENCES

Ahmad A and Abdin MZ (2000) Effect of sulphur application on lipid, RNA and fatty acid content in developing seeds of rapeseed (*Brassica campestris* L.). *Plant Science* **150:** 71—76

Angus JF, Gardner PA, Kirkegaard JA and Desmarchelier JM (1994) Biofumigation: isothiocyanates released from Brassica roots inhibit the growth of the take-all fungus. *Plant Soil* **162:** 107—112

Anon (1984) Winter oilseed rape growing systems. *Technical Note* 66 CH, The North of Scotland College of Agriculture, Aberdeen.

Anon (1985) Winter oilseed rape growing systems 1985/85. *Technical Note* 86 CH, The North of Scotland College of Agriculture, Aberdeen.

Appelqvist LA (1968) Lipids in Cruciferae, II. Fatty acid composition of *Brassica napus* seed as affected by nitrogen, phosphorus, potassium and sulphur nutrition of the plants. *Physiol Plant* **21:** 455—465

Aulakh MS, Pasricha NS and Sahota NS (1980) Yield, nutrient concentration and quality of mustard crops as influenced by nitrogen and sulphur fertilizers. *J Agric Sci (Cambridge)* **94:** 545—549

Babuchowski K (1971) The processing value of rapeseed, cooking oil and oilcake meal as affected by sulphur nutrition. *Zeszyty Naukowe Wyzszej Rolnicej w Olsztynie* **5(supplement):** 3—51

Booth E J (1994) The infulence of environment and management on the glucosinolate content of oilseed rape (*Brassica napus* L.). Ph.D. Thesis, Aberdeen University, Scotland

Booth EJ and Walker KC (1997) The effectiveness of foliar glucosinolate content raised by sulphur application on disease control in oilseed rape. In: Cram WJ, De Kok LJ, Stulen I, Brunold and C and Rennenberg H (eds) Sulphur Metabolism in Higher Plants: Molecular, Ecophysiological and Nutritional Aspects, pp 327—329. Backhuys Publishers, Leiden

Booth EJ, Walker KC and Griffiths DW (1991a) A time course study of the effect of sulphur on glucosinolates in oilseed rape from the vegetative stage to maturity. *J Sci Food Agric* **53:** 43—61

Booth EJ, Walker KC and Schnug E (1991b) The effect of site, foliar sulphur and nitrogen application on glucosinolate content and yield of oilseed rape (*Brassica napus* L.). *Proceedings of the Eighth International Rapeseed Congress*, Saskatoon, Canada, pp 567—572

Booth EJ, Coll C, Sutherland KG and Walker KC (2000) Evaluation of physiological activity of bioactives. In: High Quality Oils, Proteins and Bioactive Products for Food and Non-food Purposes Based on Biorefining of Cruciferous Oilseed Crops, Final Report. *EU project FAIR CT95-0260*

Brokenshire T, Channon AG and Wale S (1984) Recognising oilseed rape diseases, Publication 135, The Scottish Agricultural Colleges

Burton W and Marcroft S (1999) Oilseed update for the 1999 season - Walpeup. Grains Research and Development Corporation Research Update, Victoria, Australia.

De Boer and Dolf (1999) Influence of rotation and biofumigation on soil-borne diseases. *Potato Australia* **10:** 13

Eaton SV (1942) Volatile sulphur content of black mustard plants. *Bot Gazette* **104:** 82—89

Evans EJ, Bilsforrow PE, Zhao FJ and Syers JK (1991) The sulphur nutrition of winter oilseed rape in northern Britain. Proceedings of the Eighth International Rapeseed Congress, Saskatoon, Canada, pp 542—546

Fismes J, Vong P C & Guckert A (1998) Effects of nitrogen and sulphur fertiliser on yield and quality of oilseed rape (*Brassica napus* L.). *Proceedings of the European Society for Agronomy Fifth congress*, Nitra, The Slovak Republic, pp 69—70

Freeman G G & Mossadeghi N (1970). Influence of sulphate nutrition on flavour components of three cruciferous plants: radish (*Raphanus sativus*), cabbage (*Brassica oleracea capiata*) and white mustard (*Sinapis alba*). *J Sci Food Agric* **23:** 387—402

Holmes MRJ and Ainsley AM (1977) Fertilizer requirements of spring oilseed rape. *J Sci Food Agric* **28:** 301—311

Holmes MRJ and Ainsley AM (1978) Seedbed fertilizer requirements of winter oilseed rape. *J Sci Food Agric* **29:** 657—666

Holmes MRJ (1980) Sulphur. In: Holmes MRJ (ed) Nutrition of the Oilseed Rape Crop, pp101—121. Applied Science Publishers Ltd, London

Janzen HH and Bettany JR (1984) Sulfur nutrition of rapeseed. 1. Influence of fertilizer nitrogen and sulfur rates. *Soil Sci Soc Am J* **48:** 100—107

Josefsson E (1970) Glucosinolate content and amino acid composition of rapeseed (Brassica napus) meal as affected by sulphur and nitrogen nutrition. *J Sci Food Agric* **21**: 98—102

Josefsson E and Appelqvist LA (1968) Glucosinolates in seed of rape and turnip rape as affected by variety and environment. *J Sci Food Agric* **19:** 564—570

Kirkegaard JA (1999) Conservation cropping and disease control. *Research in Sustainable Agricultural Systems*, CSIRO Plant Industry

Kirkegaard JA, Matthiessen JN, Wong PTW, Mead A, Sarwar M and Smith BJ (1999) Exploiting the biofumigation potential of *Brassicas* in farming systems. *Proceeding of the Tenth International Rapeseed Congress*, Canberra, Australia

Kirkegaard JA and Sarwar M (1999) Glucosinolate profiles of Australian canola (*Brassica napus annua* L.) and Indian mustard (*Brassica juncea* L.) cultivars: implications for biofumigantion. *Aust J Agric Res* **50:** 315—24

Klessa DA and Sinclair AH (1989) Sulphur in soils, fertilisers and crops. *Technical Note* T160, Scottish Agricultural Colleges, Perth

Luong H, Booth EJ and Walker KC (1993) Utilisation of sulphur nutrition to induce the natural defence mechanisms of oilseed rape. *J Sci Food Agric* **63:** 119—120

Lakkineni KC and Abrol YP (1992) Effect of sulphur fertiliser on rapeseed mustard and groundnut. *Phyton Annales rei Botanicae. Prog Sulphur Metabolism Higher Plants* **32:** 75—78

Merrien A, Ribaillier D, Agbo P and Devineau J (1987) Impact de la fertilisation soufre sur la teneur en glucosinolates des graines chez le colza: consequences agronomiques. *Proceedings of the Seventh International Rapeseed Congress*, Poznan, Poland, pp 907—916

McGrath SP and Zhao FJ (1996) Sulphur uptake, yield responses and the interactions between nitrogen and sulphur in winter oilseed rape (*Brassica napus*). *J Agric Sci (Cambridge)* **126:** 53—62

Mithen RF, Lewis BG and Fenwick GR (1986) *In vitro* activity of glucosinolates and their products against *Leptosphaeria maculans*. *Trans British Mycol Soc* **87:** 433—440

Nehl D (2000) Cotton fungal disease on the increase. Media release, New South Wales Agriculture, Australia

Nuttall WF, Ukrainetz H, Stewart JWB and Spurr DT (1987) The effect of nitrogen, sulphur and boron on yield and quality of rapeseed (*Brassica napus* L. and *Brassica campestris* L.). *Can J Soil Sci* **67:** 545—559

Nuttall WF, Boswell CC, Sinclair AG, Moulin AP and Townley-Smith LJ (1993) The effect of time of application and placement of sulphur sources on yield of wheat, canola and barley. *Commu Soil Sci Plant Anal* **24:** 2193—2202

Pasricha NS and Aulakh MS (1991) Twenty years of sulphur research and oilseed production in Punjab, India. *Sulphur Agricu* **15:** 17—23

Peterka S and Schlosser E (1989) *In vitro* activity of glucosinolates against *Leptosphaeria maculans* in comparison to the glucosinolate content and susceptibility of seedlings of different *Brassica* spp. *Mededelingenvan de Faculteit Landbouwwetenschappen Rijksuniversiteit* Gent **54:** 439—446

Pryor DE (1940) The effect of some mineral nutrients on the development of clubroot of crucifers. *J Agric Res* **61:** 149—160

Schnug E (1989) Double low oilseed rape in West Germany: sulphur nutrition and glucosinolate levels. *Aspects App Bio* **23:** 67—82

Singh GB and Moolani MK (1970) Influence of sulphur and nitrogen levels on yield and quality of raya (*Brassica juncea*). Proceedings of Symposium on Fertilizer Use, Bulletin No. 8, Indian Society of Soil Science, pp 129—133

Sipes BS and DeFrank J (1999) Use of cover crops and biofumigation for plant-parasitic nematode control in pineapple. Final report, Hawaii IPM Program, University of Hawaii, USA

Steel C and McLeod R (1998) Control of nematode populations by cover crops. Research projects at the National Wine and Grape Industry Centre, New South Wales, Australia

Studer R (1969) Sulphur fertilizing of winter rape on rendzinas in the Champagne Berrichonne. *Comptes Rendus des Seances de l'Academie d'Agricultre de France* **55:** 316—327

Tandon HLS (1991) Crop responses to sulphur application. In: Tandon HLS (ed) Sulphur Research and Agricultural Production in India, 3rd ed, pp 45—61. The Sulphur Institute, Washington

Wetter LR, Ukrainetz H and Downey RK (1970) The effect of chemical fertilizers on the content of oil, protein and glucosinolates in *Brassica* including rapeseed. *Proceedings of the Fifth International Rapeseed Congress*, St Adele, Canada, pp 92—112

Withers PJA (1989) Effect of applied sulphur on yield and quality of winter oilseed rape. *Aspects App Biol* **23:** 191—197

Withers PJA and O'Donnell FM (1994) The response of double low winter oilseed rape to fertiliser sulphur. *J Sci Food Agric* **66:** 93—101

Zhao FJ, Syers JK, Evans EJ and Bilsborrow PE (1991) Sulphur and oilseed rape production in the UK. *Sulphur Agric* **15**: 13—16

Zhao FJ, Evans EJ, Bilsborrow PE and Syers JK (1994) Influence of nitrogen and sulphur on the glucosinolate profile of rapeseed (*Brassica napus* L.). *J Sci Food Agric* **64:** 295—304

Zhao FJ, Withers PTA, Evans EJ, Monaghan J, Salmon SE, Shewry PR and McGrath SP (1997) Sulphur nutrition: an important factor for the quality of wheat and rapeseed. *Soil Sci Plant Nut* **43:** 1137—1142

9. AFFILIATIONS

Scotaland Agriculture College (SAC), Craibstone Estate, Bucksburn, Aberdeen, Scotland, AB21 9YA, UK

For correspondence - *Keer Walker:* k.walker@as.sac.ac.uk

M.S. AULAKH

CROP RESPONSES TO SULPHUR NUTRITION

Abstract. Sulphur deficiency is wide spread in different parts of the world including India. The reasons for its fast increase to the present level of 30 % in Indian soils are discussed. Significant advancements in locating S-deficient areas and monitoring crop responses made during last 3 decades have increased our awareness about the role of S as an outstanding factor for the production and quality of crops. It has become increasingly clear that oilseeds require S in high amounts followed by pulses, forages, tuber crops and cereals. Numerous field experiments clearly showed that more than 40 crops have responded to S fertilization with mean yield increase ranging from 14 % to 74 %. The large body of evidence indicates that S requirements must be an integral part of balanced fertilization for obtaining optimum crop yields of high quality (oil, protein, fatty acids, and reduced nitrate in foliage). S fertilization is a feasible technique to enhance the uptake and fertilizer-use efficiency of N, P, K and Zn, and suppress the plant uptake of undesirable, toxic elements such as Se and Mo because of the sulphur's synergistic relationships with the former and antagonistic with the latter elements. The key measures needed for optimizing S use for sustainable crop productivity and reducing environmental risks are (i) timely prediction of S deficiency, (ii) timely sowing of crops, (iii) selection of proper method and right time of application of S from suitable sources, (iv) interrelationships of nutrients and balanced fertilization, (v) greater need for S recycling through green manures and crop residues, and (vi) minimizing desorption of SO_4^{2-} from colloidal surface and leaching by avoiding the excessive use of phosphatic fertilizers. There is an urgent need to delineate unexplored and potentially S-deficient areas; develop reliable, efficient and quick methods for diagnosing S deficiency in different soils and crops; identify the contribution of atmospheric pollution and irrigation waters towards S nutrition of crops; frame effective policies addressing the availability and distribution of S-containing fertilizers especially in S-deficient area; and to undertake systematic studies to malnourishment emerging as a consequence of wide spread S deficiency in animal and human beings.

1. INTRODUCTION

Sulphur (S) is now recognized as the 4th major plant nutrient along with nitrogen (N), phosphorus (P) and potassium (K). The deficiency of S has been reported with increasing frequency over the past several years all over the world (Scherer, 2001). Among different regions, Asia represents the region with the highest S fertilizer requirement. In Asia, India and China alone currently account for about 60 % of the total estimated deficit. Continuous mining of S from soils has led to widespread S deficiency and negative soil budget (Aulakh and Bahl, 2001). For instance, of the total 400 districts of India, more than 200 districts have varying proportion of S-deficient soils. According to Tandon (1995a, 1995b), about 51 million ha or about 30 % of its of cultivated land have various degrees of S deficiency. Extensive surveys made to delineate S deficient areas in different parts of the country revealed that S deficiency varies from 5 to 83 % with an overall average of 41 % (Aulakh and Dev, 1976a, 1976b, Singh, 2001). Most of the soils of the Indo-Gangetic alluvial plains, red and lateritic and hill soils are prone to S deficiency while coastal soils contain sufficient or even excessive amount of S as sulphide or soluble sulphates. Several parts of the country,

Yash P. Abrol and Altaf Ahmad (eds.), Sulphur in Plants, 341—358.

however, are still unexplored, which may also be potentially S-deficient areas. The increasing S deficiencies in soils are attributable to the following factors (Pasricha and Aulakh, 1986, 1991; Tandon, 1991):

- The adoption of high-yielding crop cultivars which demand a high fertility level and result in greater exploitation of soil reserve nutrients and removal of much larger quantities of nutrients in the harvested crop.
- The increased cropping intensity (intensive cultivation).
- A drastic decline in incidental additions of S through fertilizers, atmospheric SO_2 especially around industrial cities, pesticides and other agrochemicals. In developed countries, annual S deposition is measured to be 3.7 to 25.2 kg S ha^{-1} (Pasricha and Fox, 1993). In India, sulphate concentration in rainwater is usually less than 0.1 mg L^{-1}, thus very little accession of S from air pollution and acid rain (Singh, 2000).
- Another important factor is the increased use of high-analysis S-free fertilizers. In 1950s, S-containing fertilizers were common sources of N and P. At that time, most of N as ammonium sulphate and P as single superphosphate (SSP) were applied. The consumption of fertilizer N and P has increased sharply since then (Fig. 1). However, the use of N and P fertilizer compounds which contain little or no S, such as urea, diammonium phosphate (DAP), ammonium chloride, calcium ammonium nitrate, nitro-phosphates, and triple superphosphate has relatively increased. Thus, the increase in the consumption of S did not match with its removal by crops. As a result S consumption has increased from about 8000 tons in 1950-51 to about 0.8 million tons in 2000-2001 (FAI, 2001). As the total removal of S through various

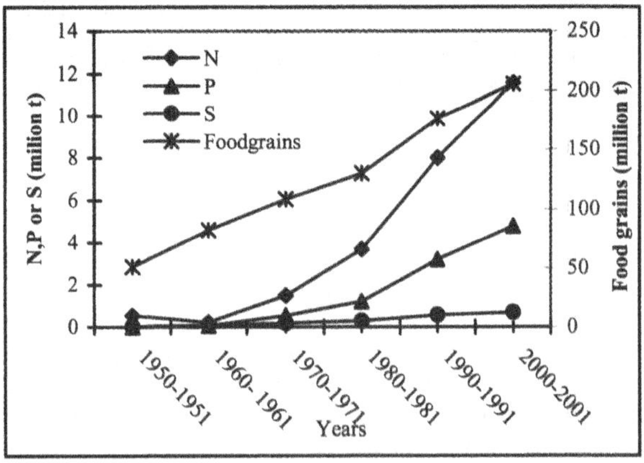

Figure 1. Trends of consumption of N, P and S fertilizers and production of foodgrains in India during last 50 years (Source: Fertiliser Statistics, 2000-2001).

sources at present level of production is 1.3 million t $year^{-1}$, a gap of 0.5 million t $year^{-1}$ suggests an extensive mining of native soil reserves. During the last three decades, researchers have conducted several laboratory, greenhouse, and field experiments to investigate various aspects of plant nutrient sulphur. This chapter synthesizes the available information on the responses of field crops, both in terms of yield and quality, to S fertilization in monoculture and cropping systems that are prevalent in different states of India, and the role of S in optimizing crop production and reducing environmental risks. Important soil and environmental factors, and management practices that affect crop responses to applied S are discussed. Interactions of S with other plant nutrients are discussed in brief (see Abdin et al., this book for more details).

2. CROP YIELD RESPONSES

A large number of studies have reported a marked influence of applied S on the yields of several cereals, pulses, oilseeds, vegetables, forages and other crops as reported in numerous reviews and recent publications (Aulakh and Pasricha, 1986, 1988, 1997, 1998; Pasricha et al., 1987; Bapat, 1989; Biswas and Tewatia, 1991; Aulakh and Chhibba, 1992; Pasricha and Aulakh, 1997; Tandon, 1991, 1995a, 1995b; Ahmad et al., 1998; Jaggi and Dixit, 1999; Sarkar, 2000; Singh, 1986, 2000, 2001). More than 40 crops have responded to S fertilization under field conditions in India (Tandon, 1995b) with mean yield increases ranging from 14 % to 74 % (Table 1). Recently, Sarkar (2000) compiled results of TSI/FAI/IFA sponsored project in six states and reported tremendous crops' response to fertilizer S.

In view of their greater S requirement, oilseed crops were the first to show deficiency of S in the country. Obviously, largest numbers of experiments were conducted on S nutrition of these crops. Among oilseeds, rapeseed-mustard crops (*Brassica campestris*, local name toria; *Brassica juncea* L., local name raya; *Brassica napus* L., local name gobhi sarson; *Brassica carinata* L., local name African sarson) have the highest requirement of S with the optimum rate varying from 20 to 60 kg S ha^{-1} depending upon the soil S status, yield potential and N level used. Yield responses to S vary from 12 to 48 % under irrigated and 25 to 124 % under dryland conditions. Tandon (1991) complied the work on S and noted that on an average of 29 studies, application of 43 kg S ha^{-1} increased mustard yield by 7.7 kg seed kg S^{-1}. In mustard (toria), 20 kg S ha^{-1} increased seed yield by 120-600 kg ha^{-1} depending upon the soil type and available S status (Pathak and Tripathi, 1979; Aulakh and Pasricha, 1997). Field experiments conducted in Punjab state of India revealed average yield responses ranging from 67 to 83 % for *Brassica juncea*, 8 to 17 % for *Brassica compestris*, and up to 7% for *Brassica napus* (Aulakh and Chhibba, 1992).

In different studies, the pod yields of groundnut (*Arachis hypogaea*) increased by 75 to 500 kg ha^{-1} with applied S (Pasricha and Aulakh, 1986). On an average, S added to S-deficient soils increased the yield at 13.0 kg sunflower (*Helianthus annuus*) seed kg^{-1} S and 9.5 kg linseed (*Linum usitatissimum*) kg^{-1} S (Tandon, 1991). In alluvial soils of Uttar Pradesh, the response of safflower (*Carthamus tinctorius*) to 30 kg S ha^{-1} was 23 %. In Punjab, average yield responses ranged from 6 to 66 % for groundnut, 14 to 19% for soybean (*Glycine max*), 9 to 12% for linseed, and 13% for sunflower (Aulakh and Chhibba, 1992). Among other pulse crops, average grain yield increase of 18% chickpea (*Cicer arietinum*), 28 % lentil (*Lens cultinaris*), 20 % mungbean (*Vigna radiata*) and blackgram (*Vigna mungo*), 22 % pigeonpea (*Cajanus cajan*), 32 %

Table 1. Crop yield response to S fertilization in various field experiments

Crop	Number of studies averaged	Yield increase due to applied S (%)	
		Range	*Mean*
Cereals			
Wheat	32	5-110	25
Rice	27	1-40	17
Maize	3	17-60	41
Sorghum	3	19-42	25
Pearl millet	1	-	15
Pulses			
Chickpea	4	16-30	18
Lentil	4	7-70	28
Mungbean	7	13-73	20
Blackgram	7	17-26	20
Pigeonpea	8	12-40	22
Fieldpea	3	8-80	32
Cowpea	1	-	33
Oilseeds			
Rapeseed-Mustard	28	4-127	33
Groundnut	23	8-107	32
Soybean	8	14-36	25
Sunflower	6	6-30	20
Sesame	3	10-33	22
Safflower	1	-	23
Linseed	5	3-31	16
Taramira	1	-	88
Forage crops			
Alfalfa	3	17-60	41
Sorghum	1	-	35
Berseem	3	20-50	32
Maize	8	6-22	14
Vegetables			
Potato	5	7-64	25
Onion	6	2-41	25
Garlic	3	27-59	40
Reddish	1	-	27
Cauliflower	1	-	45
Chillies	2	13-20	17
Ginger	3	23-32	26
French bean	2	28-48	38
Other crops			
Cotton	3	18-49	33
Sugarcane	10	11-32	22
Cassava	2	18-37	27
Jute	1	-	21
Coffee	1	-	63
Tea	2	4-144	74
Opium	1	-	19

Sources: Aulakh and Pasricha (1998); Tiwari et al. (2002)

fieldpea (*Pisum sativum*) and 33% cowpea (*Vigna unguiculata*) have been reported following S fertilization (Table 1). In recent studies conducted with two S sources (gypsum and SulFer 95), the yield responses ranged from 31-36% for groundnut, 25-44% for soybean, 30-38% for sunflower and 16-30% for linseed (Tiwari et al., 2002).

Cereals, which have relatively low S requirement, have lately started showing responses to applied S. Now a large number of experiments are available revealing average grain yield response from 15-41% for different cereals (Table 1). The high proportion of significant responses in rice and wheat is particularly noteworthy.

Besides these crops, yield of several other crops including vegetables, tuber, and root and plantation crops increased significantly with applied S (Table 1).

3. CROP QUALITY RESPONSES

Applications of S in combinations with balanced amounts of other nutrients significantly increased the oil content of groundnut (9%), rapeseed-mustard (5-6%), linseed (2%), and soybean (2%). Increased yield and oil content as a result of S applications caused an oil yield increase of 15-30% in different oilseed crops (Aulakh et al., 1980b, 1989a, 1989b, 1990, 1995; Ahmad et al., 1999a). In Punjab, a maximum of 670 kg oil ha^{-1} was obtained through a combined application of 60 kg S and 150 kg N ha^{-1} to raya (Pasricha et al., 1987). In an overall average of field studies conducted in India, application of S increased the oil content of oilseeds ranging from 3.8 % in sunflower to as high as 11.3% in groundnut. In fact, each kg of S applied can increase the supply of edible oil by 3.0 to 3.5 kg suggesting that S is a master nutrient in oil production (Tandon and Messick, 2002). For calcareous soils, high in free $CaCO_3$, biochemical oxidation of elemental S decreases soil pH and solubilizes $CaCO_3$ to make soil conditions more favourable for plant growth including the availability of other plant nutrients such as iron. Thus, application of elemental S to alkaline-calcareous soils could assist in correcting iron chlorosis (Saroha and Singh, 1980).

Sulphur fertilization significantly improves various parameters of quality within the plant. Sulphur application reduces the N:S ratio and the content of nitrate and non-protein N (Aulakh et al., 1976, 1977a, 1977b; Aulakh and Dev, 1978b), and accelerates the metabolic pathway of protein synthesis in the plant (Das et al., 1975; Dev et al., 1996; Ahmad and Abdin, 2000). Applied S increased the S-bearing amino acids, methionine and cysteine, in groundnut, *Brassica* species as well as in soybean (Table 2). In sesamum-mustard sequence, the application of 60 kg S ha^{-1} to sesamum (*Sesamum indicum*) resulted in 11% increase in protein as well as oil content of sesamum and also 5% increase in protein and 11% in oil content in the seeds of succeeding mustard (Singh and Tiwari, 1985).

Sulphur fertilization also influences the composition of oil, acetyl-CoA and acetyl-CoA carboxylase in oilseeds (Aulakh et al., 1989a, 1995; Ahmad et al., 2000). The relative concentration of different fatty acids in some oilseeds determines its use. Linseed oil with high linolenic acid and low oleic acid can be used in the manufacture of paints, oilcloths, and linoleum. Sulphur fertilization with an adequate supply of N and P accelerates the metabolic pathway of linolenic acid synthesis as it resulted in a large decrease in the percentage of stearic, oleic, and linoleic acids with concurrent increase in the content of linolenic acid (Fig. 3).

Table 2. Influence of S application on amino acids, protein and oil content of soybean

Applied S	Amino acid (mg g^{-1} of dry weight)			Protein	Oil
(mg kg^{-1})	Cystine	Cysteine	Methionine	(%)	(%)
0	1.9	1.2	7.6	40.3	19.6
40	2.4	1.6	10.5	41.0	21.0
80	2.9	1.9	13.9	41.6	20.6
120	2.9	2.0	16.4	42.2	20.8
LSD(0.05)	0.14	0.10	1.13	0.99	0.19

Source: Kumar et al. (1981)

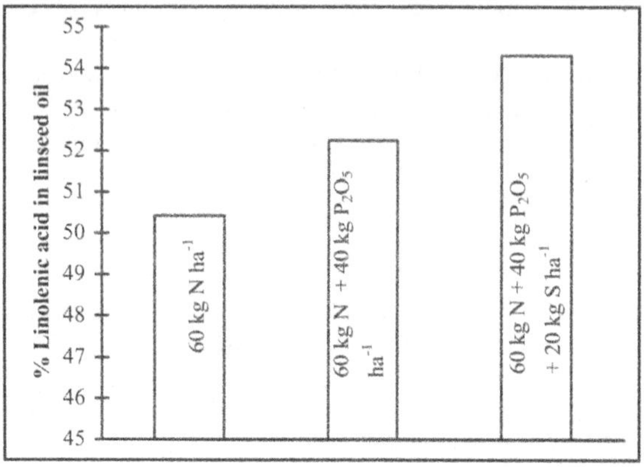

Figure 3. Influence of fertilizer S on the synthesis of linolenic acid in linseed oil. Source: Aulakh et al. (1989a)

4. CROP RESPONSES TO RESIDUAL SULPHUR IN CROPPING SEQUENCES

Fertilizer S applied to one crop could be sufficient to meet S requirement of succeeding crop. Significant residual effects of S application (20 to 53%) have also reported in groundnut-groundnut, groundnut-wheat and groundnut-mustard rotations (Pasricha and Aulakh, 1986; Biswas and Tewatia, 1991; Aulakh and Chhibba, 1992). Where direct responses with groundnut were low, residual effect with wheat was more pronounced (Pasricha and Aulakh, 1986).

In a few studies, both gypsum and pyrites had equally effective residual responses in the cropping system (Pasricha et al., 1987), In a groundnut-wheat cropping system, it was observed that in addition to significant responses of groundnut to S (450 kg ha^{-1}), the grain yield of wheat following in the cropping system increased markedly (570 kg ha^{-1}) due to residual S (Pasricha et al., 1987; Aulakh and Pasricha, 1988). The direct application of S to groundnut in a groundnut-wheat cropping sequence was more profitable than S application to wheat with residual effect on the succeeding groundnut crop. In the wheat-soybean rotation in Madhya Pradesh, soybean yield increased by 31 % when 30 kg S ha^{-1} was applied to wheat. Residual effect of 30 kg S ha^{-1} through gypsum was at par with that of 20 kg S ha^{-1} through ammonium sulphate (Bapat, 1989). Field experiments at Indore with the soybean-safflower system under rainfed conditions indicated large direct and residual response viz. 34 and 16% grain yield, 117 and 35% S uptake and 34 and 16% water-use efficiency, respectively. Similarly, experiments conducted under All India Coordinated Research Project on Micro and Secondary Nutrients and Pollutant Elements in Soils (Singh, 2001) and by Tiwari et al. (2002) showed residual effects of S in several cropping systems.

5. FACTORS FOR OPTIMIZING CROP RESPONSE TO SULPHUR

The efficiency of applied S along with other nutrients in improving crop productivity and quality is influenced by several soil, environmental factors and management practices factors. Some of the predominant factors are briefly discussed here.

5.1 Sulphur Removal by Crops and Rate of Application

S uptake by a crop is the product of S concentration in different plant parts and their biomass. However, the concentration of a nutrient in different parts of a plant of any crop species could vary with the stage of plant growth, cultivars, soil and fertilizer management and environmental factors. Crop uptake can range from 5 kg to 80 kg S ha^{-1} depending upon type of crop, yield, cropping intensity, soil S status and level of external S input and the status of other major nutrients particularly N and P (Table 3). The amount of fertilizer that needs to be applied is much larger as all fertilizers are not 100 % efficient. Depending upon the source used, this could be two to three times greater than net removal. A survey of literature revealed that optimum rates for Sulphate-S sources vary from 10 to 100 kg S ha^{-1} across crops and from 15 to 75 kg S ha^{-1} for the same crop in different environments (Tandon, 1991). In case of S applied as elemental S, pyrites etc., the amount required is often much higher ranging from 2 to 3 times of Sulphate-S sources. Some optimum rates of S application are given in Table 4. While the amount of fertilizer S needed for optimal oilseed production depends upon the source and time of application and crop species, based on the economical analysis of a large volume of crop response data, application of 20-40 kg S ha^{-1} in the SO_4^{-2}-S form is necessary to ensure an adequate supply of S to most of the field crops (Pasricha et al., 1988; Tandon, 1991, Tandon and Messick, 2002). In some years, the crops may respond to even higher levels of S due to severe S deficiencies created by high rainfall.

5.2 Sulphur Sources: Chemical Fertilizers, Organic Manures and Irrigation Water

5.2.1 Chemical fertilizers

A wide array of S-containing fertilizers is available. Sulphur sources, such as gypsum, pyrite, elemental S, and sulphuric acid have been used as S fertilizers. Fertilizers

Table 3. Average uptake of sulphur and its comparison with N, P, and K by major cereals, pulses, oilseeds and other principal crops grown in India.

Crop	Yield (kg ha^{-1})	Total nutrient uptake (kg ha^{-1})[a]			
		N	*P*	*K*	*S*
Cereals					
Wheat	3900	137	26	137	12
Rice (Summer)	2682	56	12	59	7
Rice (winter)	3131	64	11	67	10
Maize	2132	61	10	51	7
Barley	3083	23	6	43	11
Pulses					
Chickpea	1500	91	6	49	13
Lentil	2000	114	13	36	6
Mungbean	870	82	13	90	7
Blackgram	600	45	8	90	7
Pigeonpea	1200	85	8	16	9
Oilseeds					
Rapeseed & Mustard	2596	131	25	133	45
Groundnut	1900	121	19	43	15
Soybean	2500	125	43	101	22
Sunflower	2308	114	26	141	17
Safflower	1909	74	7	35	24
Sesame	1200	62	24	64	14
Linseed	1600	96	13	72	9
Other crops					
Potato	21665	95	19	140	30
Cotton	177	39	11	55	5
Sugarcane	87600	180	29	270	26
Tea	1000	110	16	37	10
Tobacco	2845	55	6	64	8
Alfalfa	91900	755	79	669	46
Forage grasses	19970	187	29	284	39
Fruits[b]	9295	55	18	86	19

[a]Nutrient uptake includes total of grain, seed, tuber or cane and their proportionate straw or leaves; [b]Average of nutrient uptake by different fruits
Source: Aulakh et al. (1985); Tandon (1991); Aulakh and Bahl (2001)

Table 4. Optimum rates of sulphur application to different groups of field crops

Crops	Cereals	Pulses	Oilseeds	Tubers	Tea	Fodders
S rate (kg ha^{-1})	*24-40*	*20-40*	*10-150*	*25-60*	*40-50*	*25-100*

Source: Tandon (1991)

containing primary nutrients, such as SSP, ammonium sulphate, potassium sulphate and sodium sulphate have also been used for this purpose. Since plants take up S in the SO_4^{-2} form, easily soluble materials containing SO_4^{-2}-S such as ammonium sulphate, potassium sulphate and sodium sulphate are considered suitable in neutral to slightly alkaline soils (Subbiah and Singh, 1970). However, SO_4^{-2}- is prone to leaching losses in coarse-textured soils. Gypsum, a less soluble material, has been found comparable to ammonium sulphate and sodium sulphate (Subbiah and Singh, 1970) or slightly less so as compared to SSP as in ^{35}S studies (Aulakh and Dev, 1977a, 1978a).

Single superphosphate is a preferred source for those crops, which need to be fertilized with S and P, instead of other P sources such as diammonium phosphate (DAP) and triple superphosphate in order to increase the crop yields. In long-term field studies with groundnut (Aulakh et al., 1980a, 1988), it was observed that the effect of SSP was more due to its S content rather than P content (Fig. 4). Recent studies carried out in All India Coordinated Research Project on Micro and Secondary Nutrients and Pollutant Elements in Soils have further shown that with the application of recommended dose of P through SSP instead of DAP resulted in an additional increase in the seed yield of 670 kg ha^{-1} in rice-wheat, 490 kg ha^{-1} in rice-lentil and 370 kg ha^{-1} in rice-linseed crop systems in Pusa, Bihar (Fig. 4).

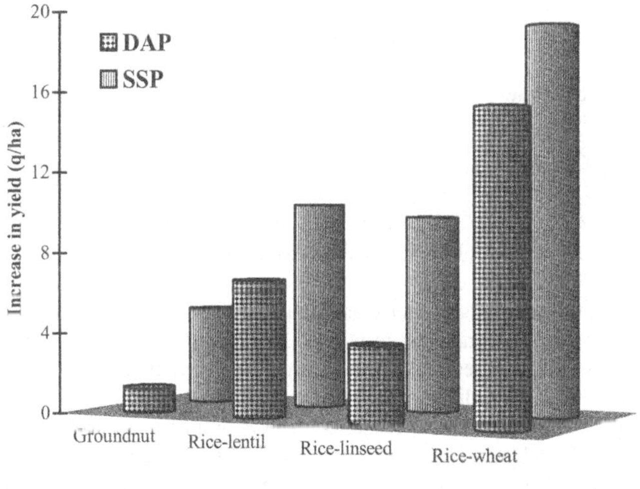

Figure 4. Increase in the productivity of groundnut (6-year average) and rice-lentil, rice-linseed and rice-wheat systems (3-year average) with the application of P through diammonium phosphate (DAP) and single superphosphate (SSP) at recommended P rate. (Source: Aulakh et al., 1988 and Singh, 2001)

Sulphuric acid (H_2SO_4) has also been used as a S-source in a few studies. Aulakh and Dev (1978a) reported H_2SO_4 as a good source of S when applied in calcareous and slightly alkaline soils because it lowers soil pH and improves the availability of other plant nutrients. The foliar spray of 0.1 % of H_2SO_4 has been found remarkably effective in increasing the yield and oil content of mustard, sunflower, sesame and soybean (Singh and Sahu, 1986).

The use of elemental sulphur (S^0) is also increasing because of several advantages such as (a) increase in the P availability in soil by lowering pH and releasing the phosphate fixed by Ca^{2+} in the neutral and alkaline soils, and P fixed by Fe and Al ions in acidic soils (Deluca et al., 1989), (b) reduction in leaching and runoff losses of SO_4^{2-} and prolonged residual effect, as it is a slow release SO_4^{2-} source, and (c) minimum handling and transportation costs as it is a concentrated source of S. Unlike sulphate fertilizers, S^0 must be converted by soil microbes to SO_4^{-2} in the soil before plants can

Table 5. Oxidation of elemental S (S^0) in acidic, neutral and alkaline soils incubated for 42 days at different temperature and moisture regimes

Treatments	**Oxidation of applied S^0 (%)**		
	Acidic soil	***Neutral soil***	***Alkaline soil***
Temperature (^{0}C)			
12	17	23	25
24	24	30	38
36	39	45	52
LSD (0.05) for Temp = 2.0, Soil = 2.0 and Temp x Soil = 4.1			
Moisture regime			
Upland-40 % WFPS[a]	18	21	27
Upland-60 % WFPS	24	30	38
Flooded-120 % WFPS	0	0	0
LSD(0.05) for Moisture = 2.6, Soil = 2.6 and Moisture x Soil = 5.3			

Sources: Jaggi et al. (1999); Aulakh et al. (2002)

period would be required for its oxidation as compared to that in acidic soils, which exist in hilly areas having low moisture. In order to fasten the availability of S, modified pyrite and elemental S products are now a days marketed in the country that are granular but quickly dispersible upon wetting such as S bentonite and SulFer-95. Micronized granular SulFer-95 contains 95 % S^0 in particles smaller than 200 mesh size. In several evaluation studies conducted in different parts of the country, it has shown promising results (Tiwari et al., 2002). In groundnut-wheat cropping in Vertisols of Maharashtra, direct effect of graded levels of applied S up to 60 kg S ha^{-1} as SulFer 95 increased the groundnut yield by 40 %. In alluvial soil at Ludhiana, Punjab, direct effect of S up to 40 kg S ha^{-1} from both gypsum and SulFer 95 increased the yield of soybean and sunflower. In Rajasthan, the S up to 60 kg ha^{-1} increased the yield of linseed by 30 %. Although its use appears to be cost effective, it is costlier than gypsum that is available as natural deposits in several parts of country.

The use of S-containing N and P fertilizers should be encouraged in order to meet N and P requirements of crops. The share of S-containing P fertilizers in total P fertilizers consumed in India has come down from 100 % in 1950-51 to around 30 % in 1970-71 to about 10% in 2000-2001 (Fertiliser Statistics 2000-2001) whereas its use is increasing and is a predominant P fertilizer in several other Asian countries. For example, in China, there are large numbers of SSP manufacturing plants dispersed in the country and about 60 % of P is met through SSP. Thus the use of SSP and ammonium sulphate in India must be encouraged. The major reasons for reduced share of SSP in meeting P requirements of crops in India include (a) reduction in SSP production, (b) subsidy based on only P content makes SSP costlier than DAP to the farmer, (c) encouraged use of DAP instead of SSP, and (d) non-printing of S content on the fertilizer bags and unawareness of farmer about its additional benefits.

5.2.2 Organic manures and crop residues

During last two decades, substantial evidence has been obtained for the effects of green manuring and crop residues on enhancing crop productivity in several crops and cropping systems (Dhillon et al., 1984; Pasricha et al., 1987; Aulakh and Pasricha, 1998; Aulakh et al., 2000, 2001). Common leguminous green manures like cowpea, sesbania, guar etc. can accumulate 45 to 100 kg N ha^{-1} and 10 to 15 kg S ha^{-1} during 40-50 days growth. Crops grown on coarse-textured soils low in organic matter respond markedly to green manuring, even in the presence of recommended dose of N, P and S fertilizers. A recent study with ^{35}S-labelled *Sesbania aculeata* green manure illustrated fast release and availability of S contained in green manure (Dhillon et al., 2001).

The recycling of S through incorporation of crop residues instead of S loss in burning of crop residues should be encouraged. At present, for instance about 70 % of the total rice straw produced from combine harvest is burnt in the open fields in Punjab. Similarly, after combine-harvesting wheat, 70 % of wheat straw is converted to '*bhusa*' for animal feed and the remaining 30 % is burnt in the field (Aulakh, 2001). Recently Singh (2002) has demonstrated that in a short period of 30-day incubation of soils amended with groundnut and gobhi sarson residues, 2 to 7 % of added residue-S was mineralized at 30° and 45°C. He observed relatively higher recovery of S added through gobhi sarson residues (C:S = 137) than groundnut residue (C:S = 210) due to its lower C:S ratio. Previous studies have shown that the incorporation of plant residues with narrow C:S ratio (<200: 1) could increase the content of SO_4^{2-}-S in soil, whereas the incorporation of those plant residues which have wide C:S ratio (>400:1) may result in a net immobilization of soil inorganic-S (Stevenson 1986).

5.2.3 Atmospheric pollution and irrigation water

Although no systematic information is available on the contribution of atmospheric pollution towards S nutrition of field crops, accretions of S from atmosphere are becoming low day by day because of the change of fuel strategy from coal and gas to electricity in the industries and due to stricter emission regulations. Sulphate concentrations of usually less than 0.1 mg L^{-1} in rain water (Singh, 2000) suggest reduced S emissions from industry and fuels.

Sulphur added through irrigation water also affects the crop response to fertilizer S. In a long-term experiment, Sahoo and Panda (1985) observed that monsoon-season rice responded significantly to S but post-rainy season rice that received 10 kg S ha^{-1}

through irrigation water did not respond to applied S. Concentration of S in tubewell water of some regions is quite high. For instance, two studies conducted in Punjab have revealed that S content in the tubewell waters ranged from 0.7 to 18 mg S L^{-1} L in Ludhiana and 0.7 to 2.7 mg S L^{-1} in Hoshiarpur (Table 6). Based on the average S content of water, the tubewell water can add, on an average, 1 to 6 kg S ha^{-1} per irrigation.

Table 6. Sulphur content of underground tubewell waters in Punjab and S added through irrigation

District (Block)	No. of tubewell sampled	S content ($mg\ L^{-1}$)		S added per irrigation[a] ($kg\ ha^{-1}$)	
		Range	**Mean**	**Range**	**Mean**
Ludhiana (Ludhiana)	50	0.75-18.00	8.04	0.56-13.50	6.03
Hoshiarpur (Bhunga)	17	0.74-2.70	1.56	0.55-2.02	1.17

[a]One irrigation = 75 x 10^4 L ha^{-1}
Sources: Cheema and Arora (1984); Singh (1987)

5.3 Time of Seeding of Crops and Fertilizer Application

5.3.1 Timely seeding of crops

Timely seeding of crops not only ensures better crop growth but also help enhance the utilization of fertilizer nutrients. For instance, advancement in the planting date of groundnut from Mid July to mid-June is better equipped to face a drought at a later stage by virtue of its having developed a deeper root system thus leading to higher response to fertilizers (Pasricha et al., 1987). Timely seeded linseed in mid-October removed significantly more S than the November-seeded crop, and consequently the optimum combination of 60 kg N, 40 kg P_2O_5 and 20 kg S ha^{-1} produced 2450 kg seed yield as compared to 1900 kg ha^{-1} produced in November-seeded crop by utilized significantly more fertilizer S.

5.3.2 Time of fertilizer S application

In general, S along with other fertilizer nutrients should be applied at the time of seeding. As the S-deficient plants have a higher propensity for S absorption that S-sufficient plants, S concentrations of deficient plants have been observed to approach or exceed normal concentrations within few days of transfer to a SO_4-containing solution. Likewise the physiological functions such as enzyme activities and photosynthesis, which are severely inhibited by S deficiency, appeared to be restored to normal rates within 5 days of transfer from S-deficient solution (Friederich and Schrader, 1978). In a study of Janzen and Bettany (1984) where ^{35}S-labelled Na_2SO_4 was applied at sowing, rosette, bud and flowering stages of rapeseed, identical grain yields were obtained with fertilizer S applied at any of the four growth stages except under conditions of severe S stress where S applied after the rosette stage did not allow recovery of maximum grain yields. Thus, if a farmer has missed an S application at sowing, he can correct S deficiency by applying S during the early growth period. In case of groundnut, if gypsum is used as a source of S, split application of S as compared to full dose of S as basal gives better return. This is because of the combined advantage of providing Ca and S during flowering and fruiting of groundnut (Chahal and Virmani, 1973; Dwivedi, 1981; Ankineedu et al., 1983; Ahmed et al., 1999b). Some experiments showed that S applied at planting, pegging, and pod formation is equally effective in obtaining yield response of groundnut (Aulakh and Pasricha, 1987; Tandon, 1991). As the most active period for S metabolism in potato is the full bloom stage, split application of S was found better than a basal application to potato. When S^0 is used as S source, its application should be 3 to 5 weeks prior to seeding a crop so that sufficient amount oxidizes to meet the plant needs. The actual period required for the oxidation and availability of S^0 to plants is determined by prevailing soil pH, moisture, and temperature (Jaggi et al. 1999; Aulakh et al., 2002).

5.3.3 Method of application of fertilizer S

While most of the Sulphate-S sources could be placed or drilled 10-15 cm deep at the time of seeding or applied as surface broadcasting during crop growing period, broadcast and mixing of elemental S and pyrite enhances soil-fertilizer contact and promote its oxidation to SO_4^{2-}.

5.4 Prediction of S Deficiency

It is very important to predict the S deficiency in time, preferentially before seeding a crop. Several indicators of S deficiency in soil and growing plants have been established.

- For testing soils, the highest correlation coefficients have been reported between plant yields and soil S extracted with 0.5M sodium acetate, 0.5M ammonium acetate, 0.15% calcium chloride, and 500 mg P L^{-1} solution of calcium phosphate (Aulakh and Chhibba, 1992; Shukla, 2001).
- During crop growth, S deficiency could be determined both from visible signs and S diagnostic indices. In general, the leaves of S-deficient plants are spindly, stunted, and pale green or yellow in color, whereas S-sufficient plants are dark green in color with vigorous growth. There are also some specific sulphur deficiency symptoms in different crops, for details refer to a recent 'Practical Sulphur Guide' by Tandon and Messick (2002).
- S content in plant foliage during active growth is a quite good parameter to ascertain S sufficiency. For example, the S content below 0.2 % in the plant tissue at the pre-flowering state of *Brassica* was found the threshold level, below which the crop yield and quality were adversely affected (Aulakh et al., 1977b). Similarly, Aulakh et al. (1980a) reported a strong relationship of groundnut pod yield with S content of kernels ($r = 0.77$, $p = 0.01$) and foliage ($r = 0.69$, $p = 0.05$). Critical level of S concentration varies in different crops and the ranges for deficiency and sufficiency for several crops are summarized in Table 7.

- The N:S ratio in the plant tissue, which serves as an index of S supply, widens with application of N alone. A critical value of N:S above which the S supply may become deficient was 15.5:1 in foliage, and 7.5:1 in grain of mustard crops (Aulakh et al., 1980b). For alfalfa, N:S ratio of 11:1 was found critical (Aulakh et al., 1976).

Table 7. Critical levels of S in select 45-55 day old plants

Crops	% S concentration in dry matter		
	Deficient	*Moderately sufficient*	*Sufficient*
Wheat, rice, maize, millets	0.10-0.20	0.20-0.30	>0.30
Groundnut, mustard, soybean, cowpea	0.10-0.25	0.25-0.40	>0.40
Sunflower, linseed brinjal, French bean, cucumber,	0.25-0.35	0.35-0.55	>0.55
Chickpea, Horse gram, pea	0.15-0.45	0.45-0.75	>0.75
Potato, cauliflower, spinach	0.30-0.40	0.40-0.75	>0.75

Source: Singh (1999)

Getting clue from these parameters, farmer can control or minimize the damage due to S deficiency by applying S during the crop growth period. For example, if a farmer has missed S application at the time of seeding, he could apply S later on during crop growth when plant analysis indices reveal S deficiency. As discussed above in sub-section 5.3.2, application of soluble S sources during early crop growth have been have been found equally effective in ameliorating S deficiency and increasing crop yield.

5.5 Crop Cultivars

Differential response among cultivars of the same crop could result in significantly different response to S as have been illustrated for four cultivars of wheat (Aulakh et al., 1977a) and three cultivars of raya (Dev et al., 1981). The rate at which optimum response was obtained varied between 28 and 39 kg S ha^{-1} for three cultivars of raya (Dev et al., 1981). Similarly there were substantial differences in S requirement between different types of coffee (Rao, 1988).

5.6 Balanced Fertilization and Interrelationships of S with Other Nutrients

Field experiments with different pulses showed significant increase in grain yield due to balanced N, P and S fertilization. The response to S over adequate rates of N + P application varied from 3 % in cowpea to as high as 20 % in lentil (Table 3). Sulphur applications also affected the patterns of crop responses to K, Ca, Mg, Zn, B, and Mo. Data from S-deficient sites showed that where no S was applied, response to these nutrients was substantially lower than when S was added (Bahl et al., 1986; Aulakh and Pasricha, 1977b, Tandon, 1991). In the event of one limiting or excessive nutrient, synergism or antagonism between different nutrients is likely to occur. Several studies have shown the interaction effects of S with N, K, Ca, and Zn are synergistic in influencing the yield, quality (oil, protein, amino acid and fatty acid synthesis) and uptake of nutrients by different crops (Aulakh et al., 1976, Jaggi et al., 1977; Aulakh and Pasricha, 1977b, 1997; Aulakh and Dev, 1978b; Dev et al., 1979, 1981; Pasricha and Aulakh, 1986, 1991; Bahl et al., 1986; Tandon, 1991; Aulakh and Chhibba, 1992). In case of SxP interaction, a differential behaviour of one nutrient in relation to the concentration or rate of application of other nutrient has been documented (Aulakh and Pasricha, 1977a, 1979; Aulakh et al., 1989b, 1990). For example, in a field study with soybean, adequate levels of 40 kg S and 80 kg P_2O_5 ha^{-1} increased the seed yield and oil content of soybean from 597 and 111 kg ha^{-1} in control to 1735 and 412 kg ha^{-1}, respectively. But application of 120 kg P_2O_5 ha^{-1} in conjunction with 20, 40 and 60 kg S ha^{-1} created an imbalance and reduced the seed yield by 5, 8 and 10 %, respectively. The reduction in oil yield was up to 15%. Adsorption and desorption of anions commonly used as fertilizers such as $H_2PO_4^-$, HPO_4^{-2}, PO_4^{-3} and SO_4^{2-} on colloidal soil surfaces occur simultaneously. Phosphate is a stronger competitor for anion adsorption sites; therefore, the large applications of P could further accentuate S deficiencies by causing concurrent desorption of SO_4^{2-}-S from soil (Pasricha and Aulakh, 1991) and its subsequent leaching with irrigation and rainwater. Such problems could occur more commonly in coarse-textured soils low in organic matter, which have little SO_4^{2-} retention but have high percolation rates. *Brassica* species and legumes grown on the soils having high levels of available Mo often accumulate relatively toxic amounts of Mo. As a result of the antagonistic effect of S on Mo, the Mo toxicity can be alleviated with the application of S as S fertilization suppresses Mo uptake and consequently improve plant growth for a crop grown on a highly Mo-polluted soils (Pasricha and Randhawa, 1972; Pasricha et al., 1977). Selenium (Se) is not an essential element for plants but its concentration in forage crops is important for animal health. Selenium poisoning of animals has been reported in the foothill areas of Shivaliks where groundwater containing appreciable amounts of soluble Se is used for irrigation. S applications alleviate the harmful effect of Se grown on Se-toxic soils (Bawa et al., 1990). Thus, S fertilization is useful not only in increasing crop production, oil content, and protein, but also in improving soil conditions for crop growth. These studies have clearly demonstrated that adequate S fertilization is a feasible technique to suppress the plant uptake of undesirable toxic elements, and to enhance the efficiency of several essential elements resulting in increased crop production and improved quality.

6. REFERENCES

Ahmad A and Abdin MZ (2000) Effect of sulphur application on lipid, RNA, protein content and fatty acid composition in developing seeds of rapeseed (*Brassica campestris* L). *Plant Sci* **150:** 71—75

Ahmad A, Abraham G, Gandotra N, Abrol, YP and Abdin MZ (1998) Interactive effect of nitrogen and sulphur on growth and yield of rapeseed-mustard (*Brassica juncea* l. Czern and Coss and *Brassica compestris* L.) genotypes. *J Agron Crop Sci* **181:** 193—199

Ahmad A, Abraham G and Abdin MZ (1999a) Physiological investigation on the impact of nitrogen and sulphur application on seed and oil yield of rapeseed (*Brassica campestris* L.) and mustard (*Brassica juncea* L Czern and Coss) genotypes. *J Agron Crop Sci* **183:** 10—25

Ahmad A, Abrol YP and Abdin MZ (1999b) Effect of split application of sulphur and nitrogen on growth and yield attributes of *Brassica* genotypes differing in their time of flowering. *Can J Plant Sci* **79:** 175—180

Ahmad A, Khan I and Abdin MZ (2000) Effect of sulphur fertilization on lipid accumulation, acetyl-CoA and acetyl-CoA carboxylase in the developing seeds of rapeseed (*Brassica campestris* L). *Aust J Agric Res* **51:** 1023—1029

Ankineedu G, Rao JV and Reddy BN (1983) Advances in fertilizer management for rainfed oilseeds. *Fert News* **28:** 76—90

Aulakh MS (2001) Impacts of integrated management of crop residues, fertilizer N and green manure on productivity, nitrate leaching, carbon sequestration, denitrification and N_2O emissions in rice-wheat system. Proceedings of 12th World Fertilizer Congress, Beijing, China

Aulakh MS and Bahl GS (2001) Nutrient mining in agro-climatic zones of Punjab. *Fert News* **46:** 47—61

Aulakh MS and Chhibba IM (1992) Sulphur in soils and responses of crops to its application in Punjab. *Fert News* **37:** 33—45

Aulakh MS and Dev G (1976a) Distribution of sulphur in the soils of the Multiple Cropping Project area in Sangrur. *J Res Punjab Agric Univ* **13:** 147—150

Aulakh MS and Dev G (1976b) Profile distribution of sulphur in some soil series of Sangrur district. Punjab. *J Indian Soc Soil Sci* **24:** 308—313

Aulakh MS and Dev G (1977a) Availability of sulphur from superphosphate and $CaSO_4.2H_2O$ to alfalfa (*Medicago sativa* L.Pers.) using tracer technique. *J Nuclear Agric Biol* **6:** 82—85

Aulakh MS and Dev G (1977b) Comparison of different plant analyses indexes in evaluating sulphur nutrition of alfalfa. *Indian J Agric Chem* **10:** 163—171

Aulakh MS and Dev G (1978a) A note on the relative efficiency of gypsum, superphosphate and sulphuric acid as source of sulphur for Lucerne. Indian *J Agric Chem* **11:** 77—80

Aulakh MS and Dev G (1978b) Interaction effect of calcium and sulphur on the growth and nutrient composition of alfalfa (*Medicago sativa* L. Pers.) using ^{35}S. *Plant Soil* **50:** 125—134

Aulakh MS and Pasricha NS (1977a) Interaction effect of sulphur and phosphorus on growth and nutrient content of moong (*Phaseolus aureus* L.). *Plant Soil* **47:** 341—350

Aulakh MS and Pasricha NS (1977b) Interrelationships between sulphur, magnesium and potassium in rapeseed. I. Yield, concentration and uptake of sulphur. *Indian J Agric Sci* **47:** 577—581

Aulakh MS and Pasricha NS (1979) Responses of gram (*Cicer arietinum* L.) and lentil (*Lens cultinaris* L.) to phosphorus as influenced by applied sulphur and its residual effect on moong (*Phaseolus aureus* L.). *Bull Indian Soc Soil Sci* **12:** 433—438

Aulakh MS and Pasricha NS (1986) Role of sulphur in the production of the grain legumes. *Fert News* **31:** 31—35

Aulakh MS and Pasricha NS (1988) Sulphur fertilization of oilseeds for yield and quality. In: Sulphur in Indian Agriculture, Section II. 3, pp 1—14. The Sulphur Institute, Washington, DC

Aulakh MS and Pasricha NS (1997) Role of balanced fertilization in oilseed based cropping systems. *Fert News* **42:** 101—111

Aulakh MS and Pasricha NS (1998) The effect of green manuring and fertilizer N application on enhancing crop productivity in mustard-rice rotation in semiarid subtropical regions. *Euro J Agron* **8:** 51—58

Aulakh MS, Dev G and Arora BR (1976) Effect of sulphur fertilization on the nitrogen-sulphur relationship in alfalfa (*Medicago sativa* L. Pers). *Plant Soil* **45:** 75—80

Aulakh MS, Pasricha NS and Dev G (1977a) Response of different crops to sulphur fertilization in Punjab. *Fert News* **22:** 32—36

Aulakh MS, Pasricha NS and Sahota NS (1977b) Nitrogen-sulphur relationship in brown sarson and Indian mustard. *Indian J Agric Sci* **47:** 249—253

Aulakh MS, Singh B and Arora BR (1977c) Effect of sulphur fertilization on the yield and quality of potatoes (*Solanum tuberosum* L.). *J Indian Soc Soil Sci* **25:** 182—185

Aulakh MS, Pasricha NS and Sahota NS (1980a) Comparative response of groundnut (*Arachis hypogaea* L.) to three phosphatic fertilizers. *J Indian Soc Soil Sci* **28:** 342—346

Aulakh MS, Pasricha NS and Sahota NS (1980b) Yield, nutrient concentration and quality of mustard crops as influenced by nitrogen and sulphur fertilizers. *J Agric Sci* (Cambridge) **94:** 545—549

Aulakh MS, Sidhu BS, Arora BR and Singh B (1985) Content and uptake of nutrients by pulses and oilseed crops. *Indian J Ecol* **12:** 238—242

Aulakh MS, Pasricha NS, Baddesha HS, Dhillon KS and Bahl GS (1988) Single superphosphate - the effective phosphorus and sulphur nutrition of groundnut. *Indian Farming* **38:** 9—10

Aulakh MS, Pasricha NS, Azad AS and Ahuja KL (1989a) Response of linseed (*Linum usitatissimum* L.) to fertilizer nitrogen, phosphorus, and sulphur, and their effect on the removal of soil sulphur. *Soil Use Management* **5:** 194—198

Aulakh MS, Pasricha NS and Bahl GS (1989b) Effect of date of sowing on efficiency of phosphorus and sulphur use by linseed (*Linum usitatissimum*). *Indian J Agric Sci* **59:** 673—675
Aulakh MS, Pasricha NS and Azad AS (1990) Phosphorus-sulphur interrelationships for soybeans on P and S deficient soil. *Soil Sci* **150:** 705—709
Aulakh MS, Pasricha NS and Ahuja KL (1995) Effect of nitrogen and sulphur application on grain and oil yield, nutrient uptake and protein content in transplanted *gobhi sarson* (*Brassica napus* L. subsp *oleifera* var *annua*). *Indian J Agric Sci* **65:** 478—482
Aulakh MS, Khera TS, Doran JW, Kuldip-Singh and Bijay-Singh (2000) Yields and nitrogen dynamics in a rice - wheat system using green manure and inorganic fertilizer. *Soil Sci Soc Am J* **64:** 1867—1876
Aulakh MS, Khera TS, Doran JW and Bronson KF (2001) Managing crop residue with green manure, urea, and tillage in a rice - wheat rotation. *Soil Sci Soc Am J* **65:** 820—827
Aulakh MS, Jaggi RC and Sharma R (2002) Mineralization-immobilization of soil organic S and oxidation of elemental S in subtropical soils under flooded and nonflooded conditions. *Biol Fertil Soils* **35:** 197—203
Bahl GS, Baddesha HS, Pasricha NS and Aulakh MS (1986) Sulphur and zinc nutrition of groundnut grown on Tolewal loamy-sand soil. *Indian J Agric Sci* **56:** 429—433
Bapat PN (1989) Response of wheat to the levels and sources of S and their residual effect on soybean in Vertisols. Proceedings of National Seminar on Sulphur in Agriculture, p 231. University of Agriculture, Bangalore and The Fertilizers and Chemicals Travancore Ltd., Udyogamandal.
Bawa SS, Dhillon SK and Dhillon KS (1990) Effect of sulphur on the absorption of selenium by different fodder crops. *Indian J Dairy Sci* **43:** 564—570
Biswas BC and Tewatia RK (1991) Results of FAO sulphur trials network in India. *Fert News* **36:** 11—35
Chahal RS and Virmani SM (1973) Preliminary study on the effect of time of application of superphosphate on the yield and nutrient uptake by groundnut. *Indian J Agric Sci* **43:** 731—733
Cheema HS and Arora CL (1984) Sulphur status of soils, tubewell waters and plants in some areas of Ludhiana under groundnut-wheat cropping system. *Fert News* **29:** 28—31
Das SK, Chabra P, Chatterjee SR, Abrol YP and Deb DL (1975) Influence of sulphur fertilization on protein quality of cereals. *Fert News* **20:** 30—35
Dev G, Jaggi RC and Aulakh MS (1979) Study of nitrate-sulphate interaction on the growth and nutrient uptake by maize using ^{35}S. *J Indian Soc Soil Sci* **27:** 302—307
Dev G, Saggar S and Bajwa MS (1981) Nitrogen-sulphur relationships in raya (*Brassica juncea* L.) as influenced by sulphur fertilization. *J Indian Society Soil Sci* **29:** 397—399
Dev G, Aulakh MS and Dhillon NS (1996) Radiotracer studies on role of sulphur in soil-plant system. In: Sachdev MS, Sachdev P and Deb DL (eds.) Isotopes and Radiations in Agriculture and Environmental Research, pp 16—33. Indian Society for Nuclear Techniques in Agriculture & Biology, New Delhi
Deluca IH, Skogley EO and Engel RE (1989) Band applied elemental sulphur to enhance the phyto-availability of phosphorus in alkaline calcareous soils. *Biol Fertil Soils* **7:** 346—350
Dhillon KS, Pasricha NS, Bahl GS and Baddesha HS (1984) Supplementation of nitrogen through green-manuring in raya (*Brassica juncea* L.). *Bull Indian Soc Soil Sci* **13:** 304—309
Dhillon SK, Dhillon KS and Bijay-Singh (2001) Dynamics of sulphur in a calcareous loam soil amended with Sulphur-35 labelled *Sesbania aculeata* green manure. *J Indian Soc Soil Sci* **49:** 259—265
Dwivedi RS (1981) Fertiliser use in groundnut-based cropping systems under different agro-climatic conditions. *Fert News* **26:** 28—34
Fertiliser Statistics 2000-2001. Fertiliser Association of India, New Delhi
FAI (2001) Frank notes: Role of sulphur in balanced fertilization. *Fert News* **46:** 11—12
Friederich JW and Schrader LE (1978) Sulphur deprivation and nitrogen metabolism in maize seedlings. *Plant Physiol* **61:** 900—903
Jaggi RC and Dixit SP (1999) Onion (*Allium cepa*) responses to sulphur in representative vegetable growing soils of Kangra valley of Himachal Pradesh. *Indian J Agric Sci* **69:** 289—91
Jaggi RC, Aulakh MS and Dev G (1977) Interaction effect of nitrogen and sulphur on growth and nutrient uptake by maize. *J Nuclear Agric Biol* **6:** 18—20
Jaggi RC, Aulakh MS and Sharma R (1999) Temperature effects on soil organic sulphur mineralization and elemental sulphur oxidation in subtropical soils of varying pH. *Nut Cycling Agroeco* **54:** 175—182
Jenzen HH and Bettany JR (1984) Sulphur nutrition of rapeseed. II. Effect of time of sulphur application. *Soil Sci Soc Am J* **48:** 107—112
Kumar V, Singh M and Singh N (1981) Effect of S, P and Mo on quality of soybean grain. *Plant Soil* **59:**3-8
Pathak RK and Tripathi RD (1979) Effect of N and S on yield and quality of rai variety Varuna. *Indian J Agric Chem* **12:** 151—157

Pasricha NS and Aulakh MS (1986) Role of S in the nutrition of groundnut. *Fert News* **31**: 17—21
Pasricha NS and Aulakh MS (1991) Twenty years of sulphur research and oilseed production in Punjab, India. *Sulphur in Agric* **15:** 17—23
Pasricha NS and Aulakh MS (1997) Sulphur - an emerging deficient nutrient. In: Kanwar JS and Katyal JC (eds) Plant Nutrient Needs, Supply, Efficiency and Policy Issues: 2000-2025, pp 265—275. National Academy of Agricultural Sciences, New Delhi
Pasricha NS and Fox RL (1993) Plant nutrient sulphur in the tropics and subtropics. *Adv Agron* **50:** 209—269
Pasricha NS and Randhawa NS (1972) Interaction effect of S and Mo on the uptake and utilization of these elements by raya (*Brassica juncea* L.). *Plant Soil* **37:** 315—320
Pasricha NS, Nayyar VK, Randhawa NS and Sinha MK (1977) Influence of sulphur fertilization on suppression of molybdenum uptake by berseem grown on molybdenum toxic soil. *Plant Soil* **46:** 245-250
Pasricha NS, Aulakh MS, Bahl GS and Baddesha HS (1987) Nutritional Requirements of Oilseed and Pulse Crops in Punjab. *Res Bull 15*. Punjab Agricultural University, Ludhiana
Pasricha NS, Aulakh MS, Bahl GS and Baddesha HS (1988) Fertilizer use research in oilseed crops. *Fert News* **33:** 15—22
Pasricha NS, Bahl GS, Aulakh MS and Dhillon KS (1991) Fertilizer Use Research in Oilseed and Pulse Crops in India. Indian Council of Agricultural Research, New Delhi
Rao WK (1988) Sulphur usage in coffee and other plantation crops. In: Sulphur in Indian Agriculture, Section II. 5, pp 1—9. The Sulphur Institute, Washington, DC and Fertiliser Association of India, New Delhi
Sahoo D and Panda N (1985) Longterm effect of intensive cropping and fertiliser use of S availability to rice in an Inceptisol. Proceedings of TNAU-FACT Seminar on Sulphur, pp 77—84. Tamil Nadu Agricultural University, Coimbatore, and The Fertilizers and Chemicals Travancore Ltd., Udyogamandal
Saroha MS and Singh HG (1980). Interrelationships between Fe and S on the prevention of chlorosis in sugarcane on alkaline calcareous soils. *Indian J Agric Sci* **50:** 34—40
Scherer NW (2001) Sulphur in crop production. *Euro J Agron* **14:** 81—111
Shukla LM (2001) Evaluation of soil test methods for sulphur in soils in India. *Fert News* **46:** 55—58
Singh M (1986) Sulphur nutrition in oilseeds and pulses and its relevance to Haryana. *Fert News* **31:** 56—57
Singh D (1987) Sulphur indexing of soils and crops in areas under maize-wheat cropping system. M.Sc. Thesis, Punjab Agricultural University, Ludhiana
Singh MV (2000) Sulphur Management for Oilseed and Pulse crops. *Ind Inst Soil Sci Bul* **13:** 1—54
Singh MV (2001) Importance of sulphur in balanced fertilizer use in India. *Fert News* **46:** 13—35
Singh B (2002) Fertilizer phosphorus management in long-term groundnut-based cropping systems. M.Sc. Thesis. Punjab Agricultural University, Ludhiana, Punjab
Singh HG and Sahu MP (1986) Response of oilseeds to sulphur. *Fert News* **31:** 23—30
Singh M and Tiwari RC (1985) Response of oilseed crops to fertilizers in dryland agriculture. Proceedings of FAI-NR Seminar, Varanasi, pp 147—159. Fertiliser Association of India, New Delhi
Sarkar MC (2000) Significant achievements of TSI/FAI/IFA sponsored collaborative project on sulphur in balanced fertilisation. In: Proc of TSI/FAI/IFA Workshop on Sulphur in Balanced Fertilisation, pp 13—23. Fertiliser Association of India, New Delhi
Stevenson FJ (ed) (1986) Cycles of Soil: Carbon, Nitrogen Phosphorus, Sulfur, Micro-nutrients. John Willey & Sons, New York
Subbiah BV and Singh N (1970) Efficiency of gypsum as a source of S to oilseed crops, studied with radioactive S and radioactive Ca. *Indian J Agric Sci* **40:** 227—234
Tandon HLS (1991) Sulphur Research and Agricultural Production in India, 3rd Edition. The Sulphur Institute, Washington DC
Tandon HLS (1995a) Sulphur in Indian agriculture: Update 1995. *Sulphur in Agric* **19:** 3—8
Tandon HLS (1995b) Sulphur Fertilisers for Indian Agriculture, 2nd Edition. Fertiliser Development and Consultation Organisation, New Delhi
Tandon HLS and Messick DL (2002) Practical Sulphur Guide. The Sulphur Institute, Washington DC
Tiwari RC, Kadlag AD, Singh D, Mathur DS, Sarkar AK and Dev G (2002) Efficiency of micronized elemental sulphur (SulFer 95) for increasing crop yields in different soils of India. Proceedings of 17th World Congress of Soil Science, Bangkok, Thailand

7. AFFILIATION

Department of Soils, Punjab Agricultural University, Ludhiana, Punjab, India
For correspondence: msaulakh@satyam.net.in

M.Z. ABDIN, A. AHMAD, N. KHAN, I. KHAN, A. JAMAL AND M. IQBAL

SULPHUR INTERACTION WITH OTHER NUTRIENTS

Abstract. Sulphur is known to interact with almost all essential macronutrients, secondary nutrients and micronutrients. These interactions can either enhance or reduce growth and yield of crops by influencing the nutrient uptake and utilization. Better understanding of these relationships can lead to more efficient crop production, higher yields, improved crop quality and harvestibility. Future improvements in crop varieties, water utilization, and general improvements of cultural techniques will require a better understanding of nutrient interaction. This chapter focuses the interaction of sulphur with other nutrients.

1. INTRODUCTION

The use of mineral fertilizers has been the most important technological factor, contributing 50% of the yield of crop (Velayutham, 1997). Developing countries are making serious efforts to increase the use of fertilizers supplying mineral nutrients. Ever increasing fuel prices coupled with dwindling petroleum reserves worldwide have a tremendous pressure on fertilizer prices. In addition, inflation and problem with the availability of foreign currency are serious constraints to the economic advancement of many developing countries including the advancement in agricultural production. One consequence of these events is the need to rationalize the use and application method of plant nutrients. The elements of rational use are mainly to increase efficiency of uptake and assimilation of mineral fertilizers by plants and prevention of losses. Among various potential fertilizer losses during production, storage, transportation, handling, distribution and use, the major losses take place in the field. Increasing the efficiency of nutrient utilization in field will not only increase the yield, but also will result in huge monitory saving (Prasad, 1998). To minimize the losses of precious nutrients, development of a sound fertilizer use and management plan is essential. The fast accumulating reports of S-deficiency from various parts of the world has triggered interest in S nutrition. The sulphur requirement of crops depends on many factors of which the balance between S and other nutrient elements is important in view of a possible synergistic or antagonistic effect. The present chapter deals with the interaction of S with other nutrients.

2. SULPHUR INTERACTION WITH NITROGEN

2.1 Sulphur and Nitrogen Interaction at Metabolic Level

Nitrogen occupies a very important place among the essential nutrients in soil. Crop productivity is determined largely by it and hence soil fertility and soil nitrogen have become almost synonymous to each other. Sulphur interacts with nitrogen in such a

Yash P. Abrol and Altaf Ahmad (eds.), Sulphur in Plants, 359—374.

way that the inadequacy of one reduces the uptake and assimilation of the other. Clarkson et al. (1989) observed a marked depression in the ability of cereal plants to take up nitrate and ammonium when plants were starved of sulphur, which was accompanied with the augmented capacity of sulphur uptake. The S and N affects enzyme activity in their respective assimilatory pathways (Reuveny et al. 1980; Barney and Bush, 1985; Brunold and Suter, 1984; Bell et al., 1995; Ahmad et al., 1999a). Under sulphur deficit conditions, reduced protein synthesis is accompanied by accumulation of organic and inorganic nitrogenous compounds. Plants starved of sulphur are accumulate arginine and asparagine with low levels of sulphur amino acids i.e. methionine and cysteine (Gilbert et al., 1997). Barney and Bush (1985), while working on tobacco plant, concluded that one nutrient accumulated when the other

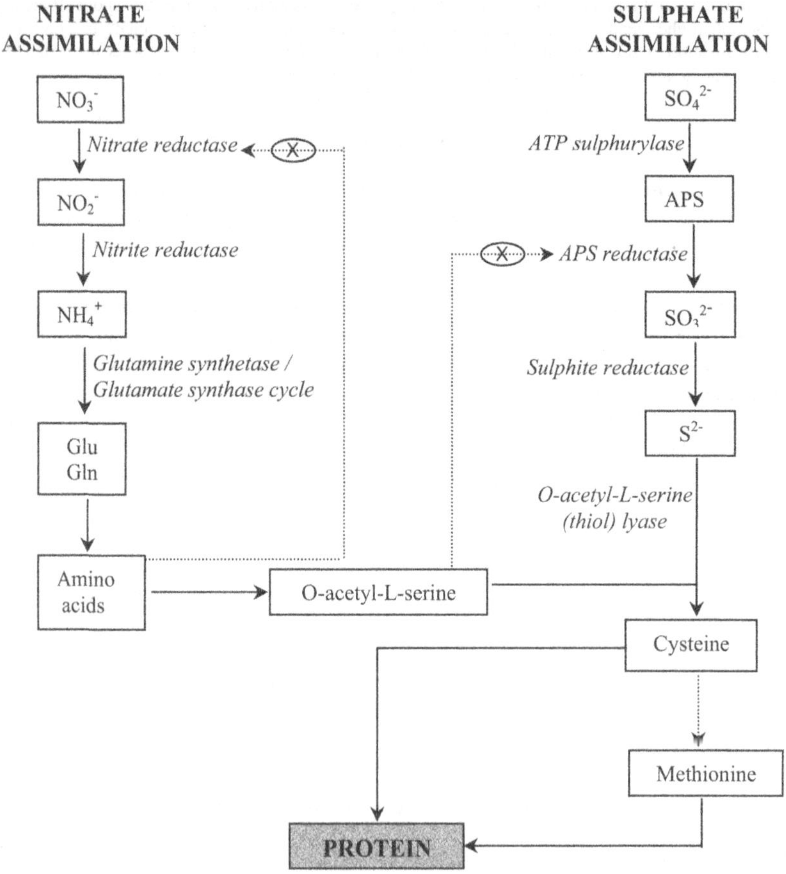

Figure 1. Metabolic coupling between sulphur and nitrogen assimilatory pathways in plants
(X) = postulated points of regulation

was limited and that the accumulated nutrient was used in protein synthesis when the treatment was reversed. S uptake and assimilation has been shown to be dependent upon a constant supply of the precursor of cysteine, O-acetylserine, which, in turn, is dependent upon adequate N availability (Kopriva and Koprivova, this book). Excess cysteine or another reduced S-compound represses the uptake and assimilation of S, when either S is in excess, or N is limiting (Zhao et al., 1999). Cysteine may be involved in the repression of the synthesis of other amino acids, ultimately repressing N-uptake. The regulatory interaction between assimilatory sulphate and nitrate reduction is believed to occur at transcriptional level (Prosser et al., 2001).

2.2 Sulphur and Nitrogen Interaction in Relation to Photosynthetic Efficiency

Photosynthesis is arguably the most important biological process on the earth. Its rate depends on (i) light harvesting capacity; (ii) the rate at which NADPH and ATP can be regenerated; and (iii) the capacity for the carboxylation of RuBP by Rubisco. The respective components responsible for these processes are chlorophyll-protein complexes (LHC), electron transport and photophosphorylation membrane complexes and photosynthetic carbon reduction (PCR) cycle. Nitrogen is an important constituent of these components. It is well documented that approximately 60% of nitrogen in a plant leaf with C_3 photosynthesis is invested in photosynthetic components. Leaves of wheat supplied with ample nitrogen fertilizer contain up to 18 g of soluble protein m^{-2} leaf. Out of this Rubisco constitutes 50-70% (Evans , 1983, Lawlor et al., 1989). Thus, nitrogen nutrition plays a crucial role in determining the photosynthetic capacity of the plant in both natural and agricultural environments (Abrol, 1993; Abrol et al., 1999; Kumar et al., 2003). While nitrogen directly affects the photosynthesis efficiency of the plant, sulphur affects the photosynthesis efficiency indirectly by improving the nitrogen utilization efficiency of the plants as evident from the relationship between nitrogen content and photosynthetic rate in the leaves of +S and –S treated plants (Ahmad and Abdin, 2000a, Fig. 2). In –S treated plants, photosynthesis is linearly related to leaf-N content only up to 1.5 g m^{-2}. At higher leaf N content, however, the relationship was non-linear. This observation was also supported by Lawlor et al. (1989), who reported a curvilinear relationship between photosynthetic rate and leaf-N content beyond 2.0 g m^{-2} in case of wheat. It is very interesting to note that in +S treated plants, the relationship between photosynthetic rate and leaf-N was linear even beyond 1.5 g m^{-2}. It has been suggested that under S deficiency (-S treatment), the extra leaf N (beyond 1.5 g m^{-2}) remains non-utilized in the form of non-protein N in the leaves of –S treated plants (Stewart and Porter, 1969; Eppendorfer, 1971; Ahmad et al., 2001) in the plants grown with insufficient sulphur. Consequently, adequate sulphur supply increased the N-utilization into protein (Rubisco) synthesis, thereby enhancing the photosynthetic rate.

2.3 Sulphur and Nitrogen Interaction in Relation to Yield and Quality of Crops

Sulphur and nitrogen relationship in terms of crop yield and quality has been established in many studies (Singh and Bairathi, 1980; Sachdev and Deb, 1990; Lakkineni and Abrol, 1992; McGrath and Zhao, 1996; Ahmad et al., 1998; Fismes et

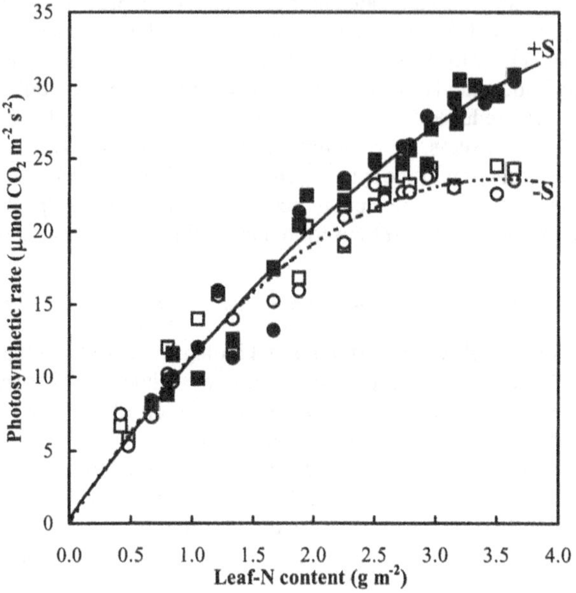

Figure 2. Relationship between Leaf-N and photosynthetic rate of Brassica genotypes grown under +S and –S conditions (Source: Ahmad and Abdin, 2000a)

al., 2000). It has been demonstrated that seed production is very sensitive to S deficiency and maximum yield responses of crop to S and N were obtained only when the availability of S and N was in approximate balance. Application of nitrogen alone suppressed the seed yields, whereas S alone produced no seed yield response. McGrath and Zhao (1996) observed an increase of 42-267% in seed yield of *Brassica napus* with the application of 40 kg S ha^{-1} with 180 and 230 kg N ha^{-1}. The seed yield declined when N was supplied at the rate of 180-230 kg ha^{-1} without S. Similar observations were reported by Ahmad et al. (1998; 1999a, 1999b) in rapeseed-mustard. In wheat, the yield increased linearly in the S and N interaction study (Reneau et al., 1986) with increased N application. Plants grown on S-limiting soils suppress the development of reproductive organs and even lead to pod abortion in rapeseed (Fismes et al., 2000). Reproductive growth and the proportion of the reproductive tissues in total dry matter was significantly increased by S during pod development (McGrath and Zhao, 1996). Under S deficient conditions, excess of unassimilated NO_3^-—N or free amino acids accumulates in leaves (Hue et al., 1991; Ahmad et al., 2001), and protein within chloroplasts is degraded (Dannehl et al., 1995; Ahmad and Abdin, 2000a). Sunrapi and Anderson (1997) have shown that high levels of N inhibit the proteolysis process in soybean and so the export of S and N from mature leaves to developing leaves or developing grains.

Interactive effect of S and N on quality aspects of various crops has been. Zhao et al. (1997) found a strong interaction between N supply and the proportion of the seed S bound in glucosinolates of rape. When S was deficient N application decreased glucosinolate concentration, but S application increased it. Also in nutrient solution experiments the glucosinolate concentration decreased when S supply was withdrawn (Blake-Kalff et al., 1998). Investigations of Fismes et al. (2000) on a rendzina soil in the Loraine, rich in organic S compounds, showed that S fertilizer application had no impact on seed yield and seed oil content, but increased the concentration of glucosinolates in seed. Ahmad and Abdin (2000b) have reported that application of S along with N improved the quality of the oil of rapeseed-mustard by increasing the linoleic acid (18:2) and oleic acid (18:1) contents, and reducing the eicosenoic acid (20:1) and erucic acid (22:1) contents. They have suggested that application of N alone probably promotes elongation of the carbon chain of linoleic acid (18:2) and oleic acid (18:1) to eicosenoic acid and erucic acid and then to linolenic acid (Ahmad and Abdin, 2000b). Since S and N serve as structural protein components, a balance between S and N in grain has been found essential for the breadmaking quality of wheat (Randall and Wrigley, 1986). Data from Eppendorfer (1971) show a linear increase in the sum of methionine, cysteine and cystine in the dry-matter of small grain cereals as N concentration increased except for the treatment without S addition.

2.4 N:S Ratio

Because of a strong interaction of N and S metabolisms, plants maintain a relatively constant ratio of organic N to organic S, particularly in their vegetative tissues, even though the ratio of total N to total S can vary widely in response to N and S supply. It has been established that for every 15 parts of N in protein there is 1 part of S which implies that the N:S ratio is fixed within a narrow range of 15:1. The N:S ratio in the whole plant in general is 20:1 (Cram, 1990). When S is deficient in relation to N, accumulation of non-protein compounds such as amides occurs, resulting in an N to S ratio greater than 15:1. On the contrary, when S supply is greater than that required for protein synthesis, sulphate accumulates in plant tissues, leading to an N to S ratio smaller than 15:1. It has been suggested that S concentration of 0.2 % and a N/S ratio of 18 in the flag leaf is sufficient for obtaining higher yields in wheat (Reneau et al., 1986), while Mahler and Maples (1987) noticed that a minimum S concentration and N:S ratio of wheat tissue for maximum yield were 1050 ppm and 16.5, respectively. Dev and Kumar (1982) reported N:S ratio of 15.6, 3.1, 14.8 and 7.1 in grain at maximum response to sulphur in maize, mustard, groundnut and wheat, respectively. Seed N:S ratio is positively correlated ($r^2 = 0.876$) with oil 22:1/18:1 ratio in rapeseed (Ahmad and Abdin, 2000a), indicating that it also affects the quality of the oil of rapeseed (Fig. 2). As the ratio of total N to total S and protein S determine the degree of availability or deficiency of S in protein, the N and S ratio is often preferred over concentration as a diagnostic criterion for S deficiency (Spencer and Freney, 1980). However, Schnug and Hanklaus (2000) argued that the use of N:S ratio as a diagnostic criterion is problematic because the same ratio of N:S can be obtained at totally different concentration levels in the tissue. Surplus with one element may, therefore, falsely be interpreted as deficiency with the other.

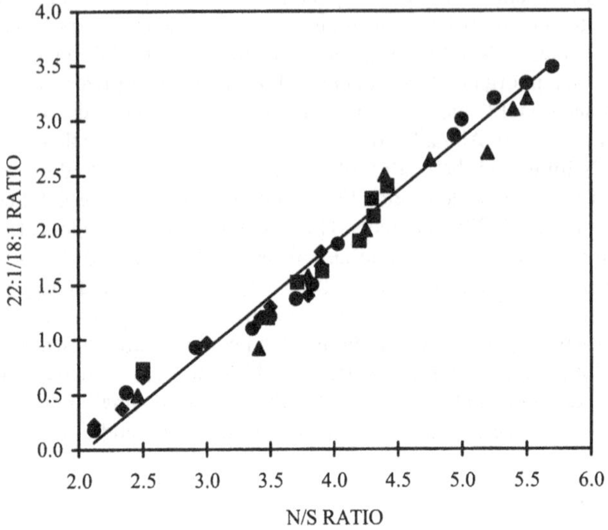

Figure 3. Inter-relationship between ratios of N:S and erucic acid:oleic acid

2.5 Iron sulphur clusters: A Coordination of Sulphur and Nitrogen Assimilation

Iron-sulphur proteins occur in all life forms. Ferredoxin and Rieske proteins each contain a (2Fe-2S) cluster whereas photosystem I (PS I) contains three (4Fe-4S) clusters (Imsande, 1999). Microbial sulphite reductases (SiRs) and related nitrite reductases (NiRs) couple a metalloporphyrin to a (4Fe-4S) cluster and carry out the six electron reduction of sulphite to sulphide or nitrite to ammonia, respectively (Crane et al., 1997a, b). These sulphite reductases (EC 1.8.1.2) contain 8 α-subunits (66 kDa each) and 4 β-subunits (64 kDa each) and rely on NADPH for reducing power. Each β-subunit contains a (4Fe-4S) cluster and a siroheme. Green plants carry out similar reactions; however, they are ferredoxin-dependent (EC 1.8.7.1) rather than NADPH-dependent. Using the PCR technique, a putative β-subunit has been cloned from *Arabidopsis* (Bruhl et al., 1996). The transcript would produce a protein of 642 amino acids, of which 66 amino acids would be a transit peptide. Thus, the mature protein would be composed of 576 amino acids with a molecular mass of 65.28 kDa. Sequence analysis revealed conservation of the cysteinly groups required for the (4Fe-4S) cluster (Bruhl et al., 1996). Genes encoding nitrite reductase have been cloned from several higher plants, including spinach, maize, bean and tobacco, and from microbes. The derived amino acid sequences of the higher plant nitrite reductases show an 80% identity and reveal a molecular mass of approximately 63 kDa. Four cysteines involved in binding the (4Fe-4S) cluster are conserved in all higher-plant nitrite reductases examined. The glutamate synthases (GltS) of several bacteria are dimers composed of an α-subunit (162 kDa) and a β-subunit (53.2 kDa) (Vanoni et al., 1998). The β-subunit of the bacterial enzyme seemingly contains the NADPH binding site and hence, is not required in photosynthetic tissue. The mature ferredoxin-dependent glutamate synthase present in green tissue of higher plants is a monomer of 165 kDa (Suzuki and Rothstein, 1997). The NAD(P)H-dependent GltS, found primarily in non-photosynthetic shoots, roots and root nodules, as a monomer of approximately 120 kDa. Both bacterial subunits and the Fe-GltS of green plants are reported to contain a $(3Fe\text{-}4S)^{+0}$ cluster (Knaff et al., 1991). Little is known about the effect of ROS on the lability of these iron-sulphur clusters. It is possible, however, that the oxidation state of plant cells could have a profound effect on these iron-sulphur clusters and consequently upon the assimilation of nitrogen and sulphur.

3. SULPHUR INTERACTION WITH PHOSPHORUS

Sulphur and phosphorus are both essential elements and are taken up by the plants in the anionic form from the soil. The requirement of plants for these elements is similar (Scherer, 2001). A proper balance of these nutrients in the soil and the plant is an important aspect in the crop production.

3.1 Phosphorus–sulphur interrelationship in soil

Interrelationship of P and S in soil is manifested mainly in adsorption studies, and formation of reaction products expressed in terms of inorganic fractionation study. Ensminger (1954) studied for the first time the possibility of replacement of sulphate by other anions of greater penetrating or coordinating ability, and Kamprath et al. (1956) established a reciprocal relationship between the concentrations of the two anions in relation to adsorption. Sulphate adsorption is completely stopped by the presence of two-thirds of the equivalent amount of phosphate compared to sulphate. It has been demonstrated that rate of downward movement of adsorbed sulphate and the extent of leaching depend on the quantity of phosphate applied, the presence or absence of plants, and on the type of the plant grown. While studying the effect of sulphate on phosphate adsorption in laterite, alluvium and black soil, Kundu (1984) reported a great extent of reduction in the value of affinity coefficient, indicating that phosphate is held less strongly in presence of sulphate. However, adsorption maxima value (from Langmuir's equation) was unaffected. In an incubation study with a calcareous soil Dawood and Kadban (1989) highlighted the importance of the source of S in adsorption relationship studies. While higher doses of elemental S (10,000 ppm) helped in decreasing the sorption of phosphorus, similar levels of S ore increased the P-sorption. Pasricha and Sparks (1990) showed that with addition of phosphorus the increased adsorption of phosphate with time resulted in a concurrent desorption of sulphate from the colloidal surfaces. The overall evolution of the adsorption relationships has raised the alarm that a continuous use of large dressings of fertilizer phosphate will aggravate the S-deficiency, specially where S is present in critical amounts. The extent of competitiveness between the two anions is related to the genesis of the soil (Metson and Blakemore, 1978). This appeared to be more serious in weakly to moderately weathered soils from sedimentary parent materials. At the other extreme, soils developed on volcanic parent material and having a high level of native adsorbed sulphate appear to have a capacity to retain added sulphate even in the presence of a similar concentration of phosphate.

The influence of S on different inorganic fractions of P is probable due to the similarity in composition of relatively insoluble Al or Fe sulphate and phosphate as shown by their formulae when written as $Al(OH)_2H_2PO_4$ and $Al_4(OH)_{10}H_2SO_4$ (Fried, 1980). The strong affinity of aqueous Al^{3+} and Fe^{3+} for $(SO_4)^{2-}$ along with $(OH)^-$ ions, suggests a keen competition among these anions for precipitation reactions with partially neutralized Al and Fe hydrous oxides. This mechanism for a positive sulphate-phosphate interaction to increase phosphorus availability by reducing the reactive phase cannot be ruled out (Kundu, 1984).

3.2 Phosphorus–sulphur interrelationship in plants

Studies on interaction of phosphorus and sulphur in plants have been more superficial than sophisticated. Results in the experiments have been expressed usually in terms of yield and concentration with relation to uptake of the nutrients by the plants. Again, the type of relationship depends on plant parameters like demand for the nutrient, rooting pattern etc. that are very different among cereals, pulses and oilseeds.

3.2.1 Cereals

Noble and Kleining (1971) studied the response of irrigated sorghum to gypsum and phosphorus. The interaction between the two can be recognized from the response of dry matter yield, which was up to 100 lb/acre level when applied along with gypsum and only 50 lb/acre level in the absence of it. Marok and Dev (1980) reported a positive interaction between P and S on dry matter production of wheat up to 50 ppm level of P and S application. At lower dose of P (12.5 ppm), the positive response was also up to 12.5 ppm, thus showing positive correlation regarding the demand for the nutrition in wheat crop. Naphada and Mutalka (1984) showed a significant increase in dry matter yield of maize over the control due to application of P fertilizers along with S. When a thiosulphate source was blended with monoammonium phosphate, total dry matter yield significantly increased in barley (Morden et al., 1986). A highly significant positive interaction was obtained between P and S in terms of P uptake leading to a high seed yield of wheat at a low rate of S application. Higher rate of S application caused antagonistic effect (Randhawa and Arora, 2000).

3.2.2 Pulses

Reports on pulses are fairly more in number but often conflicting. In a green house study on field bean, significant interaction of P and S has been reported in terms of grain yield and nutrient composition (Nayak and Dwivedi, 1990) when applied as 30 kg S ha^{-1} and 40 kg P_2O_5 ha^{-1}. Higher levels of combined application decreased the yield drastically, showing an antagonistic effect. A residual moong crop also showed a similar trend (Aulakh and Pasricha, 1979). A negative interaction between P and S was also observed by Barkas (1981) in bean grown in P and S deficient soil. Venkatalakshmi (1987) reported the beneficial effect of P-S interrelationship to be more in a low P status soil, the effect being only marginal with a high P status soil in case of gram and pea.

3.2.3 *Oilseed crops*

Oilseeds have often been used as the test crop in P-S interrelationship study, as they respond well to S application. Virmani and Gulati (1971) in their study on effect of sulphur on the response of mustard to phosphorus fertilization got highest yield at 50 ppm P and S level, with a clear indication of a beneficial effect of S at each level of P fertilization (0, 25 and 50 ppm). Joshi et al. (1973) expressed similar results showing a synergism up to 50 kg P and S ha^{-1} and an antagonism beyond that. Reports on groundnut crop show that the application of P and S significantly increased the dry matter, pod yield and shelling percentage on low and medium fertility soils (Rathee and Chahal, 1977). Soybean has most often been used as the test crop due to its high demand for both P and S corresponding to its high protein and oil content. The S and P interaction in also improves the oil yield of mustard (Jaggi and Sharma, 1999; Sharma and Jalali, 2001).

3.3 *S:P Ratio*

The S:P ratio in plant materials has often been reported in relation to P and S interaction study to find out the optimum value of the ratio corresponding to the desirable dry matter yield. The most effective S:P ratio in mustard was reported to be 0.9 - 1.4.

4. SULPHUR INTERACTION WITH POTASSIUM

The role of potassium and sulphur in augmenting the yield and improving the quality of crops is well known (Singh and Rathore 1994; Prasad et al., 1996; Razmjoo and Henderlong, 1997; Umar et al., 1997). The interaction between K and S is synergistic only up to a level, thereafter, it is antagonistic. The relative yield (Y_{K+S}) of a crop, when K and S are used together is generally determined by the following equation (Umar and Bansal, 1995):

$$\frac{1}{Y_{K+S}} = \frac{1}{A} + \frac{1}{B(K_0+K)} + \frac{1}{C(S_0+S)}$$

Where K and S are either amount of nutrient added or the per cent nutrient in the dry matter. A, B, K_0, C and S_0 are constants to be deduced from a comparison with experimental data. The values of the unknown constants are; A = 8.068; B = 62.041; C = -115.89; K_O = -2677 and S_0 = -2.044

Table 2 shows the data calculated by Umar and Bansal (1995) as per procedure given by Wallace (1990) to test the applicability of the above equation. The excellent agreement with experimental observation was found that demonstrates the success of equation. The deviation is larger in the cases of (K, S) sets (4,2) and (4, 4). This is due to the fact that the value K = 4 has been covered less while fitting the above equation to the entire collection of data. It is worth mentioning here that the values of the constants

Table 2. Relative yields with different amounts of K and S used alone and together

Nutrient application (mval)		Relative yield K & S used alone		K & S used together	a (Y_{KxS})	Y_{K+S}: Y_{KxS}	b	Fitted values of Y_{K+S}	Per cent deviation in the fitted values from the observed
K	S	Y_K	Y_s	Y_{K+S}					
3	1.5	3.41	1.48	5.27	5.05	1.04	S	5.27	-
3	3.0	3.41	1.70	6.02	5.80	1.04	S	6.06	0.66
3	4.0	3.41	1.69	5.90	5.76	1.02	S	5.90	-
3	1.5	4.02	1.48	6.58	5.95	1.10	S	6.58	-
4	3.0	4.02	1.70	8.45	6.83	1.24	S	7.87	6.86
4	4.0	4.02	1.69	8.11	6.79	1.19	S	7.59	6.41
6	1.5	6.57	1.48	6.97	9.72	0.71	A	6.91	0.86
6	3.0	6.57	1.70	8.35	11.17	0.75	A	8.35	-
6	4.5	6.57	1.69	8.04	11.10	0.72	A	8.04	-

a = according to sequential additivity; b = type of interaction; S = synergistic interaction; A = antagonistic interaction; mval = milli equivalents amounts per litre
Source: Umar and Bansal (1995)

presented in Table 1 are by no means unique even in the context of the present data. However, only slightly different values will result if some other five values of Y_{K+S} were used as input. This expectation is based on the fact that the remaining four values of Y_{K+S} are described reasonably using these values. Performing a least square search among all the data would be the ultimate. However, the form of the function does not tend itself easily to such an endeavour.

5. SULPHUR INTERACTION WITH MOLYBDENUM

Molybdenum, though required in trace amount, is essential for inorganic nitrogen metabolism as part of nitrate reductase enzyme in most plant species and nitrogenase in legumes. It is the metallic co-factor for nitrate reductase enzyme protein and helps in transfer of electrons from NADH to nitrate for its reduction to nitrite. It seems plausible that under conditions of excessive molybdenum, more than one Mo atoms may attach to the enzyme protein thus making it biologically inactive as a functional enzyme. This results in a decrease in nitrate assimilation and reduced nitrogen; consequently in proteins, nucleic acids, co-factors, chlorophyll, plant growth regulators, and ultimately in growth. Sulphur fertilization generally reduces uptake of Mo (Chaphale et al. 1991). The decrease in molybdenum content with increasing levels of sulphur for plants grown in the soil has been attributed to increased competition for exchange sites at soil-root interface by two bivalent anions of similar charge and size, i.e. sulphate and molybdate (Guyette et al., 1989). Mo when added with S improved growth parameters due to their individual effect as well as their effect on increase in nitrogen and sulphur uptake (Sairam et al., 1995). Application of Mo either alone or with sulphur significantly increased Mg uptake in wheat (Purakayastha and Nad, 1998). Combined application of Sulphur and Molybdenum was more beneficial to produce maximum yield and better quality of cabbage with fair amount of protein and ascorbic acid (Hunashikatti et al.,

2000). However, the interactive effect of sulphur and molybdenum at higher levels was not statistically significant. This may be attributed to the antagonistic effect existing between sulphur and molybdenum at higher levels as both are absorbed as anions (MoO_4^{2-} and SO_4^{2-}) and probably compete for the absorption site on plant roots.

6. SULPHUR INTERACTION WITH ZINC

Zinc is an important micronutrient, which enters the plant primarily via absorption of Zn^{2+} by roots from soil solution. Interaction of sulphur with zinc has been extensively investigated on yield of many crops. In a Zn and S deficient soil, plants responded to added S only when adequate supply of Zn to soil was maintained (Singh et al., 1997). The Zn and S interactions cause increase in seed and dry matter yield (Babhulkar, 2000), and thus, shows synergism. Indications were that S and Zn interaction perhaps occurred both at adsorption sites and within plants. S and Zn contents of plant tissue also correlated well with each other. Some reports, however, show antagonistic relationship between S and Zn (Kumar et al., 1997).

7. SULPHUR INTERACTION WITH IRON

There appears to be a close relationship between Fe and S metabolism in the plant. Application of 10 mg Fe kg^{-1} improved its availability by 10% whereas 80 mg S kg^{-1} enhances it by 49%. The combined application of both the nutrients showed tremendous boost in Fe availability giving the overall improvement of 101 % (Malewar and Ismail, 1997).

8. SULPHUR INTERACTIONS WITH SELENIUM

The antagonism between S and Se is well known. In fact, as the Se content of the fertilizer increases, the S uptake and concentration in the plant decrease. In short term experiments conducted by Pezzarossa et al. (1999) under greenhouse conditions, different genotypes of *Lycopersicon lycopersicum* (UC82B and LA2711), *Lycopersicon pennellii* (LA716) and *Lycopersicon peruvianum* (LA2157) were grown until fruit ripening. The results confirmed that sulphate in the growth solution reduced selenate uptake by plants and increased the S content of the leaves. Under low sulphate treatment there was a clear correlation ($r^2 = 0.82$) between leaf S content and shoot Se content across the genotypes, thus, indicating that the overall activity of the S transport systems also determines Se transport. Selenium was translocated from shoot to fruit, but the edible portion of the plant contained much less total Se than the inedible plant parts. The difference in Se content between the low and the high sulphate treatments was significantly higher in shoot than in root, confirming that the Se translocation from root to shoot is probably more affected by high sulphate supply than Se uptake by root.

9. SULPHUR INTERACTION WITH BORON

Available informations indicate both synergistic and antagonistic relationship between boron and sulphur. In calcareous soils (Calciorthents), application of 40 kg S ha^{-1} and

20 kg borax ha^{-1} resulted in 48.5% higher productivity of mustard, compared to any of the two applications alone. In these soils application of 40 kg S ha^{-1} and 20 kg borax ha^{-1} resulted in 591 kg ha^{-1} additional pod yield of groundnut, while application of either 40 kg S ha^{-1} or 20 kg borax ha^{-1} alone resulted in 345 and 313 kg more pod yield, respectively over the control (Sakal et al., 1996; Singh, 2000). Few reports have indicated antagonistic relationship between S and B.

10. CONCLUSION

Sulphur interactions with other nutrients are directly related to the modification of physiological and yield responses of crops, and thus required to be studied in depth. This would help to understand nutritional behaviour of S in relation to other nutrients and provide guidelines for formulating balanced fertilizer recommendations in order to optimise yield and quality of crops.

11. REFERENCES

Abrol YP (ed) (1993) Nitrogen-Soils, Physiology, Biochemistry, Microbiology, Genetics. Indian National Science Academy, New Delhi

Abrol YP, Chatterjee SR, Kumar PA and Jain V (1999) Improvement in nitrogen use efficiency: Physiological and molecular approaches. *Curr Sci* **76:** 1357—1364

Ahmad A and Abdin MZ (2000a). Photosynthesis and its related physiological variables in the leaves of *Brassica* genotypes as influenced by sulphur fertilization. *Physiol Plant* **110:** 144—149

Ahmad A and Abdin MZ (2000b) Effect of sulphur application on lipid, RNA, protein content and fatty acid composition in developing seeds of rapeseed (*Brassica campestris* L.). *Plant Sci* **150:** 71—75

Ahmad A, Abraham G, Gandotra N, Abrol YP and Abdin MZ (1998) Interactive effect of nitrogen and sulphur on growth and yield of rapeseed-mustard (*Brassica juncea* L. Czern and Coss and *Brassica campestris* L.) genotypes. *J Agron Crop Sci* **181:** 193—199

Ahmad A, Abraham G and Abdin MZ (1999a) Physiological investigation on the impact of nitrogen and sulphur application on seed and oil yield of rapeseed (*Brassica campestris* L.) and mustard (*Brassica juncea* L. Czern and Coss) genotypes. *J Agron Crop Sci* **183:** 19—25

Ahmad A, Abrol YP and Abdin MZ (1999b) Effect of split application of sulphur and nitrogen on growth and yield attributes of *Brassica* genotypes differing in their time of flowering. *Can J Plant Sci* **79:** 175—180

Ahmad A, Khan I and Abdin MZ (2001) Interactive effect of sulphur and nitrogen on N-assimilation and nitrogen harvest of rapeseed-mustard. *Indian J Plant Physiol* **6:** 46—52

Aulakh MS and Pasricha NS (1979) Responses of gram (*Cicer arietinum* L.) and lentil (*Lens cultinaris* L.) to phosphorus as influenced by applied sulphur and its residual effect on moong (*Phaseolus aureus* L.). *Bull Indian Soc Soil Sci* **12:** 433—438

Babhulkar PS, Kar D, Badole WP and Balpande SS (2000) Effect of sulphur and zinc on yield, quality and nutrient uptake by safflower in Vertisol. *J Ind Soc Soil Sci* **48:** 541—543

Barkas TD (1981) Intmercation effect of S and P on yield, quality conditions and nutrient content of bean (Phaseolus vulgare L.). Panhollenic Congress of Geothechnical Research, abstract 148.

Barney Jr PE and Bush LP (1985) Interaction of nitrate and sulphate reduction in tobacco. I. Influence of availability of nitrate and sulphate. *J Plant Nutr* **8:** 507—515

Bell CI, Clarkson DT and Cram WJ (1995) Partitioning and redistribution of sulphur during S-stress in *Macroptilium atropurpureum* cv. Siratro. *J Exp Bot* **46:** 73—81

Blake-Kalff MMA, Harrison KR, Hawkesford MJ, Zhao FJ and McGrath SP (1998) Allocation of sulfur within oilseed rape (*Brassica napus* L.) leaves in response to sulfur-deficiency. *Physiol Plant* **118:** 1337—1344

Bruhl A, Haverkamp T, Gisselmann G, Schwenn JD (1996) A cDNA clone from *Arabidopsis thaliana* encoding plastidic ferredoxin: Sulfite reductase. *Biochim Biophys Acta* **1295:** 119—124

Brunold C and Suter M (1984) Regulation of sulphate assimilation by nitrogen nutrition in the duckweek *Lemna minor* L. *Plant Physiol* **76:** 579—583

Chaphale PC, Naphade PS and Kene DR (1991) Effect of molybdenum and sulfur application on performance of mung (*Phaseolus aureus* L.) grown in black calcareous soil. *PKR Res J* **15:** 176—178

Clarkson DT, Sarker LR and Purves JV (1989) Depression of nitrate and ammonium transport in barley plants with diminished sulphate status. Evidence of co-regulation of nitrogen and sulphate intake. *J Exp Bot* **40:** 953—963

Cram WJ (1990) Uptake and transport of sulphate. In: Sulphur Nutrition and Sulphur Assimilation in Higher Plants. In: Renennberg H, Brunold C, De Kok LJ and Stulen I (eds). Sulphur Nutrition and Sulphur Assimilation in Higher Plants, pp 3—11. SPB Academic Publishers, The Hague

Crane BR, Siegel LM and Getzoff ED (1997a) Structures of the siroheme- and Fe4-S4 containing active centre of sulfite reductase in different sates of oxidation: Heme activation via reduction-gated exogenous ligand exchange. *Biochemistry* **36:** 12101—12119

Crane BR, Siegel LM and Getzoff ED (1997b) Probing the catalytic mechanism of sulfite reductase by X-ray crystallography: structures of the *Escherchia coli* hemoprotein in complex with substrates, inhibitors, intermediates, and products. *Biochemistry* **36:** 12120—12137

Dannehl HA, Herbik A and Godde D (1995) Stress-induced degradation of the photosynthetic apparatus is accompanied by changes in thylakoid protein turnover and phosphorylation. *Plant Physiol* **93:** 179—186

Dawood FM and Kadban MM (1989) Sulphur and sorbed phosphorus relationship in calcareous soil. *J Agric Water Resources Res* **8:** 89—93

Dev G and Kumer V (1982) Secondary nutrients. In: Review of Soil Research in India. Vol. I., pp 342—360. Indian Society of Soil Science, New Delhi

Ensminger LE (1954) Some factors affecting the adsorption of sulphate by Albama soil. *Soil Sci Soc Am Proc* **18:** 259—264

Eppendorfer WH (1971) Effect of sulphur, nitrogen and phosphorous on amino acid composition of field bean (*Vicia faba*) and responses of the biological value of the seed protein and sulphur-amino acid content. *J Sci Food Agric* **22:** 501—505

Evans JR (1983) Nitrogen and photosynthesis in the flag leaf of wheat (triticum aestivum L.). *Plant Physiol* **72:** 297—302

Fismes J, Vong PC, Guckert A, Frossard E (2000) Influence of sulphur on apparent N-use efficiency, yield and quality of oilseed rape (*Brassica napus* L.) grown on a calcareous soil. *Euro J Agron* **12:** 127—141

Fried A (1980) Intercqation of phosphorus with other elements in soil and plant. In: Khasawneh FE (ed) The Role of Phosphorus in Agriculture, pp 655—680. American Society of Agronomy and Soil Science Society of America, Madison, Wisconsis

Gilbert SM, Clarkson DT, Cambridge M, Lambers H and Hawkesford MJ (1997) SO_4^{2-} deprivation has an early effect on the content of rubulose-1,5-bisphosphate carboxylase/oxygenase and photosynthesis in young leaves of wheat. *Plant Physiol* **115:** 1231—1239

Guyette RP, Cutter BE and Henderson GS (1989) Long term relationships between molybdenum and sulphur concentrations in red cedar tree rings. *J Environ Quality* **18:** 385—389

Hue H, Spark D and Evan JJ (1991) Sulfur deficiency ifluences vegetative growth, chlorophyll and element concentrations and amino acids of pecan. *J Am Soc Hotri Sci* **116:** 974—980

Hunashikatti MG, Channal HT, Sarangamath PA, Manjunathaiah HM and Dharmatti PR (2000) Effect of sulphur and molybdenum on yield and quality of cabbage. *Fert News* **45:** 53—55

Imsande J (1999) Iron-sulfur clusters: formation. Perturbation and physiological functions. *Plant Physiol Biochem* **37:** 87—97

Jaggi RC and Sharma RK (1999) Sulphur-phosphorus interaction in raya (Brassica juncea var. Varuna) in acid Alfisols of western Himalaya. *Tropical Agric* **76:** 157—163

Joshi DC, Seth SP and Parekh BL (1973) Studies on S and P uptake by mustard. *J Indian Soc Soil Sci* **21:** 167—172

Kamprath EJ, Nelson WL and Fitts JW 1956. The effect of pH, sulphate, and phosphate concentrations on the adsorption of sulphate by soils. *Soil Sci Soc Am Proc* **20:** 463—466

Knaff DB, Hirasawa M, Ameyibors E, Fu W and Johnson MK (1991) Spectroscopic evidence for a (3Fe-4S) cluster in spinach glutamate synthase. *J Biol Chem* **266:** 15080—15084

Kumar M, Chattopadhyay TK, DasDK, Munsi PS and Kumar M (1997). Interaction between sulphur and zinc on the yield and yield attributes of onion (*Allium cepa* L.). *J Interacademicia* **1:** 295—300

Kumar PA, Parry MAJ, Mitchell RAC, Ahmad A and Abrol YP (2003) Photosynthesis and nitrogen use efficiency. In: Foyer C and Noctor G (eds), Photosynthetic Nitrogen Assimilation and Associated Carbon Metabolism. Advances in Photosynthesis series. Kluwer Academic Publishers, Dordrecht (in press)

Kundu S (1984) Availability and utilization of P by wheat green gram–rice cropping sequence as affected by some anions and phosphate solubilizing bacteria. Ph.D. Thesis, Indian Agricultural Research Institute, New Delhi

Lakkineni KC and Abrol YP (1992) Sulphur requirement of rapeseed-mustard, groundnut and wheat : A Comparative Assessment. *J Agron Crop Sci* **169:** 281—285

Lawlor DW, Kontturi M and Young AT (1989) Photosynthesis by flag leaves of wheat in relation to protein, ribulose *bis*phosphate carboxylase activity and nitrogen supply. *J Exp Bot* **40:** 43—52

Mahler RJ and Maples RL (1987) Sulphur fertilization of wheat grown in Arkansas. Bulletin of Arkansas Agricultural Experiment Station No. 906, pp 23

Malewar GU and S Ismail (1997) Sulphur in balanced fertilization in western India. Proceedings of the TSI/FAI/IFA Symposium on Sulphur in Balanced fertilization, New Delhi, pp S14—20

Marok AS and Dev G (1980) Phosphorus and sulphur interrelationship in wheat. *J Indian Soc. Soil Sci* **28:** 184—188

McGrath SP and Zhao FJ (1996) Sulphur uptake, yield response and the interactions between N and S in winter oilseed rape (*Brassica napus*). *J Agric Sci* (Cambridge). **126:** 53—62

Metson AJ and Blakemore LC (1978) Sulphate retention by New Zealand soil in relation to the competitive effect of phosphate. *NZ Agric Res* **21:** 243—253

Morden G, Soper R, Huzel V and Swan M (1986) The effect of thiosulphate on phosphorus availability and uptake by plants. *J Plant Nut* **9:** 1315—1321

Naphada GD and Mutalka VK (1984) Effect of phosphorus fertilizer in Saurashtra soil. Effect of sulphur and phosphatic fertilizers on the growth of groundnut and maize. *Saurashtra J Agric Soil* **7:** 5—10

Nayak GS and Dwivedi AK (1990) Interaction of sulphur and phosphorus fertilizers on yield and nutrient composition of faba bean grown on black soils of central India. *FABIS Newsletter* (ICARDA) **27:** 13—15

Noble JC and Kleining CR (1971) Attributed the positive interaction of phosphorus and gypsum to the increase in available water resulting from improved infiltration by applied gypsum. *Aust J Exp Agric Animal Husbandry* **11:** 53—57

Pasricha NS and Sparks DL (1990) Kinetics of adsorption of phosphate and sulphate in soil at two pH. *Soil Sci* **150:** 750—709

Pezzarossa, B, Piccotino D, Shennan C and Malorgio F (1999) Uptake and distribution of selenium in tomato plants as affected by genotype and sulphate supply. *J Plant Nutr* **22:** 1613—1635

Prasad R (1998) Fertilizer urea for food security, health and environment. *Curr Sci* **75:** 677—683

Prasad R, Prasad US and Sakal R (1996) Effects of potassium and sulphur on yield and quality of sugarcane grown in calcareous soils. *J Pot Res* **12:** 29—38

Prosser IM, Purves JV, Saker LR and Clarkson DT (2001) Rapid disruption of nitrogen metabolism and nitrate transport in spinach plants deprived of sulphate. *J Exp Bot* **52:** 113—121

Purakayastha TJ and Nad BK (1998) Effect of sulphur, magnesium and molybdenum on mustard (Brassica juncea L.) and wheat (*Triticum aestivum* L.) yield and uptake of macronutrients. *Indian J Plant Physiol* **3:** 112—115

Randall PJ and Wrigley CW (1986) Effect of sulfur supply on yield, composition and quality of grain from cereals, oilseed and legumes. *Adv Cereal Sci Tech* **8:** 171—206

Randhawa PS and Arora CL (2000) Phosphorus-sulphur interaction effects on dry matter yield and nutrient uptake by wheat. *J Indian Soc Soil Sci* **48:** 536—540

Rathee OP and Chahal RS (1977) Effcet of P and S application on the yield and chemical composition of groundnut in Ambala soils. *Haryana Agric Univ J Res* **VII:** 173—177

Razmjoo K and Henderlong PR (1997) Effect of potassium, sulfur, boron, and molybdenum fertilization on alfalfa production and herbage macronutrient contents. *J Plant Nutrition* **20:** 1681—1696

Reneau RB Jr, Bran DE and Donohue SJ (1986) Effect of sulphur on winter wheat grown in the coastal plain of Virginia. *Commu Soil Sci Plant Anal* **17 :** 149—158

Reuveny Z, Dougall DK and Trinity PM (1980) Regulatory coupling of nitrate and sulphate assimilation pathways in cultured tobacco cells. *Proc Natl Acad Sci USA* **77:** 6670—6672

Sachdev MS and Deb DL (1990) Nitrogen and S uptake and efficiency in the mustard - moong - maize cropping systems. *Fert News* **35:** 49—55

Sairam RK, Till AR and Blair GJ (1995) Effect of sulfur and molybldenum levels on growth, nitrate-assimilation, and nutrient conctent of *Phalaris*. *J Plant Nutr* **18:** 2093—2103

Sakal R, Sinha RB and Singh AP (1996) Research Bulletin on "Twenty five years of research on micro and secondary mutrients in soils and crops of Bihar". Department of Soil Sciences, RAU Pusa, Bihar, pp 1—98

Scherer HW (2001) Sulphur in crop production. *Euro J Agron* **14:** 81—111

Schnug E and Hankelaus S (2000) Significance of interactions between sulfur and nitrogen supply for growth and quality of crop plants.In: Brunold C, Rennenberg H, De Kok LJ, Stulen I and Davidian JC (eds), Sulfur Nutrition and Sulfur assimilation in Higher Plants: Molecular, Biochemical and Physiological Aspects, pp 345—347. Paul Haupt, Bern

Sharma MP and Jalali VK (2001) Response of rainfed mustard (*Brassica juncea*) to sulphur and phosphorus on Inceptisol in mid hill-intermediate zone of Jammu and Kashmir. *Indian J Agric Sci* **71:** 195—196

Singh KS and Bairathi RC (1980) A study on the sulphur fertilization of mustard in the semi-arid tract of Rajasthan. *Ann Arid Zone* **19:** 197—202

Singh MV (2000) Sulphur Management for Oilseed and Pulse crops. *Indian Instt Soil Sci Bull* **3:** 1—54

Singh MV (2001) Importance of sulphur in balanced fertilizer use in India. *Fert News* **46:** 13—35

Singh V and Rathore SS (1994) Effect of applied potassium and sulphur on yield, oil content and their uptake by linseed. *J Pot Res* **10:** 407—410

Singh R, Sharma PR, Singh M and Sharma R (1997) Phosphorus, sulphur and zinc interactions in barley (Hordeum vulgare L.) - concentration and uptake of sulphur and zinc. *Crop Res* (Hisar) **14:** 1 45—54

Spencer K and Freney JR (1980) Assessing of sulfur status of field grown wheat by plant analysis. Agron J **72:** 469—472

Stewart BA and Porter LK (1969) Nitrogen-sulphur relationship in wheat (*Triticum aestivum* L.), corn (*Zea mays*), and beans (*Phaseolus vulgaris*). *Agron J* **61:** 267—271

Sunarpi and Anderson JW (1997) Effect of nitrogen nutrition on the export of sulphur from leaves in soybean. *Plant Soil* **188:** 177—187

Suzuki A, Rothstein S (1997). Structure and regulation of ferredoxin- dependent glutamate synthase from *Arabidopsis thaliana*. *Euro J Biochem* **243:** 708—718

Umar S and Bansal SK (1995) Potassium and sulphur nutrition of rice and optimum K^+ content in tissue for crop production. *J Potassium Res* **11:** 307—318

Umar S, Debnath G and Bansal SK (1997) Groundnut pod yield and leaf spot disease as affected by potassium and sulphur nutrition. *Indian J Plant Physiol* **2:** 59—64

Vanoni MA, Fischer F, Ravasio S, Verzotti E, Edmondson DE, Hagen WR, Zanetti G and Curti B (1998) The recombinant α-subunit of glutamate synthesis: spectroscopic and catalytic properties. *Biochemistry* **37:** 1828—1838

Velayutham M (1997) Fertilizer use in the coming decades – some issues on research, extension, and policy matters. In: JS Kanwar and JC Katyal (eds) Plant Nutrient Needs, Supply, Efficiency and Policy Issue: 2000-2025, pp 327—329. National Academy of Agricultural Sciences, New Delhi

Venkatalakshmi G (1987) Utilization of phosphorus by some pulse and oilseed crops as influenced by S, Ca and inoculation with P solubilizing microorganisms. Ph.D. Thesis, Indian Agricultural Research Institute, New Delhi

Virmani SM and Gulati HC (1971) Effects of sulphur on the response of Indian mustard (*Brassica juncea* L. Czern and Coss) to phosphorous fertilization. *Indian J Agric Sci* **41:** 143—146

Wallace A (1990) Interaction of two parameters in crop production and in general biology: Sequential additivity, synergism, antagonism. *J Plant Nutr* **13:** 327—342

Zhao FJ, Hawkesford MJ and McGrath SP (1999) Sulphur assimilation and effects of yield and quality of wheat. *J Cereal Sci* **30:** 1—17

Zhao FJ, Withers PJA, Salmon SE, Evans EJ, Shewry PR and McGrath SP (1997) Sulphur nutrition: An important factor for the quality of wheat and rapeseed. *Soil Sci Plant Nutr* **43:** 1137—1142

12. ACKNOWLEDMENTS

Grants received from Department of Science and Technology (SERC-Fast Track Young Scientist Scheme) and Technology Mission on Oilseeds, Pulses and maize (TMOP&M), Council of Scientific and Industrial Research are gratefully acknowledged.

13. AFFILIATIONS

Department of Environmental Botany (AA, MI) and Centre for Biotechnology (MZA, NK, IK, AJ), Faculty of Science, Hamdard University, New Delh-110062, India

For correspondence — *MZ Abdin*: mzabdin@rediffmail.com

K. HAQ AND M.ALI

BIOLOGICALLY ACTIVE SULPHUR COMPOUNDS OF PLANT ORIGIN

Abstract: Sulphur containing secondary compounds of plant origin are known to have therapeutic utility. Innumerable clinical aspects of these compounds are reported. These range from anticancer, anti-HIV, antifungal to several other effects. Some sulphur metabolites have shown very promising results. Compounds like organosulphur found in family Alliaceae have cardiovascular and antineoplastic effects. Sulphur containing organoselenium compounds are reported to have antineoplastic effect. Sulphated polysaccharides showed anti-HIV activity. Sulphoraphane and glucosinolates have potent anticancer activity.

1. INTRODUCTION

The use of plants as a drug by human beings has been prevalent since time immemorial (Anthony, 1997). These have played an important role in nearly every culture on earth. During the latter half of 20th century, natural compounds were replaced gradually by synthetic compounds. However, use of the latter has been reported to have many side effects. One man's Asprin is another man's Peptic ulcer (Kitty, 1988). It has forced man to change its attitude towards compounds of natural origin. The recent popularity in use of herbal medicines can be tied to the belief that herbs can provide some benefit over and above allopathic medicine and allow users to feel that they have some control in their choice of medication (Wargovich et al., 2001). It has been estimated that up to 50% of the prescriptions presently dispensed in the USA may contain one or more drugs of natural origin (Douglas, 2000). In a recent statistical survey, it has pointed out that when compared with libraries of synthetic compounds, natural products offer the prospects of discovering a greater number of compounds with sterically more complex structures (Henkel et al., 1999).

This chapter discusses the biological profile of some of the sulphur compounds of plant origin, and rationale for their possible use in the form of medicine.

2. ORGANOSULPHUR COMPOUNDS

The antibiotic, anticancer, antithrombic, cholesterol lowering, and other beneficial health effects are widely touted in the popular and scientific/medical press associated with consumption of garlic (*Allium sativum*), onion (*A. cepa*) and other *Allium* plants such as leek (*A. possum* L.), shallot (*A. ascalonicum* auct), chive (*A. schoenopsasum* L.) and scallion (*A. fistulosum* L.) (Block, 1988). The health benefits associated with consumption of these plants are attributed to compounds found in the intact plants, flavorants formed on cutting or crushing the plants parts, substances derived from

Yash P. Abrol and Altaf Ahmad (eds.), Sulphur in Plants, 375—386.

further reaction of the flavorants and metabolic degradation products of these compounds. The details of these compounds are given below.

2.1 Allicin and related compounds

Allicin is S-allyl 2-propene sulphinothioic acid ester, extracted from garlic with ethanol at room temperature (Brodnitz et al., 1971). Allicin is formed from alliin by an enzyme allinase. Allicin itself is a precursor for array of sulphur compounds (Cavallito et al., 1945).

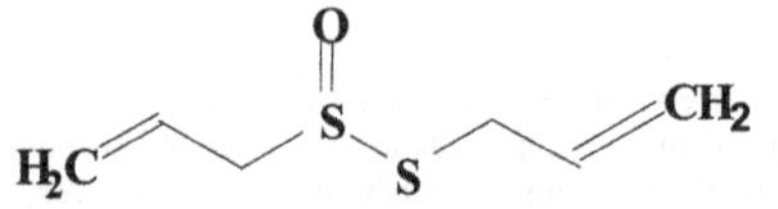

Allicin

In addition to allicin, several other volatile compounds from garlic extract have been identified as methyl allyl disulphide, diallyl disulphide, dimethyl trisulphide, allyl methyl trisulphide, diallyl trisulphide and sulphur dioxide (Freeman and Whenham, 1975). Allicin decomposes completely at 20°C within 20 h, giving diallyl disulphide as the major product and diallyl trisulphide, diallyl sulphide, sulphur oxide and many other compounds.

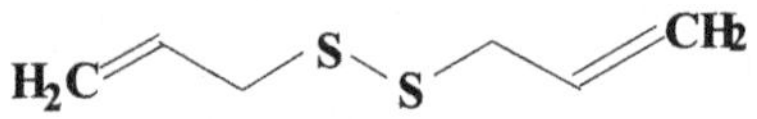

Diallyl disulphide

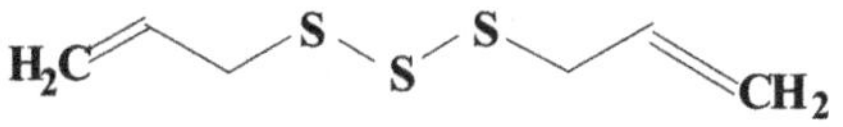

Diallyl trisulphide

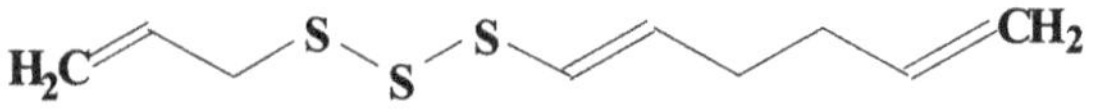

Allyl 1,5-hexadienyltrisulphide

Other compounds isolated from garlic are allyl 1,5-hexadienyltrisulphide and E- and Z- isomers of 4,5,9-trithiododeca-1,6,11-trien-9-oxide (ajoene) (Ledezma et al., 1996).

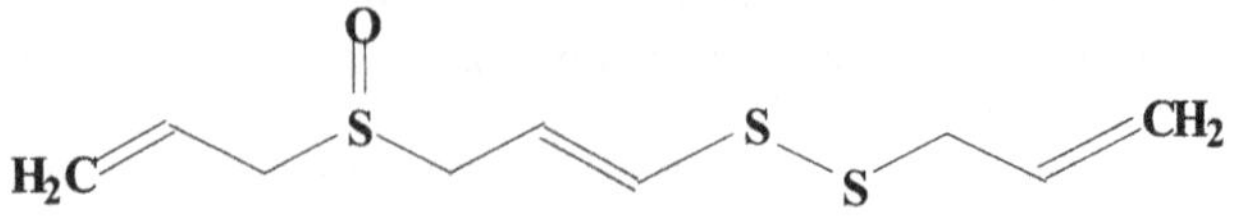

[E]-Ajoene

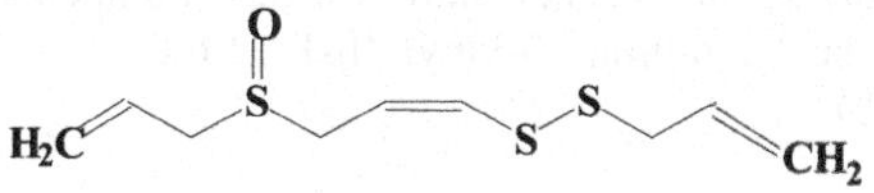

[Z]-Ajoene

2.2 Alliin and related compounds

Alliin is an amino acid derivative. Cysteine sulphoxide fraction of garlic gives 85% of alliin along with 2% S-propylcysteine sulphoxide and 13% methyl cysteine sulphoxide (Freeman and Whenham, 1975).

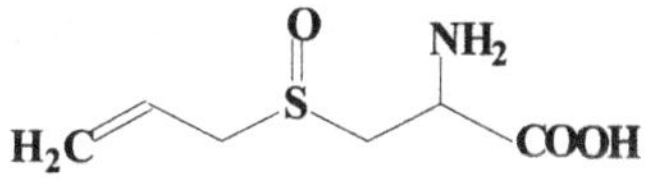

S-Allylcysteine-S-oxide (Alliin)

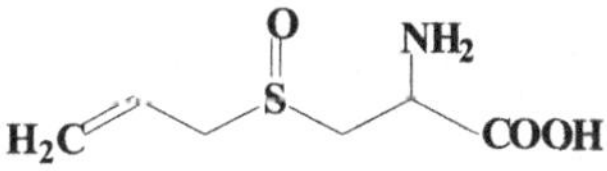

S-Allylcysteine-S-oxide

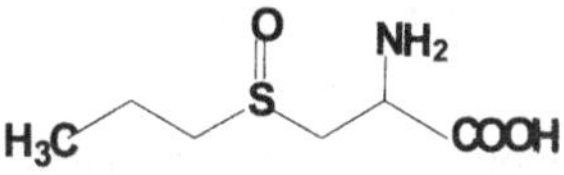

S-Methylcysteine sulphoxide

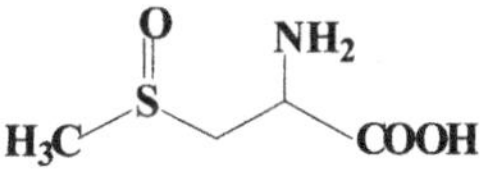

S-Propyl cysteinesulphoxide

Allinase activity on the S-substituted cysteine sulphoxide fraction yields various symmetrical or asymmetrical sulphothioic acid esters and more importantly S-allyl-L-cysteine (Suzuki et al., 1961).

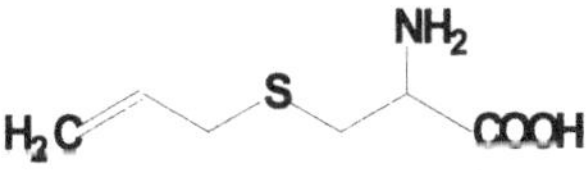

S-Allyl-L-Cysteine

Decomposition of allicin results into various compounds like 1,2-dithiin compounds, 3-vinyl-6h-1,2-dithiin, 3-vinyl-4h-1, 2-dithiin, 2-vinyl-1, 3-dithiin (Kameoka et al., 1978)

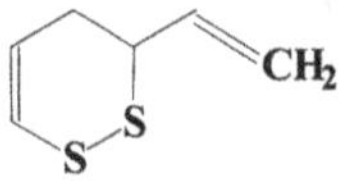

3-Vinyl-4h-1, 2-dithiin

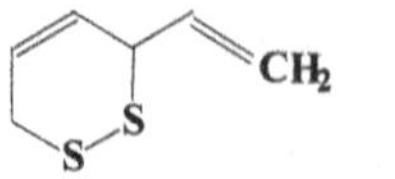

3-Vinyl-6h-1, 2-dithiin

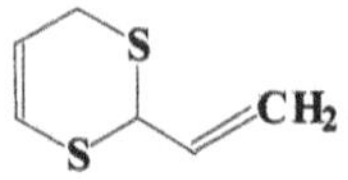

2-Vinyl-1,3-dithiin

2.3 Other compounds

Some sulphur containing acidic peptides such as γ-glutamyl-S methyl-cysteine and its sulphoxide derivative, γ-glutamyl-S-(2-carboxy-propyl)-cysteinyl-glycine, γ-glutamyl-S-allyl-cysteine, γ-glutamyl-S-propyl cysteine and γ-glutamyl-phenylalanine were also found in garlic (Suzuki et al., 1961).

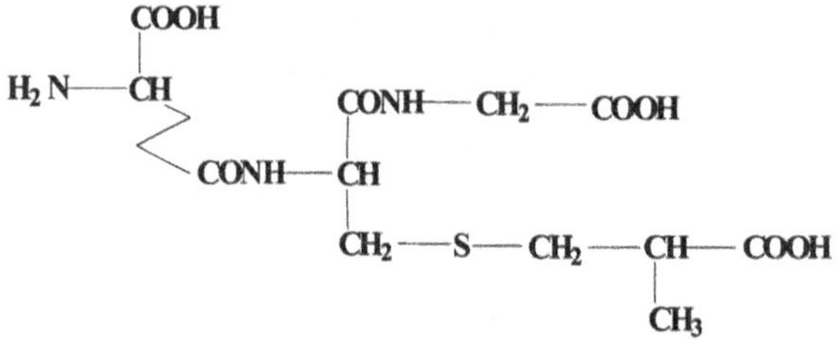

γ-Glutamyl-S-[2-Carboxy-Propyl]Cysteinyl-Glycine

Other cyclic sulphur compounds separated from benzene fraction of steam volatile oil from garlic are trithiolane derivatives, cis and trans 3-methyl-5-ethyl 1,2,4-trithiolane, and cis and trans-3,5-diethyl 1,2,4-trithiolane.

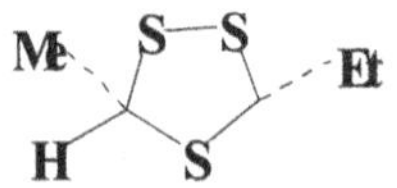

Cis-3-Methyl-5-Ethyl-1, 2,4trithiolane

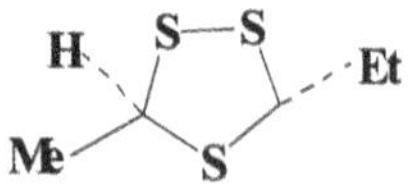

Trans-3-Methyl-5-Ethyl-1, 2,4trithiolane

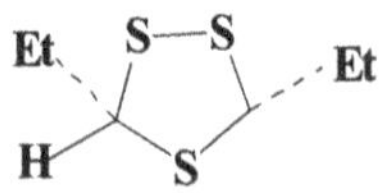

Cis-3,5 Diethyl-1, 2,4-Trithiolane

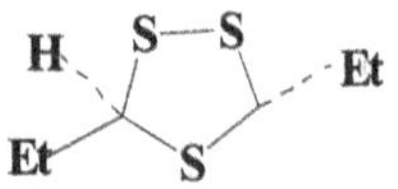

Trans-17.3, 5-Diethyl-1, 2,4trithiolane

2.4 Biological Role

Garlic constituents exhibit various biological activities such as anti-bacterial, anti-fungal, anti-viral, anti-tumor and anti-diabetic effects.

(i) Cardiovascular effects

The greatest interest of organosulphur compounds extracted from garlic has been focussed on their anti-cholesterolemic effects and anti-thrombotic activities. Applied orally, garlic and its ethanol extract reduced plasma and liver cholesterol level in male rat. It decreased serum triglycerides and decreased total serum cholesterol (Augusti, 1977). Garlic oil preparations was also found to decrease serum lipoprotein and cholesterol (Berthold et al., 1998). Allicin, methyl allyl trisulphide, ajoene, diallyltrisulphide, 2-vinyl-1,3-dithiin and allyl 1,5-hexadienyl trisulphide were found to inhibit platelet aggregation (Block et al., 1986). Aggregation of platelet induced by ADP or arachidonic acid was also inhibited when incubated with garlic. Thromboxane B_2 synthesis was suppressed completely (Ariga et al., 1981). Bordia et al. (1998) reported beneficial effect of garlic on lipids, glucose, fibrinogen and fibrinolytic activity.

Placebo-controlled trials that tested the effectiveness of oral garlic preparation in lowering cholesterol in humans have shown that total cholesterol level decreased by a statistically significant scale. Boullin (1981) demonstrated the platelet inhibitory effect of garlic and attributed the platelet inhibition to methylallyl trisulfide (MATS).ϒ-Glutamyl S-allyl cysteine inhibits blood pressure regulating angiotensin converting enzyme (ACE).

(ii) Gastrointestinal effects

Garlic relieved abdominal distension/discomfort, belching, flatulence (Ross, 1998). Ethanol-chloroform extract of fresh bulb antagonized synthesis of acetylcholine and prostaglandin E. Recent epidemiological studies suggested correlation between consumption of fresh garlic and decreased risk of gastro intestinal cancer and occurrence of colorectal polyps (Beuchat, 1994).

(iii) Anti-microbial activity

Caporaso et al. (1982) studied antifungal activity of human urine and serum from five volunteers who consumed 10-25 ml of fresh extract of garlic. It exhibited anti fungal effect on yeast and mycelial fungi species.

Freshly cut garlic applied topically acted more quickly than tolnaftate against *M. crisis* lesions on the arm of young women (Rich ,1982). Ajoene 0.4% cream was used in treatment of *Tines redis* resulting in a culture-negative cure (Ledezma et al., 1996). Garlic and onion oil were found to inhibit toxin production by *Clostridium botulinum* (DeWit et al., 1979).

(iv) Anti-neoplastic effects

Hu et al. (1997) completed a study to evaluate garlic ability to retard chemical carcinogenesis induced by benzo-[a]pyrene (BP). The BP's carcinogenic metabolites can be inactivated in vivo by conjugation with glutathione by enzyme glutathione S-transferase (GST). The investigator found that garlic organosulphides were capable of inducing GST expression in murine liver and fore-stomach as well as of preventing stomach neoplasia. Takada et al. (1994) tested effect of organosulphur components of garlic on nitrosamine-induced Hepatocarcinogenesis in rats and found no satisfactory results. They proposed that this was a result of organosulfide administration during the promotion phase rather than prior to nitroso urea administration

(v) Anti-viral effects

Garlic extract containing allicin, diallyl disulphide and diallyl trisulphide possess antiviral activity in vitro against influenza B virus and herpes simplex type I virus (Koch and Lawson, 1996).

3. ORGANOSELENIUM COMPOUNDS

Spare and Virtanen (1964) have reported selenoamino acids selenocystine [HOOC-CH $(NH_2CH_2Se)_2$] and selenomethionine [HOOC-CH $(NH_2CH_2CH_2Se)_2$] in onion on the basis of radio isotope studies. Various other organoselenium compound were discovered on the basis of various analytical techniques including GC-atomic emission detection (GC-AED), HPLC-inductively coupled Plasma – MS (HPIC – ICP – MS), and HPLC – ICP – MS (Bird et al., 1998). These compounds are Dimethyl selenide (MeSeMe), methane sulphenoselenoic acid methyl-ester (MeSe SMe), dimethyl diselenide (MeSeSeMe), (methylthio) selenide $(MeS)_2$ Se, 2-propene sulphenoselenoic acid methyl ester (MeSeSAll), and (allylthio) (methylthio) selenide (MeS SeSAll). These organoselenium compounds containing sulphur are mainly found in garlic, onion and yeast. Identification of natural organoselenium compounds is of considerable current interest in view of the discovery that Se-enriched garlic and yeast possess cancer preventive properties (Ip et al., 1992).

Lu et al. (1996) have observed that the extract of selenium enriched garlic and selenium-methyl selenocysteine caused growth inhibition, G_1 phase arrest, DNA double strand break in neoplastic murine mammary cells. Selenite produced single stranded break (an indicator of genotoxicity) and S/G_2-M phase arrest in addition to growth inhibition. Recent studies have suggested that selenocysteine provides chemoprotection against chemically induced carcinogenesis possibly by phase II Biotranformation enzyme (Hoen et al., 2002). Selenomethionine, present in garlic and yeast, have been reported to induce DNA repair and protect cell from DNA damage (Seo et al., 2002). This suggests their usefulness in prevention of cancer.

4. SULPHATED POLYSACCHARIDES

A large group of compounds, which consistently show activity in various in vitro HIV assays, are polymeric anions. The most common of these compounds are sulphated polysaccharides (Wagner, 1998). Naturally occurring sulphated polysaccharides are very common in marine algae. Carrageenins (*Condrus crispus* Linn) with molecular weights ranging from 20-500 kDa have been shown to be very potent in HIV *in vitro* assays. It is linear polysaccharide of alternating β-(1,3) and β-(1,4) galactose units (Neushul,1990). Curdlan sulphate, prepared by sulphation of the linear β-(1,3)-glucan of *Dumontiaceae* sp., has been reported to be very effective in blocking in vitro HIV effectively with reduced toxicity and low anticoagulant effect, a general side effect associated with other sulphated polysaccharides (Lednicer and Snader, 1998). Sulphation of the branched β-(1,3)-glucon from *Lentinus edodes* (fungus) and sulphated product from the immunomodulatory debranched β-(1,3)-glucon of *Schizophyllan commune* (fungus) also showed in vitro anti-HIV activity (Kornilaeva et al., 1995).

Terrestrial plants also produce numerous polysaccharides. *Viola yedoensis,* a Chinese medicinal herb, has been reported to inhibit the HIV without inducing interferon or inactivating extracellular virus (Nagan et al., 1988). This plant contains a sulphated polysaccharide of molecular weight of 10-15 kDa and composed of galactose and glucose. Nagan et al. (1988) also reported another Chinese medicinal herb, *Alternanthera philoxeroides*, in several viral systems including HIV. The activity of the aqueous extract from this drug increased exponentially. At higher concentration, reverse transcriptase activity has also been found to be inhibited. The active constituent was sulphated polysaccharide.

Lichens produce numerous polysaccharides. Ge-3-5 (a polysaccharide), a 200 KDa β-(1,6)-glucan and sulphated pustulan analogue from *Umbilicaria esculenta* has been reported to be very potent in *in vitro* HIV infected Molt-4 cells (Kornilaeva et al., 1996).

Several cholic acid sulphates have been reported to have shown reasonable anti-HIV activity with relatively lesser unwanted anticoagulant effect. Among the cholic acid sulphates specially taurolithocholic acid-3-sulphate and lithocholic acid 3-sulphate inhibited HIV replication at dose of 100 μg ml^{-1} and inhibited HIV adsorption to MT-4 cells.

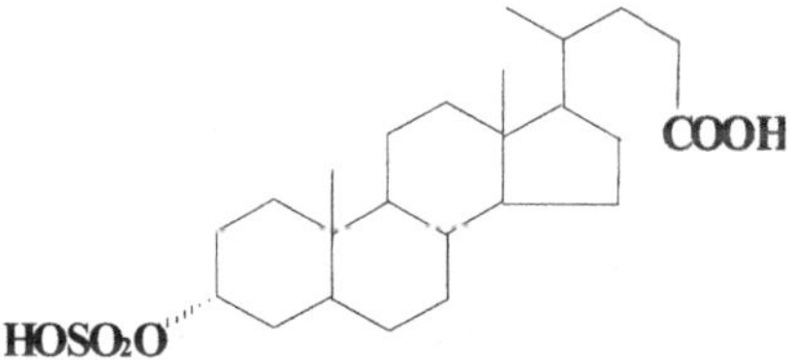

Lithocholic acid 3-sulfate

A new compound, sulphoquinovosyl-diacyglycerol, has been isolated from the cultured cyanobacteria, *Lyngbya lagerheimii* and *Phormidium tenue.* It showed positive results in HIV-1 infected T-lymphocytes in humans (Lednicer and Snader, 1998).

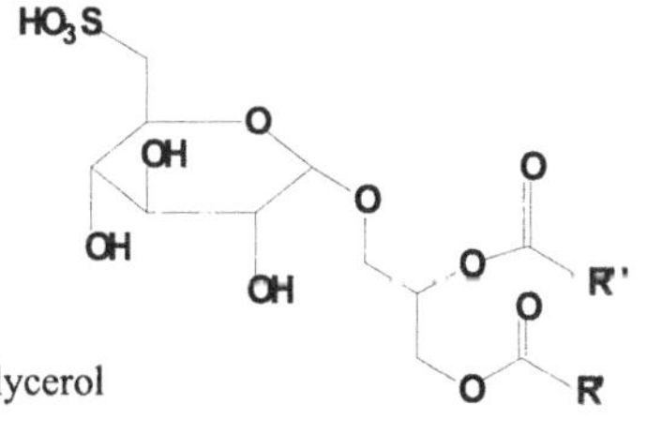

Sulfoquinovosyl-diacyglycerol

The overall results obtained were promising but it is needed to be studied further in detail to prove the worthiness of this class of compounds.

5. SULPHORAPHANE AND RELATED COMPOUNDS

Sulphoraphane [1-isothiocyanato-4-(methyl-phenyl) butane], isolated from one variety of broccoli, has been reported to be a major and very potent inducer of phase 2 detoxification enzyme in murine hepatome cells in culture (Zhang et al., 1994). Phase 2 enzyme induction is often associated with reduced susceptibility of animals and their cells to the toxic and neoplastic effects of carcinogens and other electrophiles. The analogues of sulphoraphane are exo-2-acetyl-exo-6-isothiocyanatonorbornane, endo 1-acetyl-exo-6-isothiocyanatonorbornane and exo-2-acetyl-exo-5-isothiocyanatonorbo-rnane (Zhang et al., 1993)

Sulphoraphane

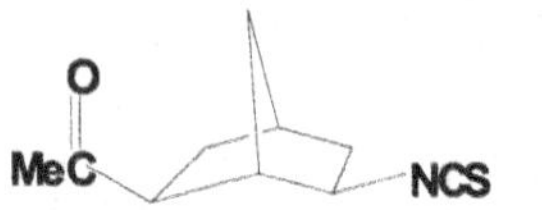

Exo-2-acetyl-exo-6-isothiocyanatonorbornane

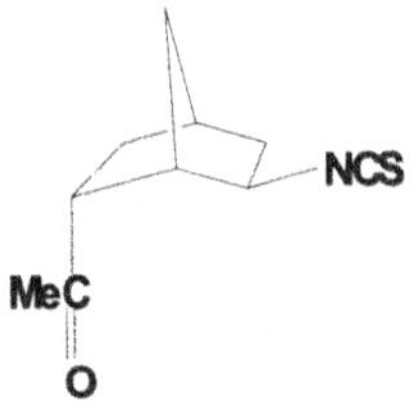

Endo 1-acetyl-exo-6-isothio cyanatonorbornane

Exo-2-acetyl-exo-5- isothio cyanatonorbornane

Sulphoraphane is of interest because of four reasons: (a) it occurs naturally in widely consumed vegetables; (b) it is a potent inducer; (c) it is a mono-functional inducer; and (d) it elevates the phase 2 detoxification enzyme without changing the synthesis of cytochrome P-450. The experiment was performed on the Sprague-Dawley rats treated with single dose of 9,10-dimethyl-1,2-benzanthracene (DMBA).The trial was designed to study the sulforaphane and its three synthetic analogue for blocking formation of mammary tumor (Zhang et al., 1994). Administration of sulphoraphane or its three synthetic analogues reduced the incidence, multiplicity and weight, and delayed the development of mammary tumor evoked by a single dose of DMBA in female Sprague-Dawley rats.

6. GLUCOSINOLATES

Glucosinolates are organic anions with S-D-thioglucose and sulphonated oxime moieties, and specific side group structure that distinguish one glucosinolates from another (for details, see Halkier, 2002 this book). Although the primary biological function of glucosinolates in plants is unknown, glucosinolate breakdown products are proposed to act as allelochemicals and play a role in plant defense against herbivores, pest and pathogens. As components of food for humans and feed for livestock, the biological activities of glucosinolates hydrolyzed products have generally considerable toxicological and pharmacological interest. Depending on glucosinolates composition and on the prevalence of hydrolyzed products, consumption of glucosinolates by mammals has been linked with goitrogenic effects (thiocyanates) or with a reduced risk of developing cancer (isothiocyanates) in experimental animals (Shapiro et al., 2001)

Natural isothiocyanates derived from aromatic and aliphatic glucosinolates are effective chemoprotective agents that block chemical carcinogenesis and prevent several types of cancer in rodent models (Gross et al., 2000). Mechanistic studies have shown that isothiocyanates target mammalian Phase 1 and Phase 2 drug-metabolizing enzymes and their coding genes, resulting in decreased carcinogen-DNA interactions and in increased carcinogen detoxification. For example, the methionine-derived isothiocyanate, sulforaphane, inhibits Phase 1 enzyme-mediated activation of pro-carcinogens, induces Phase 2 detoxification enzymes such as quinone reductase (QR) and glutathione-S transferase in hepatoma cells, and blocks mammary tumor in rats.

Induction of Phase 2 detoxification enzyme (e.g., glutathione transferase, epoxide hydrolase, NAD(P)H: quinone reductase and glucoronosyl transferases) is a powerful strategy for achieving protection against carcinogenesis, mutagenesis and other form of toxicity of electrophiles and reactive form of oxygen. Edible plants belonging to the family Brassicaceae and genus *Brassica* contain substantial amount of isothiocyanates (as glucosinolate precursor) some of which (e.g., sulphoraphane or 4-methylsulphenyl butyl isothiocyanate) are very potent inducer of Phase 2 enzymes. Unexpectedly, 3 day old sprout of cultivar of Crucifer contain 10-100 times higher level of glucoraphanin (the glucosinolate of sulforaphene) than the corresponding mature plants. These vegetables aid in the detoxification of carcinogens such as aflatoxins and polybromobiphenyl. They appear to protect against several types of cancer including rectal and colon cancer and others. Sulphoraphane is the most powerful natural inducer of chemoprotective enzyme so far reported and has become a metabolic target of breeding strategies to enhance the anti-carcinogenic potency of cruciferous vegetables (Beecher 1994). Two important mechanisms by which the effect is exerted may be antioxidant potential and altered biotransformation capacity. Result suggested that detoxification enzymes were induced and a decrease in the rate of oxidative DNA damage occurred upon consumption of Brussel Sprout (Gao et al., 2001).

Four compounds, commonly found in human diet, e.g., allyl isothiocyanate (AITC), phenyl isothiocyanate (PETC) and their parent glucosinolates sinigrin and gluconasturtin were tested for cytotoxic and genotoxic effects in a Chinese hamster ovary cell line (CHO). Isothiocyanate was found to be cytotoxic more than glucosinolate. while PITC was found inducing aberration and (SCES) sister chromatid exchange (Musk et al., 1995).

Numerous studies have indicated that hydrolytic products of at least three glucosinolates, 4 methyl sulphinyl butyl (glucoraphanin), 2 phenyl ethyl (gluconasturtiin) and 3-indolyl methyl (glucobrassicin) have anticancer activity. Indole-3-carbinol, a metabolite of glucobrassicin, has shown inhibitory effect in studies of human breast and ovarian cancer (Stoewsand ,1995). In another study Sinigrin, administered after dimethylhydrazine, suppresses induction of aberration crypt foci (ACF) (Smith et al., 1998).

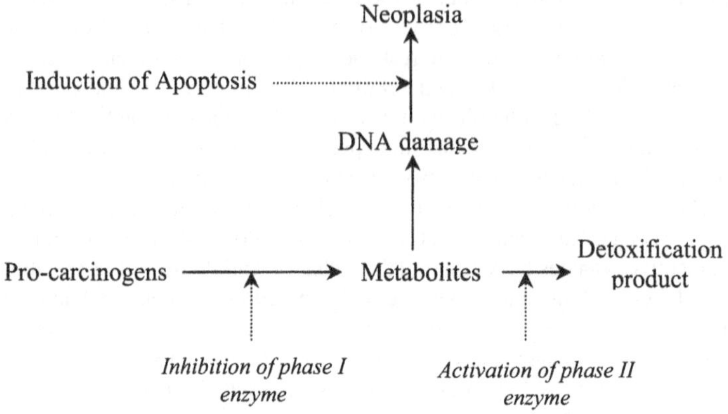

Figure 1. Targets of isothiocyanates in cancer prevention

7. MISCELLANEOUS

Methyl sulphonyl methane is naturally occurring compound and found in terrestrial plants as well as microscopic plankton in ocean (Richmond, 1986). A solution of methyl sulphonyl methane can be used in burn, rheumatoid arthritis, disc problems in back, acute injuries, tendonitis, bursitis and muscle cramp as a mean to provide extra sulphur in supplement form. It reduces pain and scarring.

Asafoetida (*Ferula foetida*), a member of Apiaceae family, is recognized by its overwhelming odour. The gum resin consists of the milky sap obtained from an incision of the green matured root. The major biochemical agent is an organic sulphur compound found as a part of essential oil which makes at very similar to the essential oil of garlic (allyl, allyl persulphide, and two terpenes). Sulphur compounds in the oil may protect against fat-induced hyperlipidemia. Two double-blind studies reported asafoetida as useful for irritable bowel syndrome (just below 5% significance level in one, near 1% in the other).

Biological compounds like polythionates, thiosulphonates and persulphides can be employed to cure cyanide poisoning. Investigation of extract of plant containing cyanogenic glycoside (e.g. cassava) is reported to cause cyanide poisoning in animals (Bhattacharya, 2000).

9. REFERENCES

Anthony A (1997) Drugs of Natural Origin: Economic and Policy Aspect of Discovery, Development and Marketing. The Pharmaceutical Press, New York

Ariga T, Oshiba S and Tamada T (1981) Platelet aggregation inhibitor in garlic. *Lancet* **1**: 150—151

Augusti KT (1977) Hypocholesterolemic effect of garlic, *Allium sativum* L. *Indian J Exp Biol* **15**: 489—490

Beecher CWW (1994) Cancer prevention properties of varieties of *Brassica oleracea*. *Am J Clin Nutr* **59**: 11166—11705

Berthold HK, Sudhop T and Von Bergmann K (1998) Effect of garlic oil preparation on serum lipoproteins and cholestrol metabolism. A randomized controlled trial. *JAMA* **279**: 1900—1902

Beuchat LR (1994) Antimicrobial properties of spices and their essential oils, Natural Antimicrobial System and Food Prevention, pp 167--169, CAB International, Wallingford

Bhattacharya R (2000) Antidotes to cyanide poisoning: Present Status. *Indian J Pharmacology* **32**: 94—101

Bird SM, Uden PC, Tyson JF, Block E and Denoyer ER (1998) High performance liquid chromatography of selenoamino acids and organoselenium compounds speciation by inductively coupled plasma mass spectrometry. *J Chromatography* **12**: 785—788

Block E (1988) The organosulphur and organoselenium components of garlic and onions. In: Weyne RB (ed) Phytochemical, pp 129—141. Technomic Publisher Lancaster, Pennysylvania

Block E, Ahmad S, Catalfamo JL, Jain MK and Apitz-Castro R (1986) Antithrombotic organosulphur compounds from garlic. Structural mechanistic and synthetic studies. *J Am Chem Soc* **108**: 7045—7055

Bordia A, Verma SK and Srivastava KC (1998) Effect of garlic (*Allium sativum*) on blood lipid, blood sugar, fibrinogen, fibrinolytic activity in patients with coronary artery disease. *Prostagland Leukotr essen Fatty Acids* **58:** 257—263

Boullin DJ (1981) Garlic as a platelet inhibitor. *Lancet* **1**: 776—777

Caporaso N, Smith SM and Eng RHK (1982) Antifungal activity in human urine and serum after ingestion of garlic. *Antimicrob Agents Chemother* **23**: 700—702

Cavallito CJ, Bailey JH and Buck JS (1945) The antibacterial priniple of *Allium sativum*. III. Its precursor and essential oil of garlic. *J Am Chem Soc* **67**: 1032—1033

De Wit JC, Noterman S, Gorin N and Kampelmacher EH (1979) Effect of garlic oil or onion oil on toxin production by *Clostridium botulinum* in meat slurry. *J Food Protect* **42**: 222—224

Douglas A (2000) Pharmacognosy in 21st centuary. *J Pharmacy Pharmacology* **53**: 135—148

Freeman GG and Whenham RJ (1975) A survey of volatile components of some *Allium* species in terms of S-alk (en)yl-L-cysteine sulphoxides present as flavour procursor. *J Sci Food Agric*. **26**: 1989—1888

Gao X, Dinkova AT and Talalay P (2001) Powerful and prolonged protection of human retinal pigment epithelial cells, keratinocytes and mouse leukemia cells against oxidative danage. Indirect antioxidant effects of sulphoraphane. *Proc Natl Acad Sci USA* **98**: 15221—15226

Gross HB, Dalebout T, Grubb CD and Abel S (2000) Functional detection of chemopreventive glucosinolates in *Arabidopsis thaliana*. *Plant Sci* **159**: 265—272

Henkel T, Brunne RM, Muller H and Reichel F (1999) Statistical investigation into the structural complementarity of natural products and synthetic compounds. *Angew Chem Int* **38**: 643—647

Hoen T, Rooseboom M, Bijsterbosch MK, van Berkel TJ, Vermeulen NP and Commandeur JN (2002) Induction of glutathione-S-transferase mRNA level by chemopreventive selenocysteine-Se conjugate. *Biochem Pharmacol* **63**: 1843—1849

Hu X, Benson PJ, Srivastava SK, Xia H, Bleicher RJ and Zaren HA (1997) Induction of glutathione-S-transferase as a bioassay for the evaluation of potency of inhibitors of benzophyrene-induced cancer in a murine model. *Int J Cancer* **73**: 897—901

Ip C, Lisk DJ and Stoewsand GS (1992) Mammary cancer prevention by regular garlic and selenium-enriched garlic. *Nutr Cancer* **17**: 279—286

Kameoka H, Demizu Y, Iwase Y and Miyazawa M (1978) Structure of cyclic sulphur compounds from *Allium* genus. *Tennen Yuki Kagobutsu Toronkai Koen Yoshihu* **21:** 199—205

Kitty C (1988) A complete guide to maintaining health and treating illness with plants. Leopard Book, London

Koch HP and Lawson LD (1996) Garlic. The science and therapeutic application of *Allium Sativum* L. and related species. 2nd Ed PP 329, Williams and Wilkins, Baltimore

Kornilaeva GA, Makarova TV, Ganzazade AI (1995) Sulphated chitosan derivatives as HIV-infection inhibitors. *Immunology* **1**: 13—16

Kornilaeva GA, Mescheryakova DV, Ganzazade AI (1996) The V3 loop and C-terminus of gp120 HIV-1 are the target of the anti-viral action of sulphated polysaccharides. Abstract of 4th International Confernce of AIDS, cancer, and human retrovirus, pp 74, Petersburg, Russia

Ledezma E, Desousa L, Jorquera A, Sanchez J, Lander A, Rodriguez E, Jain MK and Apitz-Castro R (1996) Efficiency of ajoene an organosulphur derived from garlic in the short-term therapy of *Tinia pedis*. *Mycoses* **39**: 393—395

Lednicer D and Snader KM (1998) Sulfated polysaccharides. In: Wagner H. (ed) Economic Medicinal Plant Research, Vol. 5, pp 13—16, 217—218. Academy Press,Harcourt Bracl Jovanovich Publishers, London.

Lu J, Pei H and Ip C (1996) Effect of an aqueous extract of selenium enriched garlic on in vitro markers and in vivo efficacy in cancer prevention. *Carcinogenesis* **9**: 1903—1907

Musk SR, Smith TK and Johnson IT (1995) On the cytotoxicity and genotoxicity of allyl and phenethyl isothiocyanate and their parent glucosinolates sinigrin and gluconasturtiin. *Mutat Res* **348**: 19—23

Nagan F, Chang S and Tabba H (1988) Isolation, purification and partial characterization of an active anti-HIV compund from the Chinese medicinal herb *Viole yedoensio*. *Antiviral Res* **10**: 107—116

Neushul (1990) Antiviral carbohydrate from marine red algae. *Hydrobiologia* **204/205**: 99—104

Rich GE (1982) Garlic an antibiotic. *Med J Aust* **1**: 60

Richmond VL (1986) sulphur cycle of ocean plants. *J Nutr* **116**: 54—59

Ross IA (1998) Medicinal plants of the world. Chemical constituents, traditional and modern medicinal uses. *Allium sativum*. Humana Press. Totowa, NJ

Seo YR, Sweeney C and Smith ML (2002) Selenomethione induction of DNA repair response in human fibroblast. *Oncogene* **21**: 3663—3669

Shapiro, Fahay JW, Wade KL, Stephenson KK and Talalay P (2001) Chemoprotechtive glucosinolates and isothiocyantes of Broccoli sprouts: Metablosim and extraction in Humans. *Cancer Epidemiol Biomarkers Prevention* **10**: 501—508

Smith TK, Lund EK and Johnson IT (1998) Inhibition of dimethyl hydrazine-induced aberrant crypt foci and induction of apoptosis in rat colon following oral administration of glucosinolate sinigrin. *Carcinogenesis* **19**: 267—273

Spare CG and Virtanen AI (1964) Occurrence of free selenium-containing amino acids in onion (*Allium cepa*). *Acta Chem Scand* **18**: 280—282

Stoewsand GS (1995) Bioactive organosulphur phytochemicals in *Brassica oleraceae* vegetables. *Mutat Res* **348**: 19—23

Suzuki T, Sugii M, Kakimoto T and Tsuboi N (1961) Isolation of S-allyl-L-cysteine from garlic. *Chem Pharm* 9: 251—252

Takada N, Matsuda T and Otoshi T (1994) Enhancement by organosulphur compounds from garlic and onions of diethyl nitrosamine-induced glutathione S-transferase positive foci in rat liver. *Cancer Res* **54**: 2895—2899

Wagner H (ed) (1998) Economic medicinal Plant Resaerch. Academy Press Harcort Brace Jovanovich Publishers.

Wargovich MJ, Woods C, hollis DM, Zander ME (2001) Herbals: Cancer prevention and health. *J Nutr* **131**(11 suppl): 21—22

Zhang Y, Talalay P, Cho CG and Posner GH (1993) Food and Cancer Prevention: Chemical and Biological Aspects. In: Waldron KW, Johnson IT and Fenwick (eds) pp 429. Royal Society of Chemistry, Cambridge

Zhang Y, Thomas WK, Cheon GC, Gary HP and Talalay P (1994) Anticarcinogenic activities of sulforaphane and structurally related synthetic norbornyl isothiocyanates. *Proc Natl Acad Sci USA* **91**: 3147—3150

8. ACKNOWLEDGEMENT

The patronage of Shah Alam and Shibu B Raj of Faculty of Pharmacy, Hamdard University is greatly acknowledged.

9. AFFILIATIONS

Department of Pharmacognosy and Phytochemistry, Faculty of Pharmacy, Hamdard University, New Delhi-110062, India
For correspondence - *M Ali*: mali@rediffmail.com

SUBJECT INDEX

The manufacturer's authorised representative in the EU is Springer Nature Customer Service Centre GmbH, Europaplatz 3, 69115 Heidelberg, Germany. If you have any concerns regarding our products, please contact ProductSafety@springernature.com

Printed and bound by CPI Group (UK) Ltd, Croydon, CR0 4YY
30/09/2026
02225450-0019